Fourth Edition

Using Information Technology

A Practical Introduction to Computers & Communications

Complete Version

Fourth Edition

Using Information Technology

A Practical Introduction to Computers & Communications

Complete Version

Brian K. Williams

Stacey C. Sawyer

Boston   Burr Ridge, IL   Dubuque, IA   Madison, WI   New York   San Francisco   St. Louis
Bangkok   Bogotá   Caracas   Kuala Lumpur   Lisbon   London   Madrid   Mexico City
Milan   Montreal   New Delhi   Santiago   Seoul   Singapore   Sydney   Taipei   Toronto

McGraw-Hill Higher Education

A Division of The **McGraw-Hill** *Companies*

USING INFORMATION TECHNOLOGY: A PRACTICAL INTRODUCTION TO COMPUTERS AND COMMUNICATIONS

Published by McGraw-Hill/Irwin, an imprint of the McGraw-Hill Companies, Inc. 1221 Avenue of the Americas, New York, NY, 10020. Copyright © 2001, 1999, 1997, 1995, by The McGraw-Hill Companies, Inc. All rights reserved. No part of this publication may be reproduced or distributed in any form or by any means, or stored in a data base or retrieval system, without the prior written consent of the McGraw-Hill Companies, Inc., including, but not limited to, in any network or other electronic storage or transmission, or broadcast for distance learning. Some ancillaries, including electronic and print components, may not be available to customers outside the United States.

This book is printed on acid-free paper.

international   1 2 3 4 5 6 7 8 9 0   QPD/QPD   0 9 8 7 6 5 4 3 2 1
domestic   1 2 3 4 5 6 7 8 9 0   QPD/QPD   0 9 8 7 6 5 4 3 2 1

ISBN 0-07-239803-5

Vice president/Editor-in-chief: *Robin J. Zwettler*
Executive editor: *Linda S. Schreiber*
Development editors: *Burrston House, Ltd./Craig S. Leonard/Melissa Forte*
Senior marketing manager: *Jeff Parr*
Project manager: *Christine A. Vaughan*
Lead production supervisor: *Heather D. Burbridge*
Senior designer: *Laurie J. Entringer*
Production and Quark makeup: *Stacey C. Sawyer*
Cover image: *© Masterfile*
Cover design: *Asylum Studios*
Senior photo research coordinator: *Keri Johnson*
Photo research: *Judy Mason*
Senior supplement producer: *Marc Mattson*
Media technology producer: *David Barrick*
Compositor: *GTS Graphics, Inc.*
Typeface: *10/12 Trump Mediaeval*
Printer: *Quebecor World Dubuque, Inc.*

Library of Congress Cataloging-in-Publication Data

Williams, Brian K.
    Using information technology. Complete / Brian K. Williams, Stacey C. Sawyer.—4th ed.
        p.   cm.
    ISBN 0-07-239803-5
    1. Computers.   2. Telecommunication systems.   3. Information technology.   I. Sawyer,
Stacey C. II. Title.

QA76.5 .W5332 2001
004—dc21   00-048681

INTERNATIONAL EDITION ISBN 0-07-117926-7
Copyright © 2001. Exclusive rights by The McGraw-Hill Companies, Inc., for manufacture and export. This book cannot be re-exported from the country to which it is sold by McGraw-Hill. The International Edition is not available in North America.

When ordering the title, use ISBN 0-07-239803-5

www.mhhe.com

Photo and other credits are listed in the back of the book.

Brief Contents

About the Authors

Who are **Brian Williams and Stacey Sawyer**? We are a married couple living near Lake Tahoe, Nevada, with an avid interest in seeing students become well educated—especially in information technology.

What best describes what we do? We consider ourselves **watchers and listeners.** We spend our time *watching* what's happening in business and society and on college campuses and *listening* to the views expressed by instructors, students, and other participants in the computer revolution. We then try to translate these observations into meaningful language that can be best understood by students.

Over the past two decades, we have individually or together **authored more than 20 books** (and 30 revisions), most of them on computers and information technology. Both of us have **a strong commitment to helping students succeed in college.** Brian, for instance, has also co-authored four books in the college success field: *Learning Success, The Commuter Student, The Urban Student,* and *The Practical Student.* Stacey has an interest in language education and has worked on several college textbooks in English as a Second Language (ESL) and in Spanish, German, French, and Italian. We thus bring to our information-technology books an awareness of the needs of the increasingly diverse student bodies now in our colleges.

Brian has a B.A. and M.A. from Stanford University and has held managerial jobs in education, communications, and publishing. Stacey has a B.A. from Ohio Wesleyan and the University of Freiburg, Germany, and an M.A. from Middlebury College and the University of Mainz, Germany. She has taught at Ohio State University and managed and consulted for a number of for-profit and nonprofit health, educational, and publishing organizations.

In our spare time, we enjoy travel, music, cooking, and exploring the wilds of the American West.

To the Instructor

Introduction

Information technology—"now as vital as the air we breathe," as one observer says—is the major revolution of our time. Rolling inexorably forward, it is redefining entire industries, changing the nature of work and leisure, altering conventional meanings of time and space.

The Fourth Edition of *USING INFORMATION TECHNOLOGY* puts the reader in the front row of this revolution. UIT is a concepts textbook to accompany a one-semester or one-quarter introductory course on computers or microcomputers. This eleven-chapter *Complete Version* is intended for people who will use computers as everyday tools, not those who will write programs or design computer systems. (A shorter edition—eight chapters—is available as the *Introductory Version,* which corresponds to the brief version of the previous edition of our text.)

About Our Book: A New Identity—"A Book Reborn, Not Revised"

We consider this edition of *USING INFORMATION TECHNOLOGY* to be **a book reborn,** not merely revised. The book's **new identity** is embodied in several new features—many of which were **requested by instructors**—as explained below under "What's New in This Edition?"

What Users Liked #1: Emphasis on Role of Communications in Computing

Earlier editions of this book broke new ground by emphasizing the role of the Internet and of digital convergence, the technological fusion of computers and telecommunications. This theme was enthusiastically received by both students and instructors.

But **telecommunications is now taking new forms,** which we describe in this edition.

[See examples pp. 5, 76.]

- **Moving beyond the PC:** We appear to be moving beyond the personal computer. Companies are launching a slew of simplified electronic devices that do only one or two tasks, such as checking e-mail or surfing the Web—devices uncomplicated enough to attract online users in the more than 60% of U.S. households without Internet access. Cellphones, palm computers, and information appliances are now being produced that we can use not only for communicating but also for surfing, shopping, and banking.

[See examples pp. 74, 398.]

- **The Internet becomes personal:** The World Wide Web is more and more becoming a personal resource, from getting help with advancing careers and finances to finding relationships and spirituality.

[See examples pp. 4, 278.]

- **The coming "Omninet":** Because of wireless communication, we are fast approaching the day when the Internet becomes such an all-pervasive presence that, like telephones and television, we will almost forget that it is there.

What Users Liked #2: Emphasis on Practicality

A feature that has been well received by instructors and students using past editions is that we not only cover fundamental concepts but also offer a great deal of **practical advice.** This advice, of the sort found in computer magazines and general-interest computer books, is expressed not only in the text but also in two kinds of boxes:

[See examples pp. 78, 179, 401.]

- **The Experience Box:** Appearing at the end of each chapter, the Experience Box is **optional** material that may be assigned at the instructor's discretion. However, students will find the subjects covered are of immediate value. *Examples:* "Web Research, Term Papers, & Plagiarism." "How to Buy a Notebook." "Career Strategies for the Digital Age."

[See examples pp. 44, 78.]

- **Bookmark It! Practical Action Box:** This box consists of optional material on practical matters. *Examples:* "Managing Your E-Mail." "Choosing an Internet Service Provider." "When the Internet Isn't Productive: Online Addiction & Other Time Wasters."

Ethics

What Users Liked #3: Emphasis throughout on Ethics

Many texts discuss ethics in isolation, usually in one of the final chapters. We believe this topic is too important to be treated last or lightly, and users have agreed. Thus, **we cover ethical matters throughout the book,** as indicated by the special logo shown here in the margin. *Examples:* We discuss such all-important questions as copying of Internet files, online plagiarism, privacy, computer crime, and netiquette. (A list of pages with ethics coverage appears on the inside front cover.)

[See examples pp. 22, 50, 334.]

What Users Liked #4: Emphasis on Reinforcement for Learning

Prior editions of our book offered the following features for reinforcing student learning:

- **Interesting writing:** Studies have found that textbooks **written in an imaginative style** significantly improve students' ability to retain information. Both instructors and students have commented on the distinctiveness of the writing in this book. We employ a number of journalistic devices—colorful anecdotes, short biographical sketches, interesting observations in direct quotes—to make the material as interesting as possible. We also use real anecdotes and examples rather than fictionalized ones.

[See examples pp. 118, 275.]

- **Key terms AND definitions emphasized:** To help readers avoid any confusion about which terms are important and what they actually mean, we print each key term ***bold italic underscore*** and its definition in **boldface.** *Example* (from Chapter 1): "***Data*** **consists of raw facts and figures that are processed into information.**"

[See example p. 97.]

- **Material in bite-size portions:** Major ideas are presented in **bite-size form,** with generous use of advance organizers, bulleted lists, and new paragraphing when a new idea is introduced. Most **sentences have been kept short,** the majority not exceeding 22–25 words in length.

[See example p. 92.]

A great number of brand-new features are to be found in this edition—most of which were suggested by instructors in conversations, reviews, and round-table discussions. As stated, we think they make this edition "a book reborn, not revised."

What Users Requested #1: New Organization & New Emphasis on "E-Concepts"

In accordance with instructor suggestions, **we reorganized the table of contents to permit greater treatment of "e-concepts."**

Because the Internet and the World Wide Web are now so widespread—all students have seen the terms "www" and "dot-com," and most students are already online—many instructors have told us they would like to see **"e-concepts" treated earlier and more extensively** in the book. Accordingly, we have revised the table of contents as follows:

How the Table of Contents Has Changed

Chapters in the Present Edition	Chapters in the Previous Edition
1. Introduction to Information Technology	1. The Digital Age
2. The Internet & the World Wide Web	2. Applications Software
3. Application Software	3. System Software
4. System Software	4. Processors
5. Hardware—The CPU & Storage	5. Input & Output
6. Hardware—Input & Output	6. Storage
7. Telecommunications	7. Telecommunications
8. E-Commerce, Files, & Databases	8. Communications Technology
9. The Challenges of the Digital Age	9. Files & Databases
10. The Promises of the Digital Age	10. Information Systems
11. Information Systems	11. Software Development
Appendix A: Software Development	12. Society & the Digital Age
Appendix B: Factors Affecting Communications Among Devices	

Instructors will note the following changes:

- The Internet and World Wide Web are now discussed in Chapter 2 instead of Chapter 7, reflecting their importance in students' daily lives.
- Hardware is now discussed in two chapters instead of three.
- Files and databases are discussed within the framework of e-commerce. (Storage is discussed as part of hardware.)
- The challenges and promises of the digital age, especially the "e-concepts," are discussed in two chapters instead of one.
- Software development—programming and languages—are discussed in an appendix. (Some instructors don't cover these concepts in their courses.)

What Users Requested #2: A Much More Visual Book—Many More Illustrations Designed to Aid Student Learning

In this edition we have **increased the number of visuals—particularly artwork designed to aid student learning,** so that there is a significantly increased ratio of illustrations to text.

Specifically, we offer artwork—both line art and photographs—to serve the following purposes:

To the Instructor

ix

- **Artwork to illustrate how to use software:** In this edition, we don't just describe Web browser or word processing commands, for example. We show visually how menus, toolbars, icons, and similar features work.

[See example p. 110.]

- **Artwork to illustrate how hardware works:** In Chapter 1, we ask readers to **pretend to order their own custom-built PC**—and we provide illustrations to show how the pieces go together. In Chapters 5 and 6, we present a more complex example. We show **an advertisement for a PC**—and then explain the ad by repeating its hardware features along with specific illustrations describing what they mean and how they work.

[See examples pp. 13, 18.]

[See examples pp. 176, 179.]

- **Photos to show what's exciting and unique about the Digital Age:** Many textbooks have photos showing how computers are used in ordinary ways, and we do also. But to whet student interest, we also show **how computers are used in interesting, uncommon ways.**

[See examples pp. 22, 335, 381.]

What Users Requested #3: Help Students Think Critically about Information Technology & Take Ownership over the Material

More and more instructors are becoming familiar with the taxonomy of educational objectives encompassing critical-thinking skills. These skills are organized in the following hierarchy: (a) *memorization,* (b) *comprehension,* and (c) *application, analysis, synthesis,* and *evaluation.*

Drawing on our experience in writing books to guide students to college success, we have implemented a **novel pedagogy to reinforce learning and help students take ownership over the material.** We use the following hierarchical approach in both text and supplements:

> **First level—memorization.** Tests how well reader recalls basic terms and concepts.
>
> **Second level—comprehension.** Tests how well reader understands concepts and integrates ideas.
>
> **Third level—application, analysis, synthesis, evaluation.** Tests higher-order critical-thinking skills, including the ability to solve problems and make decisions.

The following are some pedagogical features that incorporate this three-tier hierarchy of critical thinking skills.

- **Key Questions—to help students read with purpose:** We have **crafted the learning objectives as Key Questions** to help readers focus on essentials. Each Key Question appears in two places: on the first page of the chapter, and beneath the section head. Key Questions are also tied to the end-of-chapter summary, as we will explain.

[See examples pp. 33, 36.]

- **Chapter visual overview—"Graphical Interface":** Each chapter opens with a two-page **visual overview of chapter concepts,** so that students can have a "pictorial road map" of the contents of the chapter before they begin reading. This "Graphical Interface: A Visual Overview of This Chapter" may also be used as a study aid to review concepts.

[See examples pp. 34–35.]

- **Concept Checks:** Appearing periodically throughout the text, **Concept Checks** spur students to recall facts and concepts they have just read.

[See example p. 53.]

- **Visual Summary:** Each chapter ends with an innovative **Visual Summary** of important terms, with an explanation of what they are and why they are important. The terms are accompanied, when appropriate, by a picture. Each concept or term is also given a cross-reference page number that refers the reader to the main discussion within the

[See example p. 27.]

[See examples pp. 84–85.]

text. In addition, the term or concept is given a Key Question number corresponding to the appropriate Key Question (learning objective).

- **Chapter Review:** The end-of-chapter **Chapter Review** is constructed according to the three levels of educational objectives.

> **First level—memorization.** Self-Test (fill-in) Questions, Multiple-Choice Questions, and True/False Questions. Tests how well the reader recalls basic terms and concepts.
>
> **Second level—comprehension.** Short-answer questions and concept maps. Tests how well the reader understands concepts and integrates ideas.
>
> **Third level—application, analysis, synthesis, evaluation.** Knowledge in Action questions. Tests higher-order critical-thinking skills, including the ability to solve problems and make decisions.

The Student Online Learning Center

For each chapter, the student **Online Learning Center** offers both a review of the text material, in the form of an **e-learning session,** and additional exercises organized around the following themes:

- Group/team projects
- Internet/Web-related content
- Mini-case studies of actual companies
- Profiles of careers that are influenced by information technology

The content and activities further establish the three-level approach.

Resources for Instructors

We understand that, in today's teaching environment, offering a textbook alone is not sufficient to meet the needs of the many different instructors who use our books. To teach effectively, instructors must have a full complement of supplemental resources to assist them in every facet of teaching from preparing for class to conducting and lecture to assessing students' comprehension. *Using Information Technology* offers a complete, fully integrated supplements package, as described below:

Instructor's Resource Kit

The Instructor's Resource Kit contains a printed Instructor's Manual and a CD-ROM containing the Instructor's Manual in both MS Word and .pdf formats, PowerPoint slides, Brownstone's Diploma test generation software, and accompanying test item files for each chapter. The distinctive features of each component of the Instructor's Resource Kit are described below.

- **Instructor's Manual** Prepared by Kerry Thompson of Digital Light Studios, the Instructor's Manual contains learning objectives, a Chapter Outline with lecture notes, a list of the chapter competencies, tips for covering difficult material, and answers to the Concept Checks. Also included are references to corresponding topics on the Interactive Companion CD-ROM, answers to all the exercises in the Chapter Review section, and answers to the On the Web Exercises. The manual also includes a helpful introduction that explains the features, benefits, and suggested uses of the IM and an index of concepts and corresponding competencies.

- **PowerPoint Presentation** Prepared by Linda Mehlinger of Morgan State University, the PowerPoint presentation is designed to provide instructors with a comprehensive teaching resource and includes chapter learning objectives, concepts overviews, figures from the text, additional examples/illustrations, anticipated student questions with answers, discussion topics, and Concept Checks recalling key points throughout each chapter. Each chapter ends with key terms illustrating those key points.

- **Testbank** The *Using Information Technology* edition testbank contains over 3000 questions categorized by level of learning (definition, concept, and application). This is the same learning scheme that is introduced in the text to provide a valuable testing and reinforcement tool. The test questions are identified by text page number to assist you in planning your exams, and rationales for each answer are also included. Additional test questions, which can be used as pretests and posttests in class, can be found on the Online Learning Center, accessible through our supersite (www.mhhe.com/it).

Business Week Edition of Using Information Technology

An exciting new supplement with this edition of *Using Information Technology* is our *Business Week Edition*. With the purchase of a *Business Week Edition* of a McGraw-Hill/Irwin textbook, students will receive a 15-week subscription to *Business Week* for only $8.25 more than the price of the book alone. Professors who adopt the *Business Week Edition* will enjoy a complimentary subscription for a full year to *Business Week* magazine and complimentary access to the Business Week Resource Center Web site (www.resourcecenter.businessweek.com) as well as Business Week Online through the duration of their subscription.

Students will also enjoy free access to the Business Week Resource Center Web site for the duration of their magazine subscription. The Business Week Resource Center Web site contains a wealth of supplemental materials, including the Business Week Online Archives. Students will have instant access to any business topic from the past ten years of *Business Week*—from 1991 to 2000. From the Resource Center, students may also access Business Week Online (www.businessweek.com) for current issues, online-only features, and career tips. Access to these sites provides a marvelous opportunity to increase students' Internet literacy as instructors explore new ways to integrate the Web into a wide array of student exercises and research projects.

Interactive Companion CD-ROM

This free student CD-ROM, designed for use in class, in the lab, or at home by students and professors alike includes a collection of interactive tutorial labs on some of the most popular topics in information technology. By combining video, interactive exercises, animation, additional content, and actual "lab" tutorials, we expand the reach and scope of the textbook.

Digital Solutions to Help You Manage Your Course

- **PageOut**—PageOut is our Course Web Site Development Center that offers a syllabus page, URL, McGraw-Hill Online Learning Center content, online exercises and quizzes, gradebook, discussion board, and an area for student Web pages. For more information, visit the PageOut Web site (www.mhla.net/pageout).

- **Online Learning Centers**—The Online Learning Center that accompanies *Using Information Technology* is accessible through our information Technology Supersite (www.mhhe.com/it). This site provides additional learning and instructional tools developed using the same three-level approach found in the text and supplements. This offers a consistent method for students to enhance their comprehension of the concepts presented in the text.

- **Online Courses Available**—OLCs are your perfect solutions for Internet-based content. Simply put, these Centers are "digital cartridges" that contain a book's pedagogy and supplements. As students read the book, they can go online and take self-grading quizzes or work through interactive exercises. These also provide students appropriate access to lecture materials and other key supplements.

- **UIT Website**—The new Website that accompanies the fourth edition of *Using Information Technology* offers additional, text-specific resources to include: a sample chapter, an overview, Meet the Authors section, the preface, a What's New section, a Feature Summary section, and links to professional resources. In addition, the Instructor's Manual and PowerPoint Presentation slides can be downloaded from the instructor section of the site, located at www.mhhe.com/cit/uit4e.

Online Learning Centers can be delivered through any of these platforms:

McGraw-Hill Learning Architecture (TopClass)
Blackboard.com
ECollege.com (formerly Real Education)
WebCT (a product of Universal Learning Technology)

Office 2000 Application Series

Available for discount packaging, our **Advantage Series** leads students through the features of Microsoft Office 2000 with a critical-thinking, "what, why, and how" orientation. McGraw-Hill/Irwin also offers additional MS Office 2000 series to suit your teaching preference. Each MS Office series is Microsoft Office User Specialist (MOUS) certified.

Skills Assessment

McGraw-Hill/Irwin offers two innovative systems to meet your skills assessment needs. These two products are available for use with any of our applications manual series.

ATLAS (Active Testing and Learning Assessment Software)—Atlas is one option to consider for an application skills assessment tool from McGraw-Hill. Atlas allows students to perform tasks while working live within the Microsoft applications environment. ATLAS is Web-enabled, customizable, and is available for Microsoft Office 2000. **SimNet** (Simulated Network Assessment Product) is another option for a skills assessment tool that permits you to test students' software skills in a simulated environment. Sim-Net is available for Microsoft Office 97 (deliverable via a network) and Microsoft Office 2000 (deliverable via a network and the Web). Both ATLAS and SimNet provide flexibility for you in your course by offering:

- Pre-testing options
- Post-testing options
- Course placement testing
- Diagnostic capabilities to reinforce skills
- Proficiency testing to measure skills

Acknowledgments

Two names are on the front of this book, but a great many others are important contributors to its development. First, we wish to think our publisher, David Brake, for his support and his encouragement during a difficult revision and production schedule. Thanks go also to Jodi McPherson, our sponsoring editor, and Jeff Parr, our marketing manager, for their enthusiasm and new ideas. Everyone in production provided excellent support and direction: Laurie Entringer, Heather Burbridge, Christine Vaughan, Keri Johnson, Marc Mattson, and David Barrick in production and manufacturing. Gladys True, who worked with us on many of our editions over the past ten years, started the production process with us on this edition, before she left for her well-deserved but unwelcome-to-authors retirement. Outside of McGraw-Hill we were fortunate to have the invaluable services of Burrston House, which provided developmental direction and reviews; Judy Mason, photo researcher; and Bernard Gilbert, Martha Ghent, and James Minkin, who provided some of the best copyediting, proofreading, and indexing in the business. Thanks also to all the extremely knowledgeable and hard-working professionals at GTS Graphics, who provided all the prepress services.

Finally, we are grateful to the following people for their participation in manuscript reviews. Their contributions have helped us to make this the most market-driven book possible.

Robert L. Barber
Lane Community College

Vic Barlow
Purdue University

Bob Bretz
Western Kentucky University

Martin Cronlund
Anne Arundel Community College

Donald L. Dershem
Mountain View College

James Frost
Idaho State University

Susan Fry
Boise State University

Enrique Garcia
Laredo Community College

Ron Higgins
Grand Rapids Community College

Washington James
Collin County Community College/Plano

Donna Madsen
Kirkwood Community College

George P. Novotny
Ferris State University

Bettye Jewel Parham
Daytona Beach Community College

Esther Steiner
New Mexico State University

Cora Lee Whitcomb
Bentley College

Student's Guide

A One-Minute Course on How to Succeed in This Class

Got one minute to read this section? It could mean the difference between getting an A instead of a B, or a B instead of a C. Or even passing instead of failing.

Here Are the Rules

There are only four rules, and they aren't difficult.

Rule 1. You have to attend every class. (But that alone won't get you an A, as some students think.)

Rule 2. You can't put off studying, then cram the night before a test. This may work in high school, but college isn't high school.

Rule 3. You have to read or repeat material more than once. The important thing isn't reading. It's *re*reading.

Rule 4. You have to learn the secrets to using your textbook. It would be nice if all textbooks were organized the same way, but they aren't. Different texts have different features.

Getting the Most Information in the Least Time from This Book

Let's consider how you can best read *Using Information Technology.*

- Get an overview of the chapter first
- Check the key questions in each section before you read it
- Read the section, trying to answer the key question(s)
- Do the Concept Checks
- Read the Visual Summary
- Answer the questions in the Chapter Review

   A look through the next seven pages will show you what the features we discussed look like.

Get an Overview of the Chapter First

Before you set out on a trip to a place you've never been to before, you would probably look at a map so you would get a "big picture" view of the route. Reading is the same way.

Scan the first page of the chapter and look at the **Chapter Outline** and the **Key Questions.**

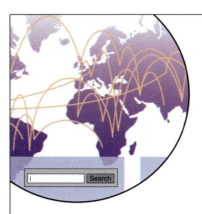

Chapter 2

The Internet & the World Wide Web

Exploring Cyberspace

Key Questions

You should be able to answer the following questions.

2.1 **Choosing Your Internet Access Device & Physical Connection: The Quest for Broadband** What are the means of connecting to the Internet, and how fast are they?

2.2 **Choosing Your Internet Service Provider (ISP)** What are the three types of Internet service provider, and what kinds of services do they provide?

2.3 **Sending & Receiving E-Mail** What are the options for obtaining e-mail software, what are the components of an e-mail address, and what are netiquette and spam?

2.4 **The World Wide Web** What are Web sites, Web pages, browsers, URLs, and search engines?

2.5 **The Online Gold Mine: More Internet Resources, Your Personal Cyberspace, E-Commerce, & the E-conomy** What are FTP, Telnet, newsgroups, real-time chat, and e-commerce?

Chapter Outline

Each chapter begins with an outline of the section headings in the chapter.

Key Questions

Use these Key Questions to help you read with purpose. Key Questions are repeated throughout the text.

Then turn the page and read through the **Graphical Interface: A Visual Overview.** These will give you a "big picture" of the chapter material.

Graphical Interface

A Visual Overview of This Chapter

Graphical Interface: A Visual Overview

Reading experts suggest you "preview" the material you are about to read. You can do this by reading the pages immediately following the chapter opening, which present the chapter's key terms and concepts. Graphics provide visual reinforcement.

Internet user

Internet service provider

1 **Choosing Your Internet Access Device & Physical Connection.** Some Internet **physical connections,** either wired or wireless, have more **bandwidth**—are able to transmit more data—than others. Data transmission is expressed in **bps** (bits per second—8 bits equal 1 character), **Kbps** (kilobits—thousands of bits per second), **Mbps** (megabits—millions), and **Gbps** (gigabits—billions). Data is **downloaded** from a remote computer to a local computer or **uploaded,** the reverse.

There are four principal types of Internet physical connections: (1) Telephone (dial-up) modem connection is low-speed but inexpensive (up to 56 Kbps). (2) High-speed phone connections are **ISDN** (up to 128 Kbps), which transmits over traditional phone lines; **DSL** (up to 8.4 Mbps), also using traditional phone lines; and **T1** (1.5 Mbps), a special trunk line. (3) **Cable modems** (10 Mbps) connect to cable TV systems. (4) Wireless systems include microwave systems, such as **communications satellites** or space stations (up to 56 Kbps).

2 **Choosing Your Internet Service Provider (ISP).** With a physical connection installed, you then need an **Internet service provider (ISP),** a company to help you connect or **log on** to the Internet. Three types of ISPs are free-service providers, which make money presenting ads; basic-service providers, or small, local companies; and full-service providers, which offer more technical support and other services—examples are AOL and MSN—and which offer **menus,** or lists of commands of services.

3 **Sending & Receiving E-Mail.** Four alternatives for getting and sending e-mail are to buy e-mail software, get the software as part of a browser or other software, get it from your ISP, and get it free (for example, from CNN.com or NetZero). People will send e-mail to you at your **domain,** a location on the Internet consisting of your user name and domain name, such as *user@domain.*

E-mail allows users to send attachments, or separate long documents, with their e-mail messages. It also allows **instant messaging (IM),** in which incoming messages are displayed at once in a **window,** a rectangular area on screen. You can exchange e-mail from people worldwide with similar interests through **list-serves,** or e-mail mailing lists.

The two basic rules of online behavior, or **netiquette,** are these: Don't waste people's time, and don't say anything online you wouldn't say to someone's face. In particular, you should always first consult **FAQs,** or Frequently Asked Questions; avoid **flaming,** such as insults or obscenities; and smooth communication using **emoticons,** or friendly graphic symbols.

To manage your e-mail, filters or instant organizers are recommended. In addition, you will need to know how to manage **spam,** or unsolicited e-mail. Finally, assume e-mail messages are not private: Anyone could read them.

Key Terms

Important terms are presented in bold type.

Check the Key Questions in Each Section before You Read It

Look at the **Key Question** near the section heading. Read this aloud (or silently) or write it down.

Bookmark It! Practical Action Box

These boxes present material on practical matters that students find useful.

PRACTICAL ACTION BOX
Choosing an Internet Service Provider

If you belong to a college or company, you may get an ISP free. Some public libraries also offer freenet connections.

If these options are not available to you, be sure to ask these questions when you're making phone calls to locate an Internet service provider:[a]

Costs

- Is there a setup fee? (Most ISPs no longer charge this, though some "free" ISPs will.)
- How much is unlimited access per month? (Most charge about $20 for unlimited usage. But inquire if there are free or low-cost trial memberships or discounts for long-term commitments.)
- If access is supposedly free, what are the trade-offs besides putting up with heavy advertising? (For instance, if the ISP closely monitors your activity in order to accurately target ads, what guarantees do you have that information about you will be kept private? What charges will you face if you try to scrap the advertising window or drop the service?)
- Is there a contract, and for what length of time? That is, are you obligated to stick with the ISP for a while even if you're unhappy with it?

Access

- ...ocal phone call? (If not, your ...ne tolls could exceed the ISP
- ...l-up number if the main num-

- Is access available when you're traveling? Your provider should offer either a wide range of local access numbers in the cities you tend to visit or toll-free 800 numbers.

Support

- What kind of help does the ISP give in setting up your connection?
- Is there free, 24-hour technical support? Is it reachable through a toll-free number?
- How difficult is it to reach tech support? (Try calling the number before you sign up for the ISP and see how long it takes to get a response. Many ISPs keep customers on hold for a long time.)

Reliability

- What is the average connection success rate for users trying to connect on the first try? (The industry average call-success rate is 93.1%. You can try dialing the number during peak hours, to see if you get a modem screech, which is good, rather than a busy signal, which is bad. You can also check Visual Networks, *www.inversenet.com*, for call-failure/call-success rates of various ISPs.)
- Will the ISP keep up with technology? (Are they planning to offer broadband technology such as DSL for speedier access?)
- Will the ISP sell your name to marketers or bombard you with junk messages (spam)?

Key Questions

Use these Key Questions presented at the start of each section to help you read with purpose.

CONCEPT CHECK

What are the measures of data transmission speed?

Explain the differences among the methods of going online.

Describe the different types of Internet service providers.

2.3 Sending & Receiving E-Mail

KEY QUESTIONS
What are the options for obtaining e-mail software, what are the components of an e-mail address, and what are netiquette and spam?

Once connected with an ISP, one of the first things most people want to do is join the millions of users who send and receive electronic mail. E-mail can be sent at any time and to several people simultaneously. You can receive e-mail wherever you are, using your user name and password to connect to the Internet. In addition, you can attach long (or short) documents or other materials to your e-mail message.

Read the Section, Trying to Answer the Key Question(s)

Read the section that follows the section heading, trying to answer the Key Question or Key Questions as you go. Make marks in the book if this helps you answer the question. In particular, look at the **key terms and definitions,** which appear in boldface. Look at the **graphics** (artwork and photos), which help to clarify the discussion.

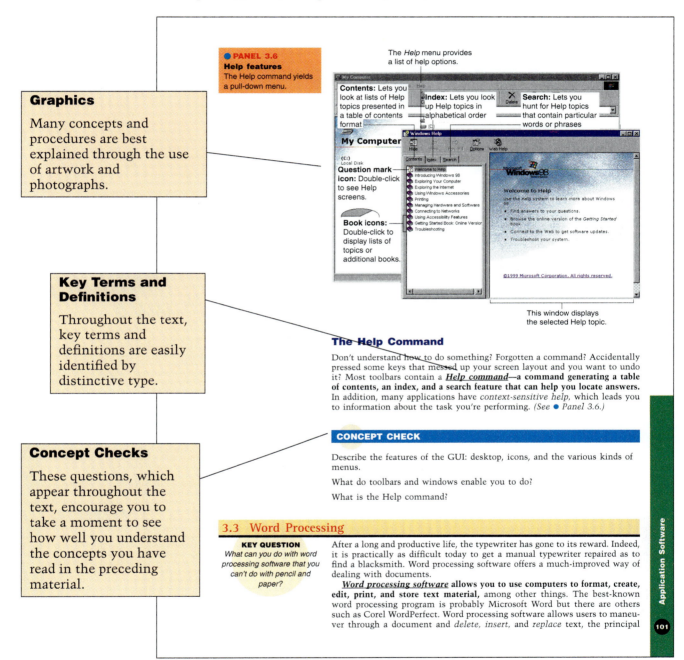

Graphics

Many concepts and procedures are best explained through the use of artwork and photographs.

Key Terms and Definitions

Throughout the text, key terms and definitions are easily identified by distinctive type.

Concept Checks

These questions, which appear throughout the text, encourage you to take a moment to see how well you understand the concepts you have read in the preceding material.

● **PANEL 3.6**
Help features
The Help command yields a pull-down menu.

The *Help* menu provides a list of help options.

Contents: Lets you look at lists of Help topics presented in a table of contents format

Index: Lets you look up Help topics in alphabetical order

Search: Lets you hunt for Help topics that contain particular words or phrases

My Computer

Question mark icon: Double-click to see Help screens.

Book icons: Double-click to display lists of topics or additional books.

This window displays the selected Help topic.

The Help Command

Don't understand how to do something? Forgotten a command? Accidentally pressed some keys that messed up your screen layout and you want to undo it? Most toolbars contain a ***Help command—a command generating a table of contents, an index, and a search feature that can help you locate answers.*** In addition, many applications have *context-sensitive help,* which leads you to information about the task you're performing. *(See ● Panel 3.6.)*

CONCEPT CHECK

Describe the features of the GUI: desktop, icons, and the various kinds of menus.

What do toolbars and windows enable you to do?

What is the Help command?

3.3 Word Processing

KEY QUESTION
What can you do with word processing software that you can't do with pencil and paper?

After a long and productive life, the typewriter has gone to its reward. Indeed, it is practically as difficult today to get a manual typewriter repaired as to find a blacksmith. Word processing software offers a much-improved way of dealing with documents.

***Word processing software* allows you to use computers to format, create, edit, print, and store text material,** among other things. The best-known word processing program is probably Microsoft Word but there are others such as Corel WordPerfect. Word processing software allows users to maneuver through a document and *delete, insert,* and *replace* text, the principal

Do the Concept Checks

The **Concept Checks** are questions that appear throughout the text that encourage you to see how well you understand the concepts you have read. Take a break from your reading and try to answer as many of these as possible. If you have trouble with the questions, you probably should go back and review the section before proceeding.

Read the Visual Summary

After you read the whole chapter, go through the **Visual Summary,** which gives the important concepts and terms of the chapter in alphabetical order and tells you why they are important.

Visual Summary

Visual Summary

Each chapter ends with an innovative Visual Summary of important terms and concepts, with an explanation of what they are and why they are important. Graphics provide visual reinforcement.

analytical graphics (p. 111, KQ 3.4) Also called *business graphics;* graphical forms that make numeric data easier to analyze than when it is organized as rows and columns of numbers. The principal examples of analytical graphics are bar charts, line graphs, and pie charts. Why it's important: *Whether viewed on a monitor or printed out, analytical graphics help make sales figures, economic trends, and the like easier to comprehend and analyze.*

Cell:
Formed by intersection of row and column (letter and number)

cell (p. 109, KQ 3.4) Place where a row and a column intersect in a spreadsheet worksheet; its position is called a *cell address.* Why it's important: *The cell is the smallest working unit in a spreadsheet. Data and formulas are entered into cells. Cell addresses provide location references for spreadsheet users.*

computer-aided design (CAD) programs (p. 122, KQ 3.6) Programs intended for the design of products, structures, civil engineering drawings, and maps. Why it's important: *CAD programs, which are available for microcomputers,* help architects design buildings and workspaces and help engineers design cars, planes, electronic devices, roadways, bridges, and subdivisions. While similar to drawing programs, CAD programs provide precise dimensioning and positioning of the elements being drawn, so that they can be transferred later to computer-aided manufacturing programs; in addition, they lack special effects for illustrations. One advantage of CAD software is that three-dimensional drawings can be rotated on screen, so the designer can see all sides of the product.

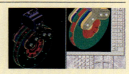

computer-aided design/computer-aided manufacturing (CAD/CAM) software (p. 122, KQ 3.6) Programs allowing products designed with CAD to be input into an automated manufacturing system that makes the products. Why it's important: *CAM systems have greatly enhanced efficiency in many industries.*

copyright (p. 91, KQ 3.1) Exclusive legal right that prohibits copying of intellectual property without the permission of the copyright holder. Why it's important: *Copyright law aims to prevent people from taking credit for and profiting from other people's work.*

cursor (p. 103, KQ 3.3) Movable symbol on the display screen that shows where you may next enter data or commands. The symbol is often a blinking rectangle or an I-beam. You can move the cursor on the screen using the keyboard's directional arrow keys or a mouse. The point where the cursor is located is called the *insertion point.* Why it's important: *All application software packages use cursors to show the current work location on the screen.*

database (p. 111, KQ 3.5) Collection of interrelated files in a computer system. These computer-based files are organized according to their common elements, so that they can be retrieved easily. Why it's important: *Businesses and organizations build databases to help them keep track of and manage their affairs. In addition, online database services put enormous resources at the user's disposal.*

database file (p. 92, KQ 3.1) File created by database management programs; it consists of organized data that can be analyzed and displayed in various useful ways. Why it's important: *Database files make up a database.*

Application Software

125

Answer the Questions in the Chapter Review

The **Chapter Review** at the end of the chapter offers a three-level process that helps you truly understand and "take ownership" of the material. Here's how it works:

- **First-Level Review—Memorization:** Level 1 questions test how well you recall basic terms and concepts. They include **Self-Test, Multiple-Choice,** and **True/False** questions.

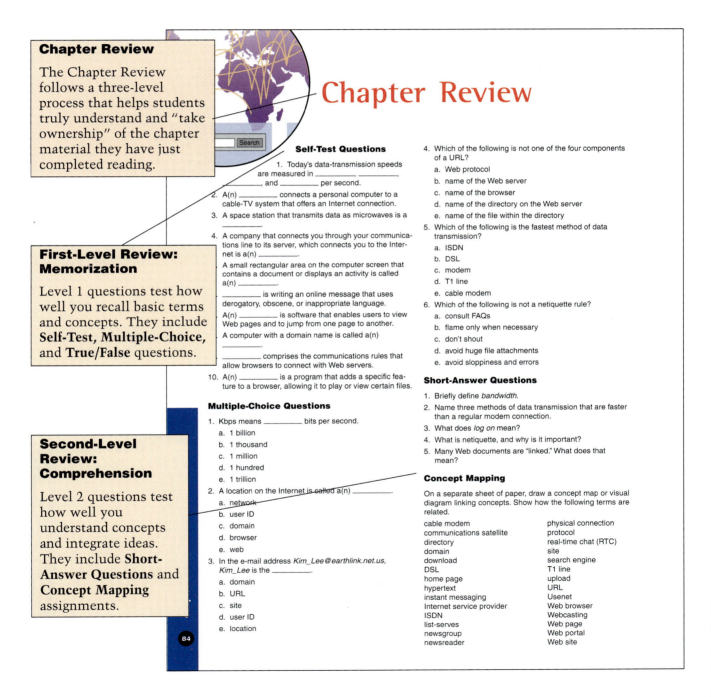

Chapter Review

The Chapter Review follows a three-level process that helps students truly understand and "take ownership" of the chapter material they have just completed reading.

First-Level Review: Memorization

Level 1 questions test how well you recall basic terms and concepts. They include **Self-Test, Multiple-Choice,** and **True/False** questions.

Second-Level Review: Comprehension

Level 2 questions test how well you understand concepts and integrate ideas. They include **Short-Answer Questions** and **Concept Mapping** assignments.

Self-Test Questions

1. Today's data-transmission speeds are measured in _____, _____, _____, and _____ per second.
2. A(n) _____ connects a personal computer to a cable-TV system that offers an Internet connection.
3. A space station that transmits data as microwaves is a _____.
4. A company that connects you through your communications line to its server, which connects you to the Internet is a(n) _____.
5. A small rectangular area on the computer screen that contains a document or displays an activity is called a(n) _____.
6. _____ is writing an online message that uses derogatory, obscene, or inappropriate language.
7. A(n) _____ is software that enables users to view Web pages and to jump from one page to another.
8. A computer with a domain name is called a(n) _____.
9. _____ comprises the communications rules that allow browsers to connect with Web servers.
10. A(n) _____ is a program that adds a specific feature to a browser, allowing it to play or view certain files.

Multiple-Choice Questions

1. Kbps means _____ bits per second.
   a. 1 billion
   b. 1 thousand
   c. 1 million
   d. 1 hundred
   e. 1 trillion
2. A location on the Internet is called a(n) _____.
   a. network
   b. user ID
   c. domain
   d. browser
   e. web
3. In the e-mail address *Kim_Lee@earthlink.net.us*, *Kim_Lee* is the _____.
   a. domain
   b. URL
   c. site
   d. user ID
   e. location
4. Which of the following is not one of the four components of a URL?
   a. Web protocol
   b. name of the Web server
   c. name of the browser
   d. name of the directory on the Web server
   e. name of the file within the directory
5. Which of the following is the fastest method of data transmission?
   a. ISDN
   b. DSL
   c. modem
   d. T1 line
   e. cable modem
6. Which of the following is not a netiquette rule?
   a. consult FAQs
   b. flame only when necessary
   c. don't shout
   d. avoid huge file attachments
   e. avoid sloppiness and errors

Short-Answer Questions

1. Briefly define *bandwidth*.
2. Name three methods of data transmission that are faster than a regular modem connection.
3. What does *log on* mean?
4. What is netiquette, and why is it important?
5. Many Web documents are "linked." What does that mean?

Concept Mapping

On a separate sheet of paper, draw a concept map or visual diagram linking concepts. Show how the following terms are related.

cable modem
communications satellite
directory
domain
download
DSL
home page
hypertext
instant messaging
Internet service provider
ISDN
list-serves
newsgroup
newsreader

physical connection
protocol
real-time chat (RTC)
site
search engine
T1 line
upload
URL
Usenet
Web browser
Webcasting
Web page
Web portal
Web site

84

- **Second-Level Review—Comprehension:** Level 2 questions test how well you understand concepts and integrate ideas. They include **Short-Answer Questions** and **Concept Mapping** assignments.
- **Third-Level Review—Application, Analysis, Synthesis, Evaluation:** Level 3 questions and assignments appear under **Knowledge in Action.** They show your mastery of the material, including your ability to use it to solve problems and make decisions.

Third-Level Review: Application, Analysis, Synthesis, Evaluation

Level 3 questions and assignments appear under **Knowledge in Action.** They test higher-order critical-thinking skills, including the ability to solve problems and make decisions.

Knowledge in Action

1. Identify the four parts of the following URL:
   http://www.nps.gov/yell/swimming.htm

2. Some Web sites go overboard with multimedia effects, while others don't include enough. Locate a Web site that you think makes effective use of multimedia. What is the purpose of the site? Why is the site's use of multimedia effective? Take notes, and repeat the exercise for a site with too much multimedia and one with too little.

3. Visit the following job-hunting Web sites:

   www.monster.com

   www.occ.com

   www.careerpath.com

   www.cweb.com

   Investigate job offerings in a field you are interested in. Which is the easiest site to use? Why? What recommendations would you make for improving the site?

4. Distance learning uses electronic links to extend college campuses to people who otherwise would not be able to take college courses. Is your school or someone you know involved in distance learning? If so, research the system's components and uses. What hardware and software do students need in order to communicate with the instructor and classmates? What courses are offered? Prepare a short report on this topic.

6. It's difficult to conceive how much information is available on the Internet and the Web. One method you can use to find information among the millions of documents is to use a search engine, which helps you find Web pages based on typed keyword or phrases. Use your browser to visit the following search sites: *www.yahoo.com* and *www.goto.com*. Click in the Search box and then type the phrase "personal computers," then hit "Go," "Find it!" or hit the enter key. When you locate a topic that interests you, print it out by choosing File, Print from the menu bar or by clicking the Print button on the toolbar. Report on your findings.

Investigating the World Wide Web

Activities for exploring the Web often appear under **Knowledge in Action** in the **Chapter Review.**

Clearly, this method takes longer than simply reading the material once and rapidly underlining it. But because it requires your involvement and understanding, it is a better way to learn.

Contents

Chapter 3

APPLICATION SOFTWARE: TOOLS FOR THINKING & WORKING MORE PRODUCTIVELY 87

Chapter 4

SYSTEM SOFTWARE: THE POWER BEHIND THE POWER 133

Chapter 5

HARDWARE—THE CPU & STORAGE: HOW TO BUY A MULTIMEDIA COMPUTER SYSTEM   167

Chapter 6

HARDWARE—INPUT & OUTPUT: TAKING CHARGE OF COMPUTING & COMMUNICATIONS   219

Chapter 7

TELECOMMUNICATIONS—NETWORKS & COMMUNICATIONS: THE "NEW STORY" IN COMPUTING 265

Chapter 8

E-COMMERCE, FILES, & DATABASES: DIGITAL ENGINES FOR THE NEW ECONOMY 311

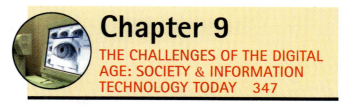

Chapter 9
THE CHALLENGES OF THE DIGITAL AGE: SOCIETY & INFORMATION TECHNOLOGY TODAY 347

Contents

xxvii

Chapter 10

THE PROMISES OF THE DIGITAL AGE: SOCIETY & INFORMATION TECHNOLOGY TOMORROW   375

Chapter 11

INFORMATION SYSTEMS: INFORMATION MANAGEMENT & SYSTEMS DEVELOPMENT   407

Contents

Appendix A

SOFTWARE DEVELOPMENT: PROGRAMMING & LANGUAGES 435

Appendix B

FACTORS AFFECTING COMMUNICATIONS AMONG DEVICES 459

Introduction to Information Technology

Mind Tools for Your Future

Key Questions

You should be able to answer the following questions.

1.1 Infotech Becomes Commonplace: Cellphones, E-Mail, the Internet, & the E-World How does information technology facilitate e-mail, networks, and the use of the Internet and the Web, and what is the meaning of the term *cyberspace*?

1.2 The "All-Purpose Machine": The Varieties of Computers What are the five sizes of computers, and what are clients and servers?

1.3 Understanding Your Computer: What If You Custom-Ordered Your Own PC? What four basic operations do all computers follow, and what are some of the devices associated with each operation? How does communications affect these operations?

1.4 Where Is Information Technology Headed? What are three directions of computer development and three directions of communications development?

A Visual Overview of This Chapter

1 **Infotech Becomes Commonplace. Information technology** merges computers and high-speed communications; new infotech phones are an example. **Computer**—a programmable, multiuse machine that processes data into information. **Communications**—electromagnetic devices and systems for communicating over distances. **Online**—using a computer to access information and services via a network.

Infotech has given us some now-commonplace technologies: **E-mail**—messages transmitted over a computer **network,** a communications system connecting computers. Major componenets of **cyberspace**—the wired and wireless world of communications—are the **Internet,** a network of 400,000 smaller networks, and its graphical subsection, the **World Wide Web,** which stores information in **multimedia** form—text, graphics, sound, video.

2 **The "All-Purpose Machine."** Computers are of five types. (1) **Supercomputers**—perform 1 trillion calculations per second. (2) **Mainframes**—for processing millions of calculations; often accessed by a **terminal,** a display screen and keyboard that can't do its own processing. (3) **Workstations**—used for scientific, mathematical, engineering, and certain manufacturing applications. (4) **Microcomputers**—personal computers, which may be **desktop PCs, tower PCs, laptops,** or **personal digital assistants (PDAs);** frequently used to connect to a **local area network (LAN),** which links equipment in an office or building. (5) **Microcontrollers**—tiny specialized computers installed in cars and appliances. **Servers** are central computers holding data and programs for many other computers.

3 **Understanding Your Computer.** You should know three things: (1) The purpose of a computer is to process **data,** raw facts, into **information,** summarized data. (2) **Hardware** consists of machinery and equipment; **software** instructs the hardware what to do. (3) Computers perform four basic operations: **input,** putting data into the system; **processing,** manipulation of data into information; **storage,** of two types—**memory,** which temporarily holds data to be processed, and **secondary storage,** which holds data or information permanently; **output,** putting out the results of processing. In addition, communication could be considered a fifth operation.

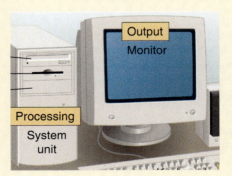

Input hardware includes the **keyboard,** which converts characters into signals readable by the processor, and the **mouse,** used to manipulate objects on the display screen.

Processing and memory hardware includes the **case (system cabinet),** which contains: a processor **chip** with miniature electronic circuits that process information in megahertz, millions of processing cycles per second; and **memory chips (RAM chips),** which hold data prior to processing. Both chips are connected to the **motherboard,** which contains **expansion slots** for plugging in additional circuit boards.

Storage hardware includes: a **floppy-disk drive,** which stores data on removable 3½-inch disks; a **hard-disk drive,** which stores data on a nonremovable disk platter; and a **CD-ROM drive,** which reads data from optical disks.

Output hardware includes: a **video card,** which converts processor information so it can be displayed as text and images on the **monitor;** a **sound card,** which outputs sound to **speakers;** and a **printer,** which produces images and text on paper.

Communications hardware includes a **modem,** which sends and receives data over phone lines to and from computers.

Software is of two types. **System software** helps the computer perform essential operating tasks. **Application software** performs specific tasks, such as word processing.

4 **Where Is Infotech Headed?** Computer developments have focused on three areas: miniaturization, speed, and affordability. Three developments in communications may be noted: **connectivity,** the ability to connect with computers via communications lines; **interactivity,** enabling a user to have a two-way dialogue; and multimedia. The melding of computers and communications has produced three developments: **convergence,** the combining of several industries through the language of computers; portability; and personalization. One result of these developments is information overload. Three ethical concerns raised by information technology are associated with speed and scale, unpredictability, and complexity.

Adds science-fiction writer Bruce Sterling, "We should never again feel all mind-boggled at anything that human beings create. No matter how amazing some machine may seem, the odds are very high that we'll outlive it."[2]

The personal computer is over two decades old. The Internet has been familiar to the public for over 10 years. It has been more than five years since the now-commonplace "www" for World Wide Web began appearing in company ads. And, like cars, elevators, air-conditioning, and television—all of which have wrought tremendous changes on society and the landscape—they are rapidly achieving what technology is supposed to do: *become ordinary.* They are becoming part of the wallpaper of our lives, almost invisible.

"The fact that technology tends to disappear ensures that we will take much of it, when it works right, for granted," says Allen. "And because of that, we stay most aware of technologies that are changing rapidly or are grossly immature. Right now we're especially aware of the Internet precisely because of its youth."

But the notion that the Internet is still a brash upstart has already pretty much faded. For instance, its two most popular aspects—e-mail and the World Wide Web—are rapidly becoming as ordinary as the telephone and the freeway. Beyond them, however, lie some even more astonishing developments, as we discuss in this book.

1.1 Infotech Becomes Commonplace: Cellphones, E-Mail, the Internet, & the E-World

KEY QUESTIONS
How does information technology facilitate e-mail, networks, and the use of the Internet and the Web, and what is the meaning of the term cyberspace?

This book is about computers, of course. But not *just* about computers. It is also about the way computers communicate with each other. When computer and communications technologies are combined, the result is **_information technology_—"infotech"—technology that merges computing with high-speed communications links carrying data, sound, and video.** Examples of information technology include personal computers, of course, but also new forms of telephones, televisions, and various handheld devices such as personal digital assistants.

Note there are two parts to this definition—*computers* and *communications*:

- **Computer technology:** You have certainly seen, and possibly used, a computer. Nevertheless, let's define what it is. **A _computer_ is a programmable, multiuse machine that accepts data—raw facts and figures—and processes, or manipulates, it into information we can use,** such as summaries, totals, or reports. Its purpose is to speed up problem solving and increase productivity.

- **Communications technology:** Unquestionably you've been using communications technology for years. **_Communications technology_, also called _telecommunications technology_, consists of electromagnetic devices and systems for communicating over long distances.** The principal examples are telephone, radio, broadcast television, and cable TV. More recently there has been the addition of communications among computers, as when people tell you they "went online" with the Internet. **_Online_ means using a computer or other information device, connected through a voice or data network, to access information and services from another computer or information device.**

As our first example, let's consider something that seems to be everywhere these days—the cellphone.

The Telephone Grows Up

Seventy-something Louis DeMartino hadn't even had a chance to unwrap the package when a ringing sound came from inside it—a call from his grand-

See-through cover, so you access services without opening up phone

Online menus, for accessing Internet services

Speaker phone, for hands-free use

Connector to your laptop or personal digital assistant for wireless access

● PANEL 1.1
The Internet phone
With an Internet-accessible phone, you have e-mail and specific Web services such as news and stock quotes.

daughter as a way of making the family's gift presentation of a cellphone all the more dramatic.[3] DeMartino may have just been getting started on using this apparatus, but Amit Sinai, 18, like many other teenagers, cannot imagine life without it. Flopping down on her towel after a quick dip in the ocean at a New York beach, for instance, the first thing she does, even before her face starts to dry, is reach for her cellphone and start dialing. "Hi, it's me," she says. "I'm on the beach. Where are you?"[4]

Cellphone mania has swept the world. All across the globe, people have acquired the portable gift of gab; some make 45 calls or more a day. It has taken more than 100 years for the telephone to get to this point—getting smaller, acquiring push buttons, losing its cord connection. In 1964, the ★ and # were added to the keypad to allow users to interact with data services and equipment. In 1973, the first cellphone call was processed. In its standard form, the phone is still so simply designed that even a young child can use it. However, it is now becoming more versatile—a way of connecting to the Internet and the World Wide Web.

Why introduce a book that is about computers with a discussion of telephones? Because Internet phones—such as AT&T's PocketNet Phone or Motorola's i1000 plus—represent another giant step for information technology. *(See ● Panel 1.1.)* Now you no longer need a personal computer to get on the Internet. These infotech phones, with their small display screens, provide a direct, wireless connection that enables you not only to make voice calls and check your daily "to-do" list but also to browse the World Wide Web and receive all kinds of information: news, sports scores, stock prices, term-paper research. And you can also send and receive e-mail.

"You've Got Mail!" E-Mail's Mass Impact

It took the telephone 40 years to reach 10 million customers, and fax machines 20 years. Personal computers made it into that many American homes 5 years after they were introduced. E-mail, which appeared in 1981, became popular far more quickly, reaching 10 million users in little more than a year.[5] No technology has ever become so universal so fast. Not surprisingly, then, one of the first things new computer users learn is how to send and receive **_e-mail_—"electronic mail," messages transmitted over a computer network,** most often the Internet. **A _network_ is a communications system connecting two or more computers;** the Internet is the largest such network.

In 1998, the volume of e-mail in the United States surpassed the volume of hand-delivered mail. By 2002, an average of 8 billion messages a day are expected to zip back and forth across the U.S. (which has some 60% of the world's e-mail accounts), up from 3.5 billion three years earlier.[6] Because of this explosion in usage, suggests a *Business Week* report, "E-mail ranks with such pivotal advances as the printing press, the telephone, and television in mass impact."[7]

How is electronic mail different from calling on a telephone or writing a conventional letter? E-mail "occupies a psychological space all its own," says one journalist. "It's almost as immediate as a phone call, but if you need to, you can think about what you're going to say for days and reply when it's

E-mailer

A college student sends e-mail.

convenient."[8] E-mail has blossomed, points out another writer, not because it gives us more immediacy but because it gives us *less.* "The new appeal of e-mail is the old appeal of print," he says. "It isn't instant; it isn't immediate; it isn't in your face." E-mail has succeeded for the same reason that the videophone—which allows callers to see each other while talking—has been so slow to catch on: because "what we actually want from our exchanges is the minimum human contact commensurate with the need to connect with other people."[9]

From this it is easy to conclude, as *New York Times* computer writer Peter Lewis did, that e-mail "is so clearly superior to paper mail for so many purposes that most people who try it cannot imagine going back to working without it."[10] What is interesting, though, is that in these times when images often seem to overwhelm words, e-mail is actually *reactionary.* "The Internet is the first new medium to move decisively backward," points out one writer, "for it is, essentially, written. . . . Ten years ago, even the most literate of us wrote maybe a half a dozen letters a year; the rest of our lives took place on the telephone."[11] E-mail has changed all that.

The Internet, the World Wide Web, & the "Plumbing of Cyberspace"

Communications is extending into every nook and cranny of civilization. It has been called the "plumbing of cyberspace." The term *cyberspace* was coined by William Gibson in his novel *Neuromancer,* to describe a futuristic computer network into which users plug their brains. In everyday use, it has a rather different meaning.

Today many people equate cyberspace with the Internet. But it is much more than that, says David Whittle in his book *Cyberspace: The Human Dimension.*[12] Cyberspace includes not only the World Wide Web, chat rooms, online bulletin boards, and member-based services like America Online—all features we explain in this book—"but also such things as conference calls and automatic teller machines," he says.[13] We may say that **cyberspace encompasses not only the online world and the Internet in particular but also the whole wired and wireless world of communications in general.**

The two most important aspects of cyberspace are the Internet and that part of it known as the World Wide Web.

- **The Internet—"the mother of all networks":** The Internet is at the heart of the "Information Age." Called "the mother of all networks," **the _Internet_ is a worldwide network that connects up to 400,000 smaller networks in more than 200 countries.** These networks link educational, commercial, nonprofit, and military entities.
- **The World Wide Web—the multimedia part of the Net:** The Internet has actually been around for more than 30 years. But what made it popular, apart from e-mail, was the development of the **_World Wide Web_—an interconnected system of computers all over the world that store information in multimedia form.** The word **_multimedia_, from "multiple media," refers to technology that presents information in more than one medium, such as text, still images, moving images, and sound.**

There is no doubt the influence of the Net and the Web is tremendous. At present, one-third to one-half of the U.S. population is online; if the Internet industry were a nation, it would be the 18th-largest economy in the world.[14] But just how revolutionary is it? Is it equivalent to the invention of television, as some technologists say? Or is it even more important—equivalent to the invention of the printing press? "Television turned out to be a powerful force that changed a lot about society," says *USA Today* technology reporter Kevin Maney. "But the printing press changed everything—religion, government, science, global distribution of wealth, and much more. If the Internet equals the printing press, no amount of hype could possibly overdo it."[15]

Perhaps in a few years we'll begin to know the answer. No massive study was ever done of the influence of the last great electronic revolution to touch us, namely, television. But the Center for Communication Policy at the University of California, Los Angeles, in conjunction with other international universities, has begun to take a look at the effects of information technology—and at how people's behavior and attitudes toward it will change over a span of years.[16]

BOOKMARK IT!

PRACTICAL ACTION BOX
Managing Your E-Mail

For many people, e-mail is *the* online environment, more so than the World Wide Web. In one study, respondents reported they received an average of 31 e-mail messages a day; users had an average of 2.8 separate e-mail accounts each, and 69% reported using free e-mail services.[a]

But then there are those such as Jeremy Gross, managing director of technology for Countrywide Home Loans, who gets 300 e-mails a day—and about 200 are junk e-mail (spam), bad jokes, or irrelevant memos (the "cc," or carbon copy).[b] Astronomer Seth Shostak gets 50 electronic messages daily, at least half requiring a reply. "If I spend five minutes considering and composing a response to each correspondence," he complains, "then two hours of my day are busied with e-mail, even when I don't initiate a single one."[c]

It's clear, then, that e-mail will increase productivity only if it is used properly. Overuse or misuse just causes more problems and wastes time. The following are some ideas to keep in mind when using e-mail:[d]

- **Do your part to curb the e-mail deluge:** Don't reply to every e-mail message you get. Avoid "cc:ing" (copying to) people unless absolutely necessary. Don't send chain letters or lists of jokes, which just clog mail systems.

- **Be helpful in sending attachments:** Attachments—computer files of long documents or images—attached to an e-mail are supposed to be a convenience, but

often they can be an annoyance. Sending a 1-megabyte file to a 500-person mailing list creates 500 copies of that file—and that many megabytes can clog or even cripple the mail system. (A one-megabyte file is about the size of a 300-page double-spaced term paper.) Ask your recipients beforehand if they want the attachment and tell them how to open it; they may need to know what word-processing program to use, for example.

- **Be careful about opening attachments you don't recognize:** Some dangerous computer viruses—renegade programs that can damage your computer—have been spread by e-mail attachments, which are activated only upon opening.

- **Use discretion about the e-mails you send:** E-mail should not be treated as informally as a phone call. Don't send a message electronically that you don't want some third party to read. E-mail messages are not written with disappearing ink; they remain in a computer system long after they have been sent. Also, recipients can easily copy and even alter your messages and forward them to others without your knowledge.

- **Don't use e-mail in place of personal contact:** Because e-mail carries no tone or inflection, it's hard to convey personal concern. Avoid criticism and sarcasm in electronic messaging. Nevertheless, you *can* use e-mail to provide quick praise, even though doing it in person will take on greater significance.

The E-World & Welcome to It

One thing we know already is that cyberspace is saturating our lives. More than 52% of American adults are Internet users, according to a recent survey.[17] While the average age of users is rising, to age 40 by late 1999, there's no doubt that young people love the Net. For instance, an amazingly high percentage of American teenagers—*81%*—use the Internet, according to Teenage Research Unlimited (TRU), a Chicago market research firm.[18] Why is that? E-mail is certainly one important reason. "It's all about staying connected," says New Yorker Kimy Tolentino, 19. His friends agree. "I'm an AOL junkie," says Molly Lowe, 15. "When I was on vacation, I asked people if I could use their laptops, just to check my e-mail." "I do that, too," says Raysa Rodriguez, 17.[19]

But it's more than just about e-mail. Teenagers are also big participants in online commerce. "When teens want information on a brand," says TRU's director of research, "they turn to the Internet first." Indeed, teens are voracious consumers of all things electronic. Unlike many adults, they are adept at "multitasking" (a term borrowed from computer jargon, as we explain later)—doing more than one task at once. Thus, they may be surfing the Web, but they're also watching TV and listening to music and maybe chatting on the phone at the same time.

The fact is that, not just for teens but for most Americans, the use of the Internet's favorite letter, "e"—as in e-business, e-commerce, e-shopping—is rapidly becoming outmoded. "E" is now part of nearly everything we do. As an executive for a marketing research firm says, "E-business is just business."[20] The electronic world is everywhere. The Net and the Web are everywhere. Cyberspace permeates everything. What editor Frederick Allen was talking about has already happened: Infotech has become . . . ordinary.

CONCEPT CHECK

What are the two parts of information technology?

How does e-mail differ from other means of communication?

What is cyberspace, and what are its two most important aspects?

1.2 The "All-Purpose Machine": The Varieties of Computers

KEY QUESTIONS
What are the five sizes of computers, and what are clients and servers?

When the alarm clock★ blasts you awake, you leap out of bed and head for the kitchen, where you plug in the coffee maker★. After using your electric toothbrush★ and showering and dressing, you stick a bagel in the microwave★, then pick up the TV remote★ and click on the TV★ to catch the weather forecast. Later, after putting dishes in the dishwasher★, you go out and start up the car★ and head toward campus or work. Pausing en route at a traffic light★, you turn on your portable CD player★ to listen to some music.

You haven't yet touched a PC, a personal computer, but you've already dealt with at least 10 computers—as you probably guessed from the ★s. All these familiar appliances rely on tiny "computers on chips" called microprocessors. "These marvels of engineering have been infiltrating our everyday lives for more than a quarter of a century," says technology writer Dan Gillmor, but "in some ways the revolution has only begun."[21]

Maybe, then, the name *computer* is inadequate. As computer pioneer John Von Neumann has said, the device should not be called a computer but rather the "all-purpose machine." It is not, after all, just a machine for doing calculations. The most striking thing about it is that it can be put to *any number of uses.*

What are the various types of computers? Let's take a look.

All Computers, Great & Small: The Categories of Machines

At one time, the idea of having your own computer was almost like having your own personal nuclear reactor. In those days, in the 1950s and '60s, computers were enormous machines affordable only by large institutions. Now they come in a variety of shapes and sizes, which can be classified according to their processing power.

Supercomputer
Blue Mountain

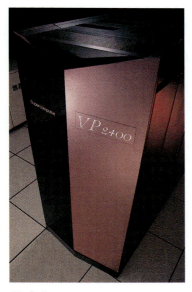

Mainframe computer

- **Supercomputers: Typically priced from $500,000 to over $35 million, <u>supercomputers</u> are high-capacity machines with hundreds of thousands of processors that can perform over 1 trillion calculations per second.** These are the most expensive but fastest computers available. "Supers," as they are called, have been used for tasks requiring the processing of enormous volumes of data, such as doing the U.S. census count, forecasting weather, designing aircraft, modeling molecules, breaking codes, and simulating explosion of nuclear bombs. More recently they have been employed for business purposes—for instance, sifting demographic marketing information—and for creating film animation. The fastest computer in the world, which cost $121.5 million and looks like 48 refrigerator-size blue boxes, is Blue Mountain, located at Los Alamos Laboratory in New Mexico.

- **Mainframe computers:** The only type of computer available until the late 1960s, <u>mainframes</u> **are water- or air-cooled computers that cost $5000–$5 million and vary in size from small, to medium, to large, depending on their use.** Small mainframes ($5000–$200,000) are often called *midsize computers;* they used to be called *minicomputers,* although today the term is seldom used. Mainframes are used by large organizations—such as banks, airlines, insurance companies, and colleges—for processing millions of transactions. Often users access a mainframe by means of a <u>**terminal,**</u> **which has a display screen and a keyboard and can input and output data but cannot by itself process data.**

- **Workstations:** Introduced in the early 1980s, <u>workstations</u> **are expensive, powerful computers usually used for complex scientific, mathematical, and engineering calculations and for computer-aided design and computer-aided manufacturing.** Providing many capabilities comparable to midsize mainframes, workstations are used for such tasks as designing airplane fuselages, prescription drugs, and movie special effects. Workstations have caught the eye of the public mainly for their graphics capabilities, which are used to breathe three-dimensional life into movies such as *Jurassic Park* and *Titanic.* The capabilities of low-end workstations overlap those of high-end desktop microcomputers.

Workstation

- **Microcomputers:** <u>*Microcomputers,*</u> **also called** <u>*personal computers,*</u> **which cost $500–$5000, can fit next to a desk or on a desktop, or can be carried around.** They are either stand-alone machines or are connected to a computer network, such as a local area network. **A** <u>***local area network (LAN)***</u> **connects, usually by special cable, a group of desktop PCs and other devices, such as printers, in an office or a building.**

Three types of micro-computers

Desktop, tower, and note-book

Microcomputers are of several types: *desktop PCs, tower PCs, laptops* (or *notebooks*), and *personal digital assistants*—handheld computers or palmtops.

<u>Desktop PCs</u> are those in which the case or main housing sits on a desk, with keyboard in front and monitor (screen) often on top. **<u>Tower PCs</u> are those in which the case sits as a "tower,"** often on the floor beside a desk, thus freeing up desk surface space.

<u>Notebook computers</u>, also called laptop computers, are lightweight portable computers with built-in monitor, keyboard, hard-disk drive, battery, and AC adapter that can be plugged into an electrical outlet; they weigh anywhere from 1.8 to 9 pounds.

<u>Personal digital assistants (PDAs)</u>, also called <u>handheld computers</u> or <u>palmtops</u>, combine personal organization tools—schedule planners, address books, to-do lists—with the ability in some cases to send e-mail and faxes. Some PDAs have touch-sensitive screens. Some also connect to desktop computers for sending or receiving information. (For now, we are using the word *digital* to mean "computer based.")

Three types of personal digital assistants

● **Microcontrollers: <u>Microcontrollers</u>, also called embedded computers, are the tiny, specialized microprocessors installed in "smart" appliances and automobiles.** These microcontrollers enable microwave ovens, for example, to store data about how long to cook your potatoes and at what temperature. Recently microcontrollers have been used to develop a new universe of experimental electronic appliances—e-pliances—for example, as tiny Web servers embedded in clothing, jewelry, and household appliances such as refrigerators.[22]

Microcontroller

Servers

The word *server* does not describe a size of computer but rather a particular way in which a computer is used. Nevertheless, because servers have become so important to telecommunications, especially with the rise of the Internet and World Wide Web, they deserve mention here. (Servers are discussed in detail in Chapter 5.)

A <u>server</u>, or network server, is a central computer that holds collections of data (databases) and programs for connecting PCs, workstations, and other devices, which are called <u>clients</u>. These clients are linked by a wired or wire-

less network. The entire network is called a *client/server network*. In small organizations, servers can store files and transmit e-mail. In large organizations, servers can house enormous libraries of financial, sales, and product information.

You may never see a supercomputer or mainframe or server or even a tiny microcontroller. But you will most certainly get to know the personal computer, if you haven't already. We consider this machine next.

CONCEPT CHECK

Describe the five sizes of computers.

What are the different types of microcomputers?

Define servers and clients.

1.3 Understanding Your Computer: What If You Custom-Ordered Your Own PC?

KEY QUESTIONS
What four basic operations do all computers follow, and what are some of the devices associated with each operation? How does communications affect these operations?

Could you build your own personal computer? Some people do, putting together bare-bones systems for just a few hundred dollars. "If you have a logical mind, are fairly good with your hands, and possess the patience of Job, there's no reason you can't . . . build a PC," says science writer David Einstein. And, if you do it right, "it will probably take only a couple of hours," because industry-standard connections allow components to go together fairly easily.[23]

Actually, probably only techies would consider building their own PC. But many ordinary users order their own *custom-built* PCs. Let's consider how you might do this.

How Computers Work: Three Key Concepts

We're not actually going to ask you to build or order a PC—just to *pretend* to do so. The purpose of this exercise is to help you understand how a computer works. That information will help you when you go shopping for a new system or, especially, if you order up a custom-built system.

Note: Complete advice about how to buy a computer system is available on our publisher's Web site (*www.mhhe.com/cit/uit4e*).

Before you begin, you will need to understand three key concepts.

FIRST: *The purpose of a computer is to process data into information.* **Data consists of the raw facts and figures that are processed into information**—for example, the votes for different candidates being elected to student-government office. **Information is data that has been summarized or otherwise manipulated for use in decision making**—for example, the total votes for each candidate, which are used to decide who won.

SECOND: You should know *the difference between hardware and software.* **Hardware consists of all the machinery and equipment** in a computer system. The hardware includes, among other devices, the keyboard, the screen, the printer, and the "box"—the computer or processing device itself. **Software, or programs, consists of all the instructions that tell the computer how to perform a task.** These instructions come from a software developer in a form (such as a CD-ROM disk) that will be accepted by the computer. Examples you may have heard of are Microsoft Windows 98 or Office 2000.

THIRD: Regardless of type and size, *all computers follow the same four basic operations:* (1) *input,* (2) *processing,* (3) *storage,* and (4) *output.* To this we will add (5) *communications.*

1. **Input operation: _Input_ is whatever is put in ("input") to a computer system.** Input can be nearly any kind of data—letters, numbers, symbols, shapes, colors, temperatures, sounds, or whatever raw material needs processing. When you type some words or numbers on a keyboard, those words are considered _input data._

2. **Processing operation: _Processing_ is the manipulation a computer does to transform data into information.** When the computer adds 2 + 2 to get 4, that is the act of processing. The processing is done by the **_central processing unit_—frequently called just the CPU—a device consisting of electronic circuitry that executes instructions to process data.**

3. **Storage operation:** Storage is of two types—temporary storage and permanent storage, or primary storage and secondary storage. **_Primary storage_, or memory, is the computer circuitry that temporarily holds data waiting to be processed.** This circuitry is inside the computer. **_Secondary storage_, simply called storage, is the area in the computer where data or information is held permanently.** A floppy disk or hard disk is an example of this kind of storage. (Storage also holds the _software_—the computer programs.)

4. **Output operation: _Output_ is whatever is output from ("put out of") the computer system, the results of processing,** usually information. Examples of output are numbers or pictures displayed on a screen, words printed out on paper in a printer, or music piped over some loudspeakers.

5. **Communications operation:** These days, most (though not all) computers have _communications_ ability, which offers an _extension_ capability—in other words, it extends the power of the computer. With wired or wireless communications connections, data may be input from afar, processed in a remote area, stored in several different locations, and output in yet other places. However, you don't need communications ability to write term papers, do calculations, or perform many other computer tasks.

These five operations are summarized in the illustration opposite. _(See ● Panel 1.2.)_

Pretending to Order a Custom-Built Desktop Computer

Now let's see how you would order a custom-built desktop PC. Remember, the purpose of this is to help you understand the internal workings of a computer so that you'll be better equipped when you go to buy one. (If you were going to build it yourself, you would pretend that someone had acquired the PC components for you from a catalog company and that you're now sitting at a workbench or kitchen table about to begin assembling them. All you would need is a screwdriver, perhaps a small wrench, and a static-electricity-free work area. You would also need the manuals that come with some of the components.) Although prices of components are always subject to change, we have indicated _general ranges_ of prices current as of fall 2000 so that you can get a sense of the relative importance of the various parts.

Input Hardware: Keyboard & Mouse

Input hardware consists of devices that allow people to put data into the computer in a form that the computer can use. At minimum, you will need two things: a _keyboard_ and a _mouse_.

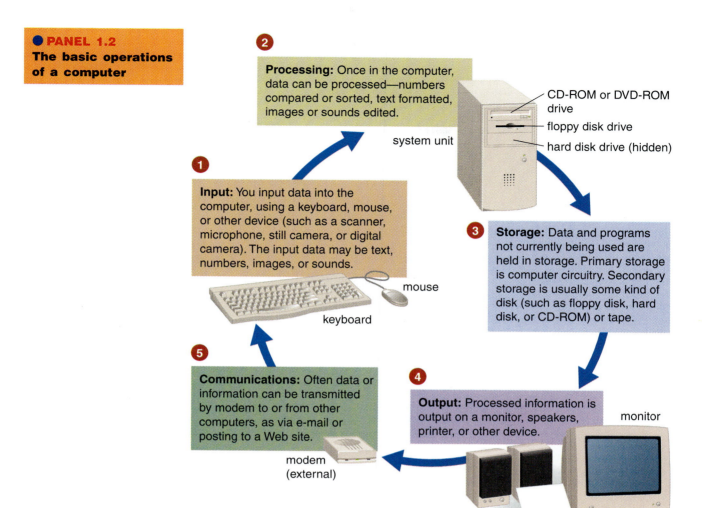

② **Processing:** Once in the computer, data can be processed—numbers compared or sorted, text formatted, images or sounds edited.

CD-ROM or DVD-ROM drive

floppy disk drive

hard disk drive (hidden)

system unit

① **Input:** You input data into the computer, using a keyboard, mouse, or other device (such as a scanner, microphone, still camera, or digital camera). The input data may be text, numbers, images, or sounds.

mouse

keyboard

③ **Storage:** Data and programs not currently being used are held in storage. Primary storage is computer circuitry. Secondary storage is usually some kind of disk (such as floppy disk, hard disk, or CD-ROM) or tape.

⑤ **Communications:** Often data or information can be transmitted by modem to or from other computers, as via e-mail or posting to a Web site.

④ **Output:** Processed information is output on a monitor, speakers, printer, or other device.

monitor

modem (external)

speakers

printer

Keyboard

Mouse

- **Keyboard:** Cost: $30–$50. On a PC, a keyboard is the primary input device. **A _keyboard_ is an input device that converts letters, numbers, and other characters into electrical signals readable by the processor.** A PC keyboard looks like a typewriter keyboard, but besides keys for letters and numbers it has several keys (such as *F* keys and *Ctrl, Alt,* and *Del* keys) intended for computer-specific tasks. After other components are assembled, the keyboard will be plugged into the back of the computer in a socket intended for that purpose.

- **Mouse:** $15–$50. **A _mouse_ is an input device that is used to manipulate objects viewed on the computer display screen.** The mouse will be plugged into the back of the computer in a socket after the other components are assembled.

Processing & Memory Hardware: Inside the System Cabinet

This is the part where most of the assembly work will have to be done. The brains of the computer are the *processing* and *memory* devices, which are housed in the case or system cabinet.

Case or system cabinet

- **Case and power supply:** About $60. Also known as the <u>**system unit**</u>, the <u>**case**</u> or <u>**system cabinet**</u> **is the box that houses the processor chip (CPU), the memory chips, and the motherboard with power supply, as well as storage devices**—floppy-disk drive, hard-disk drive, and CD-ROM or DVD drive, as we will explain. The case comes in desktop or tower models. It includes a power supply unit and a fan to keep the circuitry from overheating.

- **Processor chip:** $100–$750. It may be small and not look like much. But it could be the most expensive hardware component of a build-it-yourself PC—and doubtless the most important. **A processor <u>chip</u> is a tiny piece of silicon that contains millions of miniature electronic circuits.** The speed at which a chip processes information is expressed in *megahertz (MHz)*, millions of processing cycles per second, or *gigahertz (GHz)*, billions of processing cycles per second. For $100, you might get a 333-MHz chip, which is adequate for most student purposes, such as writing term papers. For top dollar, you might get a 1-gigahertz chip, which you would want if you're running software with gee-whiz graphics, such as those with some video games.

Processor chip

- **Memory chips:** $100–$175. These are also small. <u>**Memory chips**</u>, **also known as RAM (random access memory) chips, represent primary storage or temporary storage; they hold data prior to processing and information after processing,** before it is sent along to an output or storage device. You'll want enough memory chips to hold at least 64 megabytes, or roughly 64 million characters, of data, which is adequate for most student purposes. If you work with large graphic files, you'll need more memory capacity, 128 megabytes. (We explain the numbers used to measure storage capacities in a moment.)

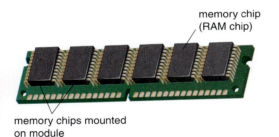

memory chip (RAM chip)

memory chips mounted on module

- **Motherboard:** About $75–$150. Also called the *system board*, **the <u>motherboard</u> is the main circuit board in the computer.** This is the big green circuit board to which everything else—such as the keyboard, mouse, and printer—attaches through connections (called *ports*) in the back of the computer. The processor chip and memory chips are also installed on the motherboard.

  Note that the motherboard has <u>**expansion slots**</u>—**for expanding the PC's capabilities—which give you places to plug in additional circuit boards,** such as those for video, sound, and communications (modem).

Now the components can be put together. As the illustration opposite shows, ① the memory chips are plugged into the motherboard. Then ② the processor chip is plugged into the motherboard. Now ③ the motherboard is attached to the system cabinet. Then ④ the power supply unit is connected to the system cabinet. Finally, ⑤ the wire for the power switch, which turns the computer on and off, is connected to the motherboard.

Motherboard, with processor chips, memory chips, and expansion slots

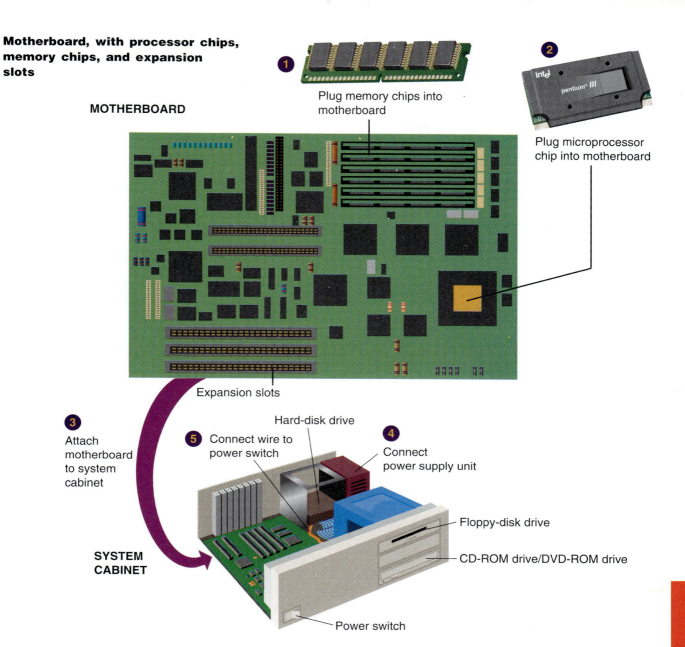

MOTHERBOARD

1 Plug memory chips into motherboard

2 Plug microprocessor chip into motherboard

Expansion slots

3 Attach motherboard to system cabinet

SYSTEM CABINET

Hard-disk drive

5 Connect wire to power switch

4 Connect power supply unit

Floppy-disk drive

CD-ROM drive/DVD-ROM drive

Power switch

Storage Hardware: Floppy Drive, Hard Drive, & CD-ROM Drive

With the motherboard in the system cabinet, the next step is installation of the storage hardware. Whereas memory chips deal with temporary storage, now we're concerned with *secondary storage* or *permanent storage,* in which you'll be able to store your data for the long haul.

For today's student purposes, you'll need at minimum a floppy-disk drive, a hard-disk drive, and a CD-ROM drive. If you work with large files, you'll also want a Zip-disk drive. These storage devices slide into the system cabinet from the front and are secured with screws. Each drive is attached to the motherboard by a flat cable (called a ribbon cable). Also, each drive must be hooked up to a plug extending from the power supply.

A computer system's data/information storage capacity is represented by bytes, kilobytes, megabytes, gigabytes, and terabytes. Roughly speaking, a *byte = 1 character* of data, a *kilobyte = 1000 characters,* a *megabyte = 1 million characters,* a *gigabyte = 1 billion characters,* and a *terabyte = 1 trillion characters.* A character could be alphabetic (A, B, or C) or numeric (1, 2, or 3) or a special character (!, ?, *, $, %).

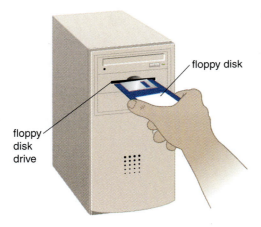

floppy disk

floppy disk drive

- **Floppy-disk and Zip drives:** $100–$110. A *floppy-disk drive* **is a storage device that stores data on removable 3.5-inch-diameter diskettes.** These diskettes don't seem to be "floppy," because they are encased in hard plastic, but the mylar disk inside is indeed flexible or floppy. Each can store 1.44 million bytes (characters) or more of data. With the floppy-disk drive installed, you'll later be able to insert a diskette through a slot in the front and remove it by pushing the eject button. **A *Zip-disk drive* is a storage device that stores data on floppy-disk cartridges with at least 70 times the capacity of the standard floppy.**

- **Hard-disk drive:** $175 for 10.8 gigabytes of storage. A *hard-disk drive* **is a storage device that stores** billions of characters of data on a nonremovable disk platter. With 10.8 gigabytes of storage, you should be able to handle most student needs.

Hard-disk drive

- **CD-ROM drive/DVD-ROM drive:** $75–$200. A *CD-ROM drive*, or its more recent variant, **a DVD-ROM drive, is a storage device that uses laser technology to read data from optical disks.** These days new software is generally supplied on CD-ROM disks rather than floppy disks. And even if you can get a program on floppies, you'll find it easier to install a new program from one CD-ROM disk rather than repeatedly inserting and removing, say, 10 or 12 floppy disks.

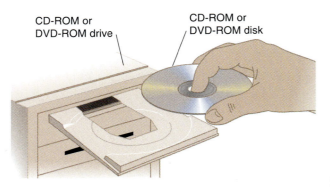

CD-ROM or DVD-ROM drive

CD-ROM or DVD-ROM disk

The system cabinet has lights on the front that indicate when these drives are in use. (You must not remove the diskette from the floppy-disk drive until the relevant light goes off, or else you risk damage to both disk and drive.) The wires for these lights need to be attached to the motherboard.

Output Hardware: Video & Sound Cards, Monitor, Speakers, & Printer

Output hardware consists of devices that translate information processed by the computer into a form that humans can understand—print, sound, graphics, or video, for example. Now a video card and a sound card need to be installed in the system cabinet. Next the monitor, speakers, and a printer are plugged in.

Incidentally, this is a good place to introduce the term *peripheral device*. **A *peripheral device* is any component or piece of equipment that expands a computer's input, storage, and output capabilities.** Examples include printers and disk drives.

Video card

- **Video card:** $50–$100. You doubtless want your monitor to display color (rather than just black-and-white) images. Your system cabinet will therefore need to have a device to make this possible. **A *video card* converts the processor's output information into a video signal that can be sent through a cable to the monitor.** Remember the expansion slots we mentioned? Your video card is plugged into one of these on the motherboard. (You can also buy a motherboard with built-in video.)

Sound card

- **Sound card:** $30–$100. You may wish to listen to music on your PC. If so, you'll need a ***sound card***, **which enhances the computer's sound-generating capabilities by allowing sound to be output through speakers.** This, too, would be plugged into an expansion slot on the motherboard. (Once again, you can buy a motherboard with built-in sound.) With the CD-ROM drive connected to the card, you can listen to music CDs.

- **Monitor:** $225 for a 17-inch model, $380 for a 19-inch model. As with television sets, the "inch" dimension on monitors is measured diagonally corner to corner. **The *monitor* is the display device that takes the electrical signals from the video card and forms an image using points of colored light on the screen.** Later, after the system cabinet has been closed up, the monitor will be connected by means of a cable to the back of the computer, using the clearly marked connector. The power cord for the monitor will be plugged into a wall plug.

Monitor

Sound speakers

- **Pair of speakers:** $30–$200. ***Speakers*—the devices that play sounds transmitted as electrical signals from the sound card—**may not be very sophisticated, but unless you're into high-fidelity recordings they're probably good enough. The two speakers are connected to a single wire that is plugged into the back of the computer once installation is completed.

- **Printer:** $100–$500. Especially for student work, you certainly need a ***printer*, an output device that produces text and graphics on paper.** There are various types of printers, as we discuss later (Chapter 5). The printer has two connections. One, which relays signals from the computer, goes to the back of the PC, where it connects with the motherboard. The other is a power cord that goes to a wall plug.

Printer

Communications Hardware: Modem

Computers can be stand-alone machines, unconnected to anything else. If all you're doing is word processing to write term papers, that may be fine. As we have seen, however, the *communications* component of the computer system *vastly* extends the range of a PC. Thus, while the system cabinet is still open there is one more piece of hardware to install.

- **Modem:** $50. A ***modem* is a device that sends and receives data over telephone lines to and from computers.** The modem is mounted on an expansion card, which is fitted into an expansion slot on the motherboard. Later you can run a telephone line from the telephone wall plug to the back of the PC, where it will connect to the modem.

Modem

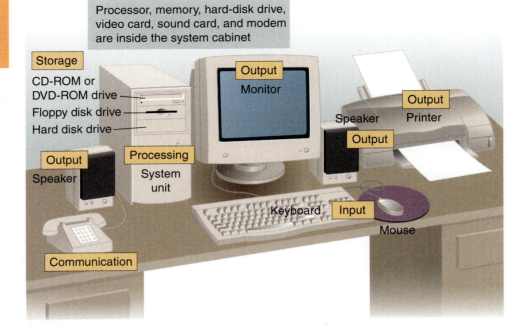

Processor, memory, hard-disk drive, video card, sound card, and modem are inside the system cabinet

Storage

CD-ROM or DVD-ROM drive

Floppy disk drive

Hard disk drive

Output

Monitor

Output

Speaker   Printer

Output

Output

Speaker

Processing

System unit

Keyboard   Input

Mouse

Communication

Now the system cabinet is closed up. The person building the system will plug in all the input and output devices and turn on the power "on" button. Your PC system will look as shown in the illustration above. *(See ● Panel 1.3.)* Are you now ready to roll? Not quite.

Software

With all the pieces put together, the assembler needs to check the motherboard manual for instructions on starting the system. One of the most important tasks is to install software, of which there are two types—system software and application software.

System software: Windows 98

Software comes in a package like this, with the software itself on a CD-ROM disk.

● **System software:** You're really interested in the type of software that allows you to write documents, do spreadsheets, and so on, which is called application software. Before you can get to that, however, system software must be installed. Examples are Windows 98 or Windows Me, which come on CD-ROM disks. The assembler will insert these into your CD-ROM drive and follow the onscreen directions for installation.

  System software **helps the computer perform essential operating tasks and enables the application software to run.** System software consists of several programs. The most important is the *operating system,* the master control program that runs the computer. Examples of operating system software for the PC are various Microsoft programs (such as Windows 95, 98, and Me and NT/2000), Unix, and Linux. The Apple Macintosh computer is another matter altogether. As we explain elsewhere, it has its own hardware components and software, which aren't directly transferable to the PC, by and large.

  After the system software is installed, setup software for the hard drive, the video and sound cards, and the modem must be installed. These setup programs (called *device drivers,* discussed in Chapter 3) will probably come on CD-ROMs (or maybe floppy disks). Once again, the installer inserts these into the drive and then follows the instructions that appear on the monitor screen.

**Application software:
Microsoft Office 2000**
The program comes on a
CD-ROM.

● **Application software:** *Now* we're finally getting somewhere! After the application software has been installed, one can actually start using the PC. ***Application software* enables you to perform specific tasks—solve problems, perform work, or entertain yourself.** For example, when you prepare a term paper on your computer, you will use a word processing program. (Microsoft Word and Corel WordPerfect are two specific brands.) Application software is specific to the system software you use. If you want to run Microsoft Word, for instance, you'll need to first have Microsoft Windows on your system, not Unix or Linux.

Application software comes on CD-ROMs or floppy disks. You insert them into your computer, and then follow the instructions on the screen for installation. Later on you may obtain entire application programs by getting them off the Internet, using your modem.

Although we have said a lot less about software than about hardware, they are equally important. We discuss software in much more detail in Chapters 3 and 4.

Is Getting a Custom-Built PC Worth the Effort?

Does the foregoing description make you want to try putting together a PC yourself? If you add up the costs of all the components (not to mention the value of your time), and then start checking ads for PCs, you might wonder why anyone would bother going to the trouble of building one. And nowadays you would probably be right. "[I]f you think you'd save money by putting together a computer from scratch," says David Einstein, "think again. You'd be lucky to match the price PC-makers are charging these days in their zeal to undercut the competition."[24]

But if you had done this for real, it would not be a futile exercise: By knowing how to build a system yourself, you'd not only be able to impress your friends, but you'd also know how to upgrade any store-bought system to include components that are better than standard. For instance, as Einstein points out, if you're into video games, knowing how to construct your own PC would enable you to make a system that's right for you. You could include the latest three-dimensional graphics video card and a state-of-the-art sound card, for example. More importantly, you'd also know how to *order* a custom-built system (as from Dell, Gateway, or Micron, the mail-order computer makers) that's right for you.

In Chapters 4 and 5, we'll expand on this discussion so that you can *really know what you're doing* when you go shopping for a PC system. However, if you're interested in tips on buying a system now, you can go to the Web site at *www.mhhe.com/cit/uit4e* for advice.

Before we end this introductory overview of information technology, let's wrap up the chapter with a look at the future.

CONCEPT CHECK

What are the three key concepts to understand about how computers work?

Name principal types of input, processing/memory, storage, output, and communications hardware.

Describe the two types of software.

Considering going into hotel work? Then you'll want to know that the hottest new job there is "computer concierge," or *compcierge* (pay: $40,000-plus a year), someone with a knowledge of computer systems who helps the 55% of guests who travel with laptop computers when they have PC or online problems.

Or what about accounting? There the newest specialty is *e-commerce accountant* (pay: up to $54,250), who offers advice on when it makes financial sense to sell goods online, how to find Web developers, and how to manage credit card and sales records.

Today, a knowledge of computers coupled with training in another field offers interesting career paths. Other examples include *bioinformaticist* (biology and computer background, $100,000-plus salary), who studies gene maps, and *virtual set designer* (training in architecture and 3-D computer modeling; pay: up to $150,000), who designs sets for TV shows by combining décors built on computers with physical spaces.[25]

Clearly, information technology is changing old jobs and inventing new ones. To prosper in this environment, you will need to combine a traditional education with training in computers and communications. And you will need to understand what the principal trends of the Information Age are. Let's consider these trends in the development of computers and communications and, most excitingly, the area where they intersect.

Three Directions of Computer Development: Miniaturization, Speed, & Affordability

One of the first computers, the outcome of military-related research, was delivered to the U.S. Army in 1946. ENIAC (short for Electronic Numerical Integrator And Calculator) weighed 30 tons and was 80 feet long and two stories high, but it could multiply a pair of numbers in the then-remarkable time of three-thousandths of a second. This was the first general-purpose, programmable electronic computer, the grandparent of today's lightweight handheld machines.

Since the days of ENIAC, computers have developed in three directions—and are continuing to do so.

- **Miniaturization:** Everything has become smaller. ENIAC's old-fashioned radio-style vacuum tubes gave way to the smaller, faster, more reliable transistor. A *transistor* is a small device used as a gateway to transfer electrical signals along predetermined paths (circuits).

  The next step was the development of tiny integrated circuits. *Integrated circuits* are entire collections of electrical circuits or

Grandpa and offspring
ENIAC (left) is the grandfather of today's handheld machines (right).

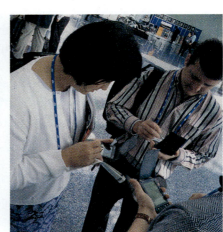

pathways that are now etched on tiny squares of silicon half the size of your thumbnail. *Silicon* is a natural element found in sand. In pure form, it is the base material for computer processing devices.

The miniaturized processor, or microprocessor, in a personal desktop computer today can perform calculations that once required a computer filling an entire room.

Microprocessor

- **Speed:** Thanks to miniaturization, computer makers can cram more hardware components into their machines, providing faster processing speeds and more data storage capacity.

- **Affordability:** Processor costs today are only a fraction of what they were 15 years ago. A state-of-the-art processor costing less than $1000 provides the same processing power as a huge 1980s computer costing more than $1 million.

These are the three major trends in computers. What about communications?

Three Directions of Communications Development: Connectivity, Interactivity, & Multimedia

Once upon a time, we had the voice telephone system—a one-to-one medium. You could talk to your Uncle Joe and he could talk to you, and with special arrangements (conference calls) more than two people could talk with one another. We also had radio and television systems—one-to-many media (or mass media). News announcers such as Dan Rather, Tom Brokaw, or Peter Jennings could talk to you on a single medium such as television, but you couldn't talk to them.

There have been three recent developments in communications:

- **Connectivity:** *Connectivity* **is the ability to connect computers to one another by communications line, so as to provide online information** access. The connectivity resulting from the expansion of computer networks has made possible e-mail and online shopping, for example.

Auto PC
The Clarion Auto PC is the first personal computer designed for a car.

- **Interactivity:** *Interactivity* **is about two-way communication; a user can respond to information he or she receives and modify the process.** That is, there is an exchange or dialogue between the user and the computer or communications device. The ability to interact means users can be active rather than passive participants in the technological process. On the television networks MSNBC or CNN, for example, you can immediately go on the Internet and respond to news from broadcast anchors. In the future, cars may respond to voice commands or feature computers built into the dashboard.

- **Multimedia:** Radio is a single-dimensional medium (sound), as is most e-mail (mainly text). As we mentioned earlier in this chapter, multimedia refers to technology that presents information in more than one medium—such as text, pictures, video, sound, and animation—in a single integrated communication. The development of the World Wide Web expanded the Internet to include pictures, sound, music, and so on, as well as text.

Exciting as these developments are, truly mind-boggling possibilities emerge as computers and communications cross-pollinate.

When Computers & Communications Combine: Convergence, Portability, Personalization—& Information Overload

Sometime in the 1990s, computers and communications started to fuse together, beginning a new era within the Digital Age. The result was three further developments, which haven't ended yet.

- **Convergence:** _**Convergence**_ **describes the combining of several industries through various devices that exchange data in the format used by computers. The industries are computers, communications, consumer electronics, entertainment, and mass media.** Convergence has led to electronic products that perform multiple functions, such as TVs with Internet access or phones with screens displaying text and pictures.

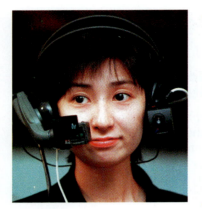

Portability

This wearable computer, from Japan, has a mini-display screen in front.

- **Portability:** In the 1980s, portability, or mobility, meant trading off computing power and convenience in return for smaller size and weight. Today, however, we are close to the point where we don't have to give up anything. As a result, experts have predicted that small, powerful, wireless personal electronic devices will transform our lives far more than the personal computer has done so far. "[T]he new generation of machines will be truly personal computers, designed for our mobile lives," wrote one journalist back in 1992. "We will read office memos between strokes on the golf course and answer messages from our children in the middle of business meetings."[26] Today such activities are commonplace. The risk they bring is that, unless we're careful, work will invade our leisure time.

- **Personalization:** Personalization is the creation of information tailored to your preferences—for instance, programs that will automatically cull recent news and information from the Internet on just those topics you have designated. Companies involved in e-commerce can send you messages about forthcoming products based on your pattern of purchases, usage, and other criteria. Or they will build products (cars, computers, clothing) customized to your heart's desire.

We might also note a fourth—and unintended—development:

- **Information overload:** At one time, most business transactions involved creating, using, sending, and storing paper. The arrival of the computer generated talk of the "paperless office," but there seems to be more paper than ever. Having more paper records means having more information, which, however, does not mean that we have more knowledge. To avoid being buried in an avalanche of unnecessary data, we must learn to distinguish what we really need from what we think we need.

"E" Also Stands for Ethics

Ethics

Every computer user will have to wrestle with ethical issues related to the use of information technology. _**Ethics**_ **is defined as a set of moral values or principles that govern the conduct of an individual or a group.** Because ethical questions arise so often in connection with information technology, we will note them, wherever they appear in this book, with the symbol shown in the margin. Here, for example, are some important ethical concerns pointed out by Tom Forester and Perry Morrison in their book _Computer Ethics._[27] These considerations are only a few of many; we'll discuss others in subsequent chapters.

- **Speed and scale:** Great amounts of information can be stored, retrieved, and transmitted at a speed and on a scale not possible before. Despite the benefits, this has serious implications "for data security and personal privacy," as well as employment, they say, because information technology can never be considered totally secure against unauthorized access.
- **Unpredictability:** Computers and communications are pervasive, touching nearly every aspect of our lives. However, at this point, compared to other pervasive technologies—such as electricity, television, and automobiles—information technology seems a lot less predictable and reliable.
- **Complexity:** Computer systems are often incredibly complex—some so complex that they are not always understood even by their creators. "This," say Forester and Morrison, "often makes them completely unmanageable," producing massive foul-ups or spectacularly out-of-control costs.

CONCEPT CHECK

Describe the three directions in which computers have developed.

Describe the three recent directions in which communications have developed.

Discuss four effects of the fusion of computers and communications.

Name three ethical considerations that result from information technology.

Onward: Are We Headed toward a Global Brain?

In the mid-1990s, the news media—newspapers, magazines, radio, television—became transfixed by the idea of cyberspace as an amazing source of knowledge, and the term information superhighway became a buzz word. More recently the term that has gained momentum is electronic commerce or e-commerce. "Talk about the information superhighway evoked images of freewheeling, wide-ranging exploration . . . a resource for learning and communication," says media critic Norman Solomon. Today, however, cyberspace "is best understood as a way to make and spend money."[28] This growing commercialization has produced a number of benefits, such as e-retailing, e-banking, and B2B (business-to-business) Internet transactions, some of which we discuss in Chapter 2. It has also flooded the Web with trivia; because nearly every company feels it has to have an "Internet strategy," many Web sites amount simply to junk.[29] (Try those listed on snack-food packages.) Indeed, critical-thinking skills are essential to sort out what is drivel and dangerous from what is reliable—for example, in evaluating medical and dietary advice on the Web.

But the universe of information technology is much greater than all this. Earlier philosophers, such as Teilhard de Chardin (1881–1955) and H. G. Wells (1866–1946), foresaw the emergence of what might be called a "global brain," or planetary intelligence, as human knowledge and wisdom expanded exponentially. Today this explosion of information—advanced by supercomputers, fiber-optic cables, satellites, and the trends we have described—might be called the Technosphere.

Despite its possible downside (overconsumption, resource depletion, privacy loss, unequal information access), suggests one futurist, the Technosphere could also offer important benefits in the 21st Century: unlimited information anywhere anytime for low cost; high-speed "tele-health" and "tele-education" services delivered to remote areas; smart-energy and smart-technology transportation, houses, and offices; and "telecommuting" workers engaged in "electronic migration" to online jobs all over the planet.[30]

Let us begin to look at the mind tools for our future.

Experience Box

Better Organization & Time Management: Dealing with the Information Deluge in College—and in Life

The Experience Box appears at the end of each chapter. Each box offers you the opportunity to acquire useful experience that directly applies to the Digital Age. This first box illustrates skills that will benefit you in college, in this course and others. (Students reading the first three editions of our book have told us they received substantial benefit from these suggestions.)

"How on earth am I going to be able to keep up with what's required of me?" you may ask yourself. "How am I going to handle the information glut?" The answer is: *by learning how to learn.* By building your skills as a learner, you certainly help yourself do better in college. More than that, however, you also train yourself to be an information manager in the future.

Using Your "Prime Study Time"

Each of us has a different energy cycle. The trick is to use it effectively. That way, your hours of best performance will coincide with your heaviest academic demands. For example, if your energy level is high during the evenings, you should plan to do your studying then.

To capitalize on your prime study time, you take the following steps: (1) Make a study schedule for the entire term and indicate the times each day during which you plan to study. (2) Find some good places to study—places where you can avoid distractions. (3) Avoid time wasters, but reward your studying with frequent rewards, such as a snack, a TV show, a conversation with a friend.

Improving Your Memory Ability

Memorizing is, of course, one of the principal requirements for succeeding in college. And it's a great help for success in life afterward. Some suggestions:

Get Rid of Distractions Distractions are a major impediment to remembering (as they are to other forms of learning). *External distractions* are those over which you have no control—hallway noises, instructor's accent, people whispering in the library. If you can't banish the distraction by moving, you might try to increase your interest in the subject you are studying. *Internal distractions* are daydreams, personal worries, hunger, illness, and other physical discomforts. Small worries can be shunted aside by listing them on a page for future handling. Large worries may require talking with a friend or counselor.

Space Your Studying, Rather Than Cramming Cramming—making a frantic, last-minute attempt to memorize massive amounts of material—is probably the least effective means of absorbing information. Research shows that it's best to space out your studying of a subject over successive days. A series of study sessions over several days is preferable to trying to do it all during the same number of hours on one day.[31] It is *repetition* that helps move information into your long-term memory bank.

Review Information Repeatedly—Even "Overlearn" It By repeatedly reviewing information—what is known as "rehearsing"—you can improve both your retention and your understanding of it. Overlearning is continuing to review material even after you appear to have absorbed it.

Use Memorizing Tricks There are several ways to organize information so that you can retain it better. For example, you can make drawings or diagrams (as of the parts of a computer system). Some methods of establishing associations between items you want to remember are given in the box. *(See ● Panel 1.4, next page.)*

- **Mental and physical imagery:** Use your visual and other senses to construct a personal image of what you want to remember. Indeed, it helps to make the image humorous, action-filled, sexual, bizarre, or outrageous in order to establish a personal connection. Example: To remember the name of the 21st president of the United States, Chester Arthur, you might visualize an author writing the number "21" on a wooden chest. This mental image helps you associate chest (Chester), author (Arthur), and 21 (21st president).

- **Acronyms and acrostics:** An acronym is a word created from the first letters of items in a list. For instance, *Roy G. Biv* helps you remember the colors of the rainbow in order: red, orange, yellow, green, blue, indigo, violet. An acrostic is a phrase or sentence created from the first letters of items on a list. For example, *Every Good Boy Does Fine* helps you remember that the order of musical notes on the stave is *E-G-B-D-F.*

- **Location:** Location memory occurs when you associate a concept with a place or imaginary place. For example, you could learn the parts of a computer system by imagining a walk across campus. Each building you pass could be associated with a part of the computer system.

- **Word games:** Jingles and rhymes are devices frequently used by advertisers to get people to remember their products. You may recall the spelling rule "I before E except after C or when sounded like A as in *neighbor* or *weigh.*" You can also use narrative methods, such as making up a story.

How to Improve Your Reading Ability: The SQ3R Method

SQ3R stands for *survey, question, read, recite,* and *review.*[32] The strategy behind it is to break down a reading assignment into small segments and master each before moving on. The five steps of the SQ3R method are as follows:

1. Survey the Chapter Before You Read It Get an overview of the chapter or other reading assignment before you begin reading it. If you have a sense what the material is about before you begin reading it, you can predict where it is going. In this text, we offer on the first page of every chapter a list of **Key Questions.** We also present "preview"-type material under the heading **Graphical Interface: A Visual Overview of This Chapter.** Finally, at the end of the chapter we offer a **Summary,** which describes what a term or concept means and why it is important.

2. Question the Segment in the Chapter before You Read It This step is easy to do, and the point, again, is to get yourself involved in the material. After surveying the entire chapter, go to the first segment—section, subsection, or even paragraph, depending on the level of difficulty and density of information. Look at the topic heading of that segment. In your mind, restate the heading as a question. In this book, following each section head we present a **Key Question.** An example in this chapter was "What are three directions of computer development and the three directions of communications development?"

After you have formulated the question, go to steps 3 and 4 (read and recite). Then proceed to the next segment and restate the heading there as a question.

3. Read the Segment about Which You Asked the Question Now read the segment you asked the question about. Read with purpose, to answer the question you formulated. Underline or color-mark sentences you think are important, if they help you answer the question. Read this portion of the text more than once, if necessary, until you can answer the question. In addition, determine whether the segment covers any other significant questions, and formulate answers to these, too. After you have read the

segment, proceed to step 4. (Perhaps you can see where this is all leading. If you read in terms of questions and answers, you will be better prepared when you see exam questions about the material later.)

4. Recite the Main Points of the Segment Recite means "say aloud." Thus, you should speak out loud (or softly) the answer to the principal question or questions about the segment and any other main points. In this book, we have placed **Concept Checks** here and there throughout that serve this function.

5. Review the Entire Chapter by Repeating Questions After you have read the chapter, go back through it and review the main points. Then, without looking at the book, test your memory by repeating the questions.

Clearly the SQ3R method takes longer than simply reading with a rapidly moving color marker or underlining pencil. However, the technique is far more effective because it requires your involvement and understanding. This is the key to all effective learning.

Learning from Lectures

Does attending lectures really make a difference? Research shows that students with grades of B or above were more apt to have better class attendance than students with grades of C− or below.[33]

Some tips for getting the most out of lectures:

Take Effective Notes by Listening Actively
Research shows that good test performance is related to good note taking.[34] And good note taking requires that you listen actively—that is, participate in the lecture process. Here are some ways to take good lecture notes:

- **Read ahead and anticipate the lecturer:** Try to anticipate what the instructor is going to say, based on your previous reading. Having background knowledge makes learning more efficient.

- **Listen for signal words:** Instructors use key phrases such as "The most important point is . . .," "There are four reasons for . . .," "The chief reason . . .,"

"Of special importance . . .," "Consequently . . ." When you hear such signal phrases, mark your notes with a ! or *. "There are four reasons for . . .," "The chief reason . . .," "Of special importance . . .," "Consequently . . ." When you hear such signal phrases, mark your notes with a ! or *.

- **Take notes in your own words:** Instead of just being a stenographer, try to restate the lecturer's thoughts in your own words, which will make you pay attention more.

- **Ask questions:** By asking questions during the lecture, you necessarily participate in it and increase your understanding.

Review Your Notes Regularly Make it a point to review your notes regularly—perhaps on the afternoon after the lecture, or once or twice a week. We cannot emphasize enough how important this kind of reviewing is.

Becoming an Effective Test Taker

Besides having knowledge of the subject matter, you can acquire certain skills that will help during the test-taking process. Some suggestions.[35]

Reviewing: Study Information That Is Emphasized & Enumerated Because you won't always know whether an exam will be an objective or essay test, you need to prepare for both. Here are some general tips.

- **Review material that is emphasized:** In the lectures, this consists of any topics your instructor pointed out as being significant or important. It also includes anything he or she spent a good deal of time discussing or specifically advised you to study. In the textbook, pay attention to key terms, which are often emphasized in boldface type, and their definitions. (In this text, we put key terms *and* their definitions in boldface.)

- **Review material that is enumerated:** Pay attention to any numbered lists, both in your lectures and in your notes. Enumerations often provide the basis for essay and multiple-choice questions.

Spend the night before the test reviewing your notes. Then go to bed without interfering with the material you have absorbed. (For example, don't watch TV.) Get up early the next morning, and review your notes again.

Test Taking: Read Directions & Budget Your Time Get to the classroom early and find a quiet spot. If you don't have a watch, sit where you can see a clock. Avoid talking with others (so as not to increase your anxiety), and do a last review of your notes. Some other test survival skills:

- **Read the test directions:** Many students don't do this and end up losing points because they didn't understand precisely what was required of them. Also, listen to any verbal directions or hints your instructor gives before the test.

- **Budget your time:** Here is an important point of test strategy: Before you start, read through the entire test and figure out how much time you can spend on each section. There is a reason for budgeting your time, of course. You would hate to find you have only a few minutes left and a long essay still to be written. Write the number of minutes allowed for each section on the test booklet or scratch sheet and stick to the schedule. The way you budget your time should correspond to how confident you feel about answering the questions.

Objective Tests: Answer Easy Questions & Eliminate Options Some suggestions for taking objective tests, such as multiple-choice, true/false, or fill-in, are as follows:

- **Answer all questions—easy ones first:** Unless the instructor says you will be penalized for wrong answers, try to answer all questions. Don't waste time stewing over difficult questions. Do the easy ones first, and come back to the hard ones later.

- **Eliminate the options:** Cross out answers you know are incorrect. Be sure to read all the possible answers, especially when the first answer is correct. (After all, other answers could also be correct, so that "All of the above" may be the right choice.) Pay attention to options that are long and detailed, since answers that are more detailed and specific are likely to be correct.

Essay Tests: First Anticipate Answers & Prepare an Outline Because time is limited, your instructor is likely to ask only a few essay questions during the exam. The key to success is to try to anticipate beforehand what the questions might be and memorize an outline for an answer. Here are the specific suggestions:

- **Anticipate ten probable essay questions:** Use the principles we discussed above of reviewing lecture and textbook material that is emphasized and enumerated. You will then be in a position to identify ten essay questions your instructor may ask. Write out these questions.

- **Prepare and memorize informal essay answers:** For each question, list the main points that need to be discussed. Put supporting information in parentheses. Circle the key words in each main point and below the question put the first letter of the key word. Make up catch phrases, using acronyms, acrostics, or word games, so that you can memorize these key words. Test yourself until you can recall the key words the letters stand for and the main points the key words represent.

Visual Summary

Note to the reader: "KQ" refers to Key Questions on the first page of each chapter. The number ties the summary term to the corresponding section in the book.

application software (p. 19, KQ 1.3) Software that has been developed to solve a particular problem, perform useful work on general-purpose tasks, or provide entertainment. *Why it's important: Application software such as word processing, spreadsheet, database manager, graphics, and communications packages are commonly used tools for increasing people's productivity.*

case (p. 14, KQ 1.3) Also known as the *system unit* or *system cabinet;* the box that houses the processor chip (CPU), the memory chips, and the motherboard with power supply, as well as storage devices—floppy-disk drive, hard-disk drive, and CD-ROM or DVD drive. *Why it's important: The case protects many important processing and storage components.*

CD-ROM drive (p. 16, KQ 1.3) Storage device that uses laser technology to read data from optical disks. *Why it's important: New software is generally supplied on CD-ROM disks rather than diskettes. And even if you can get a program on floppies, you'll find it easier to install a new program from one CD-ROM disk rather than repeatedly inserting and removing many diskettes. The newest version is called DVD-ROM.*

central processing unit (CPU) (p. 12, KQ 1.3) Device consisting of electronic circuitry that executes instructions to process data. *Why it's important: The CPU is the "brain" of the computer.*

chip (p. 14, KQ 1.3) Tiny piece of silicon that contains millions of miniature electronic circuits used to process data. *Why it's important: Chips have made possible the development of small computers.*

clients (p. 10, KQ 1.2) Computers and other devices connected to a server, a central computer. *Why it's important: Client/server networks are used in many organizations for sharing databases, devices, and programs.*

communications technology (p. 4, KQ 1.1) Also called *telecommunications technology;* consists of electromagnetic devices and systems for communicating over long distances. *Why it's important: Communications systems using electronic connections have helped to expand human communication beyond face-to-face meetings.*

computer (p. 4, KQ 1.1) Programmable, multiuse machine that accepts data—raw facts and figures—and processes (manipulates) it into useful information, such as summaries and totals. *Why it's important: Computers greatly speed up problem solving and other tasks, increasing users' productivity.*

connectivity (p. 21, KQ 1.4) Ability to connect computers to one another by communications lines, so as to provide online information access. *Why it's important: Connectivity is the foundation of the advances in the digital age. It provides online access to countless types of information and services. The connectivity resulting from the expansion of computer networks has made possible e-mail and online shopping, for example.*

convergence (p. 22, KQ 1.4) The combining of several industries through various devices that exchange data in the format used by computers. The industries are computers, communications, consumer electronics, entertainment, and mass media. *Why it's important: Convergence has led to electronic products that perform multiple functions, such as TVs with Internet access or phones with screens displaying text and pictures.*

cyberspace (p. 6, KQ 1.1) Term used to refer to the online world and the Internet in particular but also the whole wired world of communications in general. Why it's important: *More and more human activities take place in cyberspace.*

data (p. 11, KQ 1.3) Raw facts and figures that are processed into information. Why it's important: *Users need data to create useful information.*

desktop PC (p. 10, KQ 1.2) Microcomputer unit that sits on a desk, with the keyboard in front and the monitor often on top. Why it's important: *Desktop PCs and tower PCs are the most commonly used types of microcomputer.*

e-mail (electronic mail) (p. 5, KQ 1.1) Messages transmitted over a computer network, most often the Internet. Why it's important: *E-mail has become universal; one of the first things new computer users learn is how to send and receive e-mail.*

Ethics

ethics (p. 22, KQ 1.4) Set of moral values or principles that govern the conduct of an individual or a group. Why it's important: *Ethical questions arise often in connection with information technology.*

expansion slots (p. 14, KQ 1.3) Internal "plugs" used to expand the PC's capabilities. Why it's important: *Expansion slots give you places to plug in additional circuit boards, such as those for video, sound, and communications (modem).*

floppy-disk drive (p. 16, KQ 1.3) Storage device that stores data on removable 3.5-inch-diameter flexible diskettes encased in hard plastic. Why it's important: *Floppy-disk drives are included on almost all microcomputers and make many types of files portable.*

hard-disk drive (p. 16, KQ 1.3) Storage device that stores billions of characters of data on a nonremovable disk platter inside the computer case. Why it's important: *Hard disks hold much more data than diskettes do. Nearly all microcomputers use hard disks as their principal secondary-storage medium.*

hardware (p. 11, KQ 1.3) All machinery and equipment in a computer system. Why it's important: *Hardware runs under the control of software and is useless without it. However, hardware contains the circuitry that allows processing.*

information (p. 11, KQ 1.3) Data that has been summarized or otherwise manipulated for use in decision making. Why it's important: *The whole purpose of a computer (and communications) system is to produce (and transmit) usable information.*

information technology (p. 4, KQ 1.1) Technology that merges computing with high-speed communications links carrying data, sound, and video. Why it's important: *Information technology is bringing about the fusion of several important industries dealing with computers, telephones, televisions, and various handheld devices.*

input (p. 12, KQ 1.3) Whatever is put in ("input") to a computer system. Input devices include the keyboard and the mouse. Why it's important: *Useful information cannot be produced without input data.*

interactivity (p. 21, KQ 1.4) Two-way communication; a user can respond to information he or she receives and modify the process. Why it's important: *Interactive devices allow the user to actively participate in a technological process instead of just reacting to it.*

Internet (p. 6, KQ 1.1) Worldwide network that connects up to 400,000 smaller networks linking computers at academic, scientific, and commercial institutions in more then 200 countries. Why it's important: *Thanks to the Internet, millions of people around the world can share all types of information and services.*

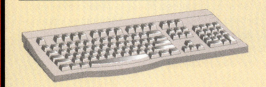

keyboard (p. 13, KQ 1.3) Input device that converts letters, numbers, and other characters into electrical signals readable by the processor. Why it's important: *Keyboards are the most common kind of input device.*

local area network (LAN) (p. 9, KQ 1.2) Network that connects, usually by special cable, a group of desktop PCs and other devices, such as printers, in an office or building. *Why it's important: LANs have replaced mainframes for many functions and are considerably less expensive.*

mainframe (p. 9, KQ 1.2) Second-largest computer available, after the supercomputer; capable of great processing speeds and data storage. Costs $5000–$5 million. Small mainframes are often called *midsize computers. Why it's important: Mainframes are used by large organizations (banks, airlines, insurance companies, universities) that need to process millions of transactions.*

memory chip (p. 14, KQ 1.3) Also known as *RAM* (for *random access memory*) chip; represents *primary storage* or *temporary storage. Why it's important: Holds data prior to processing and information after processing, before it is sent along to an output or storage device.*

microcomputer (p. 9, KQ 1.2) Also called *personal computer;* small computer that fits on or next to a desktop, or can be carried around. Costs $500–$5000. *Why it's important: The microcomputer has lessened the reliance on mainframes and has provided more ordinary users with access to computers. It can be used as a stand-alone machine or connected to a network.*

microcontroller (p. 10, KQ 1.2) Also called an *embedded computer;* the smallest category of computer. *Why it's important: Microcontrollers are built into "smart" electronic devices, such as appliances and automobiles.*

modem (p. 17, KQ 1.3) Device that sends and receives data over telephone lines to and from computers. *Why it's important: A modem enables users to transmit data from one computer to another by using standard telephone lines instead of special communications equipment.*

monitor (p. 17, KQ 1.3) Display device that takes the electrical signals from the video card and forms an image using points of colored light on the screen. *Why it's important: Monitors enable users to view output without printing it out.*

motherboard (p. 14, KQ 1.3) Main circuit board in the computer. *Why it's important: This is the big green circuit board to which everything else—such as the keyboard, mouse, and printer—is attached. The processor chip and memory chips are also installed on the motherboard.*

mouse (p. 13, KQ 1.3) Input device used to manipulate objects viewed on the computer display screen. *Why it's important: For many purposes, a mouse is easier to use than a keyboard for inputting commands. Also, the mouse is used extensively in many graphics programs.*

multimedia (p. 6, KQ 1.1) From "multiple media"; technology that presents information in more than one medium, including text graphics, animation, video, and sound. *Why it's important: Multimedia is used increasingly in business, the professions, and education to improve the way information is communicated.*

network (p. 5, KQ 1.1) Communications system connecting two or more computers. *Why it's important: Networks allow users to share applications and data and to use e-mail. The Internet is the largest network.*

notebook computer (p. 10, KQ 1.2) Also called *laptop computer;* lightweight portable computer with a built-in monitor, keyboard, hard-disk drive, battery, and adapter; weighs 1.8–9 pounds. *Why it's important: Notebook and other small computers have provided users with computing capabilities in the field and on the road.*

online (p. 4, KQ 1.1) Using a computer or other information device, connected through a voice or data network, to access information and services from another computer or information device. *Why it's important: Online communication is widely used by businesses, services, individuals, and educational institutions.*

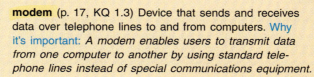

output (p. 12, KQ 1.3) Whatever is output from ("put out of") the computer system; the results of processing. Why it's important: *People use output to help them make decisions. Without output devices, computer users would not be able to view or use the results of processing.*

peripheral device (p. 16, KQ 1.3) Any component or piece of equipment that expands a computer's input, storage, and output capabilities. Examples include printers and disk drives. Why it's important: *Most computer input and output functions are performed by peripheral devices.*

personal digital assistant (PDA) (p. 10, KQ 1.2) Also known as *handheld computer* or *palmtop;* used as a schedule planner and address book and to prepare to-do lists and send e-mail and faxes. Why it's important: *PDAs make it easier for people to do business and communicate while traveling.*

Memory/RAM

primary storage (p. 12, KQ 1.3) Also called *memory;* computer circuitry that temporarily holds data waiting to be processed. Why it's important: *By holding data, primary storage enables the processor to process.*

printer (p. 17, KQ 1.3) Output device that produces text and graphics on paper. Why it's important: *Printers provide one of the principal forms of computer output.*

processing (p. 12, KQ 1.3) The manipulation the computer does to transform data into information. Why it's important: *Processing is the essence of the computer, and the processor is the computer's "brain."*

secondary storage (p. 12, KQ 1.3) Also called *storage;* devices and media that store data and programs permanently—such as disks and disk drives, tape and tape drives, CDs and CD drives. Why it's important: *Without secondary storage, users would not be able to save their work. Storage also holds the computer's software.*

server (p. 10, KQ 1.2) Computer in a network that holds collections of data (databases) and programs for connecting PCs, workstations, and other devices, which are called *clients.* Why it's important: *Servers enable many users to share equipment, programs, and data.*

software (p. 11, KQ 1.3) Also called *programs;* step-by-step instructions that tell the computer hardware how to perform a task. Why it's important: *Without software, hardware is useless.*

sound card (p. 17, KQ 1.3) Special circuit board that enhances the computer's sound-generating capabilities by allowing sound to be output through speakers. Why it's important: *Sound is used in multimedia applications. Also, many users like to listen to music CDs on their computers.*

speakers (p. 17, KQ 1.3) Devices that play sounds transmitted as electrical signals from the sound card. Speakers are connected to a single wire plugged into the back of the computer. Why it's important: *See sound card.*

supercomputer (p. 9, KQ 1.2) High-capacity computer with hundreds of thousands of processors that is the fastest calculating device ever invented. Costs $500,000–$35 million. Why it's important: *Supercomputers are used primarily for research purposes, airplane design, oil exploration, weather forecasting, and other activities that cannot be handled by mainframes and other less powerful machines.*

system software (p. 18, KQ 1.3) System software helps the computer perform essential operating tasks. *Why it's important:* Application software cannot run without system software. System software consists of several programs. The most important is the operating system, the master control program that runs the computer. Examples of operating system software for the PC are various Microsoft programs (such as Windows 95, 98, and Me and NT/2000), Unix, Linux, and the Macintosh operating system.

terminal (p. 9, KQ 1.2) Input and output device that uses a keyboard for input and a monitor for output; it cannot process data. *Why it's important:* Terminals are generally used to input data to and receive data from a mainframe computer system.

tower PC (p. 10, KQ 1.2) Microcomputer unit that sits as a "tower" often on the floor, freeing up desk space. *Why it's important:* Tower PCs and desktop PCs are the most commonly used types of microcomputer.

video card (p. 16, KQ 1.3) Circuit board that converts the processor's output information into a video signal for transmission through a cable to the monitor. *Why it's important:* Virtually all computer users need to be able to view video output on the monitor.

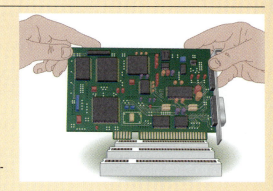

workstation (p. 9, KQ 1.2) Smaller than a mainframe; expensive, powerful computer generally used for complex scientific, mathematical, and engineering calculations and for computer-aided design and computer-aided manufacturing. *Why it's important:* The power of workstations is needed for specialized applications too large and complex to be handled by PCs.

World Wide Web (p. 6, KQ 1.1) The part of the Internet that stores information in multimedia form—sounds, photos, and video as well as text. *Why it's important:* The Web is the most widely known part of the Internet.

Zip-disk drive (p. 16, KQ 1.3) Storage device that stores data on floppy-disk cartridges with at least 70 times the capacity of the standard floppy. *Why it's important:* Zip drives are used to store large files.

Chapter Review

Self-Test Questions

1. The _____ refers to the part of the Internet that stores information in multimedia form.

2. _____ and _____ refer to the two types of microcomputer. One sits on the desktop and the other usually is placed on the floor.

3. _____ technology merges computing with high-speed communications lines carrying data, sound, and video.

4. A _____ is a programmable, multiuse machine that accepts data and processes it into information.

5. Messages transmitted over a computer network are called _____.

6. The _____ is a worldwide network that connects hundreds of thousands of smaller networks in more than 200 countries.

7. _____ refers to technology that presents information in more than one medium.

8. _____ are high-capacity machines with hundreds of thousands of processors.

9. Embedded computers, or _____, are installed in "smart" appliances and automobiles.

10. The kind of software that enables users to perform specific tasks is called _____ software.

Multiple-Choice Questions

1. Which of the following converts computer output into displayed images?
   a. printer
   b. monitor
   c. floppy-disk drive
   d. processor
   e. hard disk drive

2. Which of the following computer types is the smallest?
   a. mainframe
   b. microcomputer
   c. microcontroller
   d. supercomputer
   e. workstation

3. Which of the following is a secondary storage device?
   a. processor
   b. memory chip
   c. floppy-disk drive
   d. printer
   e. monitor

4. Since the days when computers were first made available, computers have developed in three directions. What are they?
   a. increased expense
   b. miniaturization
   c. increased size
   d. affordability
   e. increased speed

5. Which of the following operations constitute the four basic operations followed by all computers?
   a. input
   b. storage
   c. programming
   d. output
   e. processing

True/False Questions

T F   1. Mainframe computers process faster than microcomputers.

T F   2. Main memory is a software component.

T F   3. The operating system is part of the system software.

T F   4. Processing is the manipulation by which a computer transforms data into information.

T F   5. Primary storage is the area in the computer where data or information is held permanently.

Short-Answer Questions

1. What does *online* mean?

2. What is the difference between system software and application software?

3. Briefly define *cyberspace*.

4. What is the difference between software and hardware? Between system software and application software?

5. Briefly describe what a local area network is.

Knowledge in Action

1. Determine what types of computers are being used where you work or go to school. In which departments are the different types of computer used? What are they used for? How are they connected to other computers?

2. Imagine a business you could start or run at home. What type of business is it? What type of computer(s) do you think you'll need? Describe the computer system in as much detail as possible, including hardware components in the areas we have discussed so far. Keep your notes and then refine your answers after you have completed the course.

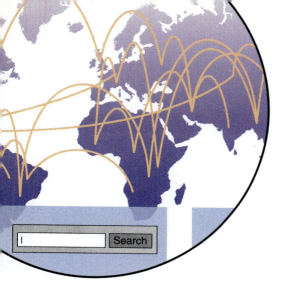

The Internet & the World Wide Web

Exploring Cyberspace

Key Questions

You should be able to answer the following questions.

2.1   **Choosing Your Internet Access Device & Physical Connection: The Quest for Broadband** What are the means of connecting to the Internet, and how fast are they?

2.2   **Choosing Your Internet Service Provider (ISP)** What are the three types of Internet service provider, and what kinds of services do they provide?

2.3   **Sending & Receiving E-Mail** What are the options for obtaining e-mail software, what are the components of an e-mail address, and what are netiquette and spam?

2.4   **The World Wide Web** What are Web sites, Web pages, browsers, URLs, and search engines?

2.5   **The Online Gold Mine: More Internet Resources, Your Personal Cyberspace, E-Commerce, & the E-conomy** What are FTP, Telnet, newsgroups, real-time chat, and e-commerce?

A Visual Overview of This Chapter

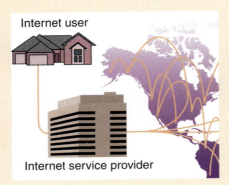

Internet user

Internet service provider

1 **Choosing Your Internet Access Device & Physical Connection.** Some Internet **physical connections,** either wired or wireless, have more **bandwidth**—are able to transmit more data—than others. Data transmission is expressed in **bps** (bits per second—8 bits equal 1 character), **Kbps** (kilobits—thousands of bits per second), **Mbps** (megabits—millions), and **Gbps** (gigabits—billions). Data is **downloaded** from a remote computer to a local computer or **uploaded,** the reverse.

There are four principal types of Internet physical connections: (1) Telephone (dial-up) modem connection is low-speed but inexpensive (up to 56 Kbps). (2) High-speed phone connections are **ISDN** (up to 128 Kbps), which transmits over traditional phone lines; **DSL** (up to 8.4 Mbps), also using traditional phone lines; and **T1** (1.5 Mbps), a special trunk line. (3) **Cable modems** (10 Mbps) connect to cable TV systems. (4) Wireless systems include microwave systems, such as **communications satellites** or space stations (up to 56 Kbps).

2 **Choosing Your Internet Service Provider (ISP).** With a physical connection installed, you then need an **Internet service provider (ISP),** a company to help you connect or **log on** to the Internet. Three types of ISPs are free-service providers, which make money presenting ads; basic-service providers, or small, local companies; and full-service providers, which offer more technical support and other services—examples are AOL and MSN—and which offer **menus,** or lists of commands of services.

3 **Sending & Receiving E-Mail.** Four alternatives for getting and sending e-mail are to buy e-mail software, get the software as part of a browser or other software, get it from your ISP, and get it free (for example, from CNN.com or NetZero). People will send e-mail to you at your **domain**, a location on the Internet consisting of your user name and domain name, such as *user@domain.*

E-mail allows users to send attachments, or separate long documents, with their e-mail messages. It also allows **instant messaging (IM),** in which incoming messages are displayed at once in a **window,** a rectangular area on screen. You can exchange e-mail from people worldwide with similar interests through **list-serves,** or e-mail mailing lists.

The two basic rules of online behavior, or **netiquette,** are these: Don't waste people's time, and don't say anything online you wouldn't say to someone's face. In particular, you should always first consult **FAQs,** or Frequently Asked Questions; avoid **flaming,** such as insults or obscenities; and smooth communication using **emoticons,** or friendly graphic symbols.

To manage your e-mail, filters or instant organizers are recommended. In addition, you will need to know how to manage **spam,** or unsolicited e-mail. Finally, assume e-mail messages are not private: Anyone could read them.

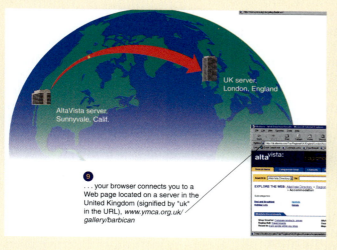

... your browser connects you to a Web page located on a server in the United Kingdom (signified by "uk" in the URL), *www.ymca.org.uk/ gallery/barbican*

4 **The World Wide Web.** What makes the Web so graphically inviting is that it is in multimedia form—graphics, video, and audio as well as text. What makes it easily navigable is that it uses **hypertext,** a system based on **hypertext markup language (HTML)** that uses "tags" or special instructions to provide links among words and phrases in many documents at many Internet sites.

A computer with a domain name (.com, .org, and the like) is called a **site,** and a **Web site** is the location of a Web domain name in a computer somewhere on the Internet. A **Web page** is a document (with text, pictures, sound) on the Web; the first page on the Web site is the **home page.** A **Web browser,** software for viewing and connecting to Web pages, is used to connect with the Web site's address, or **Universal Resource Locator (URL).** A URL, such as *http://www.nps.gov/yose/camping.htm,* consists of (1) the **protocol,** or communication rules—in particular, **HyperText Transfer Protocol (HTTP),** the protocol for connecting with Web servers—(2) the Web server name, (3) the directory, and (4) the file (perhaps with an extension, such as *htm*).

To get around the Web with a browser, you start out from the home page (which you can personalize or customize), then use directional features (Back, Forward, Home, Search), history lists (to keep track of where you've been), and bookmarks (to mark favorite URLs). To interact with a Web page, you use your mouse to click on hyperlinks, click on **radio buttons** (circles in front of options), and enter content in fill-in boxes. You can also click on **scroll arrows** to do **scrolling**—move up and down the Web page.

A starting point for obtaining information is a **Web portal,** a site (such as AOL or Yahoo!) that provides popular features such as search tools. You can check the portal's home page; use a **directory** or category of topics; or use a **keyword,** or subject word, to search for a topic. You can also use a **search engine** to find specific documents through keyword searches and menu choices. Search engines may be human-organized, computer-created, hybrid, or metacrawlers. Among the search strategies are use of quotation marks around search terms and use of operators (AND, OR, NOT, +, −).

Multimedia on the Web may require a **plug-in** (or player or viewer), a program on the browser that allows certain files to be played or viewed. Helper applications run multimedia elements separate from the browser. Web-site developers use **applets** (small multimedia programs) written in **Java,** a programming language for creating animated, interactive Web pages. **Animation** is rapid sequencing of still images. **Streaming video** transfers data in a continuous flow. **Streaming audio** lets you listen to a file as it is being downloaded. **Push technology,** such as **webcasting,** automatically downloads data to your computer—customized text, video, and audio. **Internet telephony** allows you to make phone calls on the Net.

5 **The Online Gold Mine.** Four other Internet resources are (1) **FTP,** a method for copying files; (2) **Telnet,** a means for connecting to remote computers; (2) **newsgroups,** electronic bulletin boards that take place on a special network called **Usenet,** which requires a **newsreader** (part of most browsers) to access; and (3) **real-time chat (RTC),** typed online discussions, which require a chat client (also part of most browsers) to initiate.

The Internet offers personal resources—the ability to do online matchmaking, acquire an online education through **distance learning,** get health information, and amuse yourself. It also offers **e-commerce,** or online business activities, such as retail commerce online ("e-tailing"); online auctions; online finance; online job hunting; and **B2B commerce,** for business-to-business commerce, or the exchange of goods and services directly between companies.

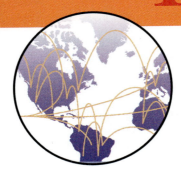

It's the biggest technological generation gap in history."[1]

At least that's the opinion of an executive whose online movie company, Sightsound.com, markets almost exclusively to college students. He's talking about the technological gap—it's not strictly a generation gap—between on-campus students provided with high-speed Internet access by their colleges and most off-campus people, whose slow online connections squeeze through plain old copper-wire phone lines.

The difference can mean receiving information for a report in 3 seconds versus twiddling your thumbs for an entire minute. But academic pursuits account for only some of the activity on campus networks. A lot of the traffic consists of data containing instant messages, music files, toll-free phone calls, online games, e-commerce orders, movies, and much more. Students are passionate users of ICQ, shorthand for a service called "I Seek You," which allows people to exchange messages across the hall (or the world) as fast as they can type. As a result, it's not uncommon to see students emerge simultaneously from their separate dormitory rooms and head toward the dining hall together, having just messaged each other that it's time for dinner.[2] Thanks to this on-campus wiring, a high proportion of Internet users are college students.

Call it the Bandwidth Gap. In general terms, **_bandwidth_ is an expression of how much data—text, voice, video, and so on—can be sent through a communications channel in a given amount of time.** A college dormitory wired with coaxial or fiber-optic cable will have more bandwidth than will a house out in the country served by conventional copper-wire telephone lines; access to information will be hundreds of times faster in the dorm. Of course, many students are not privileged to have this kind of *broadband* or high-speed access. But so significant is the difference that students apparently are making decisions about college housing—even choice of college—based on the availability of high-bandwidth. And those who are about to graduate wonder how they will ever survive in a narrow-bandwidth world.

Because of its standard interfaces and low rates, "the Internet has been the great leveler for communications—the way the PC was for computing," says Boston analyst Virginia Brooks.[3] Let's consider the Internet, a worldwide network connecting up to 400,000 smaller networks. *(See ● Panel 2.1.)* To gain access to the Internet, you need three things: (1) an *access device,* such as a personal computer with a modem; (2) a *physical connection,* such as a telephone line; and (3) an *Internet service provider (ISP).*

2.1 Choosing Your Internet Access Device & Physical Connection: The Quest for Broadband

KEY QUESTIONS

What are the means of connecting to the Internet, and how fast are they?

Can you watch movies or television on any Internet-connected PC? Perhaps you can if you're fortunate enough to be in a state-of-the-art college dorm with "24/7" (24-hours-a-day, seven-days-a-week) high-speed Internet access. But maybe you live off-campus, and while you may be able to connect to the campus network, it may mean a wait of several minutes. Or maybe you have no access to the campus network.

Let's assume you're in the last situation. What are your choices of a **_physical connection_—the wired or wireless means of connecting to the Internet?** A lot depends on where you live. As you might expect, urban and many suburban areas offer more broadband connections than rural areas do. Among the principal means of connection are: (1) telephone (dial-up) modem, (2) sev-

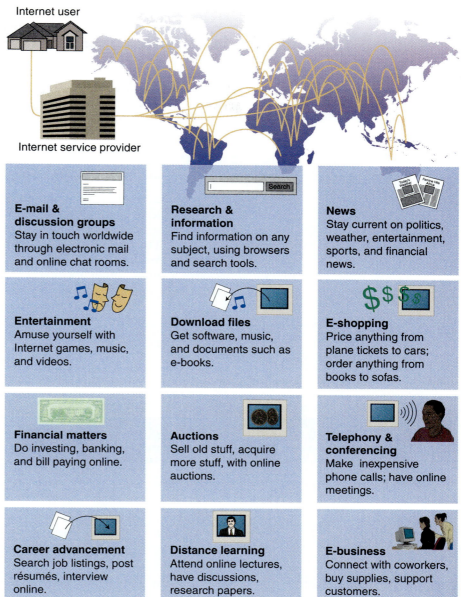

Internet user

Internet service provider

E-mail & discussion groups
Stay in touch worldwide through electronic mail and online chat rooms.

Research & information
Find information on any subject, using browsers and search tools.

News
Stay current on politics, weather, entertainment, sports, and financial news.

Entertainment
Amuse yourself with Internet games, music, and videos.

Download files
Get software, music, and documents such as e-books.

E-shopping
Price anything from plane tickets to cars; order anything from books to sofas.

Financial matters
Do investing, banking, and bill paying online.

Auctions
Sell old stuff, acquire more stuff, with online auctions.

Telephony & conferencing
Make inexpensive phone calls; have online meetings.

Career advancement
Search job listings, post résumés, interview online.

Distance learning
Attend online lectures, have discussions, research papers.

E-business
Connect with coworkers, buy supplies, support customers.

eral high-speed phone lines—ISDN, DSL, and T1, (3) cable modem, and (4) wireless—satellite and other through-the-air links.

Data is transmitted in characters, or collections of bits. Today's data-transmission speeds are measured in bits, kilobits, megabits, and gigabits per second.

- **bps:** A computer with an older modem might have a speed of 28,800 bps. The ***bps*** stands for ***bits per second***. (Eight bits equals one character. A bit, as we will discuss later, is the smallest unit of information used by computers.)

- **Kbps:** This is the most frequently used measure; ***kilobits per second***, **or *Kbps*, are 1000 bits per second.** The speed of a modem that is 28,800 bps might be expressed as 28.8 Kbps.

- **Mbps:** Faster means of connection are measured in ***megabits per second***, **or *Mbps*—1 million bits per second.**

- **Gbps:** At the extreme are ***gigabits per second, Gbps*—1 billion bits per second.**

Why is it important to know these terms? Because the number of bits affects how fast you can upload and download information from a remote computer. **_Download_ is the transmission of data from a remote computer to a local computer,** as from your college's mainframe to your own PC. **_Upload_ is the transmission of data from a local computer to a remote computer.**

The table below shows the transmission rates for various connections, as well as the approximate costs (always subject to change, of course) and their pros and cons.[4] *(See ● Panel 2.2.)* Let's consider each option in turn.

Telephone (Dial-Up) Modem: Low Speed but Inexpensive & Widely Available

The telephone line that you use for voice calls is still the cheapest means of online connections and is available everywhere. As we discussed in Chapter 1, a *modem* is a device that sends and receives data over telephone lines to and from computers. This is known as a *dial-up* connection. These days, the modem is generally installed inside your computer, but there are also external modems. (We explain modems further in Chapter 7.) The modem is attached to the telephone wall outlet. *(See ● Panel 2.3.)*

● PANEL 2.2
Methods of going online compared

Service	Cost per Month (Plus Installation, Equipment)	Maximum Speed (Download Only)	Pluses	Minuses
Telephone (dial-up) modem	$0–$30	56 Kbps	Inexpensive, available everywhere	Slow
ISDN	$40–$110	128 Kbps	Faster than dial-up, uses conventional phone lines	More expensive than dial-up
DSL	$30–$50	1.5–8.4 Mbps	Fast download, always on, higher security	Need to be close to phone company switching station, limited choice of service providers
T1 line	$1000	1.5 Mbps	24 separate circuits of 64 Kbps each, high speed both ways	Expensive, best for businesses
Cable modem	$30–$60	10 Mbps	Fast, always on, most popular broadband type	Slower service during peak times, vulnerability to hackers, limited choice of service providers
Satellite	$30–$130	400 Kbps	Wireless, fast, reliable	Slow (56 Kbps) uploads over phone lines
Radio waves	$159–$1400	155 Mbps	Wireless, fast, always on, reliable	Antenna must be line-of-sight with base station

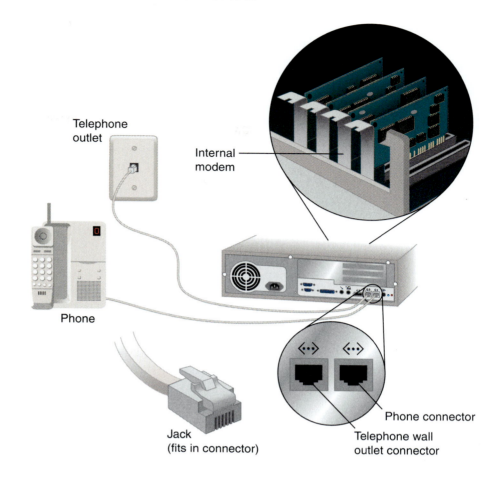

Telephone outlet

Internal modem

Phone

Jack (fits in connector)

Phone connector

Telephone wall outlet connector

Check online connections
You can check your online connection speed by going to the Windows taskbar and double-clicking the connection icon (bottom right of screen):

The result:

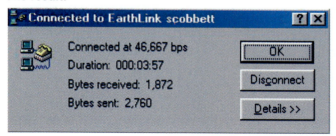

Most modems today have a maximum speed of 56 Kbps. That doesn't mean that you'll be sending and receiving data at that rate. The modem in your computer must negotiate with the modems used by your Internet service provider (ISP), the organization that actually connects you to the Internet, which may have modems operating at slower speeds, such as 28.8 Kbps. In addition, lower-quality phone lines or heavy traffic during peak hours—such as 5 p.m. to 11 p.m. in residential areas—can slow down your rate of transmission. (You can check your connection speed when you're online by using your mouse and clicking twice on the little symbol at the bottom right on your Windows screen—see illustration at left.)

One disadvantage of a telephone modem is that, while you're online, you can't use your phone to make voice calls. In addition, people who try to call you while you're using the modem will get a busy signal. (Call waiting will also interrupt an online connection, so you need to talk to your phone company about disabling it.) As we discuss in a few pages, you probably won't need to pay long-distance phone rates, since most ISPs offer local access numbers. The cost of a dial-up modem connection to the ISP is $10–$30 per month, plus a possible setup charge of $10–$25.

High-Speed Phone Lines: More Expensive but Available in Most Cities

Waiting while your computer's modem takes 25 minutes to transmit a 1-minute low-quality video from a Web site may have you pummeling the desk in frustration. To get some relief, you could enhance your POTS (for *plain old telephone system)* connection with a high-speed adaptation or get a new, dedicated line. Among the choices are ISDN, DSL, and T1, available in most major cities, though not in rural and many suburban areas.

- **ISDN line:** ___ISDN (Integrated Services Digital Network)___ **consists of hardware and software that allows voice, video, and data to be communicated over traditional copper-wire telephone lines.** Capable of transmitting up to 128 Kbps, ISDN is able to send signals over POTS lines. If you were trying to download an approximately 6-minute-long music video from the World Wide Web, it would take you about 4 hours and 45 minutes with a 28.8-Kbps modem. An ISDN connection would reduce this to an hour.

  ISDN costs $10–$40 a month to the phone company, plus perhaps another $30–$70 per month to the Internet service provider. In addition, you may need to pay your phone company $350–$700 or so to hook up an ISDN connector box, possibly run in a new phone line, and install the necessary software in your PC.

- **DSL line:** ___DSL (digital subscriber line)___ **also uses regular phone lines to transmit data in megabits per second.** (The most common type is asynchronous DSL, or ADSL, as we explain elsewhere.) Incoming data is significantly faster than outgoing data. That is, your computer can *receive* data at the rate of 1.5–8.4 Mbps, but it can *send* data at only 16–640 Kbps. This arrangement may be fine, however, if you're principally interested in obtaining very large amounts of data (video, music) rather than in sending them to others. With DSL, you can download that 6-minute music video in only 11 minutes (compared to an hour with ISDN). A big advantage of DSL is that it is always on and, unlike cable (discussed shortly), its transmission rate is consistent. One-time installation cost is $100–$200 plus $100–$300 for a modem supplied by the phone company, and the monthly cost is $30–$50.

  There is one big drawback to DSL: You have to live within 3.3 miles of a phone company central switching office, because the access speed and reliability degrade with distance. However, phone companies are building thousands of remote switching facilities to enhance service throughout their regions. Another drawback is that you have to choose from a list of Internet service providers that are under contract to the phone company you use, although other DSL providers exist.

- **T1 line:** How important is high speed to you? Is it worth $1000 a month? Then consider getting a ___T1 line___, **essentially a traditional trunk line that carries 24 normal telephone circuits and has a transmission rate of 1.5 Mbps.** Generally, T1 lines are used by corporate, government, and academic sites. (Another high-speed line, the T3 line, which transmits at 43 Mbps, costs $10,000 or more a month.)

Cable Modem: Close Competitor to DSL

If DSL's 11 minutes to move a 6-minute video sounds good, 2 minutes sounds even better. That's the rate of transmission for cable modems, which can transmit outgoing data at 500 Kbps and incoming data at 10 Mbps (and eventually, it's predicted, at 30 Mbps). **A *cable modem* connects a personal computer to a cable-TV system that offers an Internet connection.** With around 1 million subscribers, cable has been rapidly gaining market share in high-speed delivery services. Like a DSL connection, it is always on; unlike DSL, you don't need to live near a switching station. Costing $30–$60 a month plus installation and equipment ($300–$500), cable is available in most major cities.

A disadvantage, however, is that you and your neighbors are sharing the system and consequently, during peak-load times, your service may be slowed to the speed of a regular dial-up modem. (You're also more vulnerable to attacks from hackers, although there are defensive or "firewall" programs, such as Winproxy, that reduce the risk.) Finally, cable companies may force you to use their own Internet service providers. For example, if you're with Comcast, you might have to use Excite@Home although you prefer the content offered by AOL, say.

Wireless Systems: Satellite & Other Through-the-Air Connections

Suppose you live out in the country and you're tired of the molasses-like speed of your cranky local phone system. You might consider taking to the air.

- **Satellite:** With a pizza-size satellite dish on your roof, you can receive data at the rate of 400 Kbps from a **_communications satellite_, a space station that transmits radio waves called microwaves from earth-based stations.** Unfortunately, your outgoing transmission will still be only 56 Kbps, because you'll have to use your phone line for that purpose, although genuine two-way satellite service is under development (as from DirectPC). Equipment available from InfoDish or PC Connection costs about $190; installation runs $100–$250; and monthly charges (including ISP charges) are $30–$130, depending on how much time you spend online.
- **Other wireless connections:** In urban areas, some businesses are using radio waves transmitted between towers that handle cellular phone calls, which can send data at up to 155 Mbps and are not only fast and dependable but also always on. The cost is $159–$1400 a month, depending on speed. However, because your antenna must be within line of sight of the service's base station, this system is not available outside cities. Perhaps more practical and affordable is wireless technology such as that to be offered by AT&T (though only in areas not served by its TCI cable service), which enables you to send and receive data (at 512 Kbps) from a box on the outside wall of your house; it does not require line-of-sight—that is, unimpeded—connection with the base station.

Universal Broadband Is Coming

Most PC users employ the physical connections we have described, but there are other possibilities. With WebTV, for instance, you use your television to access the Internet. If your wireless aims are modest—you want to access the Net from your back porch—Apple Computer's AirPort allows you to connect to other Macs and the Internet. The Palm VII, a personal digital assistant from 3Com, enables you to get on the Net from greater distances, but the small screen best accommodates specially prepared Web pages, of which there are few so far.[5]

We are living in a time of rapid changes. Already, nearly a third of the online households in the U.S. have DSL, cable, or wireless services. Most of them, admittedly, are in and around major cities, and the companies providing connections have been slow to expand them to the rest of North America. But broadband is coming. When the telcos (telephone companies) finish scrambling to upgrade their phone lines to DSL, the cable companies make their "pipes" better handle two-way data, and the wireless companies refine their through-the-air links, ordinary Internet users will probably have connections _100 times faster_ than they are today.[6]

Suppose you have an access device such as a modem and you've signed up for a wired or wireless connection. Next, unless you're already on a college campus network, you'll need to arrange for an ***Internet service provider (ISP)**, **a company that connects you through your communications line to its server, or central computer, which connects you to the Internet.** Some well-known ISPs are America Online (AOL, which is also a portal or proprietary network, as we explain), EarthLink, Microsoft Network (MSN), AT&T World-Net, and Prodigy. There are also many local ISPs.

If you've decided simply to use the regular 56-Kbps dial-up modem in your PC connected to a plain old telephone line, you'll quickly notice the fierce competition between companies vying to become your ISP. For instance, perhaps your new PC comes with a keyboard button labeled "Internet," which, when pressed, begins the steps toward connecting you with a service provider—the provider that has come to a financial arrangement with the computer's manufacturer. Beside dealing with the blizzard of ads in magazines and on television, you may also receive promotional ISP start-up disks in the mail. In addition, your phone company probably offers an Internet service. (To do some comparison shopping, go online to *www.thelist.com*, which lists ISPs from all over the world and will guide you through the process of finding one that's best for you.)

The Three Types of ISP Services: Free, Basic, & Full

There are three types of ISPs, with varying levels of service: (1) *free service*, (2) *basic service*, and (3) *full service*.

- **Free-service providers:** One recent development in computers is the availability of things that are free, including ISPs. Among ISPs offering free services are NetZero, AltaVista (partnered with 1stUp.com), BlueLight.com (fromYahoo, Kmart, and Softbank), and Excite@Home (with 1stUp.com). The trade-off here is that in return for your free Internet service, you have to endure a barrage of ads in a banner-size box on your screen. In addition, you need to fill out questionnaires with personal data about yourself, opening yourself up to customized ads and e-mails. ISPs such as BlueLight.com have partnered with merchants in hopes of boosting e-commerce sales. Some brokerage firms (such as J.B. Oxford) use free access to retain customers.

  What's happening here is that, as high-speed Net access at $40 a month spreads, users are more reluctant to pay $20 or so a month for conventional dial-up access. Eventually, some experts think, the trend will halve the prices of conventional ISPs.[7] One new free-ISP company, Broadband Digital Group, hopes to go the competition one better by offering free DSL connections. It expects to make money by asking consumers to fill out surveys and then targeting them with ads for services such as video-on-demand, pay-per-view, online games, and Internet phone calling.[8]

- **Basic-service providers:** The trend toward free service can only be bad news for basic-service ISPs, which tend to be small, local companies offering dial-up access. In return for low fees, such companies typically have offered local access numbers, software to connect you to the Net, and technical support.

- **Full-service providers:** Full-service ISPs include AOL, EarthLink, MSN, Prodigy, and WorldNet. Compared to basic providers, full-service ISPs offer more technical support, backup access numbers in case the main number is out of service, and space for your own Web site on their server. They also offer national and even international access from local numbers so that, when you travel, you pay only the price of a local call for your connection.

The basis for AOL's early success was that it was an *online service*. As such, it was able to offer people most of the content and services they wanted within a single site: e-mail, news, chat, instant messaging, shopping, travel reservations, and financial services. On a lesser scale, MSN acts as an ISP and an entry point to mostly Microsoft-produced content.

AOL content is divided into channels such as Games, Entertainment, Computing, Health, Sports, Shopping, and International. Finding content is easy because each has its own **menu—a list of commands or options**—of services. "While sophisticates dismiss AOL as 'the Net on training wheels,'" says *Business Week* writer Stephen Wildstrom, "its simplicity and sheltered environment have helped the company become, by far, the largest Internet service provider—and have positioned it to become the globe's biggest media company."[9] That status comes thanks to AOL's merger with media giant Time Warner, announced in early 2000, which exploded the whole category of ISP into another realm. By the time you read this, other companies may have followed suit.

Some ISPs give you only one e-mail account, but AOL, Earthlink, AT&T WorldNet, and Prodigy provide five or more addresses, so that family members can share your account.[10] (See ● *Panel 2.4*.)

Once you have contacted an ISP and paid the required fee (charged to your credit card), the ISP will provide you with information about phone numbers for a local connection. This connection is called a *point of presence (POP)*—a server owned by the ISP or leased from a common carrier, such as AT&T. The ISP will also provide you with communications software for setting up your computer and modem to dial into their network of servers. For this you will be given (or give yourself) a *user name* ("user ID") and a *password*, a secret word or string of characters that enables you to **log on, or make a connection to the remote computer.** You will also need to get yourself an e-mail address, as we discuss next.

Internet Service Provider	Monthly Cost for Unlimited Hours of Use	Number of E-Mail Users per Account	Number of Megabytes for a Web Site
American Online (AOL) 800-827-6364 www.aol.com	$21.95	1 master name, 4 additional names	2 per screen name
AT&T WorldNet 800-967-5363 www.att.net	$21.95, first month free	6	5 per user
Earthlink 800-395-8425 www.earthlink.net	$19.95	5	6
Microsoft Network (MSN) 800-373-3676 www.msn.com	$19.95, first month free	1 primary, up to 5 subaccounts	12
Prodigy 800-776-3439 www.prodigy.com	$19.95	5	6

Chapter 2

PRACTICAL ACTION BOX
Choosing an Internet Service Provider

If you belong to a college or company, you may get an ISP free. Some public libraries also offer freenet connections.

If these options are not available to you, be sure to ask these questions when you're making phone calls to locate an Internet service provider:[a]

Costs

- Is there a setup fee? (Most ISPs no longer charge this, though some "free" ISPs will.)
- How much is unlimited access per month? (Most charge about $20 for unlimited usage. But inquire if there are free or low-cost trial memberships or discounts for long-term commitments.)
- If access is supposedly free, what are the trade-offs besides putting up with heavy advertising? (For instance, if the ISP closely monitors your activity in order to accurately target ads, what guarantees do you have that information about you will be kept private? What charges will you face if you try to scrap the advertising window or drop the service?)
- Is there a contract, and for what length of time? That is, are you obligated to stick with the ISP for a while even if you're unhappy with it?

Access

- Is the access number a local phone call? (If not, your monthly long-distance phone tolls could exceed the ISP fee.)
- Is there an alternative dial-up number if the main number is out of service?

- Is access available when you're traveling? Your provider should offer either a wide range of local access numbers in the cities you tend to visit or toll-free 800 numbers.

Support

- What kind of help does the ISP give in setting up your connection?
- Is there free, 24-hour technical support? Is it reachable through a toll-free number?
- How difficult is it to reach tech support? (Try calling the number before you sign up for the ISP and see how long it takes to get a response. Many ISPs keep customers on hold for a long time.)

Reliability

- What is the average connection success rate for users trying to connect on the first try? (The industry average call-success rate is 93.1%. You can try dialing the number during peak hours, to see if you get a modem screech, which is good, rather than a busy signal, which is bad. You can also check Visual Networks, *www.inversenet.com*, for call-failure/call-success rates of various ISPs.)
- Will the ISP keep up with technology? (Are they planning to offer broadband technology such as DSL for speedier access?)
- Will the ISP sell your name to marketers or bombard you with junk messages (spam)?

CONCEPT CHECK

What are the measures of data transmission speed?

Explain the differences among the methods of going online.

Describe the different types of Internet service providers.

2.3 Sending & Receiving E-Mail

KEY QUESTIONS
What are the options for obtaining e-mail software, what are the components of an e-mail address, and what are netiquette and spam?

Once connected with an ISP, one of the first things most people want to do is join the millions of users who send and receive electronic mail. E-mail can be sent at any time and to several people simultaneously. You can receive e-mail wherever you are, using your user name and password to connect to the Internet. In addition, you can attach long (or short) documents or other materials to your e-mail message.

MailStation

It's not necessary, incidentally, to have a PC for e-mail. Ameritech's e-LISTEN allows you to listen to e-mail messages read to you over the telephone or printed out on a fax machine. A single-purpose e-mail device called MailStation—about the size of a hardcover book, with laptop-style keyboard and keys about 85% of normal size plus an adjustable-angle screen—allows you to send and receive messages of up to 1000 words. Using other devices (from Sharp and JVC), you can send and receive e-mail from just about any phone—something travelers toting laptop computers may envy.

E-Mail Software & Carriers

If you aren't on a campus network, there are four ways to go about getting and sending e-mail:

- **Buy e-mail software:** Popular e-mail software programs are Eudora, Outlook Express, or Lotus Notes. However, there is probably no need for you to spend money on these programs because of the following alternatives.

- **Get e-mail program as part of other computer software:** When you buy a new computer, the system will probably include e-mail software, perhaps as part of the software (called *browsers*) used to search the World Wide Web, such as Internet Explorer or Netscape Communicator. An example is Microsoft's Outlook Express, which is part of its Explorer.

- **Get e-mail software as part of your ISP package:** Internet Service Providers—AOL, Prodigy, EarthLink, AT&T WorldNet—provide e-mail software for their subscribers.

- **Get free e-mail services:** According to one study, 69% of e-mail users reported employ free e-mail services.[11] These are available from a variety of sources, ranging from so-called portals or Internet gateways such as Yahoo!, Excite, or Lycos to cable-TV channel CNN's Web site to Juno and NetZero.

Free e-mail from Yahoo!

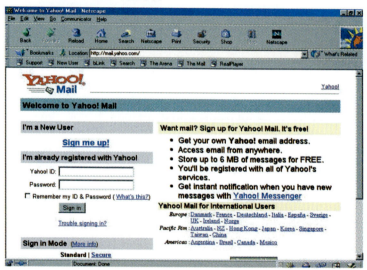

E-Mail Addresses

You'll need an e-mail address, of course, a sort of electronic mailbox used to send and receive messages. All such addresses follow the same approach: *user@domain.* (E-mail addresses are different from Web site addresses, which do not use the symbol @.) **A _domain_ is simply a location on the Internet.** Consider the following address:

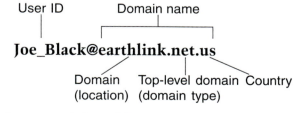

Let's look at the elements of this address.

Joe_Black The first section, the *user ID*, identifies who is at the address—in this case, *Joe_Black*. (There are many ways that Joe Black's user name might be designated, with and without capital letters: *Joe_Black, joe_black, joe.black, joeblack, jblack, joeb*, and so on.)

@earthlink The second section, the *domain name*, which is located after the @ (called "at") symbol, tells the location and type of address. Domain name components are separated by periods (called "dots"). The *domain* portion of the address (such as *Earthlink*, an Internet service provider) provides specific information about the *location*—where the message should be delivered.

.net The *top-level domain* is a three-letter extension that describes the *domain type*: .net, .com, .gov, .edu, .org, .mil, .int—network, commercial, government, educational, nonprofit, military, or international organization. *(See ● Panel 2.5.)*

.us Some domain names also include a two-letter extension for the country—for example, *.us* for United States, *.ca* for Canada, *.uk* for United Kingdom, *.jp* for Japan, *.tr* for Turkey.

Sometimes you'll see an address in which people have their own domains—for example, *Joe@Black.com*. However, you can't simply make up a domain name; it has to be registered. (You can check on whether an address is available, as well, and register it by checking *www.register.com* or *www.internicregistrations.com*)

Incidentally, many people who are unhappy with their ISPs don't change because they don't want to have to notify their friends of a new e-mail address. However, you can switch ISPs by using an e-mail forwarding service (such as Pobox.com, or a college alumni group that offers lifetime e-mail addresses). That way, you can keep one e-mail address no matter how many times you change providers.[12]

Some tips about using e-mail addresses:

- **Type addresses carefully:** You need to type the address *exactly* as it appears, including all spaces, underscores, and periods. If you type an e-mail address incorrectly, your message will be returned to you with a header of (to most people) incomprehensible strings of characters.
- **Use the "reply" command:** When responding to an e-message someone has sent you, the easiest way to avoid making address mistakes is to use the "Reply" command, which will automatically fill in the correct address in the "To" line.

● **PANEL 2.5**
The meaning of Internet top-level domain abbreviations

Domain	Description	Example
.com	Commercial businesses	Editor@mcgraw-hill.com
.edu	Educational and research institutions	Professor@stanford.edu
.gov	U.S. government agencies and bureaus	President@whitehouse.gov
.int	International organizations	Secretary_general@unitednations.int
.mil	U.S. military organizations	Chief_of_staff@pentagon.mil
.net	Internet network resources	Contact@earthlink.net
.org	Nonprofit and professional organizations	Director@redcross.org

Sending e-mail

Address Book: Lists e-mail addresses you use most; can be attached automatically to messages

Send: Command for sending messages

cc: For copying ("carbon copy") message to others

bcc: For copying others ("blind carbon copy") without the primary recipient knowing it

Message area

You can conclude every message with a custom "signature"

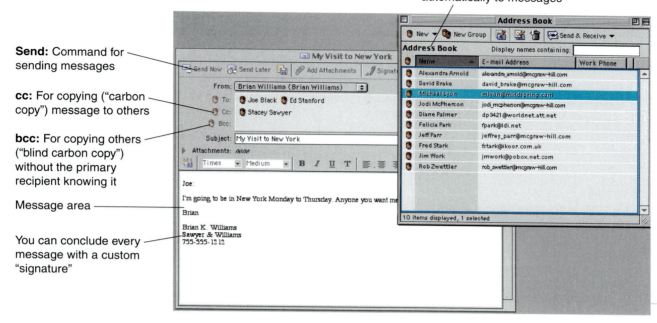

Receiving e-mail

Reply, Reply All, Forward, Delete: For helping you handle incoming e-mail

Inbox lists messages waiting in e-mailbox. (Unopened envelope icon shows unread mail.)

New message displayed here

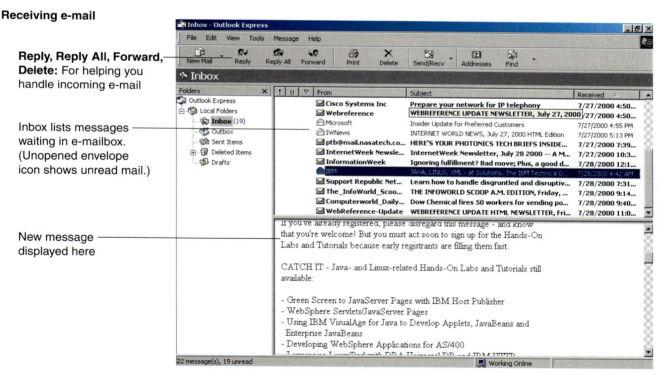

Replying to e-mail

Using the **Reply** command automatically fills in To, From, and Subject lines

- **Use the "address book" feature:** You can store the e-mail addresses of people sending you messages in your program's "address book." This feature also allows you to organize your e-mail addresses according to a nickname or the person's real name so that, for instance, you can look up your friend Joe Black under his real name, instead of under his user name, *bugsme2,* which you might not remember. The address book also allows you to organize addresses into various groups—such as your friends, your parents, club members—so you can easily send all members of a group the same message with a single command.

Attachments

You have written a great research paper and you immediately want to show it off to someone. If you were sending it via the Postal Service, you would write a cover note—"Folks, look at this great paper I wrote about term-paper cheating! See attached"—then attach it to the paper, and stick it in an envelope. E-mail has its own version of this. If the file of your paper exists in the computer from which you are sending e-mail, you can write your e-mail message (your cover note) and then use the Attach File command to attach the document. (Note: It's important that the person receiving the e-mail attachment have the exact same software that created the attached file, such as Microsoft Word 2000, or have software that can read and convert the file.)

While you could also copy your document into the main message and send it that way, e-mail tends to lose formatting options such as **bold** or *italic* text or special symbols. And if you're sending song lyrics or poetry, the lines of text may break differently on someone else's display screen than they do on yours. Thus, the benefit of the attachment feature is that it preserves all such formatting, provided the recipient is using the same word processing software that you did. (If your e-mail is written in Hypertext Markup Language, or HTML, the code designed for making Web pages, as we shall discuss, you can add special fonts, images, and colors to your messages.)

You can also attach pictures, sounds, videos, and other files to your e-mail message.

Note: Many *viruses*—those rogue programs that can seriously damage your PC or programs—ride along with e-mail as attached files. Thus, you should never open an attached file from an unknown source. This was what made the so-called May 2000 Love Bug (ILOVEYOU virus) such a disaster, as we describe in Chapter 9.

Instant Messaging

Instant messages are like a cross between e-mail and phone, allowing for communication that is far speedier than conventional e-mail. With **_instant messaging (IM),_ any user on a given e-mail system can send a message and have it pop up instantly on the screen of anyone else logged onto that system.** Then, if both parties agree, they can initiate online typed conversations in real time. The messages appear on the display screen in a small **_window_—a rectangular area containing a document or activity**—so that users can exchange messages almost instantaneously while operating other programs. Eventually, there will probably be an "open standard" so that anyone can send instant messages to anyone else, no matter what system people are on.[13]

Examples of present instant-message systems are AOL Instant Messenger (AIM—AOL pioneered the idea by allowing members to add other members' names to a "Buddy List"), ICQ ("I Seek You," also from AOL), MSN Messenger, Prodigy Instant Messaging, Tribal Voice PowWow, and Yahoo Messenger. Some of these, such as Yahoo!'s, allow voice chats among users, if their PCs are microphone-equipped.

Sending an e-mail attachment

Third, use your e-mail software's toolbar buttons or menus to attach the file that contains the attachment.

Fourth, click on *Send* to send the e-mail message and attachment.

First, address the person who will receive the attachment.

Second, write a "cover letter" e-mail advising the recipient of the attachment.

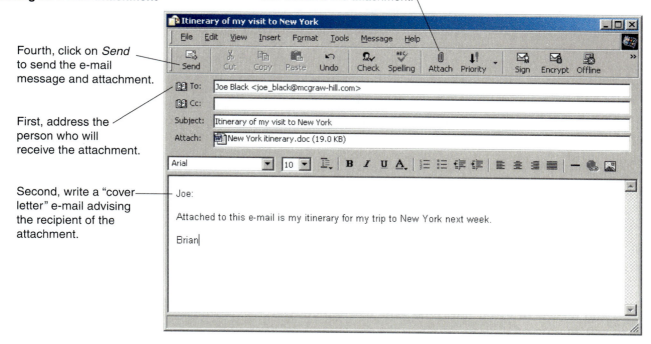

Receiving an e-mail attachment

When you receive a file containing an attachment, you'll see an icon indicating the message contains more than just text. You can click on the icon to see the attachment.

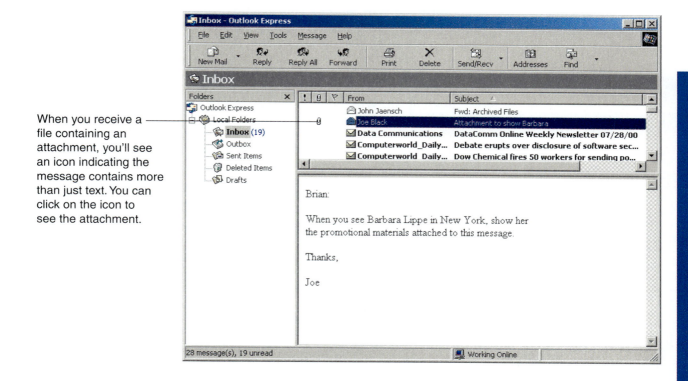

To get instant messaging, which is available for free, you download software and register with the service, providing it with a user name and password. You can then create a list of "buddies" with whom you want to communicate regularly. When your computer is connected to the Internet, the software checks in with a central server, which verifies your identity and looks to see if any of your "buddies" are also online. You can then start a conversation by sending a message to any buddy currently online.[14]

IM has become a hit with many users; indeed, users of AOL's product alone exchange 750 million messages a day. "Instant messaging has made me a happier, more adjusted person," says Randy Weston, 33, an editor who lost touch with many old friends after relocating from San Francisco to Boston. "People who slipped off my radar screen are now back on. With instant messaging, I can see which of my friends are online and I can talk with them in real-time as if we were on the phone."[15] Instant messaging is especially useful in the workplace as a way of reducing long-distance telephone bills when you have to communicate with colleagues who are geographically remote—in South America, say—but with whom you must work closely.

However, you need to be aware of a couple of drawbacks:

- **Lack of common standards:** As of this writing, none of the existing IM products can communicate with each other. If you're using AOL's IM, not only can you not communicate with a buddy on Yahoo!—you can't even communicate with a buddy on AOL's ICQ. Perhaps this will have changed by the time you read this.
- **Time wasters when you have to get work done:** An instant message "is the equivalent of a ringing phone because it pops up on the recipient's screen right away," says one writer.[16] Some analysts suggest that, because of its speed, intrusiveness, and ability to show who else is online, IM can destroy workers' concentration in some offices. You can put off acknowledging e-mail, voice mail, or faxes. But instant messaging is "the cyber-equivalent of someone walking into your office and starting up a conversation as if you had nothing better to do," says one critic. "It violates the basic courtesy of not shoving yourself into other people's faces."[17]

  You can turn off your instant messages, but that is like turning off the ringer on your phone; after a while people will wonder why you're never available. Buddy lists or other contact lists can also become very in-groupish. When that happens, people are distracted from their work as they worry about staying current with their circle (or being shut out of one). Some companies have reportedly put an end to instant messaging, sending everyone back to the use of conventional e-mail.

Mailing Lists: E-Mail–Based Discussion Groups

Want to receive e-mail from people all over the world who share your interests? You can try finding a mailing list and then "subscribing"—signing up, just as you would for a free newsletter or magazine. **_List-serves_ are e-mail mailing lists of people who regularly participate in discussion topics.** To subscribe, you send an e-mail to the list-serve moderator and ask to become a member, after which you will automatically receive e-mail messages from anyone who responds to the server. A directory of mailing lists is available at Publicly Accessible Mailing Lists *(www.neosoft.com/internet/paml)* or OneList *(www.onelist.com)*.

Ethics

Netiquette: Appropriate Online Behavior

You may think etiquette is about knowing which fork to use at a formal dinner. Basically, though, etiquette has to do with politeness and civility—with rules for getting along so that people don't get upset or suffer hurt feelings.

New Internet users, known as *newbies*, may accidentally offend other people in a discussion group or in an e-mail simply because they are unaware of **netiquette, or "network etiquette"—guides to appropriate online behavior.** In general, netiquette has two basic rules: (a) don't waste people's time, and (b) don't say anything to a person online that you wouldn't say to his or her face.

Some more specific rules of netiquette are as follows:

- **Consult FAQs:** Most online groups post **FAQs (Frequently Asked Questions) that explain expected norms of online behavior for a particular group.** Always read these first—before someone in the group tells you you've made a mistake.

- **Avoid flaming:** A form of speech unique to online communication, **flaming is writing an online message that uses derogatory, obscene, or inappropriate language.** Flaming is a form of public humiliation inflicted on people who have failed to read FAQs or otherwise not observed netiquette (although it can happen just because the sender has poor impulse control and needs a course in anger management). Something that smoothes communication online is the use of **emoticons, keyboard-produced pictorial representations of expressions.** *(See ● Panel 2.6.)*

- **Don't SHOUT:** Use of all-capital letters is considered the equivalent of SHOUTING. Avoid, except when they are required for emphasis of a word or two (as when you can't use italics in your e-messages).

- **Avoid sloppiness, but avoid criticizing others' sloppiness:** Avoid spelling and grammatical errors. But don't criticize those same errors in others' messages. (After all, they may not be English native speakers.) Most e-mail software comes with spell-checking capability, which is easy to use.

- **Don't send huge file attachments, unless requested:** Your cousin living in the country may find it takes minutes rather than seconds for his or her computer to download a massive file (as of a video that you want to share). This may tie up the system at a time when your relative badly needs to use it. Better to query in advance before sending large files as attachments. Also, whenever you send an attachment, be sure the recipient has the appropriate software to open your attachment (you both are using Microsoft Word 2000, for example).

- **When replying, quote only the relevant portion:** If you're replying to just a couple of matters in a long e-mail posting, don't send back the entire message. This forces your recipient to wade through lots of text to find the reference. Instead, edit his or her original text down to the relevant paragraph and then put in your response immediately following.

● PANEL 2.6
Some emoticons

Tilt your head to the left and take a look at this emoticon **:=)**
Do you see a smiley face with a long nose? A comment or joke followed by a smiley is often a good way to ensure that it was taken in good humor. Have sad news? Show it while you tell it. **:-(** Feeling teary-eyed? **:'(** Feeling sarcastic? **:-/** Stick your tongue out at someone **:=P** Or pucker up for a kiss **:-***
You can also send hugs. **(())** or **{{}}** And you can send roses. **@---^---** Placing **<w>** before your words signifies a whisper, and **<g>** a grin.

:-)	**Happy face**	**<g>**	**Grin**
:-(	**Sorrow or frown**	**BTW**	**By the way**
:-O	**Shock**	**IMHO**	**In my humble opinion**
:-/	**Sarcasm**	**FYI**	**For your information**
;-)	**Wink**		

Filtering e-mail

E-mail software lets you create folders for storing mail.

Using Filters to Sort Your E-Mail

The average corporate employee gets and sends 201 messages a day, according to a 1999 study by Pitney Bowes.[18] Hopefully, your volume of e-correspondence won't be anywhere near this high. However, one way to stay organized is to use *filters* or instant organizers, using the name of the person or the mailing list to put that particular mail into one folder. Then you can read e-mails sent to this folder later when you have time, freeing up your inbox for mail that needs your more immediate attention. Instructions on how to set up filters are in your e-mail program's Help section.

Spam: Unwanted Junk E-Mail

Several years ago, Monty Python, the British comedy group, did a sketch in which restaurant customers were unable to converse because people in the background (a group of Vikings, actually) kept chanting "Spam, spam, eggs and spam . . ." The term *spam* was picked up by the computer world to describe another kind of "noise" that interferes with communication. Now **spam refers to unsolicited e-mail in the form of advertising or chain letters.** Usually you won't recognize the sender on your list of incoming mail, and often the subject line will give no hint, stating something such as "The status of your application" or "It's up to you now." The solicitations can range from money-making schemes to online pornography.

Some ways to deal with this nuisance are as follows:[19]

- **Delete without opening the message:** Opening the spam message can actually send a signal to the spammer that someone has looked at the onscreen message and therefore that the e-mail address is valid—which means you'll probably get more spams in the future. If you don't recognize the name on your inbox directory or the topic on the inbox subject line, you can simply delete the message without reading it. Or you can use a preview feature in your e-mail program to look at the message without actually opening it, then delete it. (Hint: Be sure to get rid of all the deleted messages from time to time; otherwise, they will build up in your "trash" area.)

- **Never reply to a spam message!** The following advice needs to be taken seriously: *Never reply in any way to a spam message!* Replying confirms to the spammer that yours is an active e-mail address. Some spam senders will tell you that if you want to be removed from their mailing list, you should type the word REMOVE or UNSUBSCRIBE in the subject line and use the reply command to send it back. Invariably, however, all this does is confirm to the spammer that your address is valid, setting you up to receive more unsolicited messages.

  Michael Ashley Lopez, an archaeology graduate student at the University of California at Berkeley, found he had been included on an e-mail list for fans of teen idol Britney Spears. He opened the first e-mail message, wasted 13 seconds reading it, then clicked on the link to unsubscribe from it. The result was he couldn't get off. Months later, despite a determined effort to shake this nuisance, he was still receiving invitations to check out Britney's latest single or preview her latest video.[20]

- **Enlist the help of your ISP or use spam filters:** Your ISP may offer a spam filter to stop the stuff before you even see it. If it doesn't, you can sign up for a filtering service, such as ImagiNet *(www.imagin.net)* for a small monthly charge. Or there are do-it-yourself spam-stopping programs. Examples: Brightmail *(www.brightmail.com)*, Novasoft SpamKiller *(www.spamkiller.com)*, High Mountain Software SpamEater Pro *(www.hms.com)*.

Be warned, however: Even so-called spam killers don't always work. Certainly it didn't for Michael Lopez, victim of the repeated Britney Spears e-mail, even though he had signed up for an e-mail blocking service. "Nothing will work 100%, short of changing your e-mail address," says the operator of an online service called SpamCop. "No matter how well you try to filter a spammer, they're always working to defeat the filter."[21]

- **Fight back:** If you want to get back at spammers, check with abuse.net *(www.abuse.net)*, The Anti-Spam HOWTO *(zikzak.zikzak. net/~acb/features/anti-spam-howto.html)*, or Ed Falk's Spam Tracking Page *(www.rahul.net/falk)*. These will tell you where to report spammers, the appropriate people to complain to, and other spam-fighting tips.

Ethics

What about Keeping E-Mail Private?

The single best piece of advice that can be given about sending e-mail is this: *Pretend every electronic message is a postcard that can be read by anyone.* Because the chances are high that it could be. (And this includes e-mail on college campus systems as well.)

Think the boss can't snoop on your e-mail at work? The law allows employers to "intercept" employee communications if one of the parties involved agrees to the "interception." The party "involved" is the employer. And in the workplace, e-mail is typically saved on a server, at least for a while. Indeed, Federal laws require employers to keep some e-mail messages for years.

Think you can keep your e-mail address a secret among your friends? You have no control over whether they might send your e-messages on to someone else—who might in turn forward it again. (One thing you can do for them, however, is delete their names and addresses before sending one of their messages on to someone.)

Think your ISP will protect your privacy? Often service providers post your address publicly or even sell their customer lists.

Think spammers can't find you? They will if you post an e-mail to an Internet message or bulletin board, making yourself a target for pieces of software (known as "harvester bots") that scour such boards for active e-mail addresses.

And we have not even mentioned your e-mail being intercepted by those knowledgeable individuals known as hackers or crackers, which we discuss elsewhere.

If you're really concerned about preserving your privacy, you can try certain technical solutions—for instance, installing software that encodes and decodes messages (such as PGP, discussed in Chapter 9). But the simplest solution is the easiest: Don't put any sensitive or embarrassing information in your e-mail.

CONCEPT CHECK

What are the options for getting and sending e-mail?

What are some features available with e-mail?

Explain appropriate e-mail etiquette, use of filters, avoidance of spam, and why you should keep e-mail private.

KEY QUESTIONS
What are Web sites, Web pages, browsers, URLs, and search engines?

"I found my old first-grade teacher by surfing the Internet."

When people talk about the Internet in this way, they really mean the World Wide Web. After e-mail, visiting sites ("surfing") on the Web is the most popular use of the Internet. Among the forces driving its popularity are entertainment and e-commerce. *Entertainment* offerings range from listening to music to creating your own, from playing online games by yourself to playing with others, from checking out local restaurants to researching overseas travel. *E-commerce* offers online auctions, retail stores, and discount travel services as well as all kinds of "B2B," or business-to-business, connections, as when General Motors buys online from its steel suppliers.

What makes the World Wide Web so graphically inviting and easily navigable is that this international collection of servers (1) contains information in multimedia form and (2) is connected by hypertext links.

1. **Multimedia form—what makes the Web graphically inviting:**
   Whereas e-mail messages are generally text, the Web provides information in *multimedia* form—graphics, video, and audio as well as text. You can see color pictures, animation, and full-motion video. You can download music. You can listen to radio broadcasts. You can have telephone conversations with others.

2. **Use of hypertext—what makes the Web easily navigable:** Whereas with e-mail you can connect only with specific addresses you know about, with the Web you have hypertext. **Hypertext is a system in which documents scattered across many Internet sites are directly linked, so that a word or phrase in one document becomes a connection to a document in a different place.** The format, or language, used on the Web is called hypertext markup language. (It is not, however, a programming language.) **Hypertext markup language (HTML) is the set of special instructions (called "tags" or "markups") that are used to specify document structure, formatting, and links to other documents.**

For example, if you were reading this book onscreen, you could use your mouse to click on the word *multimedia*—which would be highlighted—in paragraph 1 above, and that would lead you to another location, where perhaps "multimedia" is defined.

How hypertext markup language (HTML) works
The coding in the HTML files tells your Web browser, first, how to find the files of text, graphics, and multimedia files on the server and, second, how to display them on the Web page. The browser also interprets HTML tags, or instructions, as links to other Web sites or to other Web resources, such as files to download.

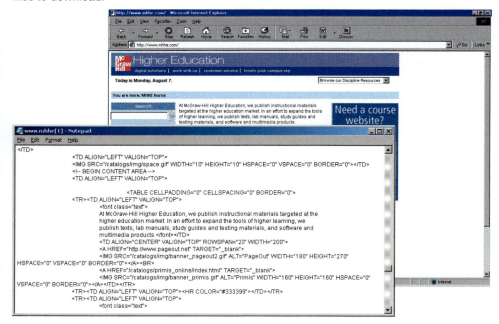

Meaning of tags: Every HTML tag is surrounded by a less-than and greater-than sign—for example, <TR>. Tags often appear as beginning and ending tags, which are identical except for a slash in the end tag—for example, <TR> Paragraph of text. </TR>.

Then you could click on a word in that definition, and that would lead you to some related words—or even some pictures.

The result is that one term or phrase will lead to another, and so you can access all kinds of databases and libraries all over the world. Among the droplets in what amounts to a Niagara Falls of information available: *Weather maps and forecasts. Guitar chords. Recipe archives. Sports schedules. Daily newspapers in all kinds of languages. Nielsen television ratings. A ZIP code guide. Works of literature. The Alcoholism Research Data Base. U.S. Government phone numbers. The Central Intelligence Agency world map. The daily White House press releases.* And on and on.

The Web & How It Works

If a Rip Van Winkle fell asleep in 1989 (the year computer scientist Tim Berners-Lee developed the Web software) and awoke today, he would be completely baffled by the new vocabulary that we now encounter on an almost daily basis: *Web site, home page, www.* Let's see how we would explain to him what these and similar Web terms mean.

- **Web site—the domain on the computer:** You'll recall we described top-level domains, such as .com, .edu, .org, and .net, in our discussion of e-mail addresses. **A computer with a domain name is called a _site_.** When you decide to buy books at the online site of bookseller Barnes & Noble, you would visit its Web site *www.barnesandnoble.com*; the **_Web site_ (often spelled "website") is the location of a Web domain name in a computer somewhere on the Internet.** That computer might be located in Barnes & Noble offices, but it might be located somewhere else entirely. (The Web site for New Mexico's Carlsbad Caverns is not located underground in the caverns, but the Web site for your college is probably on the campus.)

- **Web pages—the documents on a Web site:** A Web site is composed of a Web page or collection of related Web pages. **A _Web page_ is a document on the World Wide Web that can include text, pictures, sound, and video.** The first page you see at a Web site is like the title page of a book. This is the **_home page_, or welcome page, which identifies the Web site and contains links to other pages at the site.** If you have your own personal Web site, it might consist of just one page—the home page. Large Web sites have scores or even hundreds of pages. As of 1999, it was estimated there were around a billion pages on the World Wide Web.[22] (The contents of home pages often change. Or they may disappear, so that the connecting links to them in other Web pages become links to nowhere.)

- **Browsers—software for connecting with Web sites:** A **_Web browser_, or simply _browser_, is software that enables users to view Web pages and to jump from one page to another.** The two best known browsers are

Web site home page

Home page

Other pages on this Web site

Microsoft Internet Explorer

Netscape Navigator

Microsoft's Internet Explorer, which most users prefer, and Netscape Communicator, once the leader but now used by only about a third of consumers.[23] When you connect to a particular Web site with your browser, the first thing you will see is the home page. Then, using your mouse, you can move from one page to another by clicking on hypertext links.

● **URLs—addresses for Web pages:** Before your browser can connect with a Web site, it needs to know the site's address, the URL. **The _URL (Universal Resource Locator)_ is a string of characters that points to a specific piece of information anywhere on the Web.** A URL consists of (1) the Web _protocol_, (2) the name of the Web _server_, (3) the _directory_ (or folder) on that server, and (4) the _file_ within that directory (perhaps with an _extension_ such as _html_ or _htm_). Usually you need to type a URL _exactly_ the way it appears—not type a capital letter, for instance, if a lowercase letter is indicated.

Consider the following example of a URL for a Web site offered by the National Park Service for Yosemite National Park:

Protocol   Web server name   Directory name   File name and extension

http://www.nps.gov/yose/camping.htm

Let's look at these elements.

http://   A **_protocol_ is a set of communication rules for exchanging information.** When you see the _http://_ at the beginning of some Web addresses (as in _http://www.mcgraw-hill.com_), that stands for **_HTTP (HyperText Transfer Protocol),_ the communications rules that allow browsers to connect with Web servers.** Note: Most browsers assume that all Web addresses begin with _http://_ and so you don't need to type this part; just start with whatever follows, such as _www._

www.nps.gov/   The _Web server_ is the particular computer on which this Web site is located. The _www_ stands for "World Wide Web," of course; the _.nps_ stands for "National Park Service," and the _.gov_ is the top-level domain name indicating that this is a government Web site. The server might be physically located in Yosemite National Park in California; in the Park Service's headquarters in Washington, D.C.; or somewhere else entirely.

yose/ The *directory* name is the name on the server for the directory, or folder, from which you need to pull the file. Here it is *yose* for "Yosemite." For Yellowstone National Park, it is *yell.*

camping.htm The *file* is the particular page or document that you are seeking. Here it is *camping.htm*, because you have gone to a Web page about Yosemite's camping facilities. The *.htm* is an extension to the file name, and this extension informs the browser that the file is an HTML file.

A URL, you may have observed, is *not* the same thing as an e-mail address. Some people might type in *president@whitehouse.gov.us* and expect to get a Web site, but it won't happen. The Web site for the White House (which includes presidential information, history, a tour, and guide to federal services) is *http://www.whitehouse.gov*

Using Your Browser to Get around the Web

As stated, the World Wide Web now consists of an estimated 1 billion Web pages. Moreover, the Web is constantly changing; more sites are created and old ones are retired. Without a browser and various kinds of search tools, there would be no way any of us could begin to make any kind of sense of this enormous amount of data.

As we mentioned, a Web page may include *hyperlinks*—words and phrases that appear as underlined or color text—that are references to other Web pages. On a home page, for instance, the hyperlinks serve to connect the top page with other pages throughout the Web site. Other hyperlinks will connect to other pages on other Web sites, whether located on a computer next door or one on the other side of the world.

If you buy a new computer, it will come with a browser already installed. Most browsers have a similar look and feel. Here we show one of the popular browsers in use—Microsoft Internet Explorer.

Notice that the Web browser screen has five basic elements: *menu bar, toolbar, URL bar, workspace,* and *status bar.* To execute menu bar and toolbar commands, you use the mouse to move the pointer over the word, known as a *menu selection,* and click the left button of the mouse. This will result in a *pull-down menu* of other commands for other options. *(See ● Panel 2.7.)*

● PANEL 2.7

The commands on a browser screen

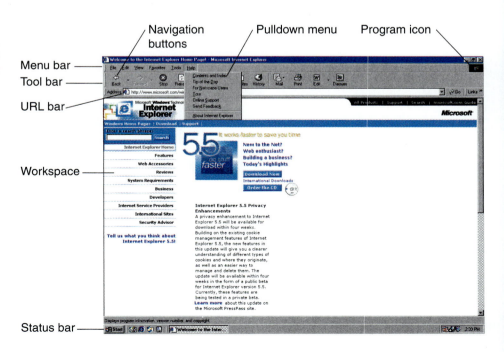

Suppose you live in North America and are planning a trip to Europe. While you're in England, you want to visit London, and you're interested in finding an inexpensive place to stay there. The World Wide Web and its hyperlinks can help you achieve this.

1 You might begin your search by going to the portal AltaVista at *www.altavista.com.* AltaVista's servers are located in Sunnyvale, California.

2 Scrolling down the AltaVista home page shows several underlined links, which contain the URLs (universal resource locators, or Web locations) for other Web pages, or documents on the Web.

When you click on the link *Europe*, . . .

3 . . . your Web browser takes you to the Web page on the AltaVista server containing lists of European countries at *http://dir.altavista.com/Top/Regional/Europe*

When you click on *United Kingdom*, . . .

4 . . . your browser takes you to another page on the server that offers features about the UK, *http://dir.altavista.com/ Top/Regional/Europe/UK*

Clicking on *England* . . .

5 . . . takes you to a Web page of locations in England, *http://dir.altavista.com/Top/ Regional/Europe/UK/England*

Clicking on *London* . . .

6 . . . takes you to a Web page of features about London, *http://dir.altavista.com/Top/ Regional/Europe/UK/England/London*

Clicking on *Accommodation* . . .

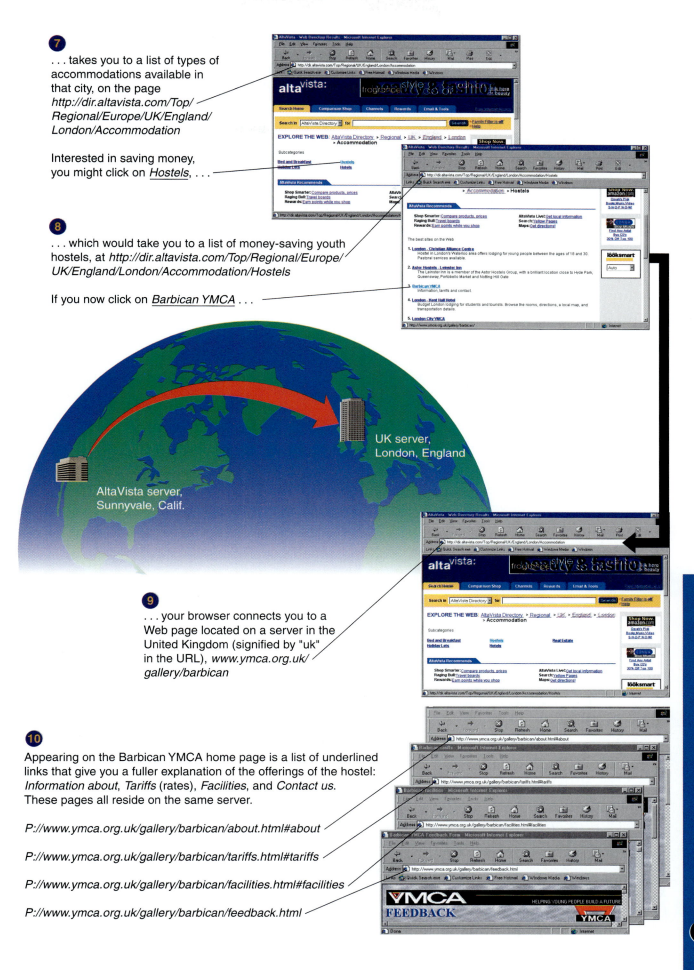

7

... takes you to a list of types of accommodations available in that city, on the page *http://dir.altavista.com/Top/ Regional/Europe/UK/England/ London/Accommodation*

Interested in saving money, you might click on *Hostels*, ...

8

... which would take you to a list of money-saving youth hostels, at *http://dir.altavista.com/Top/Regional/Europe/ UK/England/London/Accommodation/Hostels*

If you now click on *Barbican YMCA* ...

UK server,
London, England

AltaVista server,
Sunnyvale, Calif.

9

... your browser connects you to a Web page located on a server in the United Kingdom (signified by "uk" in the URL), *www.ymca.org.uk/ gallery/barbican*

10

Appearing on the Barbican YMCA home page is a list of underlined links that give you a fuller explanation of the offerings of the hostel: *Information about*, *Tariffs* (rates), *Facilities*, and *Contact us*. These pages all reside on the same server.

P://www.ymca.org.uk/gallery/barbican/about.html#about

P://www.ymca.org.uk/gallery/barbican/tariffs.html#tariffs

P://www.ymca.org.uk/gallery/barbican/facilities.html#facilities

P://www.ymca.org.uk/gallery/barbican/feedback.html

After you've been using a mouse for a while, you may find moving the pointer around somewhat time-consuming. As a shortcut, if you click on the right mouse button, you can reach many of the commands on the toolbar (*Back, Forward,* and so on) via a pop-up menu.

- **Starting out from home:** The first page you see when you start up your browser is the *home page* or *start page.* (You can also start up from just a blank page, if you don't want to wait for the time it takes to connect with a home page.) You can choose any page on the Web you want as your start page, but a good start page offers links to sites you want to visit frequently. Often you may find that the ISP with which you arrange your Internet connection will provide its own start page. However, you'll no doubt be able to customize it to make it your own personal home page.

- **Personalizing your home page:** Want to see the weather forecast for your college and/or hometown areas when you first log on? Or your horoscope, "message of the day," or the day's news (general, sports, financial, health, and so on)? Or the Web sites you visit most frequently? Or a reminder page (as for deadlines or people's birthdays)? You can probably personalize your home page following the directions provided with the first start page you encounter. Or if you have an older Microsoft or Netscape browser you can get a customizing system from either company. A customized start page is also provided by Yahoo!, Excite, AltaVista, and similar services.

- **Getting around—Back, Forward, Home, and Search features:** Driving in a foreign city (or even Boston or San Francisco) can be an interesting experience in which street names change, turns lead into unknown neighborhoods, and signs aren't always evident, so that soon you have no idea where you are. That's what the Internet is like, although on a far more massive scale. Fortunately, unlike being lost in Rome, here your browser toolbar provides navigational aids. *Back* takes you back to the previous page. *Forward* lets you look again at a page you returned from. If you really get lost, you can start over by clicking on *Home,* which returns you to your home page. *Search* lists various other search tools, as we will describe. Other navigational aides are history lists and bookmarks.

Menu bar

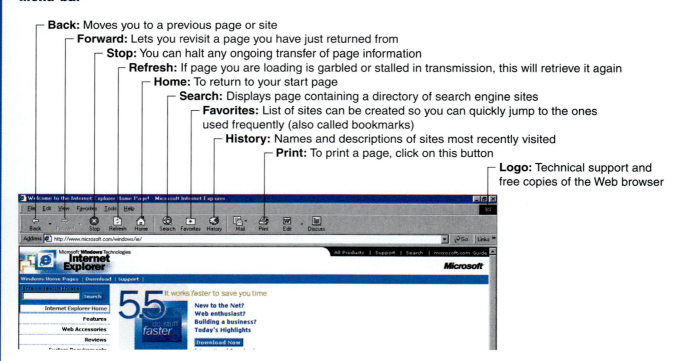

Back: Moves you to a previous page or site

Forward: Lets you revisit a page you have just returned from

Stop: You can halt any ongoing transfer of page information

Refresh: If page you are loading is garbled or stalled in transmission, this will retrieve it again

Home: To return to your start page

Search: Displays page containing a directory of search engine sites

Favorites: List of sites can be created so you can quickly jump to the ones used frequently (also called bookmarks)

History: Names and descriptions of sites most recently visited

Print: To print a page, click on this button

Logo: Technical support and free copies of the Web browser

History List

If you want to return to a previously viewed site and are using Netscape, you click on *Communicator*, then choose *History* from the menu. If you're using Internet Explorer, click on *History*.

History

Adding Bookmarks (Favorites)

If you are at a Web site you may want to visit again, you click on your *Bookmarks* (in Netscape) or *Favorites* (in Internet Explorer) button and choose *Add Bookmark* or *Add to Favorites*. Later, to revisit the site, you can go to the bookmark menu, and the site will reappear.

Favorites

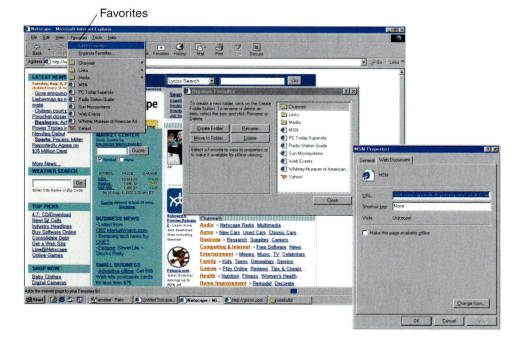

- **History lists:** If you are browsing through many Web pages, it can be difficult to keep track of the locations of the pages you've already visited. The *history list* allows you to quickly return to the pages you have recently visited.

- **Bookmarks or favorites:** One great helper for finding your way is the *bookmark* or *favorites* system, which lets you store the URLs of Web pages you frequently visit so that you don't have to remember and retype your favorite addresses. Say you're visiting a site that you really like and that you know you'd like to come back to. You click on your *Bookmark* or *Favorites* feature, which displays the URL on your screen, then click on *Add*, which automatically stores the address. Later you can locate the site name on your bookmark menu, click on it, and the site will reappear. (When you want to delete it, you can use the right mouse button and select the delete command.)

Hyperlinks

Clicking on underlined or color term transfers you to another Web page.

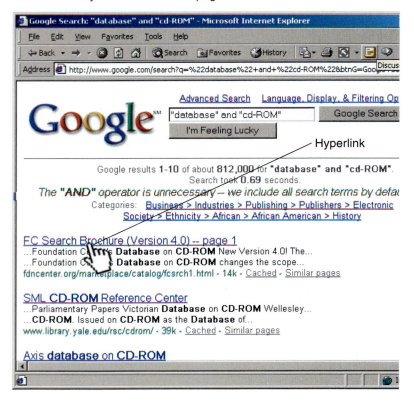

Hyperlink

Radio buttons

Act like station selector buttons on a car radio

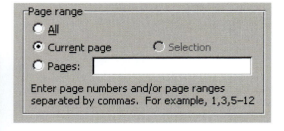

Text boxes

Require you to type in information

- **Interactivity—hyperlinks, radio buttons, and fill-in text boxes:** For any given Web page that you happen to find yourself on, there may be one of three possible ways to interact with it—or sometimes even all three on the same page.

  (1) By using your mouse to click on the hyperlinks, which will transfer you to another Web page.

  (2) By using your mouse to click on a *radio button* and then clicking on a *Submit* command. **<u>Radio buttons</u> are little circles located in front of various options; selecting an option with the mouse places a dot in the corresponding circle.**

  (3) By typing in text in a fill-in text box, then hitting the Enter key or clicking on a *Go* or *Continue* command, which will transfer you to another Web page.

- **Scrolling and frames:** To the bottom and side of your screen display, you will note **<u>scroll arrows</u>, small up/down and left/right arrows. Clicking on scroll arrows with your mouse pointer moves the screen so that you can see the rest of the Web page, a movement known as <u>scrolling</u>.** You can also use the arrow keys on your keyboard for scrolling.

  Some Web pages are divided into different rectangles known as frames, each with its own scroll arrows. **A <u>frame</u> is an independently controllable section of a Web page.** A Web page designer can divide a page into separate frames, each with different features or options.

- **Turning off images:** If you're not on a high-speed access system, you may find that it takes a while for pictures to fade in on your screen. You can save time by turning off the display of images; in their place you will see icons.

- **Looking at two pages simultaneously:** If you want to look at more than one Web page at the same time, you can position them side by side on your display screen. Select *New* from your File menu to open more than one browser window.

Web Portals: Starting Points for Finding Information

Using a browser is sort of like exploring an enormous cave with flashlight and string. You point your flashlight at something, go there, and at that location you can see another cave chamber to go to; meanwhile, you're unrolling the ball of string behind you, so that you can find your way back.

Web portals

America Online (AOL)	www.aol.com
Yahoo!	www.yahoo.com
Microsoft Network (MSN)	www.msn.com
Netscape	www.netscape.com
Lycos	www.lycos.com
Go Network	www.go.com
Excite Network	www.excite.com
AltaVista	www.altavista.com
WebCrawler	www.webcrawler.com

But what if you want to visit only the most spectacular rock formations in the cave and skip the rest? For that you need a guidebook. There are many such "guidebooks" for finding information on the Web, sort of Internet superstations known as **_Web portals_—Web sites that group together popular features such as search tools, e-mail, electronic commerce, and discussion groups.** The most popular portals are America Online, Yahoo!, Microsoft Network, Netscape, Lycos, Go Network, Excite Network, AltaVista, and WebCrawler.[24]

When you log on to a portal, you can do three things: (1) check the home page for general information, (2) use the directories to find a topic you want, and (3) use a keyword to search for a topic. (See ● Panel 2.8.)

● **Check the home page for general information:** You can treat a portal's home or start page as you would one of the mass media—something you tune in to in order to get news headlines, weather forecasts, sports scores, stock-price indexes, and today's horoscope. You might also proceed past the home page to check your e-mail, if you happen to be using the portal for this purpose.

● **PANEL 2.8**
A portal home page

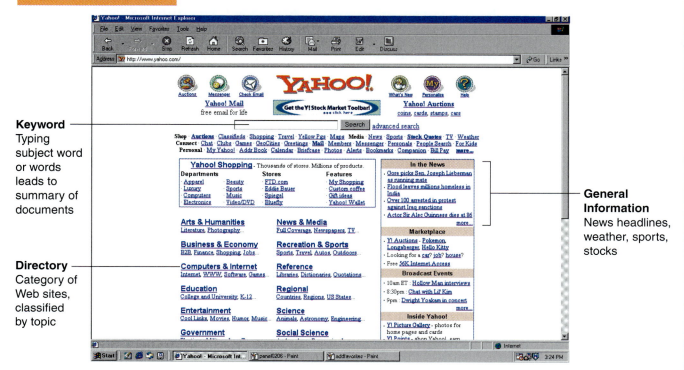

Keyword
Typing subject word or words leads to summary of documents

Directory
Category of Web sites, classified by topic

General Information
News headlines, weather, sports, stocks

- **Use the directories to find a topic:** Before they acquired their other features, many of these portals began as a type of search tool known as a _**directory**_, **providing lists of several categories of Web sites classified by topic,** such as _Business & Finance_ or _Health & Fitness._ Such a category is also called a _hypertext index,_ and its purpose is to allow you to access information in specific categories by clicking on a hypertext link.

  The initial general categories in Yahoo!, for instance, are _Arts & Humanities, Business & Economy, Computers & Internet, Education, Government, Health,_ and so on. Using your mouse to click on one general category (such as _Recreation & Sports_) will lead you to another category (such as _Sports_), which in turn will lead you to another category (such as _College & University_), and on to another category (such as _Conferences_), and so on, down through the hierarchy. If you do this long enough, you will "drill down" through enough categories that you will find a document (Web site) on the topic you want.

  Unfortunately, not everything can be so easily classified in hierarchical form. A faster way may be a _keyword search._

- **Use keyword to search for a topic:** At the top of each portal's home page is a blank space into which you can type a _**keyword**_, **the subject word or words of the topic you wish to find.** If you want a biography on former San Francisco football quarterback Joe Montana, then _Joe Montana_ is the keyword. This way you don't have to plow through menu after menu of subject categories. The results of your keyword search will be displayed in a short summary of documents containing the keyword you typed.

Many users are increasingly bypassing the better known Web portals and going directly to specialty sites or small portals, such as those featuring education, finance, and sports.[25] Examples are Webstart Communications' computer and communications site _(www.cmpcmm.com/cc)_, Travel.com's travel site _(www.travel.com/sitemap.htm)_, and the _New York Times_ home page used by the paper's own newsroom staff to find journalism-related sites _(www.nytimes.com/library/tech/reference/cynavi.html)._ Some colleges are also installing portals for their students.

Four Types of Search Engines: Human-Organized, Computer-Created, Hybrid, & Metacrawlers

At one time, as much as 40% of the Web was covered by a single search tool. By 1999, however, no one tool covered more than 16%. One study found that the top 11 search tools covered only 42% of the Web in 1999, whereas two years earlier six search tools covered 60%.[26] "As the Web billows in size," points out one writer, "search sites cover less of it, and what they cover is more likely to be popular commercial sites."[27]

When you use a keyword to search for a topic, you are using a piece of software known as a _search engine._ Whereas _directories_ are lists of Web sites classified by topic (as offered by portals), _**search engines**_ **allow you to find specific documents through keyword searches and menu choices.** The type of search engine you use depends on what you're looking for.

There are four types of such search tools: (1) human-organized, (2) computer-created, (3) hybrid, and (4) metacrawlers.[28]

- **Human-organized search sites:** If you're looking for a biography of Apple Computer founder Steve Jobs, a search engine based on human judgment is probably your best bet. Why? Because, unlike a computer-created search site, the search tool won't throw everything remotely associated with his name at you. More and more, the top five search sites on the Web (Yahoo!, AOL, MSN, Netscape, and Lycos) are going in the direction of human indexing. Unlike indexes created by computers, humans can judge data for relevance and categorize them in ways that are useful to you. Many of these sites hire people who are subject-area experts (with the idea that, for example, someone interested in gardening would be best able to organize gardening sites). Examples of human-organized search sites are Yahoo!, Open Directory, About.com, and LookSmart.

- **Computer-created search sites:** If you want to see what things show up next to Steve Jobs's name or every instance in which it appears, a computer-created search site may be best. These are assembled by software "spiders" that crawl all over the Web and send back reports to be collected and organized with little human intervention. The downside is that computer-created indexes deliver you more information than you want. Examples of this type are Northern Light, Excite, WebCrawler, FAST Search, and Inktomi.

- **Hybrid search sites:** Hybrid sites generally use humans supplemented by computer indexes. The idea is to see that nothing falls through the cracks. All the principal sites are now hybrid: AOL Search, AltaVista, Lycos, MSN Search, and Netscape Search. Others are Ask Jeeves, Direct Hit, Go, GoTo.com, Google, HotBot, and Snap. Ask Jeeves pioneered the use of natural-language queries (you ask a question as you would to a person: "Where can I find a biography of Steve Jobs?"). Google ranks listings by popularity as well as by how well they match the request. GoTo ranks by who paid the most money for top billing.

- **Metasearch sites:** Metasearch sites send your query to several other different search tools and compile the results so as to present the broadest view. Examples are Go2Net/MetaCrawler, SavvySearch, Dogpile, Inference Find, ProFusion, Mamma, The Big Hub, and C4 TotalSearch.

More information about these search tools is given in the accompanying box. *(See ● Panel 2.9, next page.)*

Tips for Smart Searching

The phrase "trying to find a needle in a haystack" will come vividly to mind the first time you type a word into a search engine and back comes a response on the order of "63,173 listings found." Clearly, it becomes mandatory that you have a strategy for narrowing your search. The following are some tips.[29]

- **Start with general search tools:** Begin with general search tools such as those offered by AltaVista, Excite, GoTo.com, HotBot, Lycos, and Yahoo! (Later, if you haven't been able to narrow your search, you can go to specific search tools, as we'll describe.)

- **Choose your search terms well and watch your spelling:** Use the most precise words possible. If you're looking for information about novelist Thomas Wolfe (author of *Look Homeward Angel*, published 1929) rather than novelist/journalist Tom Wolfe (*A Man in Full*, 1998), details are important: *Thomas*, not *Tom*; *Wolfe*, not *Wolf*. Use *poodle* rather than *dog*, *Maui* rather than *Hawaii*, *Martin guitar* rather than *guitar*, or you'll get thousands of responses that have little or nothing to do with what you're looking for. You may need to use several similar words to explore the topic you're investigating: *car racing, auto racing, drag racing, drag-racing, dragracing*, and so on.

- **Use phrases with quotation marks rather than separate words:** If you type *ski resort*, you could get results of (1) everything to do with skis on the one hand and (2) everything to do with resorts—winter, summer, mountain, seaside—on the other. Better to put your phrase in quotation marks—*"ski resort"*—to narrow your search.

● **PANEL 2.9**
Guide to search sites
Years refer to date launched.

Human-Organized Search Sites

- **Yahoo!** (*www.yahoo.com*). 1994. The most popular search site of all. Has the largest human-compiled Web directory. Supplemented by Inktomi. Users should narrow search results category before they begin searching.
- **LookSmart** (*www.looksmart.com*). 1996. One of the easiest directories to use. Supplemented by AltaVista. Used by Excite and MSN Search.
- **About** (*www.about.com*). 1997. Began as The Mining Company. Trained human "guides" cover 50,000 subjects.
- **Open Directory** (*dmoz.org*). 1998. Uses 21,500 volunteer indexers. Owned by Netscape. Used by AltaVista, AOL Search, HotBot, Lycos, and Netscape.

Computer-Created Search Sites

- **Webcrawler** (*www.webcrawler.com*). 1994. Began at University of Washington; now owned by Excite.
- **Excite** (*www.excite.com*). 1995. One of the most popular search services. Owns Magellan and WebCrawler.
- **Inktomi** (*www.inktomi.com*). 1996. Began at University of California, Berkeley; available only through partners such as AOL Search and Snap.
- **Northern Light** (*www.northernlight.com*). 1997. One of the largest indexes. Also enables users for a fee to get additional documents from sources not easily accessible, such as magazines, journals, and news wires.
- **FAST Search** (*www.alltheweb.com*). 1999. Also powers the Lycos MP3 search service. Has announced plans to index the entire Web.

Hybrid Search Sites

- **Lycos** (*www.lycos.com*). 1995. Began as search engine, shifted to human directory in 1999. Main listings come from Open Directory, secondary results from Direct Hit or Lycos' own index. Good for finding graphics, music files, and other specialized content.
- **AltaVista** (*www.altavista.com*). 1995. One of the largest search engines. Additional listings provided by Ask Jeeves and Open Directory. Multilingual searches, queries entered as simple questions.
- **Ask Jeeves** (*ask.com*). 1996. Allows users to ask questions in natural language rather than keywords. Includes Urban Cool (*www.urbancool.com*), which allows users to ask questions posed in the popular language of the street.
- **HotBot** (*www.hotbot.com*). 1996. One of the most well-rounded search engines. Owned by Lycos. First page of results from Direct Hit, secondary results from Inktomi. Directory information from Open Directory.
- **GoTo** (*www.goto.com*). 1997. Companies pay to be placed higher in search results.
- **Snap** (*www.snap.com*). 1997. Human directory of Web sites, supplemented by Inktomi. Owned by CNet and NBC.
- **Direct Hit** (*www.directhit.com*). 1998. Owned by Ask Jeeves. Refines results based on popularity. Highest-ranking results are those most frequently chosen. Used on Ask Jeeves, Lycos, and HotBot and an option on LookSmart and MSN Search.
- **Google** (*www.google.com*). 1998. Links popularity to ranking. The more sites that link to a Web page, the higher that page will rank in searches.
- **Go** (*go.com*). 1999. Owned by Disney; offers search service of former Infoseek (1995). Includes human directory. Focuses on entertainment and leisure.
- **AOL Search** (*search.aol.com*). Covers both the Web and America Online's content. Directory listings mainly from the Open Directory, with backup from Inktomi.
- **MSN Search** (*search.msn.com*). Results from LookSmart directory, secondary results from AltaVista and Direct Hit.
- **Netscape Search** (*search.netscape.com*). Results from Open Directory and Netscape's own database, secondary results from Google.

Metasearch Sites

- **Go2Net/MetaCrawler** (*www.go2net.com*). 1995. Started at University of Washington.
- **SavvySearch** (*savvysearch.com*). 1995. Started at Colorado State University, Fort Collins.
- **Dogpile** (*www.dogpile.com*). Searches a customizable list of search engines.
- **Inference Find** (*www.infind.com*). Lists results grouped by subject, rather than by search engine or in one long list.
- **ProFusion** (*www.profusion.com*).
- **The Big Hub** (*www.thebighub.com*).
- **C4 TotalSearch Technology** (*www.c4.com*).

Adapted from Elizabeth Weise, "Successful Net Search Starts with Need," *USA Today*, January 24, 2000, p. 3D.

- **Put unique words first in a phrase:** Better to have *"Tom Wolfe novels"* rather than *"Novels Tom Wolfe."* Or if you're looking for the Hoagy Carmichael song rather than the southern state, indicate *"Georgia on My Mind."*

- **Use operators—AND, OR, NOT, and + and − signs:** Most search sites use symbols called *Boolean operators* to make searching more precise. To illustrate how they are used, suppose you're looking for the song "Strawberry Fields Forever."[30]

  AND connects two or more search words and means that all of them must appear in the search results. Example: *Strawberry AND Fields AND Forever.*

  OR connects two or more search words and indicates that any of the two may appear in the results. Example: *Strawberry Fields OR Strawberry fields.*

  NOT, when inserted before a word, excludes that word from the results. Example: *Strawberry Fields NOT Sally NOT W.C.* (to distinguish from the actress Sally Field and comedian W.C. Fields).

  + (plus sign), like *AND*, precedes a word that must appear: Example: *+ Strawberry + Fields.*

  − (minus sign), like *NOT*, excludes the word that follows it. Example: *Strawberry Field − Sally.*

- **Read the Help or Search Tips section:** All search sites provide a Help section and tips. This could save you time later.

- **Try an alternate general search site or a specific search site:** If you're looking for very specific information, a general type of search site such as Yahoo! or AltaVista may not be the best way to go. Instead you should turn to a specific search site.[31] Examples: To explore public companies, try Company Sleuth *(www.companysleuth.com)*, Hoover's Online *(www.hoovers.com)*, or KnowX *(www.knowx.com)*. For news stories, try Yahoo! News *(dailynews.yahoo.com)* or Total-News *(www.totalnews.com)*. For pay-per-view information, try Dialog Web *(www.dialogweb.com)*, Lexis-Nexis *(www.lexis-nexis.com)*, and Dow Jones Interactive *(www.djnr.com)*.

Multimedia on the Web

Many Web sites (especially those trying to sell you something) are multimedia, using a combination of text, images, sound, video, or animation. While accessing Web pages with just text and images may satisfy you now, eventually you'll probably want more.

- **Plug-ins and helper applications:** In the 1990s, as the Web was evolving from text to multimedia, browsers were unable to handle many kinds of graphic, sound, and video files. To do so, external application files called plug-ins had to be loaded into the system. **A _plug-in_—also called a player or a viewer—is a program that adds a specific feature to a browser, allowing it to play or view certain files.** For example, to view certain documents, you may need to download Adobe Acrobat Reader, to listen to CD-quality video you may need to download Liquid MusicPlayer. Plug-ins are required by many Web sites if you want to fully experience their content. A variant is the *helper application*, or *add-on*, which runs multimedia elements separate from the browser.

  Recent versions of Microsoft Internet Explorer and Netscape Communicator can handle a lot of multimedia. Now if you come across a file for which you need a plug-in or add-on, the browser will ask whether you want it, then tell you how to go about downloading it, usually at no charge.

Plug in
Adobe Acrobat Reader allows you
to view or print certain Web documents.

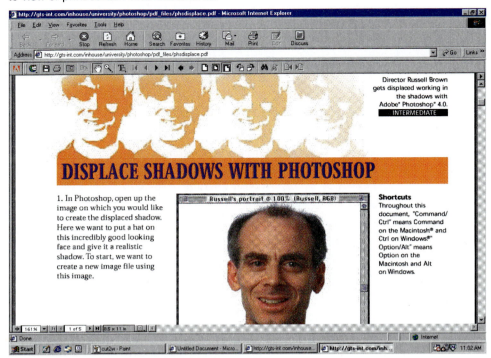

Web page combining text and images
This example shows the National Park Service's
opening screen about Mt. Rushmore in South Dakota.

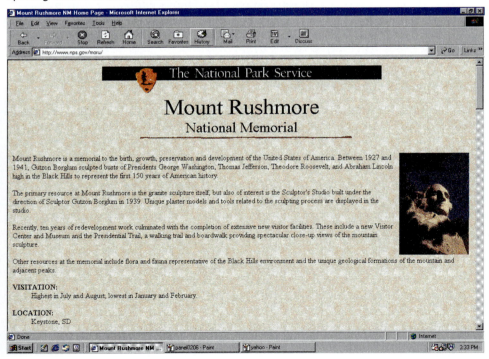

- **Developing multimedia—applets, Java, JavaScript, and ActiveX:** How do Web-site developers get all those nifty special multimedia effects? Often Web pages contain links to **_applets_, small programs that can be quickly downloaded and run by most browsers.** Applets are written in **_Java_, a complex programming language that enables programmers to create animated and interactive Web pages.** Java applets enhance Web pages by playing music, displaying graphics and animation, and providing interactive games.

If you are creating your own Web multimedia, you may want to learn techniques such as JavaScript and ActiveX, which may be used to create Web-page interest and activity—such as scrolling banners, pop-up menus, and the like—as we discuss further in Chapter 9.

Animation
This Web page shows an example of animation
in a virtual approach to Mars.

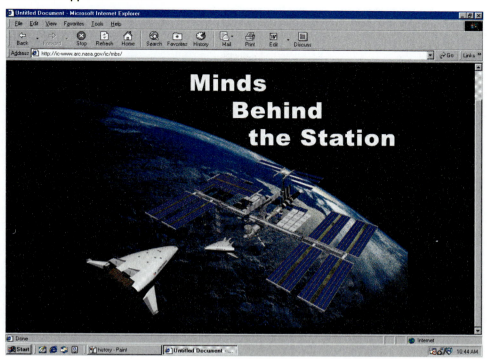

• **Text and images:** You can call up all kinds of text documents on the Web, such as newspapers, magazines, famous speeches, and works of literature. You can also view images, such as scenery, famous paintings, and photographs. Most Web pages combine both text and images.

• **Animation:** *Animation* **is the rapid sequencing of still images to create the appearance of motion,** as in a Road Runner cartoon. Animation is used in online video games as well as in moving banners displaying sports scores or stock prices.

• **Video:** Video can be transmitted in two ways. (1) A file, such as a movie or video clip, may have to be completely downloaded before you can view it. This may take several minutes in some cases. (2) A file may be displayed as *streaming video* and viewed while it is still being downloaded to your computer. *Streaming video* **is the process of transferring data in a continuous flow so that you can begin viewing a file even before the end of the file is sent.** For instance, RealPlayer offers live, television-style broadcasts over the Internet as streaming video for viewing on your PC screen. You download RealPlayer's software, install it, then point your browser to a site featuring RealVideo. That will produce a streaming-video television image in a window a few inches wide.

• **Audio:** Audio, such as sound or music files, may also be transmitted in two ways: (1) downloaded completely before they can be played or (2) downloaded as *streaming audio,* **allowing you to listen to the file while the data is still being downloaded to your computer.** A popular standard for transmitting audio is RealAudio, which compresses sound so that it can be played in real time, even though sent over telephone lines. You can, for instance, listen to 24-hour-a-day net.radio, which features "vintage rock," or English-language services of 19 shortwave outlets from World Radio Network in London. Many large radio stations outside the U.S. have net radio, allowing people around the world to listen in.

Streaming audio
RealPlayer is used for
transmitting streaming
audio, which many radio
stations now use for
live broadcasts

Click on for
live broadcast

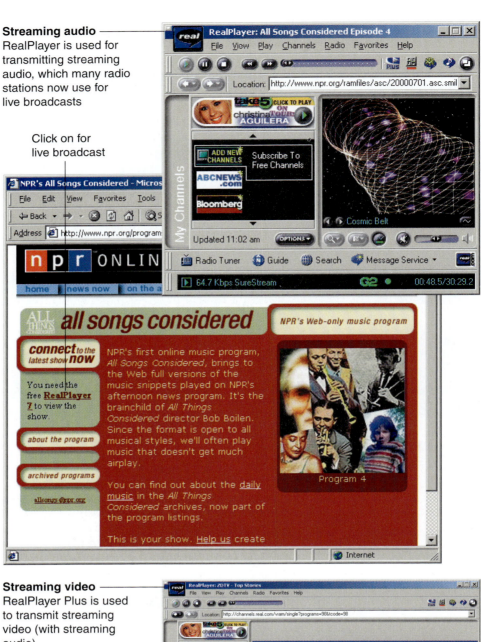

Streaming video
RealPlayer Plus is used
to transmit streaming
video (with streaming
audio)

Push Technology & Webcasting

It used to be that you had to do the searching on the World Wide Web. Now, if you wish, the Web will come searching for you. The driving force behind this is **_push technology_, software that automatically downloads information to your computer** (as opposed to "pull" technology, in which you go to a Web site and pull down the information you want).

One result of push technology is **_webcasting_, in which customized text, video, and audio are sent to you automatically on a regular basis.** The idea here is that you choose the categories, or (in Microsoft Internet Explorer) the *channels*, of Web sites that will automatically send you updated information. Thus, it saves you time because you don't have to go out searching for the information. Several services offer personalized news and information, based on a profile that you define when you register with them. Entrypoint.com, for example, will send news on a particular topic, such as all financial news about sugar beets, the weather in Omaha, or the game results for the Tennessee Titans. You can view the news as it comes in or save it to read later.

Push technology

The push technology from Entrypoint periodically delivers news that you preset according to your specifications

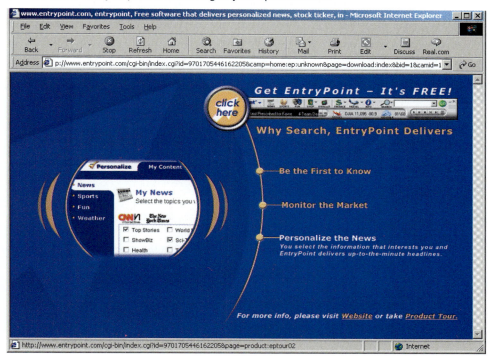

The Internet Telephone & Videophone

A few years ago, the idea of using the Internet to make phone calls was considered "a gimmicky technology for nerdy hobbyists," in the words of one writer.[32] Now Internet firms have taken aim at the traditional telephone companies.

The key element is that the Internet breaks up conversations (as it does any other transmitted data) into "information packets," which can be sent over separate lines and then regrouped at the destination, whereas conventional voice phone lines carry a conversation over a single path. Thus, the Internet can move a lot more traffic over a network than the traditional telephone link can. (We describe how telecommunications technology works in Chapter 5.)

With **_Internet telephony_—using the Net to make phone calls, either one-to-one or for audioconferencing**—you can make long-distance phone calls that are surprisingly inexpensive. Indeed, it's theoretically possible to do this without owning a computer, simply by picking up your standard telephone and dialing a number that will "packetize" your conversation. However, it's more common practice to use a PC with a sound card and a microphone, a modem linked to a standard Internet service provider, and Internet telephone software such as Netscape Conference (part of Netscape Communicator) or Microsoft NetMeeting (part of Microsoft Internet Explorer).

Besides carrying voice signals, Internet telephone software also allows videoconferencing, in which participants are linked by a videophone that will transmit their pictures, thanks a video camera attached to their PCs. It can also allow people to make sketches on "whiteboards" as they talk, as when three people meet online to discuss the floor plan for a new house.

Designing Web Pages

If you want to advertise a business online or just have your own personal Web site, you will need to design a Web page, determine any hyperlinks, and hire 24-hour-a-day space on a Web server or buy one of your own. Professional Web page designers can produce a page for you, or you can do it yourself using a menu-driven program included with your Web browser or a Web-page design software package such as Microsoft FrontPage or Adobe PageMill. After you have designed your Web page, you can put it on your ISP's server. (We describe Web authoring software in Chapter 7.)

CONCEPT CHECK

Describe how the World Wide Web works.

How do you use a browser to get around the Web

How do you use a Web portal to find information?

Describe the types of search engines and some tips for searching.

Discuss Web multimedia, push technology, webcasting, and telephony.

What are the measures of data transmission speed?

Explain the differences among the methods of going online.

2.5 The Online Gold Mine: More Internet Resources, Your Personal Cyberspace, E-Commerce, & the E-conomy

KEY QUESTIONS
What are FTP, Telnet, newsgroups, real-time chat, and e-commerce?

Deborah Thebes of San Francisco drove from store to store over a period of several months, testing out one couch after another. "I had pretty specific wants and just never saw what I wanted in a store," she said. Then someone told her about Furniture.com on the Internet, and she ended up ordering a custom-built couch from the site. It wasn't quite the size and color she had in mind, but there was no sales tax—a considerable savings on an $800 piece of furniture. "It's very comfortable, it's beautiful, and I'm not sorry I bought it," says Thebes. And at less than $800, it seemed a bargain.[33]

Is this a glimpse of our cyberfuture? Will people be ordering pianos and stoves this way? Certainly the opportunities offered by the Net seem inexhaustible. Let's consider some of them. We'll examine four Internet resources other than e-mail and the Web, we'll look at personal aspects of cyberspace, and we'll explore the ever-expanding realm of e-commerce.

Other Internet Resources: FTP, Telnet, Newsgroups, & Real-Time Chat

E-mail and the World Wide Web seem to attract all the attention. But other cyber resources are also widely used: FTP, Telnet, newsgroups, and real-time chat.

- **FTP—for copying all the free files you want:** Many Net users enjoy "FTPing"—cruising the system and checking into some of the tens of thousands of FTP sites, which predate the Web and offer interesting

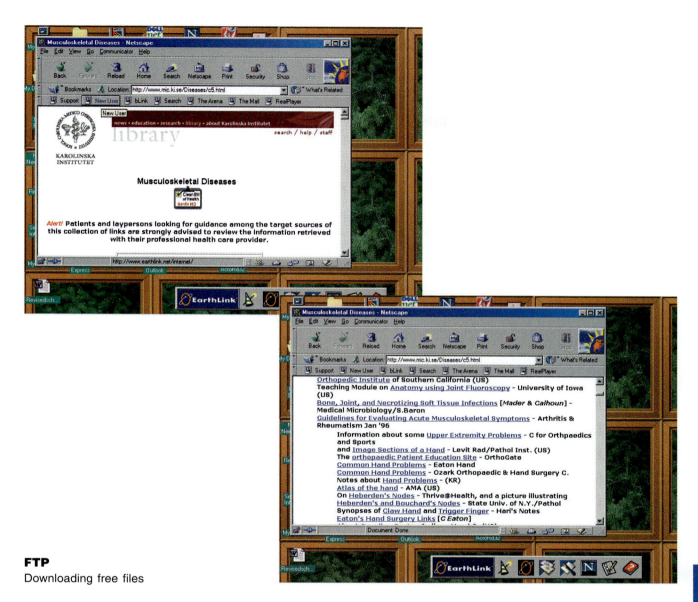

FTP
Downloading free files

free files to copy (download). ***FTP (File Transfer Protocol)* is a method whereby you can connect to a remote computer called an FTP site and transfer publicly available files to your own microcomputer's hard disk.** The free files offered cover nearly anything that can be stored on a computer: software, games, photos, maps, art, music, books, statistics.

Some FTP files are open to the public, some are not. For instance, a university might maintain an FTP site with private files (such as lecture transcripts) available only to professors and students with assigned user names and passwords. It might also have public FTP files open to anyone with an e-mail address. You can download FTP files using either your Web browser or special software (called an *FTP client program*), such as Fetch.

• **Telnet—to connect to remote computers: *Telnet* is a program or command that allows you to connect to remote computers on the Internet.** This feature, which allows microcomputers to communicate successfully with mainframes, enables you to tap into Internet computers and access public files as though you were connected directly instead of, for example, through your ISP site.

The Telnet feature is especially useful for perusing large databases at universities, government agencies, or libraries. As an electronic version of a library card catalog, Telnet can be used to search most major public and university library catalogs. (See, for example, Internet Public Library, *www.ipl.org*, and Library Spot, *www.libraryspot.com*)

- **Newsgroups—for online typed discussions on specific topics:** A ***newsgroup* is a giant electronic bulletin board on which users conduct written discussions about a specific subject.** There are more than 30,000 newsgroup forums—which charge no fee—and they cover an amazing array of topics. Some examples are *rec.arts.startrek.info*, *soc.culture.african.american*, and *misc.jobs.offered*. Newsgroups take place on a special network of computers called ***Usenet*, a worldwide network of servers that can be accessed through the Internet.** To participate, you need a ***newsreader*, a program included with most browsers that allows you to access a newsgroup and read or type messages.** (Messages, incidentally, are known as *articles*.)

  One way to find a newsgroup of interest to you is to use a portal such as Yahoo!, Excite, or Lycos to search Usenet for specific topics. Or you can use the search engine Dejanews (*www.dejanews.com*), which will present the newsgroups matching the topic you specify. About a dozen major topics, identified by abbreviations ranging from *alt* (alternative topics) to *talk* (opinion and discussion), are divided into hierarchies of subtopics.

- **Real-time chat—typed discussions among online participants:** With newsgroups (and mailing lists, which we described under e-mail), participants may contribute to a discussion, then go away and return hours or days later to catch up on others' typed contributions. With ***real-time chat (RTC)*, participants have a typed discussion ("chat") while online at the same time,** just like a telephone conversation except that messages are typed rather than spoken. Otherwise the format is much like a newsgroup, with a message board to which participants may send ("post") their contributions. To start a chat, you use what is known as a *chat client*, a program available on your browser that will connect you to a chat server. One of the most popular chat clients is Internet Relay Chat (IRC).

  Unlike instant messaging (discussed under e-mail), which tends to involve one-on-one conversation, real-time chat usually involves several participants. As a result, RTC "is often like being at a crowded party," says one writer. "There are any number of people present and many threads of conversation occurring all at once."[34]

Your Personal Cyberspace

As we mentioned in Chapter 1, information technology has become more personal as it has evolved. Unlike the generally impersonal mass media, the Internet allows you to pursue your personal interests in the areas of relationships, education, health, and entertainment, for example.

- **Relationships—online matchmaking:** It's like walking into "a football stadium full of single people of the gender of your choice," says Trish McDermott, an expert for Match.com, a San Francisco online-dating service. People who connect online before meeting in the real world, she points out, have the chance to base their relationship on personality, intelligence, and sense of humor rather than purely physical attributes. "Online dating allows people to take some risks in an anonymous capacity," she adds. "When older people look back at their lives, it's the risks that they didn't take that they most regret."[35]

(Still, there *are* some risks in trying to establish intimacy through online means because people pretend to be quite different from who they really are.)

People can also use search sites such as Infospace.com, Switchboard.com, and GTE Superpages.com to try to track down old friends and relatives.[36] Others find common bonds by joining online communities such as The WELL *(www.well.com)*, the women's site Ivillage.com *(www.ivillage.com)*, the older people's site Third Age *(www.thirdage.com)*, and the gardening site GardenWeb *(www.gardenweb.com)*.[37] Finally, the Net is no longer dominated by English; 43% of the users are non-English speakers, and that number was expected to pass 50% in 2000. After English, the most common languages among Internet users are (in order) Japanese, Spanish, and German, with French and Chinese tied for fifth.[38]

Distance learner

- **Education—the rise of distance learning:** Sally Wells of Oregon has four children, a full-time job, and 14 cows to milk. She'd like to get a master's degree, but with no time to drive an hour to campus she takes four marketing courses online.[39] Adult learners—defined by educators as those over age 24—aren't the only ones involved in ***distance learning*, the name given to online education programs.** Younger college students also like it because they don't have to spend time commuting, the scheduling is flexible, and they often have a greater selection of course offerings. Although for instructors an online class is more labor-intensive than a regular chalk-and-talk class, they often find there is better interaction with students.[40]

- **Health—patient self-education:** Health is one of the most popular subject areas of research on the Web, although it can be difficult to get accurate information. (Many sites are trying to sell you something.) Among the sites offering reputable advice are Intelihealth *(www.intelihealth.com)*, Mayo Clinic Health Oasis *(www.mayohealth.org)*, The Physician and Sportsmedicine Online *(www.physportsmed.com)*, Cyber Diet *(cyberdiet.com)*, Phys.com *(phys.com)*, the American College of Physicians *(acponline.org)*, and Medline *(www.nlm.nih.gov/medlineplus)*.[41]

- **Entertainment—amusing yourself:** Two-thirds of all Internet users in the U.S. seek out entertainment on the Web, according to a 1999 survey.[42] No wonder so many major media companies have created Web sites to try to help promote or sell movies, music, TV shows, and the like. Of course, there are many other types of entertainment sites, devoted to games, hobbies, jokes, and so on. If you want to see animated hamsters dancing, for instance, try visiting *www.hamster dance.com.*

E-Commerce

"What's your opinion on the Internet vs. the real world?" a *USA Today* reader asked. "[T]he Internet *is* the real world," replied reporter Lorrie Grant, who had been assigned to spend a month using the Web for shopping, working, banking, and other activities. "I had everything at my fingertips: office supplies, groceries, stocks, banking and bill payment, apparel, flowers, music, gifts, greeting cards, and more. Just point and click, *voilà.*"[43]

The explosion in ***electronic commerce (e-commerce)*—conducting business activities online**—is not only widening consumers' choice of products and services but also creating new businesses and compelling established businesses to develop Internet strategies. Let's look at some of the developments.

- **E-tailing—retail commerce online:** Is the Internet spawning an entirely new way of doing business? Certainly many so-called *brick-and-mortar* retailers—those operating out of physical buildings—have been surprised at the success of such online companies as Amazon.com, seller of books, CDs, and other products. As a result, traditional retailers from giant Wal-Mart to funky little Buch Spieler Music in Montpelier, Vermont, have rushed to put their products online—and it has helped to revive some small-town main streets that had suffered from plant closings and competition from mega-malls.[44]

  Retail goods can be classified into two categories—hard and soft. *Hard goods* are those that can be viewed and priced online, such as computers, clothes, furniture, and—yes—even groceries, but are then sent to buyers by mail or truck. *Soft goods* are those that can be downloaded directly from the retailer's site, such as music, software, and greeting cards.

- **Auctions—linking individual buyers and sellers:** Today millions of buyers and sellers are linking up at online auctions, where everything is available from comic books to wines. More than 500 Web sites now wield an electronic gavel and within a few years are expected to lure about 10% of all online purchasers. The Internet is also changing the tradition-bound art and antiques business (dominated by such venerable names as Sotheby's, Christie's, and Butterfield & Butterfield). There are generally two types of auction sites: (1) person-to-person auctions, such as eBay *(www.ebay.com)*, that connect buyers and sellers for a listing fee and a commission on sold items, and (2) vendor-based auctions, such as OnSale *(www.onsale.com)*, that buy merchandise and sell them at discount. Some auctions are specialized, such as Priceline *(www.priceline.com)*, an auction site for airline tickets and other items.

- **Online finance—trading, banking, and e-money:** The Internet has changed the nature of stock trading. For the first time, says technology observer Denise Caruso, "anyone with a computer, a connection to the global network, and the requisite ironclad stomach for risk has the information, tools, and access to transaction systems required to play the stock market, a game that was once the purview of an elite few."[45] Companies such as E*Trade and Ameritrade are building one-stop financial supermarkets offering a variety of money-related services, including home mortgage loans and insurance. More than 1000 banks have Web sites, offering services that include account access, funds transfer, bill payment, loan and credit card applications, and investments. You can, for instance, apply for a Visa card called NextCard and get approved (or turned down) in about two minutes.

- **Online job hunting:** There are 2500 Web sites that promise to match job hunters with an employer. Some are specialty "boutique" sites looking for, say, scientists or executives. Some are general sites, the leaders being Monster.com, CareerPath.com, Headhunter.net, CareerMosaic.com, www.usajobs.opm.gov, and CareerBuilder.com.[46] Job sites can help you keep track of job openings and applications by downloading them to your own computer.

- **B2B commerce:** Of course, every kind of commerce has taken to the Web, ranging from travel bookings to real estate. One of the most important variations is **_B2B (business-to-business) commerce_, the electronic sales or exchange of goods and services directly between companies, cutting out traditional intermediaries.** Expected to grow even more rapidly than other forms of e-commerce, B2B commerce covers an extremely broad range of activities, such as supplier-to-buyer display of inventories, provision of wholesale price lists, and sales of closed-out items and used materials—usually without agents, brokers, or other third parties. Companies say e-purchasing, for example, slashes up to 20% off what they buy from each other.[47]

CONCEPT CHECK

Describe FTP, Telnet, newsgroups, and real-time chat.

What are some ways the Internet can be of personal use and of e-commerce use?

Onward: Are We in a New Kind of E-conomy?

At the beginning of this new century, the American economy was sizzling: the stock market at a record high, corporate profits soaring, unemployment at a 30-year low, the rate of inflation barely perceptible. How did we arrive at this wonderland? One possible answer: information technology. Suggests one analysis: "The spread of computers, telecommunications equipment, and the like—which control manufacturing, supply-chain management, and business-to-business e-commerce—is helping companies hold down inventories, thus eliminating a prime cause of past recessions, when businesses found themselves overstocked with products they had to sell at discounts or simply write off their books. Fewer recessions mean greater certainty of earnings growth."[48]

New kinds of companies now dominate the landscape. For example, at the end of 1999, the valuation of Yahoo!, only a few years old, was about $90 billion—twice that of venerable General Motors, although its earnings were about one-hundredth of GM's. America Online was worth more than GM, Ford, Sears, and Disney combined.[49]

Call it the "electronic economy"—or *e-conomy*. While the ups and downs of the business cycle have yet to be repealed, the Internet and computers have dramatically changed the nature of enterprise.

Experience Box
Web Research, Term Papers, & Plagiarism

No matter how much students may be able to rationalize cheating in college—for example, trying to pass off someone else's term paper as their own (plagiarism)—ignorance of the consequences is not an excuse. Most instructors announce the penalties for cheating at the beginning of their course—usually a failing grade in the course and possible suspension or expulsion from school.

Even so, probably every student becomes aware before long that the World Wide Web contains sites that offer term papers, either for free or for a price. Some dishonest students may download papers and just change the author's name to their own. Others are more likely just to use the papers for ideas. Perhaps, suggests one article, "the fear of getting caught makes the online papers more a diversion than an invitation to wide-scale plagiarism."[50]

How the Web Can Lead to Plagiarism

Two types of term-paper Web sites are as follows:

- **Sites offering papers for free:** Such a site requires users to fill out a membership form, then provides at least one free student term paper. (Quality is a crapshoot, since free paper mills often subsist on the submissions of poor students, whose contributions may be sub-literate.)

- **Sites offering papers for sale:** Commercial sites may charge $6–$10 a page, which users may charge to their credit card. (Expense is no guarantee of quality. Moreover, the term-paper factory may turn around and make your $350 custom paper available to others—even fellow classmates working on the same assignment—for half the price.)

How Instructors Catch Cheaters

How do instructors detect and defend against student plagiarism? Leland says professors are unlikely to be fooled if they tailor term-paper assignments to work done in class, monitor students' progress—from outline to completion—and are alert to papers that seem radically different from a student's past work.[52]

Eugene Dwyer, a professor of art history at Kenyon College, requires that papers in his classes be submitted electronically, along with a list of World Wide Web site references. "This way I can click along as I read the paper. This format is more efficient than running around the college library, checking each footnote."[53]

Just as the Internet is the source of cheating, it is also a tool for detecting cheaters. Search programs make it possible for instructors to locate texts containing identified strings of words from the millions of pages found on the Web. Thus, a professor can input passages from a student's paper into a search program that scans the Web for identical blocks of text. Indeed, some Web sites favored by instructors build a database of papers over time so that students can't recycle work previously handed in by others. One system can lock on to a stolen phrase as short as eight words. It can also identify copied material even if it has been changed slightly from the original.[54]

How the Web Can Lead to Low-Quality Papers

William Rukeyser, coordinator for Learning in the Real World, a nonprofit information clearinghouse, points out another problem: The Web enables students "to cut and paste together reports or presentations that appear to have taken hours or days to write but have really been assembled in minutes with no actual mastery or understanding by the student."[55]

Philosophy professor David Rothenberg, of New Jersey Institute of Technology, reports that as a result of students doing more of their research on the Web he has seen "a disturbing decline in both the quality of the writing and the originality of the thoughts expressed."[56] How does an instructor spot a term paper based primarily on Web research? Rothenberg offers four clues:

- **No books cited:** The student's bibliography cites no books, just articles or references to Web sites. Sadly, says Rothenberg, "one finds few references to careful, in-depth commentaries on the subject of the paper, the kind of analysis that requires a book, rather than an article, for its full development."

- **Outdated material:** A lot of the material in the bibliography is strangely out of date, says Rothenberg. "A lot of stuff on the Web that is advertised as timely is actually at least a few years old."

- **Unrelated pictures and graphs:** Students may intersperse the text with a lot of impressive-looking pictures and graphs that actually bear little relation to the precise subject of the paper. "Cut and pasted from the vast realm of what's out there for the taking, they masquerade as original work."

- **Superficial references:** "Too much of what passes for information [online] these days is simply *advertising* for information," points out Rothenberg. "Screen after screen shows you where you can find out more, how you can connect to this place or that." Other kinds of information are detailed but often superficial: "pages and pages of federal documents, corporate propaganda, snippets of commentary by people whose credibility is difficult to assess."

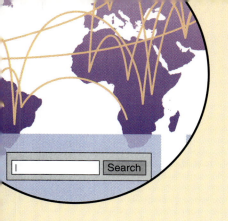

Visual Summary

animation (p. 69, KQ 2.4) The rapid sequencing of still images to create the appearance of motion, as in a cartoon. Why it's important: *Animation is a component of multimedia; it is used in online video games as well as in moving banners displaying sports scores or stock prices.*

applets (p. 68, KQ 2.4) Small programs that can be quickly downloaded and run by most browsers. Why it's important: *Web pages contain links to applets, which add multimedia capabilities.*

B2B (business-to-business) commerce (p. 76, KQ 2.5) Electronic sales or exchange of goods and services directly between companies, cutting out traditional intermediaries. Why it's important: *Expected to grow even more rapidly than other forms of e-commerce, B2B commerce covers an extremely broad range of activities, such as supplier-to-buyer display of inventories, provision of wholesale price lists, and sales of closed-out items and used materials—usually without agents, brokers, or other third parties.*

bandwidth (p. 36) Expression of how much data—text, voice, video, and so on—can be sent through a communications channel in a given amount of time. Why it's important: *Different communications systems use different bandwidths for different purposes. The wider the bandwidth, the faster data can be transmitted.*

bps (p. 37, KQ 2.1) Bits per second. Why it's important: *Data transfer speeds are measured in bits per second.*

cable modem (p. 40, KQ 2.1) Device connecting a personal computer to a cable-TV system that offers an Internet connection. Why it's important: *Cable modems transmit data faster than standard modems.*

communications satellite (p. 41, KQ 2.1) Space station that transmits radio waves called *microwaves* from earth-based stations. Why it's important: *An orbiting satellite contains many communications channels and receives signals from ground microwave stations anywhere on earth.*

directory (p. 64, KQ 2.4) Search tool that provides lists of several categories of Web sites classified by topic, such as Business & Finance or Health & Fitness. Such a category is also called a *hypertext index,* and its purpose is to allow you to access information in specific categories by clicking on a hypertext link. Why it's important: *Directories are useful for browsing—looking at Web pages in a general category and finding items of interest. Search engines may be more useful for hunting specific information.*

distance learning (p. 75, KQ 2.5) Online education programs. Why it's important: *Distance learning provides educational opportunities for people who are not able to get to a campus; also distance-learning students don't have to spend time commuting, the scheduling is flexible, and they often have a greater selection of course offerings.*

domain (p. 45, KQ 2.3) A location on the Internet. Why it's important: *A domain name is necessary for sending and receiving e-mail and for many other Internet activities.*

download (p. 38, KQ 2.1) To transmit data from a remote computer to a local computer. Why it's important: *Downloading enables users to save files on their own computers for later use, which reduces the time spent online and the corresponding charges.*

DSL (digital subscriber line) (p. 40, KQ 2.1) A hardware and software technology that uses regular phone lines to transmit data in megabits per second. Why it's important: *DSL connections are much faster than regular modem connections.*

e-commerce (electronic commerce) (p. 75, KQ 2.5) Conducting business activities online. *Why it's important:* E-commerce is not only widening consumers' choice of products and services but is also creating new businesses and compelling established businesses to develop Internet strategies.

emoticons (p. 51, KQ 2.3) Keyboard-produced pictorial representations of expressions. *Why it's important: Emoticons can smooth online communication.*

:-)	Happy face
:-(	Sorrow or frown
:-O	Shock
:-/	Sarcasm
;-)	Wink

FAQs (Frequently Asked Questions) (p. 51, KQ 2.3) Guides that explain expected norms of online behavior for a particular group. *Why it's important: Users should read a group's/site's FAQs to know how to proceed properly.*

flaming (p. 51, KQ 2.3) Writing an online message that uses derogatory, obscene, or inappropriate language. *Why it's important: Flaming should be avoided. It is a form of public humiliation inflicted on people who have failed to read FAQs or otherwise not observed netiquette (although it can happen just because the sender has poor impulse control and needs a course in anger management).*

frame (p. 62, KQ 2.4) An independently controllable section of a Web page. *Why it's important: A Web page designer can divide a page into separate frames, each with different features or options.*

FTP (File Transfer Protocol) (p. 73, KQ 2.5) Method whereby you can connect to a remote computer called an *FTP site* and transfer publicly available files to your own microcomputer's hard disk. *Why it's important: The free files offered cover nearly anything that can be stored on a computer: software, games, photos, maps, art, music, books, statistics.*

gigabits per second (Gbps) (p. 37, KQ 2.1) 1 billion bits per second. *Why it's important: Gbps is a common measure of data transmission speed.*

home page (p. 55, KQ 2.4) Also called *welcome page;* Web page that, somewhat like the title page of a book, identifies the Web site and contains links to other pages at the site. *Why it's important: The first page you see at a Web site is the home page.*

hypertext (p. 54, KQ 2.4) System in which documents scattered across many Internet sites are directly linked, so that a word or phrase in one document becomes a connection to a document in a different place. *Why it's important: Hypertext links many documents by topics, allowing users to find information on topics they are interested in.*

hypertext markup language (HTML) (p. 54, KQ 2.4) Set of special instructions (called "tags" or "markups") used to specify Web document structure, formatting, and links to other documents. *Why it's important: HTML enables the creation of Web pages.*

HyperText Transfer Protocol (HTTP) (p. 56, KQ 2.4) Communications rules that allow browsers to connect with Web servers. *Why it's important: Without HTTP, files could not be transferred over the Web.*

instant messaging (IM) (p. 48, KQ 2.3) Any user on a given e-mail system can send a message and have it pop up instantly on the screen of anyone else logged onto that system. *Why it's important: If both parties agree, they can initiate online typed conversations in real time. As they are typed, the messages appear on the display screen in a small window.*

Internet service provider (ISP) (p. 42, KQ 2.2) Company that connects you through your communications line to its server, or central computer, which connects you to the Internet. *Why it's important: Unless they subscribe to an online information service (such as AOL) or have a direct network connection (such as a T1 line), microcomputer users need an ISP to connect to the Internet.*

Internet service provider

Telephony & conferencing
Make inexpensive phone calls; have online meetings.

E-business
Connect with coworkers, buy supplies, support customers.

Internet telephony (p. 71, KQ 2.4) Using the Net to make phone calls, either one-to-one or for audioconferencing. Why it's important: *Long-distance phone calls by this means are surprisingly inexpensive.*

ISDN (Integrated Services Digital Network) (p. 40, KQ 2.1) Hardware and software that allows voice, video, and data to be communicated over traditional copper-wire telephone lines (POTS). Why it's important: *ISDN provides faster data transfer speeds than do regular modem connections.*

Java (p. 68, KQ 2.4) Complex programming language that enables programmers to create animated and interactive Web pages using applets. Why it's important: *Java applets enhance Web pages by playing music, displaying graphics and animation, and providing interactive games.*

keyword (p. 64, KQ 2.4) The subject word or words of the topic you wish to find in a Web search. Why it's important: *The results of your keyword search will be displayed in a short summary of documents containing the keyword you typed.*

kilobits per second (Kbps) (p. 37, KQ 2.1) 1000 bits per second. Why it's important: *Kbps is a common measure of data transfer speed. The speed of a modem that is 28,800 bps might be expressed as 28.8 Kbps.*

list-serves (p. 50, KQ 2.3) E-mail mailing lists of people who regularly participate in discussion topics. To subscribe, the user sends an e-mail to the list-serve moderator and asks to become a member, after which he or she automatically receives e-mail messages from anyone who responds to the server. Why it's important: *Anyone connected to the Internet can subscribe to list-serve services. Subscribers receive information on particular subjects and can post e-mail to other subscribers.*

log on (p. 43, KQ 2.2) To make a connection to a remote computer. Why it's important: *Users must be familiar with log-on procedures to go online.*

megabits per second (Mbps) (p. 37, KQ 2.1) 1 million bits per second. Why it's important: *Mbps is a common measure of data transmission speed.*

menu (p. 43, KQ 2.2) List of commands or options offered by software programs. Why it's important: *Menus make software easier to use; the user can choose from a list instead of having to remember specific commands and addresses.*

netiquette (p. 51, KQ 2.3) "Network etiquette"—guides to appropriate online behavior. Why it's important: *In general, netiquette has two basic rules: (a) Don't waste people's time, and (b) don't say anything to a person online that you wouldn't say to his or her face.*

newsgroup (p. 74, KQ 2.5) Giant electronic bulletin board on which users conduct written discussions about a specific subject. Why it's important: *There are more than 30,000 newsgroup forums—which charge no fee—and they cover an amazing array of topics.*

newsreader (p. 74, KQ 2.5) Program included with most browsers that allows users to access a newsgroup and read or type messages. Why it's important: *Users need a newsreader to participate in a newsgroup.*

physical connection (p. 36, KQ 2.1) The wired or wireless means of connecting to the Internet. Why it's important: *Without physical connections, the Internet would be impossible, as would telephone and other types of communication connections.*

plug-in (p. 67, KQ 2.4) Also called a *player* or a *viewer*, program that adds a specific feature to a browser, allowing it to play or view certain files. Why it's important: *To fully experience the contents of many Web pages, you need to use plug-ins.*

protocol (p. 56, KQ 2.4) Set of communication rules for exchanging information. Why it's important: *HyperText Transfer Protocol (HTTP) provides the communications rules that allow browsers to connect with Web servers.*

push technology (p. 71, KQ 2.4) Software that automatically downloads information to your computer, as opposed to "pull" technology, in which you go to a Web site and pull down the information you want. Why it's important: *With little effort, users can obtain information that is important to them.*

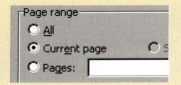

radio buttons (p. 62, KQ 2.4). An interactive tool displayed as little circles in front of options; selecting an option with the mouse places a dot in the corresponding circle. Why it's important: *Radio buttons are one way of interacting with a Web page.*

real-time chat (RTC) (p. 74, KQ 2.5) Typed discussion ("chat") among participants who are online at the same time; it is just like a telephone conversation, except that messages are typed rather than spoken. Why it's important: *RTC provides a means of immediate electronic communication.*

scroll arrows (p. 62, KQ 2.4) Small up/down and left/right arrows located to the bottom and side of your screen display. Why it's important: *Clicking on scroll arrows with your mouse pointer moves the screen so that you can see the rest of the Web page, or the content displayed on the screen.*

scrolling (p. 62, KQ 2.4) Moving quickly upward or downward through text or other screen display, using the mouse and scroll arrows (or the arrow keys on the keyboard). Why it's important: *Normally a computer screen displays only part of, for example, a Web page. Scrolling enables users to view an entire document, no matter how long.*

search engine (p. 64, KQ 2.4) Search tool that allows you to find specific documents through keyword searches and menu choices, in contrast to directories, which are lists of Web sites classified by topic. Why it's important: *Search engines enable users to find Web sites of specific interest of use to them.*

site (p. 55, KQ 2.4) Computer with a domain name. Why it's important: *Sites provide Internet and Web content.*

spam (p. 52, KQ 2.3) Unsolicited e-mail in the form of advertising or chain letters. Why it's important: *Spam filters are available that can spare users the annoyance of receiving junk mail, ads, and other unwanted e-mail.*

streaming audio (p. 69, KQ 2.4) Process of downloading audio in which you can listen to the file while the data is still being downloaded to your computer. Why it's important: *Users don't have to wait until the entire audio is downloaded to hard disk before listening to it.*

streaming video (p. 69, KQ 2.4) Process of downloading video in which the data is transferred in a continuous flow so that you can begin viewing a file even before the end of the file is sent. Why it's important: *Users don't have to wait until the entire video is downloaded to hard disk before watching it.*

T1 line (p. 40, KQ 2.1) Traditional trunk line that carries 24 normal telephone circuits and has a transmission rate of 1.5 Mbps. Why it's important: *High-capacity T1 lines are used at many corporate, government, and academic sites; these lines provide greater data transmission speeds than do regular modem connections.*

Telnet (p. 73, KQ 2.5) Program or command that allows you to connect to remote computers on the Internet. Why it's important: *This feature, which allows microcomputers to communicate successfully with mainframes, enables users to tap into Internet computers and access public files as though they were connected directly instead of, for example, through an ISP site.*

upload (p. 38, KQ 2.1) To transmit data from a local computer to a remote computer. Why it's important: *Uploading allows users to easily exchange files over networks.*

URL (Universal Resource Locator) (p. 56, KQ 2.4) String of characters that points to a specific piece of information anywhere on the Web. A URL consists of (1) the Web protocol, (2) the name of the Web server, (3) the directory (or folder) on that server, and (4) the file within that directory (perhaps with an extension such as html or htm). Why it's important: *URLs are necessary to distinguish among Web sites.*

Usenet (p. 74, KQ 2.5) Worldwide network of servers that can be accessed through the Internet. Why it's important: *Newsgroups take place on Usenet.*

Web browser (browser) (p. 55, KQ 2.4) Software that enables users to view Web pages and to jump from one page to another. Why it's important: *Users can't surf the Web without a browser. The two most well known browsers are Microsoft's Internet Explorer and Netscape Communicator.*

webcasting (p. 71, KQ 2.4) Service, based on push technology, in which customized text, video, and audio are sent to the user automatically on a regular basis. Why it's important: *Users choose the categories, or the channels, of Web sites that will automatically send updated information. Thus, it saves time because users don't have to go out searching for the information.*

Web page (p. 55, KQ 2.4) Document on the World Wide Web that can include text, pictures, sound, and video. Why it's important: *A Web site's content is provided on Web pages. The starting page is the home page.*

Web portal (p. 63, KQ 2.4) Web site that groups together popular features such as search tools, e-mail, electronic commerce, and discussion groups. The most popular portals are America Online, Yahoo!, Microsoft Network, Netscape, Lycos, Go Network, Excite Network, AltaVista, and WebCrawler. Why it's important: *Web portals provide an easy way to access the Web.*

Web site (website) (p. 55, KQ 2.4) Location of a Web domain name in a computer somewhere on the Internet. Why it's important: *Web sites provide multimedia content to users.*

window (p. 48, KQ 2.3) Rectangular area containing a document or activity. Why it's important: *This feature enables different outputs to be displayed at the same time on the screen. For example, users can exchange messages almost instantaneously while operating other programs.*

Self-Test Questions

1. Today's data-transmission speeds are measured in _____, _____, _____, and _____ per second.

2. A(n) _____ connects a personal computer to a cable-TV system that offers an Internet connection.

3. A space station that transmits data as microwaves is a _____.

4. A company that connects you through your communications line to its server, which connects you to the Internet is a(n) _____.

5. A small rectangular area on the computer screen that contains a document or displays an activity is called a(n) _____.

6. _____ is writing an online message that uses derogatory, obscene, or inappropriate language.

7. A(n) _____ is software that enables users to view Web pages and to jump from one page to another.

8. A computer with a domain name is called a(n) _____.

9. _____ comprises the communications rules that allow browsers to connect with Web servers.

10. A(n) _____ is a program that adds a specific feature to a browser, allowing it to play or view certain files.

Multiple-Choice Questions

1. Kbps means _____ bits per second.
   a. 1 billion
   b. 1 thousand
   c. 1 million
   d. 1 hundred
   e. 1 trillion

2. A location on the Internet is called a(n) _____.
   a. network
   b. user ID
   c. domain
   d. browser
   e. web

3. In the e-mail address *Kim_Lee @earthlink.net.us*, *Kim_Lee* is the _____.
   a. domain
   b. URL
   c. site
   d. user ID
   e. location

4. Which of the following is not one of the four components of a URL?
   a. Web protocol
   b. name of the Web server
   c. name of the browser
   d. name of the directory on the Web server
   e. name of the file within the directory

5. Which of the following is the fastest method of data transmission?
   a. ISDN
   b. DSL
   c. modem
   d. T1 line
   e. cable modem

6. Which of the following is not a netiquette rule?
   a. consult FAQs
   b. flame only when necessary
   c. don't shout
   d. avoid huge file attachments
   e. avoid sloppiness and errors

Short-Answer Questions

1. Briefly define *bandwidth*.

2. Name three methods of data transmission that are faster than a regular modem connection.

3. What does *log on* mean?

4. What is netiquette, and why is it important?

5. Many Web documents are "linked." What does that mean?

Concept Mapping

On a separate sheet of paper, draw a concept map or visual diagram linking concepts. Show how the following terms are related.

cable modem	physical connection
communications satellite	protocol
directory	real-time chat (RTC)
domain	site
download	search engine
DSL	T1 line
home page	upload
hypertext	URL
instant messaging	Usenet
Internet service provider	Web browser
ISDN	Webcasting
list-serves	Web page
newsgroup	Web portal
newsreader	Web site

Knowledge in Action

1. Identify the four parts of the following URL:
   http://www.nps.gov/yell/swimming.htm

2. Some Web sites go overboard with multimedia effects, while others don't include enough. Locate a Web site that you think makes effective use of multimedia. What is the purpose of the site? Why is the site's use of multimedia effective? Take notes, and repeat the exercise for a site with too much multimedia and one with too little.

3. Visit the following job-hunting Web sites:

   www.monster.com

   www.occ.com

   www.careerpath.com

   www.cweb.com

   Investigate job offerings in a field you are interested in. Which is the easiest site to use? Why? What recommendations would you make for improving the site?

4. Distance learning uses electronic links to extend college campuses to people who otherwise would not be able to take college courses. Is your school or someone you know involved in distance learning? If so, research the system's components and uses. What hardware and software do students need in order to communicate with the instructor and classmates? What courses are offered? Prepare a short report on this topic.

6. It's difficult to conceive how much information is available on the Internet and the Web. One method you can use to find information among the millions of documents is to use a search engine, which helps you find Web pages based on typed keyword or phrases. Use your browser to visit the following search sites: *www.yahoo.com* and *www.goto.com*. Click in the Search box and then type the phrase "personal computers," then hit "Go," "Find it!" or hit the enter key. When you locate a topic that interests you, print it out by choosing File, Print from the menu bar or by clicking the Print button on the toolbar. Report on your findings.

Application Software

Tools for Thinking &
Working More Productively

Key Questions

You should be able to answer the following questions.

3.1   **Application Software: Getting Started** What are five ways of obtaining application software, tools available to help you learn them, three common types of files, and the types of software?

3.2   **Common Features of Software** What are some common features of the graphical software environment?

3.3   **Word Processing** What can you do with word processing software that you can't do with pencil and paper?

3.4   **Spreadsheets** What can you do with an electronic spreadsheet that you can't do with pencil and paper and a standard calculator?

3.5   **Database Software** What is database software, and what is personal information management software?

3.6   **Specialty Software** What are the principal uses of presentation graphics, financial, desktop publishing, drawing and painting, project management, and computer-aided design software?

A Visual Overview of This Chapter

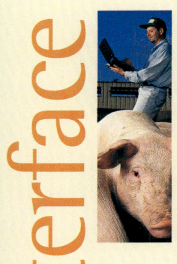

1 **Application Software.** There are five ways of obtaining application software. (1) You can use commercial software, which is **copyrighted** and is available for a fee under **software license,** meaning it may not be duplicated without permission. (2) You can freely duplicate **public-domain software,** which is not copyrighted. (3) You can use copyrighted **shareware,** which is distributed free but requires a fee for continued use. (4) You can use **freeware,** which is copyrighted but distributed free. (5) You can lease **rentalware,** the concept behind ASPs. To learn software you can use step-by-step **tutorials** (lessons) or **documentation** (reference guides).

The purpose of application software is to manipulate raw data into files of information. A **file** is a named collection of data or a program existing in secondary storage. Three types are **document files** (created by word processing), **worksheet files** (created by spreadsheets), and **database files** (created by database management programs). Files can be **imported** or acquired from other programs and **exported** or sent to other programs.

Productivity software, which is designed to make users more productive, may exist in stand-alone form, such as word processing or spreadsheet programs. Or several programs may be combined in an **office suite.** Some productivity software exists as **groupware,** which several users may share online.

2 **Common Features of Software.** The **user interface** is the display screen that enables user interaction with the computer via keyboard or via the mouse, which has an onscreen **pointer.** Today's screens have a **graphical user interface (GUI),** in which a **desktop,** or main interface screen, allows you to select from **icons** (little pictorial symbols) or **menus** (lists of activities). Most icons have a **rollover** feature that pops up an explanation when a mouse rolls over it. Menus may be **pull-down** (from the screen top), **fly-out** (explode out to the right), **pull-up** (from screen bottom), and **pop-up** (anywhere on screen). **Toolbars** (top of screen) and **taskbars** (bottom) display frequently used icons and menus. The data and programs appear in a frame called a **window,** which can be resized or repositioned on screen. Most toolbars contain a **Help command** to provide answers to questions; for some specific tasks, context-sensitive help is available.

3 **Word Processing.** Perhaps the most useful productivity program is **word processing software** (Microsoft Word, Corel WordPerfect), which allows you to create, edit, format, print, and store text material, using mouse and keyboard. A computer keyboard contains **special-purpose keys** to enter, delete, and edit data and execute commands and **function keys** (**F1, F2,** etc.) for executing commands specific to the software. Several keystrokes may be combined in one or two keystroke commands—a **macro.**

Three features that help you create documents are the **cursor,** the movable symbol on the display screen; **scrolling,** the ability to move up, down, or sideways through the text; and **word wrap,** which continues text on the next line automatically when you reach the end of a line. An **outline feature** enables you to show the hierarchy of headings within a document. Features for editing documents are insert and delete, undelete, find and replace, cut/copy and paste, **spelling checker, grammar checker,** and **thesaurus** (for presenting alternate words).

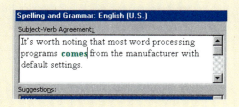

Formatting, or determining the appearance of a document, is made easier by **templates,** preformatted documents, and **wizards,** which answer your questions and format a document. Aspects of formatting are **fonts,** or typefaces and type sizes; spacing and columns; margins and justification (spacing of words in a line); page numbers and page headers/footers (repeated text at top/bottom); and other formatting such as use of clip art (ready-made pictures). The manufacturer usually specifies automatically standardized format settings, or **default settings.** Most programs give several options for printing out documents. Documents may be **saved,** or preserved, in secondary storage, such as hard disk.

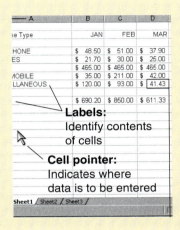

Labels:
Identify contents
of cells

Cell pointer:
Indicates where
data is to be entered

4 **Spreadsheets.** The electronic spreadsheet, now simply called the **spreadsheet** (Microsoft Excel, Corel Quattro Pro, Lotus 1-2-3), allows you to create tables and financial schedules by entering data and formulas into rows and columns arranged as a grid. A spreadsheet file contains worksheets, or single tables, with several related worksheets collected into a workbook.

Spreadsheet grids are organized with column headings across the top, row headings down the left side, and various labels or descriptive text. Columns and rows intersect in a **cell,** and its position is called a *cell address;* several adjacent cells constitute a *range.* A number entered in the cell is called a **value,** and its location is indicated by a cell pointer or spreadsheet cursor. **Formulas,** or instructions for calculations, are used to manipulate data; built-in formulas are called *functions.* Values can be changed and then recomputed; such **recalculation** is an important reason for the popularity of the spreadsheet, since it allows you to do **what-if analysis**—to see how changing numbers can change outcomes. For specialized needs, worksheet templates, custom-designed forms, are available. A nice feature of spreadsheets is the ability to create **analytical graphics,** graphical forms—such as bar charts, line graphs, and pie charts—that make numeric data easier to analyze.

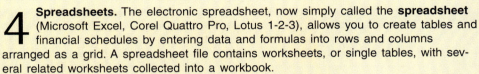

5 **Database Software.** A **database** is a collection of interrelated files, and **database software** (Microsoft Access, Corel Paradox, Lotus Approach) is a program that controls the structure of a database and access to the data. The most widely used form is the **relational database,** in which data is organized into related tables. Each table contains records (rows) and fields (columns). The records within the various tables in a database are linked by a **key field,** a common identifier that is unique, such as a Social Security number.

You can find what you want with a query, locating and displaying records. Records can also be sorted, as alphabetically, numerically, or geographically. Search results can then be saved, or they can be put into different formats, printed out, copied and placed in other documents, or transmitted (as via e-mail) to someone else. A specialized type of database software is a **personal information manager (PIM),** which helps you manage addresses, appointments, and to-do lists.

6 **Specialty Software.** Among the many thousands of specialized productivity programs available, the following are important.

Presentation graphics software (Microsoft PowerPoint, Corel Presentations, Lotus Freelance Graphics) uses graphics, animation, sound, and data to make visual presentations, commonly called *slide shows.* Design templates and content templates are available to help users get started. Material may be viewed from several perspectives (Outline, Slide, Notes Page, Slide Sorter, and Slide Show views). Presentations may be dressed up with clip art, textures, audio clips, and the like.

Financial software includes personal-finance managers, entry-level accounting programs, and business financial-management packages. **Personal-finance managers** (Quicken, Microsoft Money) help users track income and expenses, write checks, do online banking, and plan financial goals. Tax programs help with tax preparation and filing. Some financial programs automate bookkeeping and payroll tasks. Others help in business start-ups or in making investments.

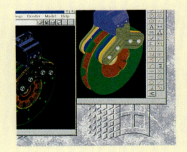

Desktop-publishing (DTP) software (QuarkXPress, PageMaker, Microsoft Publisher) enables users to mix text and graphics to produce high-quality output for commercial printing. Users can choose various type and layout styles and can use files (text, graphics) from other programs. **Drawing programs** allow users to design and illustrate objects and products; **painting programs** allow them to simulate painting on screen.

Project management software helps users plan and schedule the people, costs, and resources required to complete a project on time.

Computer-aided design (CAD) programs are used to design products, structures, engineering drawings, and maps. **CAD/CAM software**—for **computer-aided design/computer-aided manufacturing**—allows products designed with CAD to be input automatically into an automated manufacturing system that makes the products.

"Productivity is no abstract number," says economist Alan Blinder. "In the long run, it is the name of the game."[1]

The game is important—nothing less than our standard of living. Improvement in productivity is the principal reason for the economic good times of the past decade.

Productivity is a straightforward concept: It rises when a typist can produce three times as many pages this year compared to last—the result, say, of using a word processor instead of a typewriter.

Until recently, however, it wasn't clear that computers were providing those kinds of efficiencies. In 1987, it was famously observed that "You can see the computer age everywhere but in the productivity statistics."[2] Not any more. Two Federal Reserve economists found that computers—both the manufacture of them and their use, particularly the large-scale networking of computers since 1995—accounted for *two-thirds* of recent productivity growth rates.[3]

But here's a paradox. While, overall, computers help us work more productively, they can also be profoundly frustrating and even nonproductive. Experienced computer users, for instance, learn to become acquainted with "crashes": In a crash, whatever you're working on suddenly disappears from the display screen, to be replaced by what's called "the blue screen of doom."

"The appliances we use at home do not crash," says technology columnist Dan Gillmor. "We would never buy a TV that forced us to reboot [restart] the set once a month, let alone once a week or every other day."[4] Yet these are the irritations we endure in using computers in their present stage of development.

But just as automobiles evolved from finicky, unreliable contraptions into reasonably predictable, smooth-running machines, so will computers. Even in their present quirky state, however, they are essential tools for boosting our effectiveness.

3.1 Application Software: Getting Started

KEY QUESTIONS

What are five ways of obtaining application software, tools available to help you learn them, three common types of files, and the types of software?

Application software is software that has been developed to solve a particular problem, to perform useful work on specific tasks, or to provide entertainment. *System software,* which we discuss in detail in Chapter 4, enables the application software to interact with the computer and helps the computer manage its internal and external resources. New microcomputers are usually equipped not only with system software but also with some application software.

Application Software: For Sale, for Free, or for Rent?

At one time, just about everyone paid for their PC application software. You bought it as part of a computer or in a software store, or you downloaded it online with a credit card charge. Now, as we suggested earlier in describing applications services providers (ASPs), this business model may be changing. Let's look at the alternatives: *commercial software, public-domain software, shareware, freeware,* and *rentalware. (See ● Panel 3.1.)*

- ● **Commercial software:** *Commercial software,* also called *proprietary software,* is software that's offered for sale, such as Microsoft Word or Office 2000. Although such software may not show up on the bill of sale when you buy a new PC, you've paid for it as part of the purchase.

Types	Definition
Commercial software	Copyrighted. If you don't pay for it, you can be prosecuted
Public-domain software	Not copyrighted. You can copy it for free without fear of legal prosecution.
Shareware	Copyrighted. Available free, but you should pay to continue using it
Freeware	Copyrighted. Available free
Rentalware	Copyrighted. Lease for a fee

And, most likely, whenever you order a new game or other commercial program, you'll have to pay for it. This software is copyrighted. **A _copyright_ is the exclusive legal right that prohibits copying of intellectual property without the permission of the copyright holder.**

Software manufacturers don't sell you their software; rather, they sell you a license to become an authorized user of it. What's the difference? In paying for a _software license_, **you sign a contract in which you agree not to make copies of the software to give away or for resale.** That is, you have bought only the company's permission to use the software and not the software itself. This legal nicety allows the company to retain its rights to the program and limits the way its customers can use it. The small print in the licensing agreement usually allows you to make one copy (backup copy or archival copy) for your own use. (Each software company has a different license; there is no industry standard.)

Every year or so, software developers find ways to enhance their products and put forth new versions or new releases. A _version_ is a major upgrade in a software product, traditionally indicated by numbers such as 1.0, 2.0, 3.0. More recently, other notations have been used. After 1995, Microsoft labeled its Windows and Office software versions by year instead of by number, as in Microsoft's Office 97, Office 2000, and so forth. A _release,_ which now may be called an "add" or "addition," is a minor upgrade. Often this is indicated by a change in number after the decimal point. (For instance, 3.0 may become 3.1, 3.11, 3.2, and so on.) Some releases are now also indicated by the year in which they are marketed. And, unfortunately, some releases are not clearly indicated at all. (These are "patches," which may be downloaded from the software maker's Web site.)

- **Public-domain software:** _Public-domain software_ **is not protected by copyright and thus may be duplicated by anyone at will.** Public domain programs—usually developed at taxpayer expense by government agencies—have been donated to the public by their creators. They are often available through sites on the Internet. You can duplicate public domain software without fear of legal prosecution.

- **Shareware:** _Shareware_ **is copyrighted software that is distributed free of charge but requires users to make a monetary contribution in order to continue using it.** Shareware is distributed primarily through the Internet, but because it is copyrighted, you cannot use it to develop your own program that would compete with the original product.

- **Freeware:** _Freeware_ **is copyrighted software that is distributed free of charge,** today most often over the Internet. Why would any software creator let his or her product go for free? Sometimes developers want to see how users respond, so that they can make improvements in a later version. Sometimes it is to further some scholarly or humanitarian purpose—for instance, to create a standard for software on which people are apt to agree. (Linux is such a program.) In its most recent form, freeware is made available by companies trying to make money some other way—actually, by attracting viewers to their advertising. (The Web browsers Internet Explorer and Netscape Navigator are of this type.) Freeware developers generally retain all rights to their programs; technically, you are not supposed to duplicate and distribute them further.

- **Rentalware:** **_Rentalware_ is software that users lease for a fee.** This is the concept behind application services providers (ASPs), firms that lease software over the Internet. Users download programs whenever they are needed.

Tutorials & Documentation

How are you going to learn a given software program? Most commercial packages come with tutorials and documentation.

- **Tutorials:** A **_tutorial_ is an instruction book or program that helps you learn to use the product by taking you through a prescribed series of steps.** For instance, our publisher offers several how-to books, known as the *Advantage Series,* that enable you to learn different kinds of software. Tutorials may also form part of the software package.
- **Documentation:** **_Documentation_ is a user guide or reference manual that provides a narrative and graphical description of a program.** While documentation may be print-based, today it is usually available on CD-ROM, as well as via the Internet. Documentation may be instructional, but features and functions are usually grouped by category for reference purposes. For example, in word processing documentation, all features related to printing are grouped together so you can easily look them up.

Files of Data—& the Usefulness of Importing & Exporting

There is only one reason for having application software: to take raw data and manipulate it into useful files of information. A **_file_ is a named collection of (1) data or (2) a program that exists in a computer's secondary storage,** such as a floppy disk, hard disk, or CD-ROM.

Three well-known types of data files are as follows:

- **Document files: _Document files_ are created by word processing programs and consist of documents such as reports, letters, memos, and term papers.**
- **Worksheet files: _Worksheet files_ are created by electronic spreadsheets and consist of collections of (usually) numerical data such as budgets, sales forecasts, and schedules.**
- **Database files: _Database files_ are created by database management programs and consist of organized data that can be analyzed and displayed in various useful ways.** Examples are student names and addresses that can be displayed according to age, grade-point average, or home state.

Other common types of files (such as graphics, audio, and video files) are discussed in Chapter 8.

It's useful to know that often files can be exchanged—that is, *imported* and *exported*—between programs.

- **Importing:** _**Importing**_ **is defined as getting data from another source and then converting it into a format compatible with the program in which you are currently working.** For example, you might write a letter in your word processing program and include in it—that is, import—a column of numbers from your spreadsheet program.
- **Exporting:** _**Exporting**_ **is defined as transforming data into a format that can be used in another program and then transmitting it.** For example, you might work up a list of names and addresses in your database program, then send it—export it—to a document you are writing in your word processing program.

The Types of Software

Software can be classified in many ways—for entertainment, personal, education/reference, productivity, and specialized uses. _(See_ ● _Panel 3.2.)_

CONCEPT CHECK

Distinguish among the following kinds of software: commercial, public-domain, shareware, freeware, and rentalware.

Describe tutorials and documentation, and discuss three common types of files.

Distinguish importing from exporting.

In the rest of this chapter we will discuss types of _**productivity software**_— **such as word processing, spreadsheets, and database managers—whose purpose is to make users more productive at particular tasks.** Some productivity software comes in the form of an _**office suite**_, **which bundles several applications together into a single large package.** Microsoft Office 2000, for example, includes (among other things) Word, Excel, and Access—word processing, spreadsheet, and database programs, respectively. Other productivity software, such as Lotus Notes, is sold as _**groupware**_—**online software that allows several people to collaborate on the same project.**

We will now consider the three most important types of productivity

● **PANEL 3.2**
Types of software

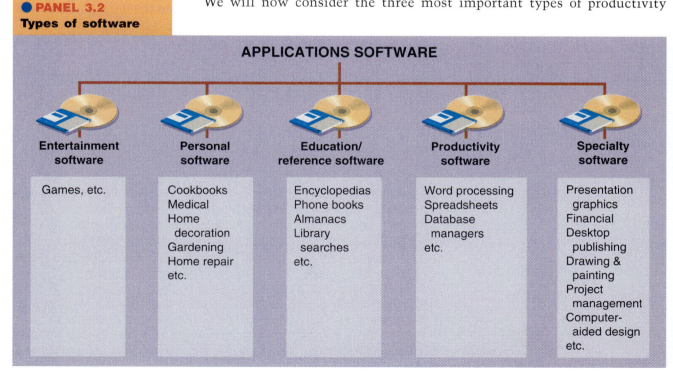

APPLICATIONS SOFTWARE

Entertainment software	Personal software	Education/ reference software	Productivity software	Specialty software
Games, etc.	Cookbooks Medical Home decoration Gardening Home repair etc.	Encyclopedias Phone books Almanacs Library searches etc.	Word processing Spreadsheets Database managers etc.	Presentation graphics Financial Desktop publishing Drawing & painting Project management Computer- aided design etc.

software: word processing, spreadsheet, and database software (including personal information managers). We will then discuss more specialized software: presentation graphics, financial, desktop-publishing, drawing and painting, project management, and computer-aided design software. But first, let's look at some common features of software.

3.2 Common Features of Software

The first thing you look at when you call up any application software on the screen is the ___user interface___—**the user-controllable display screen that allows you to communicate, or interact, with the computer.** Like the dashboard on a car, the user interface has gauges that show you what's going on and switches and buttons for controlling what you want to do. From this screen, you choose the application programs you want to run or the files of data you want to open.

You can interact with this display screen using the keys on your keyboard, but you will also frequently use your mouse. The mouse allows you to direct an on-screen pointer to perform any number of activities. The ___pointer___ **usually appears as an arrow, although it changes shape depending on the application.** The mouse is used to move the pointer to a particular place on the display screen or to point to little symbols, or icons. You can activate the function corresponding to the symbol by pressing ("clicking") buttons on the mouse. Using the mouse, you can pick up and slide ("drag") an image from one side of the screen to the other or change its size. *(See ● Panel 3.3.)*

The GUI: The Computer's Dashboard

In the beginning, personal computers had *command-driven* interfaces, which required you to type in strange-looking instructions (such as "copy a:\filename c:\" to copy a file from a floppy disk to a hard disk). In the next version, they had *menu-driven interfaces*, in which you could use the arrow keys on your keyboard (or a mouse) to choose a command from a menu, or list of activities. Today the computer's "dashboard" is usually a ___graphical user interface (GUI)___ **(pronounced "gooey"), which allows you to use a mouse or keystrokes to select icons (little symbols) and commands from menus (lists of activities).** The GUIs on the PC and on the Apple Macintosh (which was the first easy-to-use personal computer available on a wide scale) are somewhat similar. Once you learn one version, it's fairly easy to learn the other. However, the best-known GUI is that of Microsoft Windows, the latest version of which is shown on the opposite page. *(See ● Panel 3.4, p. 96.)* We consider Windows further in the following chapter.

Desktop, Icons, & Menus

Three features of a GUI are the *desktop, icons,* and *menus.*

- **Desktop:** After you turn on the computer, the first screen you will encounter is the *desktop*, a term that embodies the idea of folders of work (memos, schedules, to-do lists) on a businessperson's desk. **The ___desktop___, which is the system's main interface screen, displays pictures (icons) that provide quick access to programs and information.**

- **Icons and rollovers:** We're now ready to give a formal definition: ___Icons___ **are small pictorial figures that represent programs, data files, or procedures.** For example, a trash can represents a place to dispose of a file you no longer want. If you click your mouse pointer on a little picture of a printer, you can print out a document. One of the most important icons is the *folder*, a representation of a manila folder; folders are the collections of files in which you store your documents and other data.

Icons representing folders

To-do list

Class schedules

Term paper

Finances

Letters

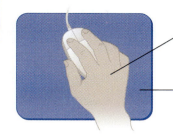

Resting your hand on the mouse, use your thumb and outside two fingers to move the mouse on your desk or mouse pad. Use your first two fingers to press the mouse buttons.

Mouse pad provides smooth surface.

Left button. Click once on an item on screen to select it. Click twice to perform an action.

Right button. Click on an object to display a shortcut list of options.

Term	Action	Purpose
Point	Move mouse across desk to guide pointer to desired spot on screen. The pointer assumes different shapes, such as arrow, hand, or I-beam, depending on the task you're performing.	To execute commands, move objects, insert data, or similar actions on screen
Click	Press and quickly release left mouse button.	To select an item on the screen
Double-click	Quickly press and release left mouse button twice.	To open a document or start a program
Drag and drop	Position pointer over item on screen, press and hold down left mouse button while moving pointer to location in which you want to place item, then release.	To move an item on the screen
Right-click	Press and release right mouse button.	To display a shortcut list of commands, such as a pop-up menu of options

Outlook Express: Part of Microsoft's browser, Internet Explorer, that enables you to use e-mail.

Microsoft Network: Click here to connect to Microsoft Network (MSN), the company's online service.

Norton Protected: Click here to activate anti-virus software.

Network Neighborhood: If your PC is linked to a network, click here to get a glimpse of everything on the network.

My Documents: Where your documents are stored unless you specify otherwise.

My Computer: Gives you a quick overview of all the files and programs on your PC.

Documents: Multitasking capabilities allow users to smoothly run more than one program at once.

Start menu: After clicking on the start button, a menu appears, giving you a quick way to handle common tasks. You can launch programs, call up documents, change system settings, get help, and shut down your PC.

Start button: Click for an easy way to start using the computer.

Taskbar: Gives you a log of all programs you have opened. To switch programs, click on the icon buttons on the taskbar.

Multimedia: Windows 98 features sharper graphics and improved video capabilites compared to Windows 95.

● **PANEL 3.4**
A graphical user interface
This is for Windows 98.

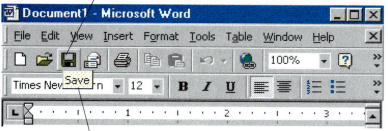

Icon: Symbol representing a program, data file, or procedure. Icons are designed to communicate their function, such as a floppy disk for saving.

Rollover: When you roll your mouse cursor over an icon or graphic, a small box with text appears that briefly explains its function.

Of course, you can't always be expected to know what an icon or graphic means. A ***rollover*** feature, a **small text box explaining the icon's function, appears when you roll the mouse pointer over the icon.** A rollover may also produce an animated graphic.

- **Menus:** Like a restaurant menu, **a *menu* offers you a list of options to choose from—in this case, a list of commands for manipulating data,** such as Print or Edit. Menus are of several types. Resembling a pull-down window shade, **a *pull-down menu*, also called a *drop-down menu*, is a list of options that pulls down from the top of the screen.** For example, if you use the mouse to "click on" (activate) a command (for example, File) on the menu bar, you will see a pull-down menu offering further commands. Choosing one of these options may produce further menus called ***fly-out menus*, menus that seem to explode out to the right.**

A ***pull-up menu* is a list of options that pulls up from the bottom of the screen.** In Windows 98, a pull-up menu appears in the lower left-hand corner when you click on the Start button.

A ***pop-up menu* is a list of command options that can "pop up" anywhere on the screen when you click the right mouse button.** In contrast to pull-down or pull-up menus, pop-up menus are not connected to a toolbar (explained later).

Pull-down menu: When you click the mouse on the menu bar, a list of options appears or pulls down like a shade.

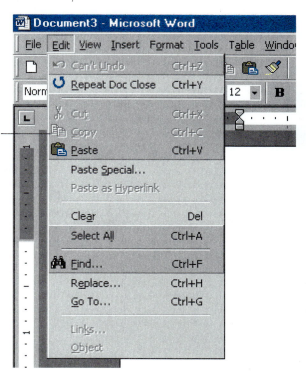

Pull-up menu: When you click the mouse cursor on the *Start* button, it produces a pull-up menu offering access to programs and documents.

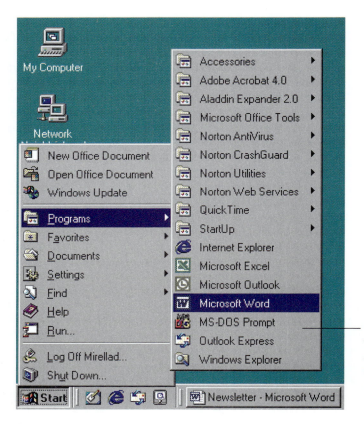

Fly-out menu: Moving the mouse cursor to an option on the pull-up menu produces a flyout menu with more options.

Documents, Toolbars, & Windows

Suppose you want to go to a document—say, a term paper you've been working on. There are two ways to begin working from the Windows 98 desktop: (1) You can click on the Start button at lower left and then make a selection from the pull-up menu that appears. Or (2) you can click on one of the icons on the desktop, probably the most important of which is the *My Computer* icon, and pursue the choices offered there. Either way, the result is the same: The document will be displayed on the window. *(See ● Panel 3.5.)*

Once you're past the desktop, the GUI's opening screen, clicking on the *My Computer* icon reveals toolbars and windows.

From Start menu

Click on *Start* button to produce Start menu, then go to *Documents* option.

From My Computer icon

Click on *My Computer* icon, which opens a window that provides access to information on your computer.

Click on C, which opens a window that provides access to information stored on your hard disk.

Click on *My Documents* icon, which opens a window providing access to document files and folders.

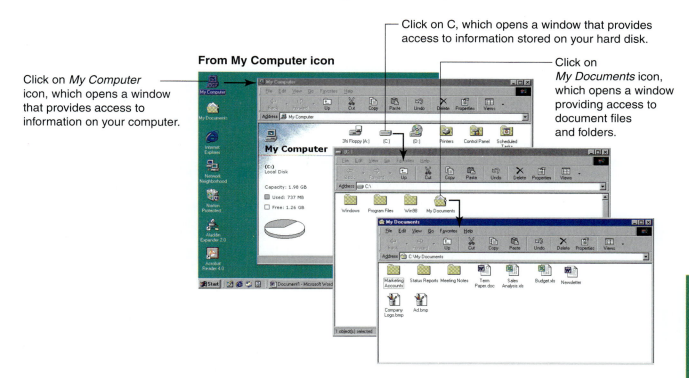

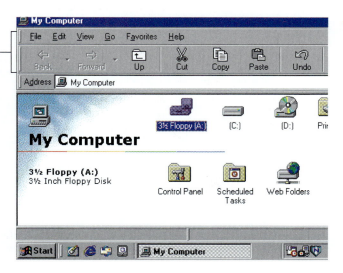

Toolbar: Bar across top of window, which contains buttons or icons representing frequently used commands

- **Toolbar:** A *toolbar* is a bar across the top of the display window. It displays menus and icons representing frequently used options or commands. Examples of menus are File, Edit, View, Favorites, and Help. An example of an icon is the picture of a printer, which issues a print command. **In Windows, the toolbar graphic at the bottom of the screen, which shows the applications that are running, is called a *taskbar*.**

- **Windows:** When spelled with a capital "W," Windows is the name of Microsoft's system software (Windows 95, 98, Me, and so on). When spelled with a lowercase "w," **a _window_ is a rectangular frame on the computer display screen. Through this frame you can view a file of data—such as a document, spreadsheet, or database—or an application program.**

In the right-hand corner of the Windows 98 toolbar are three icons that represent _Minimize, Maximize,_ and _Close._ By clicking on these icons, you can _minimize_ the window (shrink it down to an icon at the bottom of the screen), _maximize_ it (enlarge it), or _close_ it (exit the file and make the window disappear).

You can also _move_ the window around the desktop, using the mouse.

Finally, you can create _multiple windows_ to show operations going on concurrently. For example, one window might show the text of a paper you're working on, another might show the reference section for the paper, and a third might show something you're downloading from the Internet.

Minimize: Click here to shrink window so it collapses to an icon on the desktop.

Maximize: Click here to enlarge window so it fills the screen.

Close: Click here to exit a file and remove its window from your display.

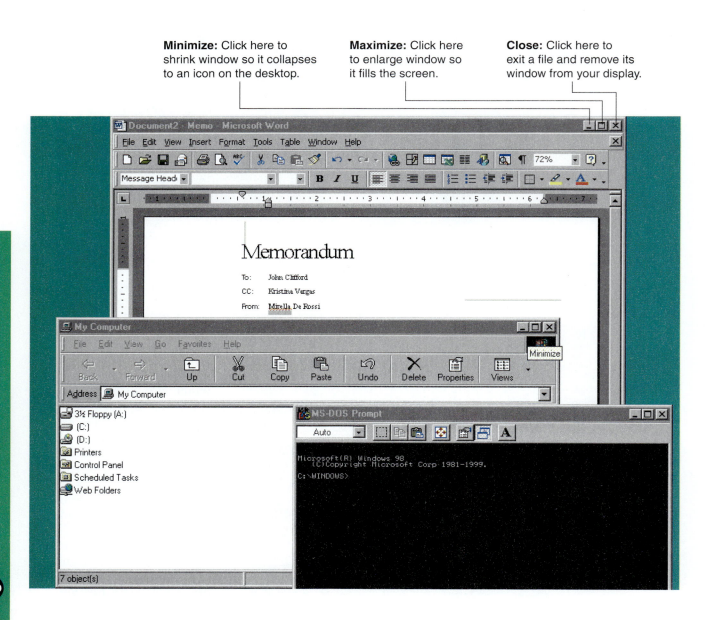

● PANEL 3.6
Help features
The Help command yields
a pull-down menu.

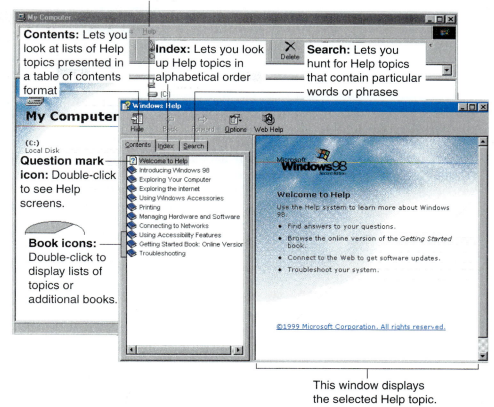

The *Help* menu provides
a list of help options.

Contents: Lets you look at lists of Help topics presented in a table of contents format

Index: Lets you look up Help topics in alphabetical order

Search: Lets you hunt for Help topics that contain particular words or phrases

Question mark icon: Double-click to see Help screens.

Book icons: Double-click to display lists of topics or additional books.

This window displays the selected Help topic.

The Help Command

Don't understand how to do something? Forgotten a command? Accidentally pressed some keys that messed up your screen layout and you want to undo it? Most toolbars contain a ***Help command—a command generating a table of contents, an index, and a search feature that can help you locate answers.*** In addition, many applications have *context-sensitive help*, which leads you to information about the task you're performing. *(See ● Panel 3.6.)*

CONCEPT CHECK

Describe the features of the GUI: desktop, icons, and the various kinds of menus.

What do toolbars and windows enable you to do?

What is the Help command?

3.3 Word Processing

After a long and productive life, the typewriter has gone to its reward. Indeed, it is practically as difficult today to get a manual typewriter repaired as to find a blacksmith. Word processing software offers a much-improved way of dealing with documents.

***Word processing software* allows you to use computers to format, create, edit, print, and store text material,** among other things. The best-known word processing program is probably Microsoft Word but there are others such as Corel WordPerfect. Word processing software allows users to maneuver through a document and *delete, insert,* and *replace* text, the principal correction activities. It also offers such additional features as *creating, editing, formatting, printing,* and *saving.*

Features of the Keyboard

Besides the mouse, the principal tool of word processing is the keyboard. As well as letter, number, and punctuation keys and often a calculator-style numeric keypad, computer keyboards have special-purpose and function keys. *(See ● Panel 3.7.)* Sometimes, for the sake of convenience, strings of keystrokes are combined in combinations called *macros.*

- **Special-purpose keys:** ***Special-purpose keys*** **are used to enter, delete, and edit data and to execute commands.** An example is the *Esc* (for "Escape") key, which tells the computer to cancel an operation or leave ("escape from") the current mode of operation. The *Enter,* or *Return,* key, which you will use often, tells the computer to execute certain commands and to start new paragraphs in a document. Commands are instructions that cause the software to perform specific actions.

  Special-purpose keys are generally used the same way regardless of the application software package being used. Most keyboards include the following special-purpose keys: *Esc, Ctrl, Alt, Del, Ins, Home, End, PgUp, PgDn, Num Lock,* and a few others. (*Ctrl* means Control, *Del* means Delete, *Ins* means Insert, for example.)

- **Function keys:** ***Function keys,*** **labeled F1, F2, and so on, are positioned along the top or left side of the keyboard. They are used to execute commands specific to the software being used.** For example, one application software package may use *F6* to exit a file, whereas another may use *F6* to underline a word.

● PANEL 3.7
Common keyboard layout

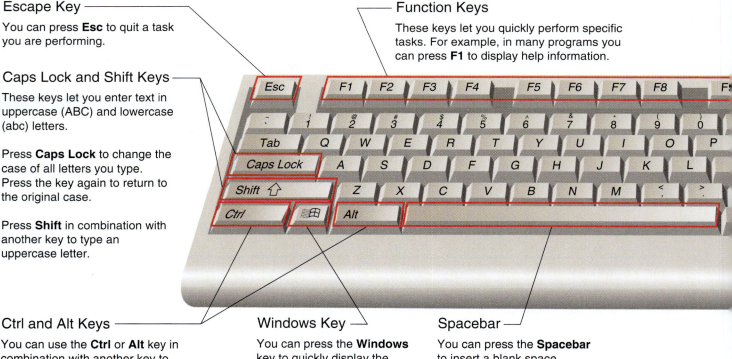

Escape Key

You can press **Esc** to quit a task you are performing.

Caps Lock and Shift Keys

These keys let you enter text in uppercase (ABC) and lowercase (abc) letters.

Press **Caps Lock** to change the case of all letters you type. Press the key again to return to the original case.

Press **Shift** in combination with another key to type an uppercase letter.

Function Keys

These keys let you quickly perform specific tasks. For example, in many programs you can press **F1** to display help information.

Ctrl and Alt Keys

You can use the **Ctrl** or **Alt** key in combination with another key to perform a specific task. For example, in some programs, you can press **Ctrl** and **S** to save a document.

Windows Key

You can press the **Windows** key to quickly display the Start menu when using Windows 95, 98, or NT operating systems.

Spacebar

You can press the **Spacebar** to insert a blank space.

- **Macros:** Sometimes you may wish to reduce the number of keystrokes required to execute a command. To do this, you use a macro. A *macro*, **also called a *keyboard shortcut*, is a single keystroke or command—or a series of keystrokes or commands—used to automatically issue a longer, predetermined series of keystrokes or commands.** Thus, you can consolidate several activities into only one or two keystrokes. The user names the macro and stores the corresponding command sequence; once this is done, the macro can be used repeatedly. (To set up a macro, pull down the Help menu and type in "macro.")

 Although many people have no need for macros, individuals who find themselves continually repeating complicated patterns of keystrokes say they are quite useful.

Creating Documents

Creating a document means entering text using the keyboard. Word processing software has three features that affect this process—the *cursor, scrolling,* and *word wrap.*

- **Cursor: The *cursor* is the movable symbol on the display screen that shows you where you may next enter data or commands.** The symbol is often a blinking rectangle or I-beam. You can move the cursor on the screen using the keyboard's directional arrow keys or a mouse. The point where the cursor is located is called the *insertion point.*
- **Scrolling: *Scrolling* means moving quickly upward, downward, or sideways through**

Scrolling

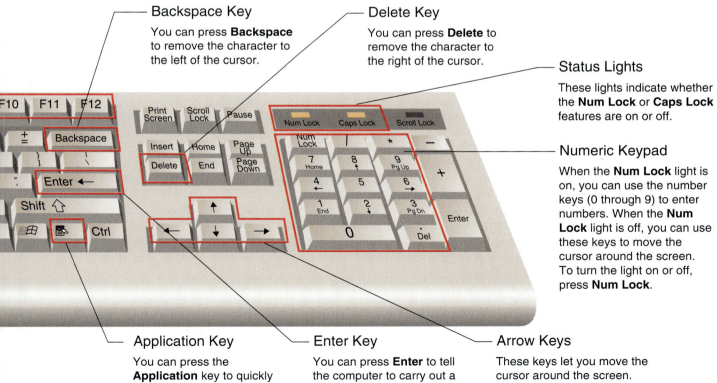

Backspace Key

You can press **Backspace** to remove the character to the left of the cursor.

Delete Key

You can press **Delete** to remove the character to the right of the cursor.

Status Lights

These lights indicate whether the **Num Lock** or **Caps Lock** features are on or off.

Numeric Keypad

When the **Num Lock** light is on, you can use the number keys (0 through 9) to enter numbers. When the **Num Lock** light is off, you can use these keys to move the cursor around the screen. To turn the light on or off, press **Num Lock**.

Application Key

You can press the **Application** key to quickly display the shortcut menu for an item on your screen.

Enter Key

You can press **Enter** to tell the computer to carry out a task. In a word processing program, press this key to start a new paragraph.

Arrow Keys

These keys let you move the cursor around the screen.

the text or other screen display. A standard computer screen displays only 20–22 lines of standard-size text. Of course, most documents are longer than that. Using the directional arrow keys, or the mouse and a scroll bar located at the side of the screen, you can move ("scroll") through the display screen and into the text above and below it.

- **Word wrap:** *Word wrap* automatically continues text on the next line when you reach the right margin. That is, the text "wraps around" to the next line. You don't have to hit a "carriage return" key or Enter key, as you do with a typewriter.

To help you organize term papers and reports, the *outline feature* puts tags on various headings to show the hierarchy of heads—for example, main head, subhead, and sub-subhead. The basics of word processing are shown in the accompanying illustration. *(See ● Panel 3.8.)*

Editing Documents

Editing is the act of making alterations in the content of your document. Some features of editing are *insert and delete, undelete, find and replace, cut/copy and paste, spelling checker, grammar checker,* and *thesaurus.* Some of these commands are in the Edit pull-down menu and icons on the toolbar.

- **Insert and delete:** *Inserting* is the act of adding to the document. You simply place the cursor wherever you want to add text and start typing; the existing characters will be pushed along. (The *Insert* key toggles between inserting and writing over.)

  Deleting is the act of removing text, usually using the *Delete* or *Backspace* keys.

  The *Undelete command* allows you to change your mind and restore text that you have deleted. Some word processing programs offer as many as 100 layers of "undo," so that users who delete several paragraphs of text, but then change their minds, can reinstate the material.

- **Find and replace:** The *Find,* or *Search, command* allows you to find any word, phrase, or number that exists in your document. The *Replace command* allows you to automatically replace it with something else.

- **Cut/Copy and paste:** Typewriter users who wanted to move a paragraph or block of text from one place to another in a manuscript used scissors and glue to "cut and paste." With word processing, it only takes a few keystrokes. You select (highlight) the portion of text you want to copy or move. Then you use the *Copy* or *Cut command* to move it to the *clipboard,* a special holding area in the computer's memory. From there, you can "paste," or transfer, the material to any point (indicated with the cursor) in the existing document or in a new document.

- **Spelling checker:** Most word processors have a **_spelling checker,_ which tests for incorrectly spelled words.** As you type, the spelling checker indicates (perhaps with a squiggly line) words that aren't in its dictionary and thus may be misspelled. *(See ● Panel 3.9, page 106.)* Special add-on dictionaries are available for medical, engineering, and legal terms. In addition, programs such as Microsoft Word have an Auto Correct function that automatically fixes such common mistakes as transposed letters—replacing "teh" with "the," for instance.

Toolbar:
Allows quick access to frequently used commands

Menu bar:
Allows access to all commands

Title bar:
Shows name of document you're working on

Spelling and Grammar button:
Click on to check for misspelled words and incorrect grammar.

Text alignment buttons:
Click on to align text to be left, right, center, or full justified.

Style button:
Click on to access variety of format styles.

Ruler:
Shows tabs and margins

Insertion point:
Blinking symbol shows where the next character you type will appear

Status bar:
Shows details about the document you're working on

Window controls:
Lets you enlarge a window, restore its previous position, or hide it from view

Mouse pointer:
Use the mouse to move the insertion point, to click on icons, or to select text for editing.

Scroll bars:
Lets you scroll the document to reveal hidden portions

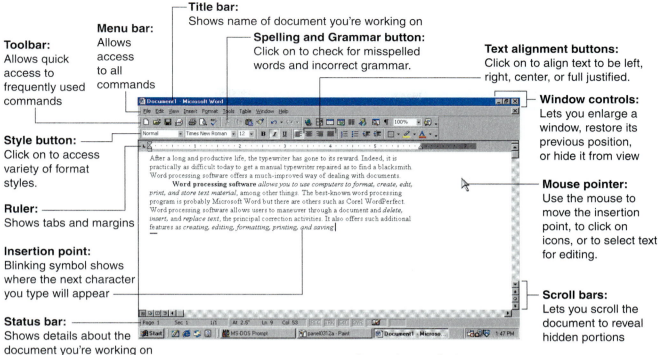

Cut-and-paste feature:
Enables you to move blocks of text. First highlight the text. On Edit menu, select Cut option. Then, on Edit Menu, select Paste option. (You can also use the icons in the toolbar.)

Outline feature:
Enables you to view headings in your document

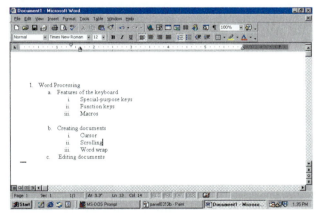

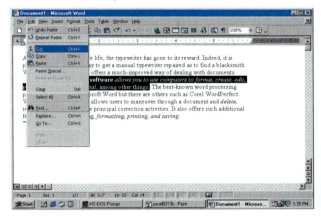

Font button

Font size button

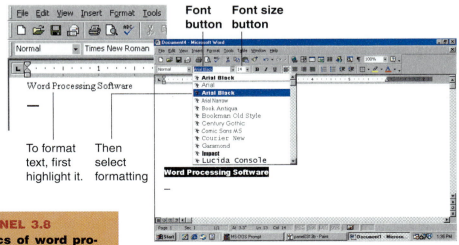

To format text, first highlight it.

Then select formatting

Formatting feature:
Enables you to change type font and size. First highlight the text. Then click next to Font button for pull-down menu of fonts. Click next to Font Size button for menu of type sizes.

● **PANEL 3.8**
Basics of word processing

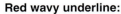

Red wavy underline:
Indicates spelling checker
doesn't recognize the word.
You have two options.

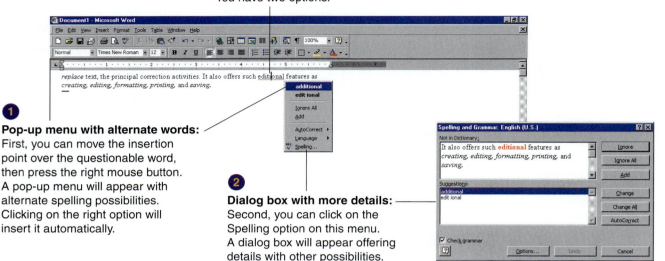

①

Pop-up menu with alternate words:
First, you can move the insertion
point over the questionable word,
then press the right mouse button.
A pop-up menu will appear with
alternate spelling possibilities.
Clicking on the right option will
insert it automatically.

②

Dialog box with more details:
Second, you can click on the
Spelling option on this menu.
A dialog box will appear offering
details with other possibilities.

● **PANEL 3.9**
Spelling checker
How a word processing
program checks for mis-
spelled words and offers
alternatives

- **Grammar checker:** A *grammar checker* **highlights poor grammar, wordiness, incomplete sentences, and awkward phrases.** The grammar checker won't fix things automatically, but it will flag (perhaps with a different color of squiggly line) possible incorrect word usage and sentence structure. *(See* ● *Panel 3.10.)*

- **Thesaurus:** If you find yourself stuck for the right word while you're writing, you can call up an on-screen *thesaurus,* **which will present you with the appropriate word or alternative words.**

Formatting Documents with the Help of Templates & Wizards

In the sense in which we use it here, *formatting* **means determining the appearance of a document.** To this end, word processing programs provide two helpful devices—templates and wizards. **A** *template* **is a preformatted document that provides basic tools for shaping a final document**—the text, layout, and style for a letter, for example. **A** *wizard* **answers your questions and uses the answers to lay out and format a document.** In Word, you can use the Memo Wizard to create professional-looking memos or the Resume Wizard to create a resume.

Among the many aspects of formatting are the following:

- **Font:** You can decide what *font*—**typeface and type size**—you wish to use. For instance, you can specify whether it should be Arial, Courier, or Times New Roman. You can indicate whether the text should be, say, 10 points or 12 points in size and the headings should be 14 points or 16 points. (There are 72 points in an inch.) You can specify what parts of it should be underlined, *italic,* or **boldface.**

- **Spacing and columns:** You can choose whether you want the lines to be *single-spaced* or *double-spaced* (or something else). You can specify whether you want text to be *one column* (like this page), *two columns* (like many magazines and books), or *several columns* (like newspapers).

- **Margins and justification:** You can indicate the dimensions of the *margins*—left, right, top, and bottom—around the text. You can specify the text *justification*—how the letters and words are spaced in

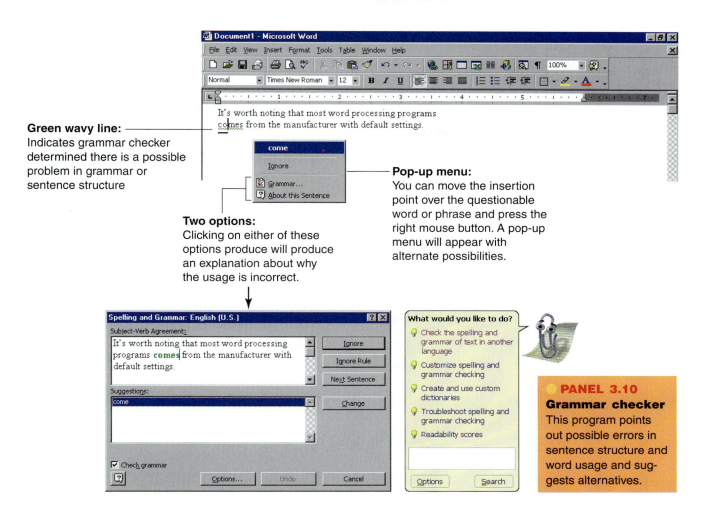

Green wavy line:
Indicates grammar checker determined there is a possible problem in grammar or sentence structure

Two options:
Clicking on either of these options produce will produce an explanation about why the usage is incorrect.

Pop-up menu:
You can move the insertion point over the questionable word or phrase and press the right mouse button. A pop-up menu will appear with alternate possibilities.

● **PANEL 3.10**
Grammar checker
This program points out possible errors in sentence structure and word usage and suggests alternatives.

each line. To *justify* means to align text evenly between left and right margins, as most newspaper columns and in this text. To *left-justify* means to align text evenly on the left. (It is "ragged right," like many business letters or this paragraph.)

● **Pages, headers, footers:** You can indicate page numbers and headers or footers. A *header* is common text (such as a date or document name) printed at the top of every page. A *footer* is the same thing printed at the bottom of every page.

● **Other formatting:** You can specify *borders* or other decorative lines, *shading, tables,* and *footnotes.* You can even import *graphics* or drawings from files in other software programs, including *clip art*—collections of ready-made pictures and illustrations available online or on CD-ROM disks.

It's worth noting that word processing programs (and indeed most forms of application software) come from the manufacturer with *default settings.* **<u>*Default settings*</u> are the settings automatically used by a program unless the user specifies otherwise, thereby overriding them.** Thus, for example, a word processing program may automatically prepare a document single-spaced, left-justified, with 1-inch right and left margins, unless you alter these default settings.

Printing, Faxing, or E-Mailing Documents

Most word processing software gives you several options for printing. For example, you can print *several copies* of a document. You can print *individual pages* or a *range of pages.* You can even preview a document before printing it out. *Previewing (print previewing)* means viewing a document on screen to see what it will look like in printed form before it's printed. Whole pages are displayed in reduced size.

You can also send your document off to someone else by fax or e-mail attachment, if your computer has the appropriate communications link.

Saving Documents

Saving means storing, or preserving, a document as an electronic file permanently—on floppy disk or hard disk, for example. Saving is a feature of nearly all application software, but anyone accustomed to writing with a typewriter will find this activity especially valuable. Having the document stored in electronic form spares you the tiresome chore of retyping it from scratch whenever you want to make changes. You need only retrieve it from the storage medium and make the changes you want. Then you can print it out again.

CONCEPT CHECK

What are the important features of the keyboard?

Describe the role of the cursor, scrolling, and word wrap in creating documents.

What word processing features are available to help edit documents?

What assistance is available to help you format documents, and what aspects of formatting should be of concern to you?

3.4 Spreadsheets

KEY QUESTION

What can you do with an electronic spreadsheet that you can't do with pencil and paper and a standard calculator?

What is a spreadsheet? Traditionally, it was simply a grid of rows and columns, printed on special light-green paper, that was used to produce financial projections and reports. A person making up a spreadsheet spent long days and weekends at the office penciling tiny numbers into countless tiny rectangles. When one figure changed, all others numbers on the spreadsheet had to be recomputed. Ultimately, there might be wastebaskets full of jettisoned worksheets.

In the late 1970s, Daniel Bricklin was a student at the Harvard Business School. One day he was staring at columns of numbers on a blackboard when he got the idea for computerizing the spreadsheet. He created the first electronic spreadsheet, now called simply a spreadsheet. **The _spreadsheet_ allows users to create tables and financial schedules by entering data and formulas into rows and columns arranged as a grid on a display screen.** Before long, the electronic spreadsheet was the most popular small-business program. Unfortunately for Bricklin, his version (called VisiCalc) was quickly surpassed by others. Today the principal spreadsheets are Microsoft Excel, Corel Quattro Pro, and Lotus 1-2-3.

You can put your checkbook register on a spreadsheet, and then use it to compute totals and compare income and expenses from one month to the next. In addition, within a spreadsheet file you may have workbooks containing worksheets. A *worksheet* is a single table. A *workbook* is a collection of related worksheets. Thus, within your Microsoft Excel spreadsheet file, you might have one workbook headed *Checkbook*, which would contain worksheets with the history of your checking account for the years 2000, 2001, and so on. You might have another workbook headed *Credit cards*, containing worksheets for each year.

The Basics: How Spreadsheets Work

The arrangement of a spreadsheet is as follows. *(See ● Panel 3.11, next page.)*

- **How a spreadsheet is organized—column headings, row headings, and labels:** In the worksheet's frame area (work area), lettered *column headings* appear across the top ("A" is the name of the first column, "B" the second, and so on). Numbered *row headings* appear down the left side ("1" is the name of the first row, "2" the second, and so forth). *Labels* are any descriptive text, such as APRIL, RENT, or GROSS SALES. You use your computer's keyboard to type in the various headings and labels.

- **Where columns and rows meet—cells, cell addresses, ranges, and values:** A _cell_ **is the place where a row and a column intersect; its position is called a cell address.** For example, "A1" is the cell address for the top left cell, where column A and row 1 intersect. A *range* is a group of adjacent cells—for example, A1 to A5. **A number or date entered in a cell is called a _value_.** The values are the actual numbers used in the spreadsheet—dollars, percentages, grade points, temperatures, or whatever. Headings and labels also go into cells. A *cell pointer*, or *spreadsheet cursor*, indicates where data is to be entered. The cell pointer can be moved around like a cursor in a word processing program.

- **Why the spreadsheet has become so popular—formulas, functions, recalculation, and what-if analysis:** Now we come to the reason the electronic spreadsheet has taken offices by storm. **_Formulas_ are instructions for calculations.** For example, a formula might be @SUM(A5:A15), meaning *Sum* (that is, add) *all the numbers in the cells with cell addresses A5 through A15.*

  Functions are built-in formulas that perform common calculations. For instance, a function might average a range of numbers or round off a number to two decimal places.

  After the values have been entered into the worksheet, the formulas and functions can be used to calculate outcomes. However, what was revolutionary about the electronic spreadsheet was its ability to easily do recalculation. **_Recalculation_ is the process of recomputing values,** either as an ongoing process as data is entered or afterward, with the press of a key. With this simple feature, the hours of mind-numbing work required to manually rework paper spreadsheets became a thing of the past.

  The recalculation feature has opened up whole new possibilities for decision making. In particular, **_what-if analysis_ allows the user to see how changing one or more numbers changes the outcome of the recalculation.** That is, you can create a worksheet, putting in formulas and numbers, and then ask, "What would happen if we change that detail?"—and immediately see the effect on the bottom line.

- **Using worksheet templates—pre-arranged forms for specific tasks:** You may find that your spreadsheet software makes worksheet templates available for specific tasks. *Worksheet templates* are forms containing formats and formulas custom-designed for particular kinds of work. Examples are templates for calculating loan payments, tracking travel expenses, monitoring personal budgets, and keeping track of time worked on projects. Templates are also available for a variety of business needs—providing sales quotations, invoicing customers, creating purchase orders, and writing a business plan.

Spreadsheet
A farmer uses a notebook computer with spreadsheet software to record data about pigs, such as age, weight, and eating patterns.

Toolbar: Allows quick access to frequently used commands

Menu bar: Allows access to all commands

Title bar: Shows name of document you're working on

Formula bar: Shows contents of cell and lets you enter data or formulas into a cell

Window controls: Lets you enlarge a window, restore its previous position, or hide it from view

Cell location: Displays column (A) and row (12) of current cell

Column headings: Lets you select an entire column with a mouse click

Row headings: Lets you select an entire row with a mouse click

Cell: Formed by intersection of row and column (letter and number)

Status bar: Shows details about the document you're working on

Labels: Identify contents of cells

Cell pointer: Indicates where data is to be entered

Values: Numbers are called *values.*

Worksheet area: This area contains the worksheet itself.

Scroll bars: Lets you scroll the document to reveal hidden portions

Recalculation and what-if analysis: *Recalculation* is the process of recomputing values. *What-if analysis* is changing one or more values to see what would happen with recalculation.

① What if March expenses for Miscellaneous changed from $41.43 to $120.75?

② Then total expenses would change from $2,151.53 to $2,230.85. Moreover, all the percentages would change.

Formulas and functions: Instructions for calculations are called *formulas.* Built-in formulas that perform common calculations such as addition are called *functions.*

Sheet tab: Lets you select a worksheet

Worksheets: A *spreadsheet file* can contain several related *worksheets,* each covering a different topic. This allows pertinent data or formulas to be easily accessed and applied when needed.

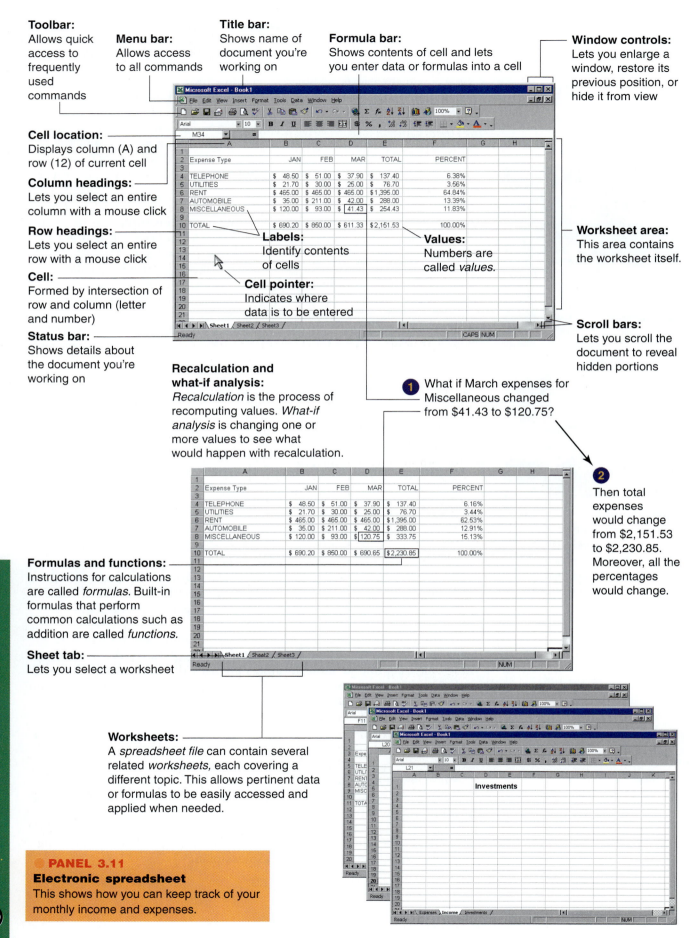

● **PANEL 3.11**
Electronic spreadsheet
This shows how you can keep track of your monthly income and expenses.

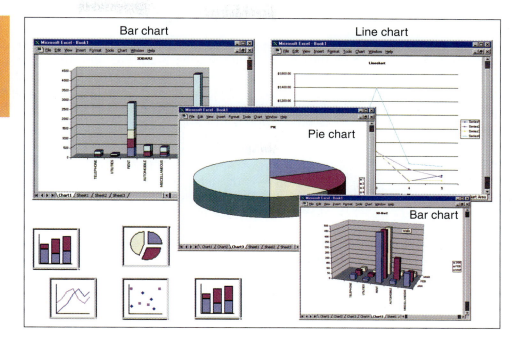

● **PANEL 3.12**

Analytical graphics

Bar charts, line graphs, and pie charts are used to display numbers in graphical form.

Bar chart

Line chart

Pie chart

Bar chart

Analytical Graphics: Creating Charts

A nice feature of spreadsheet packages is the ability to create analytical graphics, or charts. **_Analytical graphics,_ or business graphics, are graphical forms that make numeric data easier to analyze** than when it is organized as rows and columns of numbers. Whether viewed on a monitor or printed out, analytical graphics help make sales figures, economic trends, and the like easier to comprehend and analyze.

The principal examples of analytical graphics are *bar charts, line graphs*, and *pie charts. (See* ● *Panel 3.12.)* If you have a color printer, these charts can appear in color. In addition, they can be displayed or printed out so that they look three-dimensional. Spreadsheets can even be linked to more exciting graphics, such as digitized maps.

CONCEPT CHECK

What is a spreadsheet? A worksheet? A workbook?

What are the components of a spreadsheet?

What is the significance of recalculation and what-if analysis?

What's useful about worksheet templates and analytical graphics?

3.5 Database Software

KEY QUESTIONS

What is database software, and what is personal information management software?

In its most general sense, a database is any electronically stored collection of data in a computer system. In its more specific sense, **a _database_ is a collection of interrelated files** in a computer system. These computer-based files are organized according to their common elements, so that they can be retrieved easily. (Databases are covered in detail in Chapter 8.) Sometimes called a *database manager* or *database management system (DBMS)*, **_database software_ is a program that sets up and controls the structure of a database and access to the data.**

The Benefits of Database Software

When data is stored in separate files, the same data will be repeated in many files. In the old days, each college administrative office—registrar, financial aid, housing, and so on—might have a separate file on you. Thus, there was *redundancy*—your address, for example, was repeated over and over. The advantage of database software is that data is not in separate files. Rather, it is *integrated*. Thus, your address need only be listed once, and all the separate administrative offices will have access to the same information. For that reason, information in databases is considered to have more *integrity*. That is, the information is more likely to be accurate and up to date.

Databases are a lot more interesting than they used to be. Once they included only text. Now they can also include pictures, sound, and animation. It's likely, for instance, that your personnel record in a future company database will include a picture of you and perhaps even a clip of your voice. If you go looking for a house to buy, you will be able to view a real estate agent's database of video clips of homes and properties without leaving the realtor's office.

Today the principal microcomputer database programs are Microsoft Access, Corel Paradox, and Lotus Approach. (In larger systems, Oracle is a major player.) These programs also allow users to attach multimedia—sound, motion, and graphics—to forms.

The Basics: How Databases Work

Let's consider some basic features of databases:

- **How a relational database is organized—tables, records, and fields:** The most widely used form of database, especially on PCs, is the **_relational database_, in which data is organized into related tables.** Each *table* contains rows and columns; the rows are called records, and the columns are called fields. An example of a record is a person's address—name, street address, city, and so on. An example of a field is that person's last name; another field would be that person's first name, a third field would be that person's street address, and so on. (See ● *Panel 3.13.*)

  Just as a spreadsheet may include a workbook with several worksheets, so a relational database might include a database with several tables. For instance, if you're running a small company, you might have one database headed *Employees*, containing three tables—*Addresses, Payroll,* and *Benefits.* You might have another database headed *Customers*, with *Addresses, Orders,* and *Invoices* tables.

- **How various records can be linked—the key field:** The records within the various tables in a database are linked by a **_key field_, a field that can be used as a common identifier because it is unique.** The most frequent key field used in the United States is the Social Security number, but any unique identifier could be used, such as employee number.

- **Finding what you want—querying and displaying records:** The beauty of database software is that you can locate records quickly. For example, several offices at your college may need access to your records, but for different reasons: the registrar, financial aid, student housing, and so on. Any of these offices can *query records—locate and display records*—by calling them up on a computer screen for viewing and updating. Thus, if you move, your address field will need to be corrected for all relevant offices of the college. A person making a search might make the query, "Display the address of [your name]." Once a record is displayed, the address field can be changed. Thereafter, any office calling up your file will see the new address.

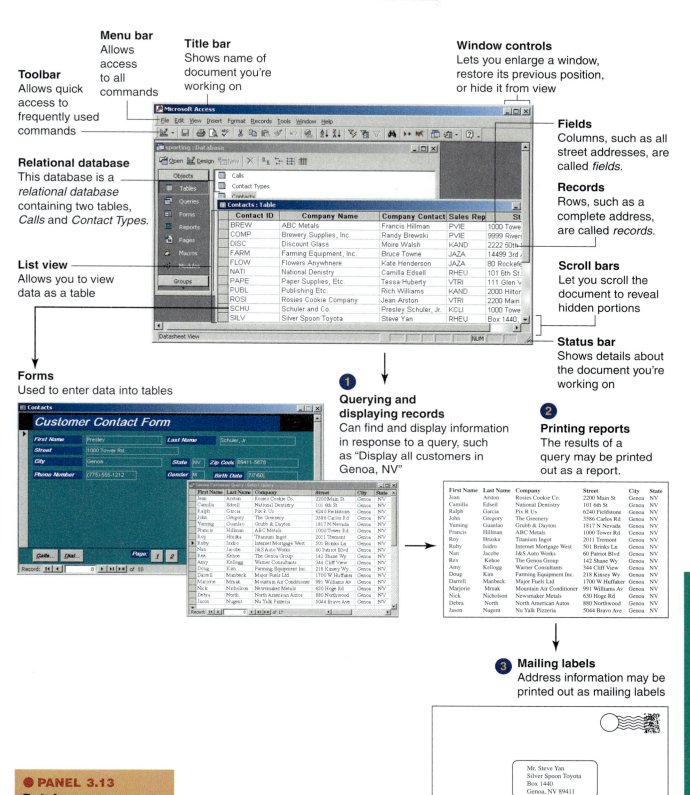

Menu bar
Allows access to all commands

Toolbar
Allows quick access to frequently used commands

Title bar
Shows name of document you're working on

Window controls
Lets you enlarge a window, restore its previous position, or hide it from view

Relational database
This database is a *relational database* containing two tables, *Calls* and *Contact Types*.

Fields
Columns, such as all street addresses, are called *fields*.

Records
Rows, such as a complete address, are called *records*.

List view
Allows you to view data as a table

Scroll bars
Let you scroll the document to reveal hidden portions

Status bar
Shows details about the document you're working on

Forms
Used to enter data into tables

❶ Querying and displaying records
Can find and display information in response to a query, such as "Display all customers in Genoa, NV"

❷ Printing reports
The results of a query may be printed out as a report.

❸ Mailing labels
Address information may be printed out as mailing labels

● **PANEL 3.13**
Database

- **Sorting and analyzing records and applying formulas:** With database software you can easily find and change the order of records in a table. Normally, records are displayed in a database in the same order in which they are entered—for instance, by the date a person registered to attend college. However, all these records can be *sorted* in different ways—arranged alphabetically, numerically, geographically, or in some other order. For example, they can be rearranged by state, by age, or by Social Security number.

  In addition, database programs contain built-in mathematical formulas so that you can analyze data. This feature can be used, for example, to find the grade-point averages for students in different majors or in different classes.

- **Putting search results to use—saving, formatting, printing, copying, or transmitting:** Once you've queried, sorted, and analyzed the records and fields, you can simply save them to your hard disk or to a floppy disk. You can format them in different ways, altering headings and type styles. You can print them out on paper as reports, such as an employee list with up-to-date addresses and phone numbers. A common use is to print out the results as names and addresses on mailing labels—adhesive-backed stickers that can be run through your printer, then stuck on envelopes. You can use the copy command to copy your search results and then paste them into a paper produced on your word processor. You can also cut and paste data into an e-mail message or make the data an attachment file to an e-mail, so that it can be transmitted to someone else.

Personal Information Managers

Pretend you are sitting at a desk in an old-fashioned office. You have a calendar, a Rolodex-type address file, and a notepad. Most of these items could also be found on a student's desk. How would a computer and software improve on this arrangement?

Many people find ready uses for specialized types of database software known as personal information managers. **A *personal information manager (PIM)* is software to help you keep track of and manage information you use on a daily basis, such as addresses, telephone numbers, appointments, to-do lists, and miscellaneous notes.** Some programs feature phone dialers, outliners (for roughing out ideas in outline form), and ticklers (or reminders). With a PIM, you can key in notes in any way you like and then retrieve them later based on any of the words you typed.

Popular PIMs are Microsoft Outlook, Lotus Organizer, and Act. Microsoft Outlook, for example, has sections labeled Inbox, Calendar, Contacts, Tasks (to-do list), Journal (to record interactions with people), Notes (scratchpad), and Files. *(See ● Panel 3.14.)*

CONCEPT CHECK

What is a database, and what are its benefits?

Describe the basic features of a relational database.

How might a PIM help you?

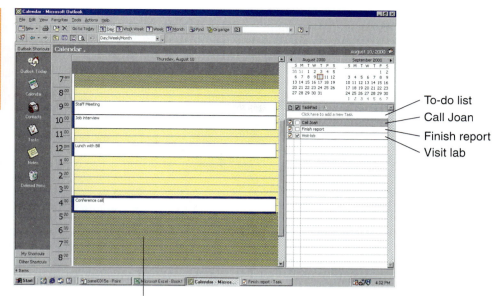

To-do list
Call Joan
Finish report
Visit lab

Appointment calendar

3.6 Specialty Software

After learning some of the productivity software just described, you may wish to become familiar with more specialized programs. For example, you might first learn word processing and then move on to desktop publishing, or first learn spreadsheets and then learn personal-finance software. We will consider the following kinds of software, although they are but a handful of the thousands of specialized programs available: *presentation graphics, financial, desktop-publishing, drawing and painting, project management,* and *computer-aided design* software.

Presentation Graphics Software

You may already be accustomed to seeing presentation graphics because many college instructors now use such software to accompany their lectures. ***Presentation graphics software* uses graphics, animation, sound and data or information to make visual presentations.** Well-known presentation graphics packages include Microsoft PowerPoint, Corel Presentations, and Lotus Freelance Graphics. *(See ● Panel 3.15, next page.)*

Visual presentations are commonly called *slide shows,* although they can consist not only of 35-mm slides but also of paper copies, overhead transparencies, video, animation, and sound. Presentation graphics packages often come with slide sorters, which group together a dozen or so slides in miniature. The person making the presentation can use a mouse or keyboard to bring the slides up for viewing or even start a self-running electronic slide show.

Let's examine the process of using presentation software.

● **Using templates to get started:** Just as word processing programs offer templates for faxes, business letters, and the like, presentation-graphics programs offer templates to help you organize your presentation, whether it's for a roomful of people or over the Internet. Templates are of two types: design and content. *Design templates* offer formats, layouts, background patterns, and color schemes that can apply to general forms of content material. *Content templates* offer formats for specific subjects; for instance, PowerPoint offers templates for "Selling Your Ideas," "Facilitating a Meeting," and "Motivating a Team." The software offers wizards that walk you through the process of filling in the template.

1 **Outline View**
This view helps you organize the content of your material in standard outline form.

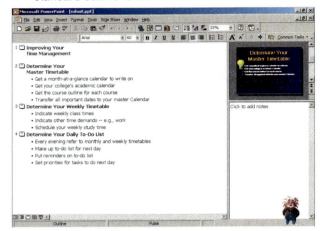

2 **Dressing up your presentation**
PowerPoint offers professional design templates of text format, background, and borders. You place your text for each slide into one of these templates. You can also import a graphic from the clip art that comes with the program.

3 **Slide View**
This view allows you to see what a single slide will look like. You can use this view to edit the content and looks of each slide.

4 **Notes Page View**
This view displays a small version of the slide plus the notes you will be using as speaker notes.

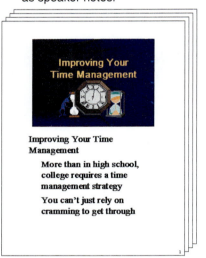

View icons
Clicking on these offers different views: *Slide, Outline, Slide Sorter, Notes Page,* and *Slide Show*

5 **Slide Sorter View**
This view displays miniatures of each slide, enabling you to adjust the order of your presentation.

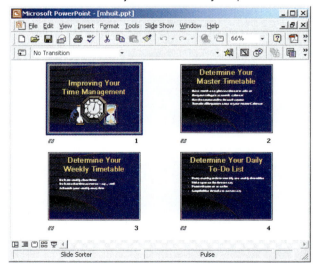

● **PANEL 3.15**
Presentation graphics
Microsoft PowerPoint 2000 helps you prepare and make visual presentations.

● **Getting assistance on content development and organization:** To provide assistance as you're building your presentation, PowerPoint displays three windows on your screen at the same time—the *Outline View,* the *Slide View,* and the *Notes Page View.* This enables you to add new slides, create and edit the text on the slides, and create notes (to use as lecture or speech notes) while developing your presentation.

The *Outline View* helps you organize the content of your material in standard outline form. The text you enter into the outline is automatically formatted into slides according to the template you selected. If you wish, you can pull in (import) your outline from a word processing document.

The *Slide View* helps you see what a single slide will look like. The outline text appears as slide titles and subtitles in subordinate order.

The *Notes Page View* displays the notes you will be using as speaker notes. It includes a small version of the slide.

Two other views are helpful in organizing and practicing. The *Slide Sorter View* allows you to view a number of slides (12 or more) at once, so you can see how to order and reorder them. The *Slide Show View* presents the slides in the order in which your audience will view them, so you can practice your presentation.

- **Dressing up your presentation:** Presentation software makes it easy to dress up each visual page ("slide") with artwork by pulling in clip art from other sources. Although presentations may make use of some basic analytical graphics—bar, line, and pie charts—they usually look much more sophisticated. For instance, they may utilize different texture (speckled, solid, cross-hatched), color, and three-dimensionality. In addition, you can add audio clips, special visual effects (such as blinking text), animation, and video clips.

CONCEPT CHECK

What are the benefits of presentation graphics software?

What kind of help is available with a presentation graphics package?

Financial Software

Financial software **is a growing category that ranges from personal-finance managers to entry-level accounting programs to business financial-management packages.**

Consider the first of these, which you may find particularly useful. *Personal-finance managers* **let you keep track of income and expenses, write checks, do online banking, and plan financial goals.** Such programs don't promise to make you rich, but they can help you manage your money. They may even get you out of trouble. Many personal-finance programs, such as Quicken and Microsoft Money, include a calendar and a calculator, but the principal features are the following:

- **Tracking of income and expenses:** The programs allow you to set up various account categories for recording income and expenses, including credit card expenses.
- **Checkbook management:** All programs feature checkbook management, with an on-screen check writing form and check register that look like the ones in your checkbook. Checks can be purchased to use with your computer printer.
- **Reporting:** All programs compare your actual expenses with your budgeted expenses. Some will compare this year's expenses to last year's.
- **Income tax:** All programs offer tax categories, for indicating types of income and expenses that are important when you're filing your tax return.
- **Other:** Some of the more versatile personal-finance programs also offer financial-planning and portfolio-management features.

Besides personal-finance managers, financial software includes small business accounting and tax software programs, which provide virtually all the

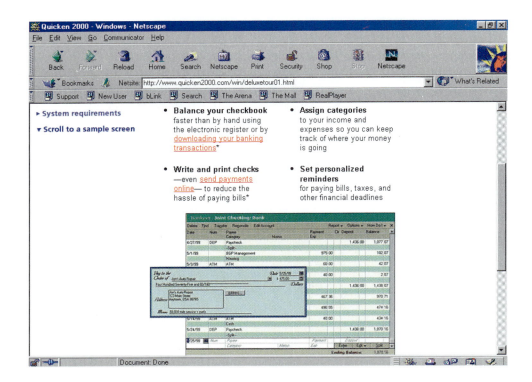

forms you need for filing income taxes. Tax programs such as TaxCut and Turbo Tax make complex calculations, check for mistakes, and even unearth deductions you didn't know existed. Tax programs can be linked to personal finance software to form an integrated tool.

Many financial software programs may be used in all kinds of enterprises. For instance, accounting software automates bookkeeping tasks, while payroll software keeps records of employee hours and produces reports for tax purposes.

Some programs go beyond financial management and tax and accounting management. For example, Business Plan Pro, Management Pro, and Performance Now can help you set up your own business from scratch.

Finally, there are investment software packages, such as StreetSmart from Charles Schwab and Online Xpress from Fidelity, as well as various retirement planning programs.

CONCEPT CHECK

What is financial software?

What functions does financial software perform?

Desktop Publishing

When Margaret Trejo, then 36, was laid off from her job because her boss couldn't meet the payroll, she was stunned. "Nothing like that had ever happened to me before," she said later. "But I knew it wasn't a reflection on my work. And I saw it as an opportunity."[5]

Today Trejo Production is a successful desktop-publishing company in Princeton, New Jersey, using Macintosh equipment to produce scores of books, brochures, and newsletters. "I'm making twice what I ever made in management positions," says Trejo, "and my business has increased by 25% every year."

Not everyone can set up a successful desktop-publishing business, because many complex layouts require experience, skill, and knowledge of graphic design. Indeed, use of these programs by nonprofessional users can lead to rather unprofessional-looking results. Nevertheless, the availability of microcomputers and reasonably inexpensive software has opened up a career area formerly reserved for professional typographers and printers.

Desktop publishing (DTP) involves mixing text and graphics to produce high-quality output for commercial printing, using a microcomputer and mouse, scanner, laser or ink-jet printer, and DTP software. Often the printer is used primarily to get an advance look before the completed job is sent to a typesetter for even higher-quality output. Professional DTP programs are

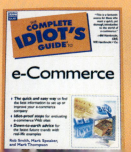
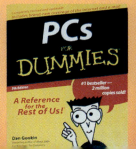

QuarkXPress and PageMaker. Microsoft Publisher is a "low-end," consumer-oriented DTP package. Some word processing programs, such as Word and WordPerfect, also have many DTP features, though at nowhere near the level of the specialized DTP packages.

Desktop publishing has the following characteristics:

- **Mix of text with graphics:** Desktop-publishing software allows you to precisely manage and merge text with graphics. As you lay out a page on-screen, you can make the text "flow," liquid-like, around graphics such as photographs. You can resize art, silhouette it, change the colors, change the texture, flip it upside down, and make it look like a photo negative.

- **Varied type and layout styles:** As do word processing programs, DTP programs provide a variety of fonts, or typestyles, from readable Times Roman to staid Tribune to wild Jester and Scribble. Additional fonts can be purchased on disk or downloaded online. You can also create all kinds of rules, borders, columns, and page numbering styles.

- **Use of files from other programs:** It's usually not efficient to do word processing, drawing, and painting with the DTP software. As a rule, text is composed on a word processor, artwork is created with drawing and painting software, and photographs are input using a scanner and then modified and stored using photo-manipulation software. Prefabricated art to illustrate DTP documents may be obtained from disks containing clip art, or "canned" images. The DTP program is used to integrate all these files. You can look at your work on the display screen as one page or as two facing pages (in reduced size). Then you can see it again after it has been printed out. *(See ● Panel 3.16.)*

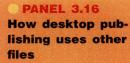

● **PANEL 3.16**
How desktop publishing uses other files

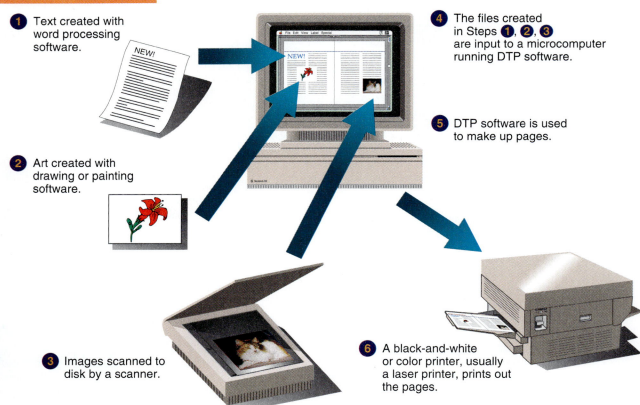

1. Text created with word processing software.

2. Art created with drawing or painting software.

3. Images scanned to disk by a scanner.

4. The files created in Steps **1**, **2**, **3** are input to a microcomputer running DTP software.

5. DTP software is used to make up pages.

6. A black-and-white or color printer, usually a laser printer, prints out the pages.

Drawing & Painting Programs

John Ennis was trained in realistic oil painting, and for years he used his skill creating illustrations for book covers and dust jackets. Now he "paints" using a computer, software, and mouse. The greatest advantage, he says, is that if "I do a brush stroke in oil and it's not right, I have to take a rag and wipe it off. With the computer, I just hit the 'undo' command."[6]

It may be no surprise to learn that commercial artists and fine artists have begun to abandon the paintbox and pen and ink for software versions of palettes, brushes, and pens. The surprise, however, is that an artist can use mouse and pen-like stylus to create computer-generated art as good as that achievable with conventional artist's tools. More surprising, even *nonartists* can produce good-looking work with these programs.

There are two types of computer art programs: drawing and painting.

- **Drawing programs: A _drawing program_ is graphics software that allows users to design and illustrate objects and products.** Some drawing programs are CorelDRAW, Adobe Illustrator, Macromedia Freehand, and Sketcher.

    Drawing programs create *vector images*—images created from mathematical calculations.

Vector image

- **Painting programs: _Painting programs_ are graphics programs that allow users to simulate painting on screen.** A mouse or a tablet stylus is used to simulate a paintbrush. The program allows you to select "brush" sizes, as well as colors from a color palette. Examples of painting programs are MetaCreations' Painter, Adobe Photoshop, Corel PhotoPaint, and JASC's PaintShop Pro.

    Painting programs produce *raster images* made up of little dots.

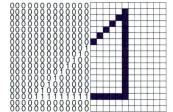

Raster image

CONCEPT CHECK

What can you do with desktop publishing software?

What are some characteristics of DTP software?

What are some features of drawing programs? Of painting programs?

Project Management Software

As we have seen, a personal information manager (PIM) can help you schedule your appointments and do some planning. That is, it can help you manage your own life. But what if you need to manage the lives of others in order to accomplish a full-blown project, such as steering a political campaign or handling a nationwide road tour for a band? Strictly defined, a *project* is a one-time operation involving several tasks and multiple resources that must be organized toward completing a specific goal within a given period of time. The project can be small, such as an advertising campaign for an in-house advertising department, or large, such as construction of an office tower or a jetliner.

Project management software **is a program used to plan and schedule the people, costs, and resources required to complete a project on time.** For instance, the associate producer on a feature film might use such software to keep track of the locations, cast and crew, materials, dollars, and schedules needed to complete the picture on time and within budget. The software would show the scheduled beginning and ending dates for a particular task—such as shooting all scenes on a certain set—and then the date that task was actually completed. Examples of project management software are Harvard Project Manager, Microsoft Project, Suretrack Project Manager, and ManagerPro.

Computer-Aided Design

Computers have long been used in engineering design. ***Computer-aided design (CAD) programs* are intended for the design of products, structures, civil engineering drawings, and maps.** CAD programs, which are available for microcomputers, help architects design buildings and workspaces and help engineers design cars, planes, electronic devices, roadways, bridges, and subdivisions. CAD and drawing programs are similar. However, CAD programs provide precise dimensioning and positioning of the elements being drawn, so that they can be transferred later to computer-aided manufacturing (CAM) programs. Also, CAD programs lack the special effects for illustrations that come with drawing programs. One advantage of CAD software is that the product can be drawn in three dimensions and then rotated on the screen so the designer can see all sides. *(See* ● *Panel 3.17.)* Examples of CAD programs for beginners are Autosketch and CorelCAD.

A variant of CAD is *CADD,* for *computer-aided design and drafting,* software that helps people do drafting. CADD programs include symbols (points, circles, straight lines, and arcs) that help the user put together graphic elements, such as the floor plan of a house. An example is Autodesk's Auto-CAD.

***CAD/CAM (computer-aided design/computer-aided manufacturing)* software allows products designed with CAD to be input into an automated manufacturing system that makes the products.** For example, CAD/CAM systems brought a whirlwind of enhanced creativity and efficiency to the fashion industry. Some CAD systems, says one writer, "allow designers to electronically drape digital-generated mannequins in flowing gowns or tailored suits that don't exist, or twist imaginary threads into yarns, yarns into weaves, weaves into sweaters without once touching needle to garment."[7] The designs and specifications are then input into CAM systems that enable robot pattern-cutters to automatically cut thousands of patterns from fabric with only minimal waste. Whereas previously the fashion industry worked about a year in advance of delivery, CAD/CAM has cut that time to 8 months—a competitive edge for a field that feeds on fads.

CONCEPT CHECK

Describe what project management software can do.

What is the purpose of CAD software? CAD/CAM software?

● **PANEL 3.17**
CAD
Example of computer-aided design

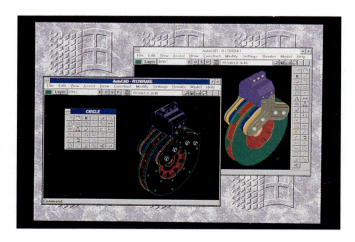

Experience Box
How to Buy Software

You can often buy music CDs at the 7-Eleven and videotapes at the gas minimart. Computer software isn't sold in convenience stores yet, but it's getting close. You *can* buy software at Target, Best Buy, Circuit City, and Office Depot. But should you? Here are some tips on buying software.

Getting Ready to Buy

Whatever type of software you're interested in, you need to be clear on a few things before you buy:

Do You Know Your Needs? Before talking to anyone about software, you should figure out what you want your computer to do for you.

Are you mainly writing research papers? Keeping track of performance of employees reporting to you? Projecting sales figures? Building a mailing list and launching a fund-raising campaign? Publishing a newsletter? Teaching children about computers? You want the machine to serve you, not vice versa.

Do You Know What Software You Want? The safest course is to pick software used successfully by people you know. Or look for ratings in the leading computer magazines. Brands that consistently get high ratings in magazine reviews are likely to be reliable.

Do You Know the Latest Version & Release? If you know the name of the software you're interested in, so much the better. If you have a particular brand and type in mind, make sure it's the most recent version and release.

A new *version* of a software package resembles a model change on a car. It adds new features, generally making the software more powerful and versatile. Versions are usually numbered in ascending order: 1.0, 2.0, 3.0, 3.1, 3.11, and so on. Thus, Office 97.2 is the second version.

A *release number* identifies a specific version of a program. Microsoft Works 4.5 is the fifth release of the fourth version, Pagemaker 6.5 is the fifth release of the sixth version. New releases generally incorporate routine enhancements and correct the annoying errors called software bugs.

Most software manufacturers are continually upgrading their products. Often the upgrades are made available online, over the Internet, in what are known as "patches"; if you find a problem with your software, you may call the developer and be told to download a patch.

Do You Know If an Upgrade Is Coming Out? Find out if a new version of the software is just around the corner. You may want to hold off buying until it's available.

Upgrades are to software as new models are to the auto industry. Every year or so, software manufacturers bring out a new version or release featuring incremental improvements, just as car makers do.

Will the Salespeople Speak Your Language?

Selling ice cream does not require a lot of product knowledge. Selling software does (or should). Some salespeople know their wares but try to intimidate newcomers with their knowledge. Others have only the barest familiarity with their products, although they may be patient with novices' questions. You hope, of course, you'll get someone who is both knowledgeable and helpful.

Some questions to ask are given in the box on the next page. (See ● *Panel 3.18.*)

Software Sellers: The Range of Outlets

The types of software sellers are as follows:

Computer Retail Stores A retail store selling both computer hardware and software may be a good place to go if you need to buy a PC as well as software. Prices won't beat those of discounters and superstores, but you may be able to find a well-informed, knowledgeable staff. Small retail stores, which often double as repair shops, may be in dealer chains, such as MicroAge Computer Centers, or may be home-grown independents, such as Lake Tahoe's Wired Solutions in Incline Village, Nevada. College bookstores are also a good source and may offer significant student discounts.

Computer Superstores Computer superstores range up to 30,000 square feet in size, versus, say, 2000 square feet for small retail stores. Not only do they carry all kinds of hardware, including computer furniture, but also upwards of 2000–3000 software titles. Like smaller stores, they offer computers for trying out software. Unlike some smaller stores, they also offer classes to train you in the use of particular software packages. Finally, they have extensive technical departments for installing software and readying and repairing hardware. Examples of chains with superstores are CompUSA and ComputerLand.

Electronics, Office, & Department Stores & Warehouse Clubs Walk into electronics stores like Radio Shack, Circuit City, and Best Buy and you'll find software there, too. Even large office-supply stores such as Staples sell software. Some department stores such as Sears, Montgomery Ward, and Dayton Hudson carry both hardware and software. It's doubtful, however, that the selection and prices are as competitive as those in specialist stores. Discount warehouse clubs such as CostCo, Office Depot, and Price Club provide computer hardware and software at steeply discounted prices. Some drawbacks are that these stores may not have repair services, customer support, or salespeople with deep product knowledge.

Online Sellers Flip through the ads in the back of a computer magazine and you may find all kinds of online sellers, such as SoftMan Products Co. which advertises "the original cheap software" (at *www.buycheapsoftware.com*), Egghead. com *(www.egghead.com)*, MicroWarehouse *(www.warehouse. com/pm)*, or TigerDirect *(www.tigerdirect.com)*. All these organizations also have toll-free numbers; you charge orders to your credit card and the product is delivered to you by UPS, FedEx, or Priority Mail. As long as you know what you want and pick a reputable company, buying software online is effective—and cost-effective.

Visual Summary

analytical graphics (p. 111, KQ 3.4) Also called *business graphics;* graphical forms that make numeric data easier to analyze than when it is organized as rows and columns of numbers. The principal examples of analytical graphics are bar charts, line graphs, and pie charts. Why it's important: *Whether viewed on a monitor or printed out, analytical graphics help make sales figures, economic trends, and the like easier to comprehend and analyze.*

Cell:
Formed by intersection of row and column (letter and number)

cell (p. 109, KQ 3.4) Place where a row and a column intersect in a spreadsheet worksheet; its position is called a *cell address.* Why it's important: *The cell is the smallest working unit in a spreadsheet. Data and formulas are entered into cells. Cell addresses provide location references for spreadsheet users.*

computer-aided design (CAD) programs (p. 122, KQ 3.6) Programs intended for the design of products, structures, civil engineering drawings, and maps. Why it's important: *CAD programs, which are available for microcomputers, help architects design buildings and workspaces and help engineers design cars, planes, electronic devices, roadways, bridges, and subdivisions. While similar to drawing programs, CAD*

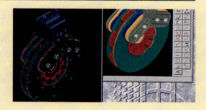

programs provide precise dimensioning and positioning of the elements being drawn, so that they can be transferred later to computer-aided manufacturing programs; in addition, they lack special effects for illustrations. One advantage of CAD software is that three-dimensional drawings can be rotated on screen, so the designer can see all sides of the product.

computer-aided design/computer-aided manufacturing (CAD/CAM) software (p. 122, KQ 3.6) Programs allowing products designed with CAD to be input into an automated manufacturing system that makes the products. Why it's important: *CAM systems have greatly enhanced efficiency in many industries.*

copyright (p. 91, KQ 3.1) Exclusive legal right that prohibits copying of intellectual property without the permission of the copyright holder. Why it's important: *Copyright law aims to prevent people from taking credit for and profiting from other people's work.*

To clean your printer, first open the top by pressing the button on the left side near the top. Swing the lid

Cursor

cursor (p. 103, KQ 3.3) Movable symbol on the display screen that shows where you may next enter data or commands. The symbol is often a blinking rectangle or an I-beam. You can move the cursor on the screen using the keyboard's directional arrow keys or a mouse. The point where the cursor is located is called the *insertion point.* Why it's important: *All application software packages use cursors to show the current work location on the screen.*

database (p. 111, KQ 3.5) Collection of interrelated files in a computer system. These computer-based files are organized according to their common elements, so that they can be retrieved easily. Why it's important: *Businesses and organizations build databases to help them keep track of and manage their affairs. In addition, online database services put enormous resources at the user's disposal.*

Contact ID	Company Name	Co
BREW	ABC Metals	Fran
COMP	Brewery Supplies, Inc.	Ran
DISC	Discount Glass	Moi
FARM	Farming Equipment, Inc.	Bru
FLOW	Flowers Anywhere	Kat
NATI	National Denistry	Car
PAPE	Paper Supplies, Etc.	Tes
PUBL	Publishing Etc.	Rich
ROSI	Rosies Cookie Company	Jea
SCHU	Schuler and Co.	Pre
SILV	Silver Spoon Toyota	Ste

database file (p. 92, KQ 3.1) File created by database management programs; it consists of organized data that can be analyzed and displayed in various useful ways. Why it's important: *Database files make up a database.*

database software (p. 111, KQ 3.5) Application software that sets up and controls the structure of a database and access to the data. Why it's important: *Database software allows users to organize and manage huge amounts of data.*

default settings (p. 107, KQ 3.3) Settings automatically used by a program unless the user specifies otherwise, thereby overriding them. Why it's important: *Users need to know how to change default settings in order to customize documents.*

desktop (p. 94, KQ 3.2) The operating system's main interface screen. Why it's important: *The desktop displays pictures (icons) that provide quick access to programs and information.*

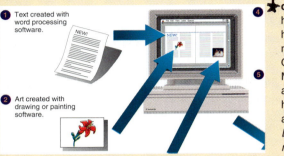

① Text created with word processing software.

② Art created with drawing or painting software.

★**desktop publishing (DTP)** (p. 119, KQ 3.6) Application software and hardware system that involves mixing text and graphics to produce high-quality output for commercial printing, using a microcomputer and mouse, scanner, laser or ink-jet printer, and DTP software (such as QuarkXPress and PageMaker or, at a more consumer-oriented level, Microsoft Publisher). Often the printer is used primarily to get an advance look before the completed job is sent to a typesetter for even higher-quality output. Some word processing programs, such as Word and WordPerfect, have rudimentary DTP features. Why it's important: *Desktop publishing has reduced the number of steps, the time, and the money required to produce professional-looking printed projects.*

document file (p. 92, KQ 3.1) File created by word processing programs; it consists of documents such as reports, letters, memos, and term papers. Why it's important: *Document files are probably the most common type of file users deal with.*

documentation (p. 92, KQ 3.1) User guide or reference manual that provides a narrative and graphical description of a program. While documentation may be print-based, today it is usually available on CD-ROM, as well as via the Internet. Why it's important: *Documentation helps users learn software commands and use of function keys, solve problems, and find information about system specifications.*

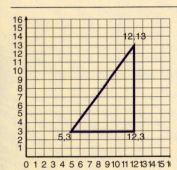

drawing program (p. 121, KQ 3.6) Graphics software that allows users to design and illustrate objects and products. Why it's important: *Drawing programs are vector-based and are best used for straightforward illustrations based on geometric shapes.*

exporting (p. 93, KQ 3.1) Transforming data into a format that can be used in another program and then transmitting it. Why it's important: *Users need to know how to export many types of files.*

file (p. 92, KQ 3.1) A named collection of data or a program that exists in a computer's secondary storage, such as on a floppy disk, hard disk, or CD-ROM disk. Why it's important: *Dealing with files is an inescapable part of working with computers. Users need to be familiar with the different types of files.*

financial software (p. 117, KQ 3.6) Applications software that ranges from personal-finance managers to entry-level accounting programs to business financial-management packages. Why it's important: *Financial software provides users with powerful management tools (personal-finance managers) as well as small business programs. Moreover, tax programs provide virtually all the forms needed for filing income taxes, make complex calculations, check for mistakes, and even unearth deductions you didn't know existed. Tax programs can also be integrated with personal finance software to form an integrated tool. Accounting software automates bookkeeping tasks, while payroll software keeps records of employee hours and produces reports for tax purposes. Some programs allow users to set up a business from scratch. Financial software also includes investment software packages and various retirement planning programs.*

fly-out menu (p. 97, KQ 3.2) Menu that seems to explode out to the right. Why it's important: *Menus make software easier to use.*

font (p. 106, KQ 3.3) A particular typeface and type size. Why it's important: *Fonts influence the appearance and effectiveness of documents, brochures, and other publications.*

formatting (p. 106, KQ 3.3) In word processing and desktop publishing, determining the appearance of a document. Why it's important: *The document format should match its users' needs. Ways to format a document include using different fonts, boldface, italics, variable spacing, columns, and margins.*

formulas (p. 109, KQ 3.4) In a spreadsheet, instructions for calculations entered into designated cells. Why it's important: *When spreadsheet users change data in one cell, all the cells linked to it by formulas automatically recalculate their values.*

freeware (p. 91, KQ 3.1) Copyrighted software that is distributed free of charge, today most often over the Internet. Why it's important: *Freeware saves users money.*

function keys (p. 102, KQ 3.3) Keys labeled F1, F2, and so on, positioned along the top or left side of the keyboard. Why it's important: *They are used to execute commands specific to the software being used.*

grammar checker (p. 106, KQ 3.3) Word processing feature that highlights poor grammar, wordiness, incomplete sentences, and awkward phrases. The grammar checker won't fix things automatically, but it will flag (perhaps with a different color of squiggly line) possible incorrect word usage and sentence structure. Why it's important: *Grammar checkers help users produce better-written documents.*

graphical user interface (GUI) (p. 94, KQ 3.2) User interface in which icons and commands from menus may be selected by means of a mouse or keystrokes. Why it's important: *GUIs are easier to use than command-driven interfaces.*

groupware (p. 93, KQ 3.1) Online software that allows several people to collaborate on the same project. Why it's important: *Groupware improves productivity by keeping users continually notified about what their colleagues are thinking and doing.*

Help command (p. 101, KQ 3.2) Command generating a table of contents, an index, and a search feature that can help users locate answers to questions about the software. Why it's important: *Help features provide a built-in electronic instruction manual.*

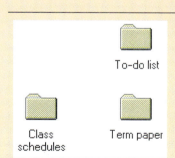

icons (p. 94, KQ 3.2) Small pictorial figures that represent programs, data files, or procedures. Why it's important: *Icons have simplified the use of software. The feature represented by the icon can be activated by clicking on the icon.*

importing (p. 93, KQ 3.1) Getting data from another source and then converting it into a format compatible with the program in which you are currently working. Why it's important: *Users will often have to import files.*

key field (p. 112, KQ 3.5) Field that can be used as a common identifier because it is unique. The most frequent key field used in the United States is the Social Security number, but any unique identifier could be used, such as an employee number. Why it's important: *Key fields are needed to identify and retrieve specific items in a database.*

macro (p. 103, KQ 3.3) Also called a keyboard shortcut; a single keystroke or command—or a series of keystrokes or commands—used to automatically issue a longer, predetermined series of keystrokes or commands. Why it's important: *Users can consolidate several activities into only one or two keystrokes. The user names the macro and stores the corresponding command sequence; once this is done, the macro can be used repeatedly.*

menu (p. 97, KQ 3.2) Displayed list of options—such as commands— to choose from. Why it's important: *Menus are a feature of GUIs that make software easier to use.*

office suite (p. 93, KQ 3.1) A single large software package that bundles several applications together. Why it's important: *Office suites cost less than do the applications purchased separately.*

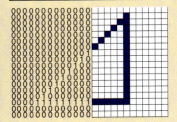

painting program (p. 121, KQ 3.6) Graphics program that allows users to simulate painting on screen. A mouse or a tablet stylus is used to simulate a paintbrush. The program allows you to select "brush" sizes, as well as colors from a color palette. Why it's important: *Painting programs, which produce raster images made up of little dots, are good for creating art with soft edges and many colors.*

personal-finance manager (p. 117, KQ 3.6) Application software that lets you keep track of income and expenses, write checks, do online banking, and plan financial goals. Why it's important: *Personal-finance software can help people manage their money more effectively.*

personal information manager (PIM) (p. 114, KQ 3.5) Software to help you keep track of and manage information you use on a daily basis, such as addresses, telephone numbers, appointments, to-do lists, and miscellaneous notes. Some programs feature phone dialers, outliners (for roughing out ideas in outline form), and ticklers (or reminders). Why it's important: *PIMs can help users better organize and manage daily business activities.*

pointer (p. 94, KQ 3.2) Indicator that usually appears as an arrow, although it changes shape depending on the application. The mouse is used to move the pointer to a particular place on the display screen or to point to little symbols, or icons. Why it's important: *It is often easier to manipulate the pointer on the screen by means of the mouse than to type commands on a keyboard.*

pop-up menu (p. 97, KQ 3.2) List of command options that can "pop up" anywhere on the screen. In contrast to pull-down or pull-up menus, pop-up menus are not connected to a toolbar. Why it's important: *Pop-up menus make programs easier to use.*

presentation graphics software (p. 115, KQ 3.6) Software that uses graphics, animation, sound, and data or information to make visual presentations. Why it's important: *Presentation graphics software provides a means to produce sophisticated graphics.*

productivity software (p. 93, KQ 3.1) Application software such as word processing, spreadsheets, and database managers. Why it's important: *Productivity software makes users more productive at particular tasks.*

project management software (p. 121, KQ 3.6) Program used to plan and schedule the people, costs, and resources required to complete a project on time. Why it's important: *Project management software increases the ease and speed of planning and managing complex projects.*

public-domain software (p. 91, KQ 3.1) Software, often available on the Internet, that is not protected by copyright and thus may be duplicated by anyone at will. Why it's important: *Public domain software offers lots of software options to users who may not be able to afford much commercial software. Users may download such software from the Internet for free and make as many copies as they wish.*

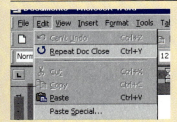

pull-down menu (p. 97, KQ 3.2) Also called a *drop-down menu;* list of options that pulls down from the menu bar at the top of the screen. Why it's important: *Like other menu-based and GUI features, pull-down menus make software easier to use.*

pull-up menu (p. 97, KQ 3.2) List of options that pulls up from the menu bar at the bottom of the screen. Why it's important: *See* pull-down menu.

recalculation (p. 109, KQ 3.4) Recomputing values in a spreadsheet, either as an ongoing process as data is entered or afterward, with the press of a key. Why it's important: *With this simple feature, the hours of mind-numbing work required to manually rework paper spreadsheets became a thing of the past.*

relational database (p. 112, KQ 3.5) Database in which data is organized into related tables. Each table contains rows and columns; the rows are called records, and the columns are called fields. An example of a record is a person's address— name, street address, city, and so on. An example of a field is that person's last name; another field would be that person's first name, a third field would be that person's street address, and so on. Why it's important: *The relational database is a common type of database.*

rentalware (p. 92, KQ 3.1) Software that users lease for a fee. Why it's important: This is the concept behind application services providers (ASPs).

rollover (p. 97, KQ 3.2) Icon feature in which a small text box explaining the icon's function appears when you roll the mouse pointer over the icon. A rollover may also produce an animated graphic. Why it's important: *The rollover gives the user an immediate explanation of an icon's meaning.*

saving (p. 108, KQ 3.3) Storing, or preserving, a document as an electronic file permanently—on diskette, hard disk, or CD-ROM, for example. Why it's important: *Saving is a feature of nearly all application software. Having the document stored in electronic form spares users the tiresome chore of retyping it from scratch whenever they want to make changes. Users need only retrieve it from the storage medium and make the changes, then resave it and print it out again.*

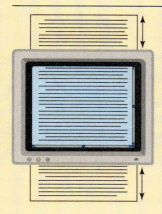

scrolling (p. 103, KQ 3.3) Moving quickly upward, downward, or sideways through the text or other screen display. Why it's important: *A standard computer screen displays only 20–22 lines of standard-size text; however, most documents are longer than that. Using the directional arrow keys, or the mouse and a scroll bar located at the side of the screen, users can move ("scroll") through the display screen and into the text above and below it.*

shareware (p. 91, KQ 3.1) Copyrighted software that is distributed free of charge but requires users to make a monetary contribution in order to continue using it. Shareware is distributed primarily through the Internet. Because it is copyrighted, you cannot use it to develop your own program that would compete with the original product. Why it's important: *Like public domain software and freeware, shareware offers an inexpensive way to obtain new software.*

software license (p. 91, KQ 3.1) Contract by which users agree not to make copies of software to give away or for resale. Why it's important: *Software manufacturers don't sell people software so much as licenses to become authorized users of the software.*

special-purpose keys (p. 102, KQ 3.3) Keys used to enter, delete, and edit data and to execute commands. For example, the Esc (for "Escape") key tells the computer to cancel an operation or leave ("escape from") the current mode of operation. The Enter, or Return, key tells the computer to execute certain commands and to start new paragraphs in a document. Why it's important: *Special-purpose keys are essential to the use of software.*

spelling checker (p. 104, KQ 3.3) Word processing feature that tests for incorrectly spelled words. As you type, the spelling checker indicates (perhaps with a squiggly line) words that aren't in its dictionary and thus may be misspelled. Special add-on dictionaries are available for medical, engineering, and legal terms. Why it's important: *Spelling checkers help users prepare accurate documents.*

spreadsheet (p. 108, KQ 3.4) Application software that allows users to create tables and financial schedules by entering data and formulas into rows and columns arranged as a grid on a display screen. Why it's important: *When data is changed in one cell, values in other cells in the spreadsheet are automatically recalculated.*

taskbar (p. 99, KQ 3.2) Graphic toolbar that appears at the bottom of the Windows screen. Why it's important: *The taskbar presents the applications that are running.*

template (p. 106, KQ 3.3) In word processing, a preformatted document that provides basic tools for shaping a final document—the text, layout, and style for a letter, for example. Why it's important: *Templates make it very easy for users to prepare professional-looking documents, because most of the preparatory formatting is done.*

thesaurus (p. 106, KQ 3.3) Word processing feature that will present you with the appropriate word or alternative words. Why it's important: *The thesaurus feature helps users prepare well-written documents.*

toolbar (p. 99, KQ 3.2) Bar across the top of the display window. It displays menus and icons representing frequently used options or commands. Why it's important: *Toolbars make it easier to identify and execute commands.*

tutorial (p. 92, KQ 3.1) Instruction book or program that helps you learn to use the product by taking you through a prescribed series of steps. Why it's important: *Tutorials enable users to practice using new software in a graduated fashion and learn the software in an effective manner.*

user interface (p. 94, KQ 3.2) Display screen that allows users to communicate, or interact, with the computer. The three types of user interface are command-driven, menu-driven, and graphical (GUI), which is now most common. Why it's important: *Without user interfaces, no one could operate a computer system.*

value (p. 109, KQ 3.4) A number or date entered in a spreadsheet cell. Why it's important: *Values are the actual numbers used in the spreadsheet—dollars, percentages, grade points, temperatures, or whatever.*

what-if analysis (p. 109, KQ 3.4) Spreadsheet feature that employs the recalculation feature to investigate how changing one or more numbers changes the outcome of the calculation. Why it's important: *Users can create a worksheet, putting in formulas and numbers, and then ask, "What would happen if we change that detail?"—and immediately see the effect.*

window (p. 100, KQ 3.2) Rectangular frame on the computer display screen. Through this frame you can view a file of data—such as a document, spreadsheet, or database—or an application program. Why it's important: *Using windows, users can display at the same time portions of several documents and/or programs on the screen.*

What would you like to do?
Check the spelling and grammar of text in another language
Customize spelling and grammar checking
Create and use custom dictionaries
Troubleshoot spelling and grammar checking

wizard (p. 106, KQ 3.3) Word processing software feature that answers your questions and uses the answers to lay out and format a document. Why it's important: *Wizards make it easy to prepare professional-looking memos, faxes, resumes, and other documents.*

word processing software (p. 101, KQ 3.3) Application software that allows you to use computers to format, create, edit, print, and store text material, among other things. Why it's important: *Word processing software allows users to maneuver through a document and delete, insert, and replace text, the principal correction activities. It also offers such additional features as creating, editing, formatting, printing, and saving.*

worksheet file (p. 92, KQ 3.1) File created by electronic spreadsheets; it consists of a collection of (usually) numerical data such as budgets, sales forecasts, and schedules. Why it's important: *Worksheet files are one common type of file users will have to deal with.*

Chapter Review

Self-Test Questions

1. _____ is the term for programs designed to perform specific tasks for the user.

2. _____ software allows you to create and edit documents.

3. _____ is the activity of moving upward or downward through the text or other screen display.

4. Name four editing features offered by word processing programs: _____, _____, _____, _____.

5. In a spreadsheet, the place where a row and a column intersect is called a _____.

6. A(n) _____ is a keyboard shortcut used to automatically issue a longer, predetermined series of keystrokes or commands.

7. The _____ is the movable symbol on the display screen that shows you where you may next enter data or commands.

8. When you buy software, you pay for a _____, a contract by which you agree not to make copies of the software to give away or for resale.

9. Records in a database are linked by a _____, which can be used as a common identifier because it is unique.

10. _____ involves mixing text and graphics to produce high-quality output for commercial printing.

Multiple-Choice Questions

1. Which of the following is not an advantage of using database software?
   a. integrated data
   b. improved data integrity
   c. lack of structure
   d. elimination of data redundancy

2. Which of the following is not a type of menu?
   a. fly-out menu
   b. pop-in menu
   c. pop-out menu
   d. pull-down menu
   e. pull-out menu

3. Which of the following is not a feature of word processing software?
   a. spelling checker
   b. cell address
   c. formatting
   d. cut and paste
   e. find and replace

True/False Questions

T F 1. Spreadsheet software enables you to perform what-if calculations.

T F 2. *Font* refers to a preformatted document that provides basic tools for shaping the final document.

T F 3. Rentalware is software that users lease for a fee.

T F 4. Public-domain software is protected by copyright and so is offered for sale by license only.

T F 5. The records within the various tables in a database are linked by a key field.

Short-Answer Questions

1. What is the difference between a command-driven interface and a graphical user interface (GUI)?

2. What are the following types of application software used for?
   a. project management software
   b. desktop-publishing software
   c. database software
   d. spreadsheet software
   e. word processing software

3. Which program is more sophisticated, analytical graphics or presentation graphics? Why?

4. How are the following different? Pop-up menu; pull-down menu; fly-out menu.

5. What is *importing*? *Exporting*?

Concept Mapping

On a separate sheet of paper, draw a concept map, or visual diagram, linking concepts. Show how the following terms are related.

CAD/CAM	recalculation
cell	relational
database	spelling checker
documentation	spreadsheet
file	template
formula	thesaurus
grammar checker	toolbar
GUI	tutorial
icon	value
key field	what-if analysis
menu	word processing
productivity software	worksheet

Knowledge in Action

1. If you were in the market for a new microcomputer today, what application software would you want to use on it? What system software? Why?

2. Several Web sites include libraries of shareware programs. Visit the *www.winfiles.com* site, click on the Windows shareware icon, and identify three shareware programs that interest you. State the name of each program, the operating system it runs on, and its capabilities. Also, describe the contribution you must make to receive technical support.

3. What is your opinion of using MP3 to download free music from the Web to play on your own PC and/or CDs? Much attention has been given lately to music downloading and copyright infringement. Research this topic in library magazines and newspapers or on the Internet, and take a position in a short report.

4. How do you think you could use desktop publishing at home? For personal items? Family occasions? Holidays? What else? What hardware and software would you have to buy?

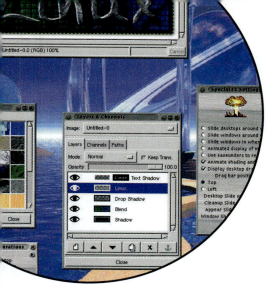

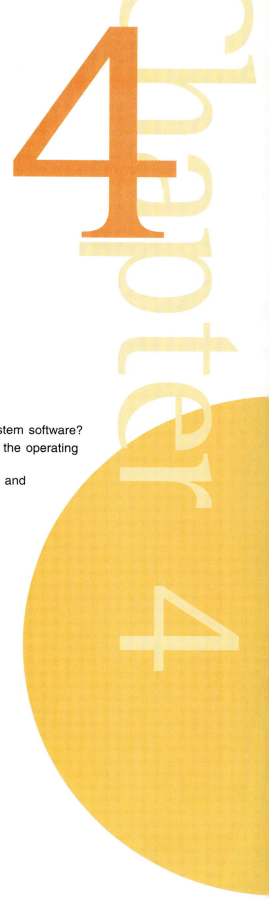

System Software

The Power behind the Power

Key Questions

You should be able to answer the following questions.

4.1   **The Components of System Software** What are three components of system software?

4.2   **The Operating System: What It Does** What are the principal functions of the operating system?

4.3   **Common Operating Systems** What are some common desktop, network, and portable OSs?

4.4   **Other System Software: Device Drivers & Utility Programs** What are the characteristics of device drivers and utility programs?

4.5   **The OS of the Future: "The Network Is the Computer"** What are some future directions operating systems might take?

4.6   **Online Software & Application Software Providers: Turning Point for the Software Industry?** What are some recent trends in online software?

A Visual Overview of This Chapter

1 **The Components of System Software.** Three basic components of system software are operating systems, device drivers, and utility programs. An operating system is the principal component of system software. Device drivers allow input/output devices to communicate with the rest of the computer system. Utility programs provide functions (such as data recovery) not supplied by other system software.

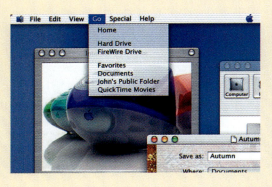

2 **The Operating System: What It Does.** The **operating system (OS)** consists of the master system of programs that manage the basic operations of the computer. Features of the OS are booting, CPU management, file management, task management, and formatting. (1) In **booting,** the OS is loaded into the computer's main memory. (2) The **supervisor** or **kernel** manages the CPU. The OS also manages memory, by portioning, dividing memory into foreground/background areas, and arranging programs in queues to be processed. (3) In file management, the OS records the storage location of files. (4) Task management includes **multitasking,** executing more than one program concurrently; **multiprogramming,** concurrent execution of different users' programs; **time sharing,** round-robin processing of programs of several users; and **multiprocessing,** simultaneous processing of two or more programs by multiple computers. (5) **Formatting,** or **initializing,** consists of preparing a disk to store data or programs.

3 **Common Operating Systems.** There are three categories of platforms, or particular combinations of processors and OSs—for desktops/laptops, for networks, and for handhelds.

Principal desktop/laptop OSs are DOS, Macintosh OS, and the Microsoft Windows series. **DOS** was Microsoft's original OS. **Macintosh operating system** runs only on Apple Macintoshes. **Microsoft Windows** (versions 3.1, **95, 98,** and most recently **Me**) is the most popular OS for desktops and portables. The forthcoming **Microsoft Whistler** targets home users but is based on Windows NT. **Emulation software** allows software designed for Windows to run on Macs.

Principal network server OSs are **NetWare** from Novell; **Windows NT** and its successor **Windows 2000** from Microsoft; **Unix,** available in several versions, including Sun's **Solaris** and **BSD;** and **Linux,** a free version of Unix and a kind of **open-source software** modifiable by anyone.

Principal OSs for handhelds are **Palm OS,** which runs the Palm and the Visor, and **Windows CE** (a slimmed-down version of Windows 95), which became **Pocket PC,** a simpler version.

4 **Other System Software: Device Drivers & Utility Programs. Device drivers** are specialized software programs that allow input and output devices to communicate with the rest of the computer system. **Utilty programs** perform tasks related to the control and allocation of computer resources. They enhance existing functions or provide services not supplied by other system software programs. Tasks performed by utilities include the following. (1) A **backup utility** is used to make a duplicate copy of the information on your hard disk. (2) A **data-recovery utility** is used to restore data that has been physically damaged or corrupted. (3) **Antivirus software** is a utility program that scans hard disks, floppy disks, and memory to detect viruses. (4) **Data compression utilities** remove redundant elements, gaps, and unnecessary data from a computer's storage space so that less space (fewer bits) is required to store or transmit data. (5) **Fragmentation** is the scattering of portions of files about the disk in nonadjacent areas, thus greatly slowing access to the files. A **defragmenter** utility program will find all the scattered files on your hard disk and reorganize them as contiguous files.

5 **The OS of the Future: "The Network Is the Computer."** The concept has been put forth of an "operating system" that extends over all kinds of networks.

Three expressions of a possible Internet-wide operating system are as follows: (1) **Microsoft.Net** is Microsoft's platform for an operating system for the entire Internet, designed to link unrelated Web sites so that people can organize all the information in their lives, using PCs and smaller devices, such as cellphones, handheld computers, and set-top boxes. Underlying .Net is a commitment to XML (for extensible markup language), an "open standards" protocol that makes it easy for machines to read Web sites by enabling Web developers to add more "tags" to a Web page. Windows Whistler is supposed to feature .Net-based services. (2) **E-speak** is Hewlett-Packard's version of an Internet operating system, or "universal language," that allows different Web sites to communicate with one another. E-speak also uses the programming standard XML. (3) Sun's **Jini** is a small layer of software designed to let all types of electronic gadgets on a wired or wireless network communicate with each other. Jini builds on another Sun technology called Java.

The opposite possibility is that no one company's operating system will dominate. Rather, there might be "massively distributed computing." A **distributed system** is a noncentralized network of several computers and other devices that can communicate with one another. Instead of an increasingly Webcentric model, computing might become decentralized, with information distributed among millions of computers.

6 **Online Software & Application Software Providers. Application service providers (ASPs)** are firms that lease software over the Internet. ASPs fit the strategy of users of **network computers**—thin clients, or inexpensive, stripped-down computers that connect to networks and run applications tied to servers. ASPs were anticipated by **enterprise resource planning (ERP) software,** which consists of large client/server software applications that help companies organize and operate their businesses. With ASPs, however, clients can rent instead of buy software to run off of servers.

What we need is a science called *practology*, a way of thinking about machines that focuses on how things will actually be used."

So says Alan Robbins, a professor of visual communications, on the subject of *machine interfaces*—the parts of a machine that people actually manipulate.[1] An interface is a machine's "control panel," ranging from the volume and tuner knobs on an old radio to all the switches and dials on the flight-deck of a jetliner. You may have found, as Robbins thinks, that on too many of today's machines—digital watches, VCRs, even stoves—the interface is often designed to accommodate the machine or some engineering ideas rather than the people actually using them. Good interfaces are intuitive—that is, based on prior knowledge and experience—like the twin knobs on a 1950s radio, immediately usable by both novices and sophisticates. Bad interfaces, such as a software program with a bewildering array of menus and icons, force us to relearn the required behaviors every time. Of course, you can prevail over a bad interface if you repeat the procedures often enough.

How well are computer hardware and software makers doing at giving us useful, helpful interfaces? The answer is: getting better all the time, but they still have some leftovers from the past to get rid of. For instance, PC keyboards still come with a SysRq (for "System Request") key, which was once used to get the attention of the central computer but now is rarely used. (The Scroll Lock key is also seldom used.)

In time, as interfaces are refined, it's possible computers will become no more difficult to use than a car. Until then, however, for smoother computing you need to know something about how system software works. Today people communicate one way, computers another. People speak words and phrases; computers process bits and bytes. For us to communicate with these machines, we need an intermediary, an interpreter. This is the function of system software. We interact mainly with the applications software, which interacts with the system software, which controls the hardware.

4.1 The Components of System Software

KEY QUESTION

What are three components of system software?

As we've said, software is of two types. *Application software* is software that can perform useful work on general-purpose tasks, such as word processing or spreadsheets, or that is used for entertainment. Hundreds of application software packages are available for personal computers. *System software*, which you will find already installed if you buy a new computer, enables the application software to interact with the computer and helps the computer manage its internal and external resources. There are only a handful of systems software packages for personal computers.

There are three basic components of system software that you need to know about. *(See ● Panel 4.1.)*

- **Operating systems:** An *operating system* is the principal component of system software in any computing system.
- **Device drivers:** *Device drivers* help the computer control a peripheral device.
- **Utility programs:** *Utility programs* are generally used to support, enhance, or expand existing programs in a computer system.

A fourth type of system software, *language translators,* is described in the Appendix.

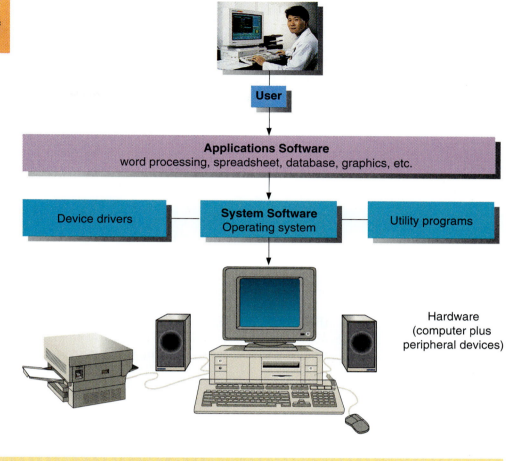

4.2 The Operating System: What It Does

KEY QUESTION
*What are the principal
functions of the operating
system?*

The <u>***operating system (OS)***</u> **consists of the master system of programs that
manage the basic operations of the computer.** These programs provide
resource management services of many kinds. In particular, they handle the
control and use of hardware resources, including disk space, memory, CPU
time allocation, and peripheral devices. The operating system allows you to
concentrate on your own tasks or applications rather than on the complexities of managing the computer.

Different sizes and makes of computers have their own operating systems.
For example, Cray supercomputers use UNICOS and COS, IBM mainframes
use MVS and VM, Data General minicomputers (midsize computers) use
AOS and DG, and DEC minicomputers use VAX/VMS. Pen-based computers
have their own operating systems—PenRight, PenPoint, Windows for Pen
Computing—that enable users to write scribbles and notes on the screen. In
general, an operating system written for one kind of hardware will not be
able to run on another kind of machine. In other words, *different operating
systems are mutually incompatible.*

Microcomputer users may readily experience the aggravation of such
incompatibility when they acquire a new or used microcomputer. Do they
get an Apple Macintosh with Macintosh Systems Software, which won't run
IBM-compatible programs? Or do they get an IBM or IBM-compatible (such
as Compaq or Dell), which won't run Macintosh programs?

Before we try to sort out these perplexities, we should have an idea of
what operating systems do. We will consider:

- Booting
- CPU management
- File management

- Task management
- Formatting

Booting

The work of the operating system begins as soon as you turn on, or "boot," the computer. ***Booting* is the process of loading an operating system into a computer's main memory.** This loading is accomplished by programs stored permanently in the computer's electronic circuitry. When you turn on the machine, programs called *diagnostic routines* test the main memory, the central processing unit, and other parts of the system to make sure they are running properly. Next, BIOS (for basic input/output system) programs are copied to main memory and help the computer interpret keyboard characters or transmit characters to the display screen or to a diskette. Then the boot program obtains the operating system, usually from hard disk, and loads it into the computer's main memory, where it remains until you turn the computer off. *(See ● Panel 4.2.)*

CPU Management

Suppose you are writing a report using a word processing program and want to print out a portion of it while continuing to write. How does the computer manage both tasks?

Like a police officer directing traffic, the ***supervisor*, or kernel, manages the CPU. It remains in main memory while the computer is running and directs other "nonresident" programs (programs that are not in main memory) to perform tasks that support application programs.**

Thus, if you enter a command to print your document, the operating system will select a printer (if there is more than one). It will then notify the computer to begin executing instructions from the appropriate program (known as a printer driver, because it controls, or "drives," the printer). Meanwhile, many operating systems allow you to continue writing. Were it not for this supervisor program, you would have to stop writing and wait for your document to print out before you could resume.

The operating system also manages memory—it keeps track of the locations within main memory where the programs and data are stored. It can swap portions of data and programs between main memory and secondary storage, such as your computer's hard disk. This capability allows a computer to hold only the most immediately needed data and programs within main memory. Yet it has ready access to programs and data on the hard disk, thereby greatly expanding memory capacity.

There are several ways operating systems can manage memory:

- **Partitioning:** In *partitioning,* the OS divides memory into separate areas called partitions, each of which can hold a program or data.

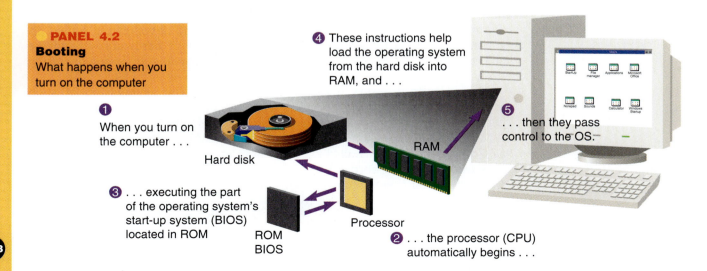

● **PANEL 4.2**
Booting
What happens when you turn on the computer

❶ When you turn on the computer . . .

Hard disk

❸ . . . executing the part of the operating system's start-up system (BIOS) located in ROM

ROM BIOS

❷ . . . the processor (CPU) automatically begins . . .

Processor

❹ These instructions help load the operating system from the hard disk into RAM, and . . .

RAM

❺ . . . then they pass control to the OS.

- **Foreground/background:** Some computer systems divide memory into *foreground* and *background* areas. Foreground programs have higher priority, and background programs have lower priority. When you're working at your microcomputer, the foreground program is the one you are currently working with, such as word processing. The background program might be regulating the flow of print images to your printer.
- **Queues:** Programs that are to be executed wait on disk in *queues* (pronounced "Qs"). A queue is a temporary holding place for programs or data.

File Management

A *file* is a named collection of related information. A file can be a program, such as a word processing program. Or it can be a data file, such as a word processing document, a spreadsheet, images, songs, and the like. We discuss files in more detail later in the chapter.

Files containing programs and data are located in many places on your hard disk and other secondary-storage devices. The operating system records the storage location of all files. If you move, rename, or delete a file, the operating system manages such changes and helps you locate and gain access to it. For example, you can *copy,* or duplicate, files and programs from one disk to another. You can *back up,* or make a duplicate copy of, the contents of a disk. You can *erase,* or remove, from a disk any files or programs that are no longer useful. You can *rename,* or give new file names to, the files on a disk.

Task Management

A computer is required to perform many different tasks at once. In word processing, for example, it accepts input data, stores the data on a disk, and prints out a document—seemingly simultaneously. Some computers' operating systems can also handle more than one program at the same time—word processing, spreadsheet, database searcher. Each program is displayed in a separate window on the screen. Others can accommodate the needs of several different users at the same time. All these examples illustrate *task management.* A "task" is an operation such as storing, printing, or calculating.

Among the ways operating systems manage tasks in order to run more efficiently are *multitasking, multiprogramming, time-sharing,* and *multiprocessing.* (Not all operating systems can do all these things.)

The high-tech army
Sergeant Scott Decker tests high-tech equipment for the U.S. Army. The lightweight computer in Decker's backpack feeds satellite data to the eyepiece display over his right eye, showing his precise geographical location. His rifle includes a digital compass and a digital video camera that can send images back to base. The helmet-mounted microphone and earphone connect to a radio system to help fellow soldiers identify each other.

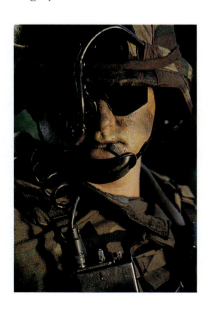

- **Multitasking—executing more than one program concurrently:** Earlier microcomputers could do only *single-tasking,* whereby an OS could run only one application program at a time. Thus, users would have to shut down the application program they were working in before they opened another application, which was inconvenient. Today, multitasking operating systems are used.

  Multitasking **is the execution of two or more programs by one user concurrently on the same computer with one central processor.** You may be writing a report on your computer with one program while another program plays a music CD. How does the computer handle both programs at once?

  The answer is that the operating system directs the processor to spend a predetermined amount of time executing the instructions for each program, one at a time. Thus, a small part of the first program is processed, and then the processor moves to the remaining programs, one at a time, processing small parts of each. The cycle is repeated until processing is complete. Because the processor is usually very fast, it may appear that all the programs are being executed at the same time. However, the processor is still executing only one instruction at a time.

- **Multiprogramming—concurrent execution of different users' programs:** *Multiprogramming* **is the execution of two or more programs concurrently on a multiuser operating system.** As with multitasking, the processor spends a certain amount of time executing each user's program. Once again, because the processor works so quickly, it seems as though all the programs are being run at the same time.

- **Time-sharing—round-robin processing of programs for several users:** In *time-sharing,* **a single computer processes the tasks of several users at different stations in round-robin fashion.** Time-sharing is used when several users are linked by a communications network to a single computer. The computer will first work on one user's task for a fraction of a second, then go on to the next user's task, and so on.

  This is accomplished through *time slicing.* Because computers operate so quickly, they can alternately apportion slices of time (fractions of a second) to various tasks. Thus, the computer may rapidly switch back and forth among different tasks, just as a hairdresser or dentist works with several clients or patients concurrently. Users are generally unaware of the switching process.

  Multitasking and time-sharing differ slightly. With multitasking, the processor directs the programs to take turns accomplishing small tasks or events, such as making a calculation, searching for a record, or printing out part of a document. Each event may take a different amount of time to complete. With time-sharing, the computer spends a fixed amount of time with each program before going on to the next one.

- **Multiprocessing—simultaneous processing of two or more programs by multiple computers:** *Multiprocessing* **is processing done by two or more computers or processors linked together to perform work simultaneously—that is, at precisely the same time.** This can entail processing instructions from different programs or different instructions within the same program at once.

  As in multitasking, which involves only a single processor, the processing should be so fast that, by spending a little bit of time working on each program in turn, several programs can be run at the same time. With both multitasking and multiprocessing, the operating system keeps track of the status of each program so that it knows where it left off and where to continue processing. But an operating system capable of multiprocessing is much more sophisticated than that required for multitasking.

Two possible approaches to multiprocessing are coprocessing and parallel processing. In *coprocessing,* the controlling CPU works together with specialized microprocessors called *coprocessors,* each of which handles a particular task, such as creating display-screen graphics or performing high-speed mathematical calculations. Many microcomputer systems have coprocessing capabilities.

In *parallel processing,* several full-fledged processors work together on the same tasks, sharing memory. Parallel processing is often used in large computer systems designed to keep running if one of the CPUs fails. These are called *fault-tolerant* systems; they have many processors and redundant components, such as memory and input, output, and storage devices. Fault-tolerant systems are used, for example, in airline reservation systems.

Comparison of Task Management

	Definition: Processing of Two or More Programs . . .	Number of Users	Number of Processors	Order of Processing
Multitasking	by one user concurrently on one processor	One	One	Concurrently
Multiprogramming	by multiple users concurrently on one processor	Multiple	One	Concurrently
Time sharing	by multiple users in round-robin fashion on one processor	Multiple	One	Round robin
Multiprocessing	by one or more users simultaneously on two or more processors	One or more	Two or more	Simultaneously

Operating system functions are summarized below. *(See ● Panel 4.3.)*

● PANEL 4.3
Basic operating system functions

Booting	Managing Storage Media	Managing Computer Resources	Managing Files	Managing Tasks
Uses diagnostic routines to test system for equipment failure	Formats diskettes	Via the supervisor, manages the CPU and directs other programs to perform tasks to support applications programs	Copies files/programs from one disk to another	May be able to perform multitasking, multiprogramming, time-sharing, or multiprocessing
Stores BIOS programs in main memory	Displays information about operating system version	Keeps track of locations in main memory where programs and data are stored (memory management)	Backs up files/programs	
Loads operating system into computer's main memory	Displays disk space available	Moves data and programs back and forth between main memory and secondary storage (swapping)	Erases (deletes) files/programs	
			Renames files	

Formatting

Formatting, or _initializing_, a disk is the process of preparing that disk so that it can store data or programs. Today it is easier to buy preformatted diskettes, which bear the label "Formatted IBM" (for floppies designed to run on PCs) or "Formatted Macintosh." However, it's useful to know how to format a blank floppy disk or reformat a diskette that wasn't intended for your machine.

How to Format a Floppy Disk on the PC

1. Insert unformatted (blank) disk in the floppy disk drive.

2. In My Computer, click once on the icon for the disk you want to format (the floppy disk).

3. On the File menu, click Format.

Notes:

- Be aware that when you format a disk, it removes all information already on the disk, such as previous files stored there.

- You can't format a disk if files are already open on that disk.

CONCEPT CHECK

What distinguishes the three types of system software?

Explain the important features of the OS: booting; CPU, file, and task management; and formatting.

4.3 Common Operating Systems

KEY QUESTION

What are some common desktop, network, and portable OSs?

The _platform_ is the particular processor model and operating system on which a computer system is based. For example, there are "Mac platforms" (Apple Macintosh) and "Windows platforms" or "PC platforms" (for personal computers such as Dell, Compaq, Gateway, Hewlett-Packard, or IBM that run Microsoft Windows). Sometimes the latter are called "_Wintel_ platforms," for "Windows + Intel," because they often combine the Windows operating system with the Intel processor chip. (We discuss processors in Chapter 5.)

Despite the dominance of the Windows platform, many so-called _legacy systems_ are still in use. A legacy system is an older, outdated, yet still functional technology, such as the DOS operating system. You may find yourself having to use DOS at some point.

Desktop & Laptop Operating Systems: DOS, Macintosh, & Windows

Let us quickly describe the principal platforms used on desktop computers: DOS, the Macintosh OS, and the Windows series (3.1, 95, 98, and Me). Desktop operating systems are used mainly on single-user computers (both desktops and laptops) rather than on mainframes or servers. We discuss operating systems for servers and for portable information appliances shortly.

The Apple iMac

- **DOS—the old-timer: _DOS_ (rhymes with "boss")—for _Disk Operating System_—was the original operating system produced by Microsoft and had a hard-to-use command-driven user interface.** Its initial 1982 version was designed to run on the IBM PC as PC-DOS. Later Microsoft licensed the same system to other computer makers as MS-DOS. With the growing popularity of cheaper PCs produced by these companies, MS-DOS came to dominate the industry. Two years before the advent of Windows 95, which eventually grew out of DOS, there were reportedly more than 100 million users of DOS, which at that time made it the most popular software ever adopted—of any sort.

- **The Macintosh Operating System—for the love of Mac:** The **_Macintosh operating system (Mac OS)_, which runs only on Apple Macintosh computers, set the standard for icon-oriented,**

easy-to-use graphical user interfaces. The software generated a strong legion of fans shortly after its launch in 1984 and inspired rival Microsoft to upgrade DOS to the more user-friendly Windows operating systems. Much later, in 1998, Apple introduced its iMac computer (the "i" stands for Internet), which added capabilities such as small-scale networking.

The newest version of the operating system, Mac OS X (called "ten"), breaks with 15 years of Mac software to offer a dramatic new look and feel.[2] (See ● Panel 4.4.) The user interface known as Aqua has Hollywood-like tricks (thanks to Apple chairman Steve Jobs' experience running Pixar Animation Studios, makers of *Toy Story* and other animated films). "Jelly-colored onscreen buttons pulse as if alive," says one description. "Menu borders are translucent, allowing you to see the documents under them. Sliders glow luminously."[3] In addition, Apple claims that OS X won't allow software conflicts, a frequent headache with Microsoft's Windows operating systems. For example, you might install a game and find that it interferes with the device driver for a sound card. Then, when you uninstall the game, the problem persists. With Mac OS X, when you try to install an application program that conflicts with any other program, the Mac simply won't allow you to run it.

Ultimately, Apple's strategy is to marry its easy-to-use operating system with its iMac (discussed in Chapter 4) to offer free Web services for everything from your photos to your home page.[4]

Can you run Windows-based application software on a Macintosh? Indeed, you can, using **_emulation software_, which allows software designed for one computer to run on another.** For instance, Virtual Windows by Connectix lets you run Windows programs on an iMac or Macintosh G3.[5] The only drawback is that you have to sacrifice a little speed because it takes time to translate Windows-related instructions into instructions the Mac OS can understand.

An alternative is *hardware emulation,* in which you install a circuit board with a PC (Intel) processor and other hardware compatible with Windows into the Macintosh system unit. Essentially this gives you two computers, PC and Mac, in a Macintosh machine.

Macintosh is still considered king in areas such as desktop publishing. However, programs for games and for common business uses such as word processing and spreadsheets are widely available. For very specialized applications, most programs are written for the Windows platform.

- **Microsoft Windows 3.1, 95, & 98:** In the 1980s, taking its cue from the popularity of Mac's easy-to-use GUI, Microsoft began working on Windows—to make DOS more user-friendly. Early attempts (Windows 1.0, 2.0, 3.0) did not catch on. However, in 1992, Windows 3.1 emerged as the preferred system among PC users. (Technically, Windows 3.1 wasn't a full operating system; it was simply a layer or shell over DOS.) Later, this version evolved into the Windows 95 operating system, which was succeeded by Windows 98.

  **Microsoft Windows 95/98** **is today's most popular operating system for desktop and portable microcomputers, supporting the most hardware and the most application software.** *(See ● Panel 4.5, repeated from Chapter 3.)* Among other improvements over their predecessors, Windows 95 and 98 adhere to a standard called Universal Plug and Play, which is supposed to let a variety of electronics seamlessly network with each other. _**Plug and play**_ **is defined as the ability of a computer to automatically configure a new hardware component that is added to it.** When capitalized or when abbreviated PnP, Plug and Play refers to the specifications developed by Microsoft to allow users of Windows PCs to "plug" in peripherals such as monitors, modems, and printers and have them "play" or operate automatically. In reality, Plug and Play has not performed as smoothly as promised. In Windows Me, however, PnP is further developed.

- **Microsoft Windows Me and Whistler—especially for multimedia mavens:** _**Microsoft Windows Millennium**_**, or** _**WinMe**_ **(the Me stands for Millennium Edition), is the successor to Windows 95 and 98, designed to support desktop and portable computers.** It is especially a boon to multimedia mavens because of its ability to handle still pictures, digital video, and audio files.

  Among the system software's improvements are the following: [6]

  WinMe claims to have reduced the problem of frequent "crashes," notorious with earlier versions of Windows. (The "blue screen of death" error is the ominous nickname given to the event in which the computer just stops and reverts to a blue screen.) When your computer crashes, all the elements on the screen simply freeze up, and you're obliged to turn off the computer and reboot. Now the operating system tries to protect core system files from being altered when you load third-party programs. If they do go haywire, it provides you with a "system restore" tool to roll back the computer to the time when it was running smoothly.

  Another benefit of WinMe is its handling of still pictures. You can easily download pictures from a digital camera into a PC; you can preview images while they're still in the camera, then store those you want to keep on your computer and share them with others online.

  You can also more easily transfer digital video to your PC, and the WinMe Movie Maker will automatically divide the video into segments organized around scene changes. This allows you to get rid of flawed shots such as those of ground and sky, then you can rearrange the remaining clips into a timeline of your choosing, adding transitions and special effects to create your own movie.

  The WinMe Media Player lets you listen to Internet radio, get information on favorite artists, organize tunes in your PC, and transfer music from your PC to portable digital audio players.

  In 2000, Microsoft announced that in late 2001 it would make available new system software code-named Whistler.[7] _**Microsoft Whistler**_ **targets home users but is based on Windows NT,** as we

Outlook Express: Part of Microsoft's browser, Internet Explorer, that enables you to use e-mail.

Microsoft Network: Click here to connect to Microsoft Network (MSN), the company's online service.

Norton Protected: Click here to activate anti-virus software.

Network Neighborhood: If your PC is linked to a network, click here to get a glimpse of everything on the network.

My Documents: Where your documents are stored unless you specify otherwise.

My Computer: Gives you a quick overview of all the files and programs on your PC.

Documents: Multitasking capabilities allow users to smoothly run more than one program at once.

Start menu: After clicking on the start button, a menu appears, giving you a quick way to handle common tasks. You can launch programs, call up documents, change system settings, get help, and shut down your PC.

Start button: Click for an easy way to start using the computer.

Taskbar: Gives you a log of all programs you have opened. To switch programs, click on the icon buttons on the taskbar.

Multimedia: Windows 98 features sharper graphics and improved video capabilites.

● PANEL 4.5
Windows 98 screen

discuss next. With this new version, Microsoft will finally be giving up the last of the Windows software carried forward from the aging DOS programming technology.

Network Operating Systems: NetWare, Windows NT/2000, Unix, & Linux

The operating systems described so far were principally designed for use with stand-alone desktop machines. Now let's consider the important operating systems designed to work with networks: NetWare, Windows NT/2000, Unix/Solaris, and Linux.

- **Novell's NetWare—PC networking software:** <u>*NetWare*</u> **has long been a popular network operating system for coordinating microcomputer-based local area networks (LANs) throughout a company or a campus.** LANs allow PCs to share programs, data files, and printers and other devices. Novell, the maker of NetWare, thrived as corporate data managers realized that networks of PCs could exchange information more cheaply than the previous generation of mainframes and midrange computers.

  In 1999, NetWare's share of the server market was 19%, down from earlier times, owing to the rise of the Internet and competition from Microsoft, Unix/Solaris, and Linux. Nevertheless, Ford, Wal-Mart, and other large companies have been eager to use Novell's directory software, which runs on Windows 2000, Solaris, and Linux servers as well as on NetWare, to keep track of computers, programs, and people on a network. This innovation has kept Novell ahead.[8] In addition, in 2000 Novell announced a new strategy of packaging its Web e-business software into a bundle to be sold to application service providers, the ASPs we discussed at the beginning of this chapter. "By packaging its software this way," says one analyst, "Novell can focus its sales efforts on 500 ASPs instead of tens of thousands of corporations and smaller businesses."[9]

- **Windows NT and 2000—the challenge from Microsoft:** Windows 95 and 98 can be used to link PCs in small networks in homes and offices. However, something more powerful was needed to run the huge networks linking a variety of computers—PCs, workstations, mainframes—used by many companies, universities, and other organizations, which previously were served principally by Unix and NetWare operating systems. <u>*Windows NT*</u> **(the NT stands for New Technology), later upgraded to** <u>*Windows 2000*</u>**, is Microsoft's multitasking operating system designed to run on network servers. It allows multiple users to share resources such as data and programs.**

Sending impressions overseas

Japanese and American members of the Mormon church journey via wagon train from Nebraska to Utah, re-enacting a trip taken by Mormons to escape religious persecution more than 150 years ago. Using a solar-powered laptop with a wireless modem, Osamu Sekiguchi of Tokyo posts his impressions of the journey to a Japanese Web site.

When it first appeared, in 1993, the system came in two versions. The *Windows NT Workstation* version enabled graphic artists, engineers, and others using stand-alone workstations to do intensive computing at their desks. The *Windows NT Server* version was designed to benefit multiple users tied together in client/server networks. In 1999, Windows NT had 38% of the server market. In early 2000, Microsoft rolled out its updated version, *Windows 2000,* to replace Windows NT version 4.0.

The choice of the name Windows 2000 is somewhat unfortunate. This is not the successor to Windows 95 and 98 for home and non-network use. If you're interested in arcade-style games, for instance, 2000 won't work, since it was specifically designed to eliminate crashes and reboots and as a result is incompatible with a whole class of programs and devices (though they will work with Windows 98 and Me). Windows 2000 has better Plug-and-Play hardware installation and works better for portables, giving slightly better battery life than either NT or 98. One weakness is that Windows 2000 still can't run the most powerful servers.[10]

- **Unix, Solaris, and BSD—first to exploit the Internet:** Unix (pronounced "*Yu*-niks") was developed at AT&T's Bell Laboratories in 1969 as an operating system for minicomputers. Today **_Unix_ is a multitasking operating system for multiple users that has built-in networking capability and versions that can run on all kinds of computers.** Government agencies, universities, research institutions, large corporations, and banks all use Unix for everything from designing airplane parts to currency trading. Unix is also used for Web-site management. Indeed, the developers of the Internet built their communications system around Unix because it has the ability to keep large systems (with hundreds of processors) churning out transactions day in and day out for years without fail. In 1999, Unix had 15% of the server market (down from 19% in 1998, probably because of inroads made by Linux, as we'll discuss).

  Sun Microsystems' **_Solaris_ is a super-reliable version of Unix that seems to be most popular for handling large e-commerce servers and large Web sites.** Solaris 8 can handle servers with as many as 64 microprocessors, compared with 32 for Windows 2000. In addition, eight computers can be clustered together to work as one, compared with four for Windows 2000.[11] As we mentioned, Sun is also an ASP, offering over-the-Net application software StarOffice and StarPortal.

  Another interesting variant is **_BSD_, free software derived from Unix.** BSD began in the 1970s in the computer science department of the University of California, Berkeley, when students and staff began to develop their own derivative of Unix, known as the Berkeley Software Distribution, or BSD. There are now three variations, which are distributed online and on CD-ROM. (1) The *FreeBSD Project* concentrates on standard PC hardware and is the most widely deployed version. For instance, Yahoo! has been relying on FreeBSD since 1995. (2) *NetBSD* focuses on putting BSD on a wide range of platforms, ranging from old Atari game machines to state-of-the-art Compaqs. (3) *OpenBSD* gives top priority to security.[12]

- **Linux—software built by a community:** It began in 1991 when programmer Linus Torvalds, a graduate student in Finland, posted his free Linux operating system on the Internet. With 25% of the server market in 1999 (up from 16% the year before), Linux (pronounced "*Linn*-uks") is the rising star of network software. **_Linux_ is a free version of Unix, and its continual improvements result from the efforts of tens of thousands of volunteer programmers.** Whereas Windows is Microsoft's proprietary product, Linux is **_open-source software_—**

meaning any programmer can download it from the Internet for free and modify it with suggested improvements. The only qualification is that changes can't be copyrighted; they must be made available to all and remain in the public domain. From these beginnings, Linux has attained cult-like status. "What makes Linux different is that it's part of the Internet culture," says an IBM general manager. "It's essentially being built by a community."[13] (The People's Republic of China announced in July 2000 that it was adopting Linux as a national standard for operating systems because it feared being dominated by the OS of a company of a foreign power—namely, Microsoft.)

Linux screen

If Linux belongs to everyone, how do companies like Red Hat Software and VA Linux Systems—two companies that base their business on Linux—make money? Their strategy is to give away the software but then sell services and support. Red Hat, for example, makes available an inexpensive ($149) application-software package that offers word processing, spreadsheets, and the like.

Because it was built for use on the Internet, Linux is more reliable than Windows for online applications. Hence, it is better suited to run Web sites and e-commerce software. Its real growth, however, may come as it reaches outward to other applications. IBM, Red Hat, and 45 other companies have formed the Embedded Linux Consortium, which envisions a time when microchips running Linux are built into everything from Internet-linked microwave ovens and refrigerators to wireless phones and TV set-top cable boxes.[14]

Operating Systems for Handhelds: Palm OS & Windows CE/Pocket PC

Maybe you're not one of the estimated (at the end of 2000) 7.7 million owners of a handheld computer or personal digital assistant (PDA). But perhaps you've seen people poking through calendars and address books, beeping through games, or (in the latest versions) checking e-mail or the Web on these palm-size devices. Handhelds have gained headway in the corporate world and on college campuses. Because of their small size, they rely on specialized operating systems, including the Palm OS and Windows CE.

Visor handheld computer

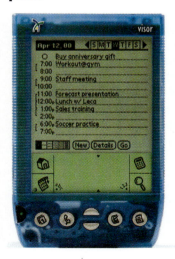

- **Palm OS—the dominant OS for handhelds:** In 1994, Jeff Hawkins took blocks of mahogany and plywood into his garage and emerged with a prototype for the PalmPilot. Two years later, Hawkins and business brain Donna Dubinsky pulled one of the most successful new-product launches in history. Today the company, Palm Computing, with 80% of the handheld market, sells the popular Palm III, V, and VII. However, Hawkins and Dubinsky left to form another company, Handspring, whose product, Visor, acts like a Palm but costs less and has an expansion slot that can transform the device into a cell phone, MP3 music player, or two-way pager.

  The _Palm OS_, **which runs the Palm and Visor, is the dominant operating system for handhelds,** with about 85% of the market, and is being licensed to competitors, such as IBM and Nokia. With a Palm OS, you have access to thousands of application programs over the

Internet, many of them free. (For instance, one $10 program, PayUp, lets you track 19 people's dinner orders and calculate an exact bill for each, including tax and tip.)[15]

- **Windows CE/Pocket PC—Microsoft Windows for handhelds:** In 1996, Microsoft released ***Windows CE*, a greatly slimmed-down version of Windows 95 for handheld computing devices,** such as those made by Casio, Compaq, and Hewlett-Packard. Windows CE had some of the familiar Windows look and feel and included rudimentary word processing, spreadsheet, e-mail, Web browsing, and other software.

  But whereas Palm concentrated on simplicity, reliability, compactness, and long battery life, Microsoft and its allies focused on what had worked in the conventional-PC marketplace: piling on new features and options. Consumers rejected it. "Windows CE is a bear to use," wrote one reviewer about the Compaq Aero, the leader in the CE market, which accounted for only 7% of handheld dollar sales in 1999. "Microsoft is essentially cramming Windows down our throats."[16]

  Microsoft abandoned Windows CE because too many buyers and developers regarded it as a loser. Its latest version, ***Pocket PC* for handhelds, is simpler and less cluttered than CE and looks and feels a lot less like desktop Windows.** Based on research findings that customers want their handhelds to be more than just electronic organizers, however, Pocket PC offers pocket versions of Word and Excel that let users read standard word processing and spreadsheet files sent as e-mail attachments from their PCs. Microsoft has also announced plans to offer a special version of Internet Explorer that would automatically reformat Web pages so they could be read on the small screens of handhelds.[17]

Besides Palm OS and Windows CE/Pocket PC, other operating systems are being developed for the handheld market. For example, the British Psion Revo, a handheld, runs a swift Java-language-compatible operating system.[18] Another operating system called BeIA, an open-source system from Be Inc. in Menlo Park, California, forms the basis of a number of Internet appliances.[19] Indeed, in what is coming to be called "the post-PC era," companies are rushing to design not only handhelds but also *intelligent linking technology*, which will connect embedded chips in all kinds of household appliances.

CONCEPT CHECK

What are the principal desktop operating systems and their features?

What are the common network operating systems and their features?

Describe the leading operating systems for handhelds.

4.4 Other System Software: Device Drivers & Utility Programs

KEY QUESTIONS
What are the characteristics of device drivers and utility programs?

We said that the three principal parts of system software are operating system, device drivers, and utility program. Let's now consider the last two.

Device Drivers: Running Peripheral Hardware

***Device drivers* are specialized software programs that allow input and output devices to communicate with the rest of the computer system.** Many basic device drivers come with system software when you buy a computer. If, however, you buy a new peripheral device, such as a mouse, scanner, or printer, the package will include a device driver (probably on a CD-ROM). You'll need to install the driver on your computer's hard-disk drive (by following the manufacturer's instructions) before the device will operate.

Utilities: Service Programs

"Utility programs do the everyday chores around your computer," suggests one homely analogy, "sweeping, taking out the trash, making sure the electronic doors are locked."[20]

Utility programs, also known as _service programs_, perform tasks related to the control and allocation of computer resources. They enhance existing functions or provide services not supplied by other system software programs. Most computers come with built-in utilities as part of the system software. (Windows 95/98/Me offers several of them.) However, they may also be bought separately as external utility programs (such as Norton Desktop, 911 Utilities, and PC Tools).

Among the tasks performed by utilities are the following:

- **Backup:** Suddenly your hard-disk drive fails, and you have no more programs or files. Fortunately, we hope, you have used a **_backup utility_ to make a backup, or duplicate copy, of the information on your hard disk.** Examples of freestanding backup utilities are Norton Backup (from Symantec) and Colorado Scheduler.

- **Data recovery:** One day in the 1970s, so the story goes, programming legend Peter Norton was working at his computer and accidentally deleted an important file. This was, and is, a common enough error. However, instead of re-entering all the information, Norton decided to write a computer program to recover the lost data. He called the program The Norton Utilities. Ultimately it and other utilities made him very rich.

  A **_data-recovery utility_ is used to restore data that has been physically damaged or corrupted.** Data can be damaged by viruses (see following), bad software, hardware failure, and power fluctuations that occur while data is being written/recorded.

- **Virus protection:** If there's anything that can make your heart sink faster than the sudden failure of your hard disk, it may be the realization that your computer system has been invaded by a virus. A *virus* consists of hidden programming instructions that are buried within an applications or systems program. Sometimes they copy themselves to other programs, causing havoc. Sometimes the virus is merely a simple prank that pops up a message. Other times, however, it can destroy programs and data and wipe your hard disk clean. Viruses are spread when people exchange floppy disks or download (make copies of) information from computer networks.

  Fortunately, antivirus software is available. **_Antivirus software_ is a utility program that scans hard disks, floppy disks, and memory to detect viruses.** Some utilities destroy the virus on the spot. Others notify you of possible viral behavior. Because new viruses are constantly being created, you need the type of antivirus software that can detect unknown viruses.

  Examples of antivirus software are Norton AntiVirus, Dr. Solomon's Anti-Virus Toolkits, McAfee's VirusScan, and Webscan. New viruses appear every day, so it's advisable to look for an antivirus utility that offers frequent updates without additional cost.

  Although it's a good idea to install an antivirus utility on your computer, virus risks are sometimes exaggerated. With few exceptions, if you don't boot your computer with a diskette in the drive, directly run programs downloaded from a network, open files attached to

PRACTICAL ACTION BOX

What to Do If the Disk with the Only Copy of Your Novel Fails

Always make a back-up copy of your files.

That's Rule No. 1.

Rule No. 2: See Rule No. 1.

While we're at it, let's mention Rule No. 3—that it's a good idea to *make a back-up copy of your files.*

What this means is that, besides saving to your hard-disk drive (or floppy disk), you should regularly—anywhere from every 10 minutes to once a day—take the important files of data you're currently working on and *copy them* on to other disks or a backup tape, if you have one. The principle is that you should never have just *one* copy of anything.

Now suppose you're like Soo-Yin Jue and didn't follow Rules 1 through 3.[a] For nine years, Jue worked on her first novel, traveling to Asia and gathering notes, which she entered onto her faithful Macintosh. Then, after she finished writing a first draft of her novel, she went to hit "Save."

The diskette on which she had been saving all her work began to spin.

And then the Mac made an odd grinding noise.

Nothing she did would bring any of her data back to the screen.

And she had not saved any of nearly a decade's worth of original material, either on another diskette or on the hard-disk drive of some other computer.

Enter DriveSavers of Novato, California, one of a handful of companies authorized by diskmakers to do rescue work. Whether the data loss results from an unrealistic faith in the invulnerability of technology or from damage by spilled coffee, floods, or even fires, DriveSavers specializes in data-recovery miracles.

The small staff of engineers uses a variety of repair techniques, software, cleansers, and even a sterile "clean room" free of dust particles to resurrect data from drives bound for the junk yard. Usually this can be done in about two days, for a charge of around $800.[b] The company claims a 95% success rate.

Soo-Yin Jue was one of the lucky ones. Data-recovery engineer John Christopher pointed out to her that he could rescue the file for her novel, though he couldn't save any formatting, such as paragraph marks or page breaks.

Jue was jubilant. "As long as you have the words, I don't care about anything else," she said.

What about those 5% of cases in which DriveSavers is unsuccessful?

The company gets paid no matter what, charging about 10% of its normal fee when the data is irretrievable. For those cases in which there is no hope, DriveSavers employs a "data crisis counselor," Nikki Stange.

"People express panic, guilt, anger, and fear," says Stange. "I use techniques I developed when working on a suicide hotline."

Say, what was that Rule No. 1 again?

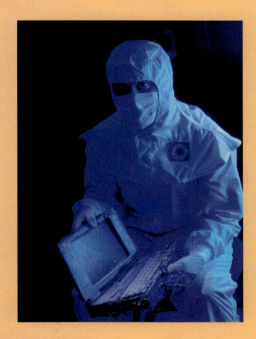

Is this data dead?
Saved by firefighters from a burning house, this laptop was melted shut and written off for dead. Then "disk doctors" came to the rescue. Chris Bioss, a recovery engineer, specializes in retrieving data from damaged hard drives.

System Software

- Never start your computer from an unknown floppy disk. Always make sure your floppy-disk drive is empty before turning on or restarting your computer.
- Run virus-scanning software on a new floppy disk before executing, installing, or copying its files into your system.
- If you download or install software from a network server (including the Internet), bulletin board, or online service, always run scanning software on the directory you place the new files in before executing them.
- Create a start-up diskette containing the scan program. Make sure this floppy disk is write-protected so that it can't become infected.
- Scan files attached to e-mail before you open them.

e-mail, or use illegally copied program diskettes, your risk of virus infection is low. *(See ● Panel 4.6.)*

We discuss viruses further in Chapter 9.

- **Data compression:** As you continue to store files on your hard disk, it will eventually fill up. You then have several choices: You can delete old files to make room for the new. You can buy a new hard-disk cartridge drive and some cartridges and transfer the old files and programs to those. Or you can use a data compression utility.

  **Data compression utilities** **remove redundant elements, gaps, and unnecessary data from a computer's storage space so that less space (fewer bits) is required to store or transmit data.** With a data compression utility, files can be made more compact for storage on your hard-disk drive. Given today's huge-capacity hard drives, you may never fill yours up. Still, data compression remains an issue.

  With the increasing use of large graphic, sound, and video files, data compression is necessary both to reduce the storage space required and to reduce the time required to transmit such large files over a network. You may also want to compress a file to fit on a floppy disk, for portability.

  As the use of sophisticated multimedia becomes common, compression and decompression will be increasingly taken over by built-in hardware boards that specialize in this process. That will leave the main processor free to work on other things, and compression/decompression software utilities will become obsolete.

- **File defragmentation:** Over time, as you delete old files from your hard disk and add new ones, something happens: the files become *fragmented.* _**Fragmentation**_ **is the scattering of portions of files about the disk in nonadjacent areas, thus greatly slowing access to the files.**

  When a hard disk is new, the operating system puts files on the disk contiguously (next to one another). However, as you update a file over time, new data for that file is distributed to unused spaces. These spaces may not be contiguous to the older data in that file. It takes the operating system longer to read these fragmented files. **A** _**defragmenter**_ **utility program, commonly called a "defragger," will find all the scattered files on your hard disk and reorganize them as contiguous files.** Defragmenting the file will speed up the drive's operation.

Many other utilities exist, such as those for transferring files back and forth between a desktop microcomputer and a laptop. Generally, the companies selling utilities do not manufacture the operating system. OS developers may eventually incorporate the features of a proven utility as part of their product. (Note: Independent, or external, utilities must be compatible with your system software; check the software packaging and user documentation.)

4.5 The OS of the Future: "The Network Is the Computer"

KEY QUESTION
What are some future directions operating systems might take?

What will happen to operating systems in the future? Or will there even be any operating systems at all?

For over a decade, Sun Microsystems has subscribed to the notion that "the network is the computer." In this view, the future of computing is no longer bound up with the programs locked into stand-alone hardware. Rather, it is in the software that will run networked machines. The concept of an "operating system" that extends over all kinds of networks, including the Internet, has also been put forth by Novell, Apple, Cisco, Hewlett-Packard, Oracle, and Microsoft. Let us see where this idea seems to be leading.

Internet "Operating Systems": Microsoft.Net, E-speak, & Jini

As a company, Microsoft has managed two difficult transitions in business strategy. The first was its decision to develop its Windows graphical user interface. Around 1985 this resulted in a new face for the DOS operating system, in which typed commands were replaced by pull-down menus, icons, and toolbars. The second was in 1995, when Microsoft decided to "embrace and extend" the Internet. (By incorporating its own Web browser, Internet Explorer, as a free component of its operating system, Microsoft within a short period of time drastically reduced the sales of its browser competitor Netscape—inviting a lawsuit from the U.S. Justice Department for anitcompetitive behavior.) Now it is turning to a new bet-the-company strategy called Microsoft.Net.

- **Microsoft.Net—Windows everywhere on the Net:** Pronounced "dot-net," **.Net is Microsoft's platform for an operating system for the entire Internet, designed to link unrelated Web sites so that people can organize all the information in their lives, using PCs and smaller devices,** such as cellphones, handheld computers, and set-top boxes. Some of the .Net software will reside on Windows, some on the Internet, and some on the devices themselves. It will incorporate new technology being explored by Microsoft, including speech recognition, handwriting recognition, real-time video, and intelligent browsers. The focus of the company's billion-dollar initiative, which has gone under the code name of Next Generation Windows Services, is to get its Windows-branded software into as many places on the Internet as possible.

  How will .Net be used? The idea is that one mouse click might set off a cascade of actions without your having to open new programs or visit additional Web sites. In one writer's example, "Booking an airline ticket online . . . might trigger a link that updates your calendar with the flight times. That might alert another site that will later check the status of your flight and e-mail your spouse if you're delayed."[21]

  Until now, Web sites and operating systems have had different interfaces. There is no reason why they should. A key to Microsoft's strategy is to alter the look and feel of Windows so that the GUI will probably look less like the current Windows desktop, with folders and

System Software

153

icons, and more like, say, Amazon.com's home page, with links and tabs. Clicking on some tabs would lead you to programs and data on your PC. Clicking on others would connect you to Web-based services, such as news sites or financial data.

Underlying .Net is a commitment to *XML* (for *extensible markup language*), an "open standards" protocol that makes it easy for machines to read Web sites by enabling Web developers to add more "tags" to a Web page. (We describe XML in the Appendix.) Because .Net uses XML, it can operate with software made by other companies. Even so, Microsoft says that using Windows will help make Web services run more smoothly. The next major overhaul of Windows, code-named *Whistler*, will feature .Net-based services. Still, with its new strategy, Microsoft will increasingly need to shift its revenue sources to services funded by subscriptions or advertising as its PC software and licensing fees begin inevitably to decline.[22]

- **E-speak—Hewlett-Packard's universal Internet language:** In 1999, Hewlett-Packard Co. introduced what it called "Chapter II of the Internet." **_E-speak_ is H-P's version of an Internet operating system, or "universal language," that allows different Web sites to communicate with one another.** Like Microsoft.Net, the purpose of e-speak is to make the Internet easier and more useful. Also similarly, e-speak uses the programming standard XML, which helps computers communicate and sort through the millions of offerings on the Web. An important difference, however, is that, whereas Microsoft seems to have a grand vision for its .Net, Hewlett-Packard sees e-speak as simply a tool to help computers talk about the details of transactions. In addition, H-P apparently fumbled its first-year marketing effort, so that few users seem to understand what e-speak is supposed to do.[23]

- **Sun's Jini:** In early 1999, Sun Microsystems formally announced a new technology called Jini (the genie in the tale *Aladdin*) and a lineup of partners to support it, including IBM, Sony, and Eastman Kodak.[24] **_Jini_ is a small layer of software designed to let all types of electronic gadgets on a wired or wireless network communicate with each other.** Jini builds on another Sun technology called *Java* (discussed in the Appendix). The difference is that, while Java allows separate devices on a network to run different applications, Jini enables these devices to actually "talk" to each other automatically, so that there is no need to install special software drivers.

Thus, for example, Jini allows a cellular phone to grab an incoming number from a pager and dial it automatically. Or a user of a notebook PC connected to a network in an airport business lounge could immediately have access to a scanner or printer. Or cars might be outfitted with voice-activated e-mail and navigation. Or the systems in a house might be networked so that families could control electrical and heating systems from their computers or TVs, as well as share files, applications, printers, videos, and Internet access.

The Opposite Possibility: No One Company Dominates

A great deal of computing horsepower sits idle most of the day. This is the unused capacity of thousands or even millions of home and office PCs, which are frequently left unattended or shut down when not in use. Couldn't something useful be done with this all this downtime?

Indeed, it could—and has been. SETI@home (SETI stands for Search for Extraterrestrial Intelligence) is a project in which thousands of users have contributed the equivalent, it's said, of more than 300,000 years of computer time, using their personal computers tied to the Internet.[25] Although SETI has not been able to find intelligence elsewhere in the universe, it has been stunningly successful at demonstrating what might be called "massively distributed computing." **A _distributed system_ is a noncentralized network consisting of several computers and other devices that can communicate with one another.** Such a system appears to users, in one description, "as parts of a single, large, accessible 'storehouse' of shared hardware, software, and data."[26] Now other organizations (such as Popular Power at _www.popularpower.com_ and Distributed Science at _www.distributedscience.com_) are also attempting distributed-computing setups among PC users on the Internet.

Yale University computer scientist David Gelernter (who came to the attention of the public in 1993 when he was critically injured by a mailbomb from the Unabomber) believes that Microsoft is misjudging the future of computing. Computing, he suggests, is becoming decentralized, with information distributed among millions of computers. Microsoft, he thinks, is behind the wave in trying to maintain a central role for its software with the Microsoft.Net strategy.[27]

Others also believe that computing will move on without Microsoft, relying on the open-source model, as Linux does.[28] Indeed, this is the strategy underlying many of the new Linux-based Internet appliances, such as i-Opener from Netpliance and a new Web-access device from Gateway. "The Internet challenge to Microsoft's Windows franchise has only begun," suggests Randall Stoss, author of _The Microsoft Way_. "The diffusion of inexpensive broadband service and small Internet-aware hardware devices will make the threat more manifest."[29]

CONCEPT CHECK

What are three possible Internet "operating systems"?

What is distributed computing?

4.6 Online Software & Application Software Providers: Turning Point for the Software Industry

What are some recent trends in online software?

Everything else seems to be going on the Internet. Why not word processors, spreadsheets, and other software? Indeed, even now, if you're tired of using a paper calendar or datebook, you can go online to jump.com, yahoo.com, or when.com and start using a Web calendar to keep track of appointments, deadlines, and birthdays. (Cautionary note: If, however, there comes a day when the Web site doesn't come up on your computer screen, as happened to _Time_ magazine technology writer Anita Hamilton, you'll be back to jotting down your appointments on Post-It notes.)[30] Other Web-based software has also been offered free for some time—browsers, e-mail, and address books, for example—in an attempt to keep users coming back to a particular portal such as AltaVista or Yahoo!

But for businesses in particular—and ultimately probably for you—it does make sense to rent software rather than buy it. Let's take a look at this concept as we begin our discussion of software.

Online Software & the Application Service Provider

Jim Obsitnik, 30, is a software salesman and avid jogger—so avid, in fact, that he keeps track of his miles and running times on an Excel spreadsheet. Business travel used to be inconvenient because he had to record his jogging data on slips of paper (which could get lost), then enter it into his PC when he got home. No more. Now when he's out of town he can log onto a free, password-protected runner's Web site called Desktop.com and download his spreadsheet. He doesn't even need his own computer; he can use someone else's.[31]

Is this a picture of the future? Every month you pay the phone bill, the electric bill, maybe the ISP (Internet service provider) bill. What about paying an ASP—an application software provider—as well as an ISP? Or even getting both ISP and ASP for free?

An _ASP (application service provider)_ is a firm that leases software over the Internet to customers. You no longer have to buy software in a store, in shrink-wrapped packages. Instead, you can simply download a particular program when you need it, for as long as you need it.

Could this be a major turning point for computers and communications? "The [ASP] trend is rocking the $8 billion software industry," says one analyst, "threatening titans such as Microsoft, IBM," and others.[32] Among the purveyors of Web-based software are Damango, Desktop.com, Halfbrain.com, iAmaze.com, Mi8, myWebOS, Sun Microsystems, ThinkFree.com, and Visto.com. _(See ● Panel 4.7.)_ Mi8, for example, offers Microsoft Office and Lotus Notes—which are well-known business application programs—for $21.95 a month. The ASP called myWebOS offers a free word processor, e-mail, and calendar; free spreadsheet and presentation software is in the works. Sun Microsystems (_www.sun.com_) offers StarOffice and plans to offer StarPortal. Microsoft also plans to provide its Office program as a monthly service. In addition, there are ASPs that make available online specialized software to handle tasks for small businesses, such as office services (HotOffice.com), expense tracking services (TimeBills.com), or human resources (Employease).[33]

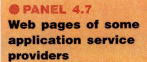

● PANEL 4.7
Web pages of some application service providers

Network Computers Revisited: "Thin Clients" versus "Fat Clients"

We can link the growth in ASPs to the concept of a *network computer*, proposed a few years ago by Larry Ellison, CEO of database-software maker Oracle Corp. **A _network computer_ is an inexpensive, stripped-down computer that connects people to networks and runs applications tied to servers.**

Out of this grew a distinction between "thin clients" and "fat clients." A *client*, you'll recall from Chapter 1, is one part of a client/server network. The *server* is the central computer that holds collections of data and programs; the clients are the PCs, workstations, and other computers on the network that use the data and programs from the server.

A *fat client* is a regular computer, perhaps a PC, that is on a network. Often it contains software with a great many features. Such "bloatware" is the result of software makers constantly trying to top themselves when they issue new versions. To run these programs efficiently, a computer requires lots of main memory and hard-disk storage capacity, as well as a powerful microprocessor. A *thin client,* by contrast, is a network computer—a stripped-down computer without hefty microprocessors or even much storage—that is supposed to operate as an inexpensive terminal tied to a server. (See ● *Panel 4.8.*)

● PANEL 4.8
Old and new ways of getting software

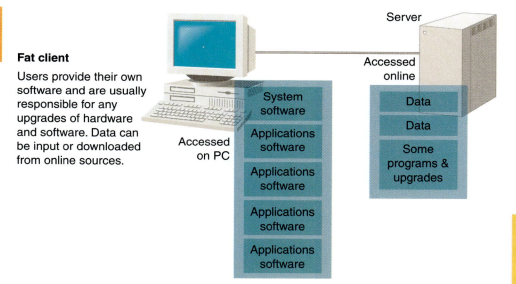

Fat client

Users provide their own software and are usually responsible for any upgrades of hardware and software. Data can be input or downloaded from online sources.

Accessed on PC

System software

Applications software

Applications software

Applications software

Applications software

Server

Accessed online

Data

Data

Some programs & upgrades

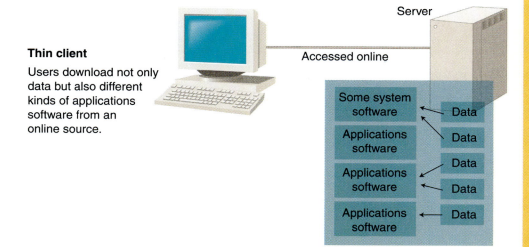

Thin client

Users download not only data but also different kinds of applications software from an online source.

Accessed online

Some system software

Applications software

Applications software

Applications software

Data

Data

Data

Data

Server

Contrary to Ellison's expectations, the widespread appearance of network computers/thin clients was delayed because the prices of regular, fat-client PCs dropped sharply, below $1000. But in the meantime, another movement advanced the cause of online software—the development of enterprise resource planning software.

From ERP to ASP: The Evolution of "Rentalware"

About 10 years ago, businesses began to adopt client/server arrangements. The tasks once performed by mainframes and minicomputers were being divided between desktop computers and servers. To ease this transition, companies relied on business and accounting programs called ERP software. **_ERP (enterprise resource planning) software_ consists of large client/server software applications that help companies organize and operate their businesses.** The makers of such software include SAP, Oracle, and PeopleSoft, as well as Microsoft.

ERP is expensive—a corporation might well spend $30 million on such a system. And the risk is all on the buyer. Even if the system doesn't work as well as it should, the buyer will be inclined to keep it, because it represents such a huge investment.

Enter the ASP model. Instead of buying an ERP system, a company can "rent" the same thing, and the installation and management of all the equipment and software becomes someone else's headache.[34] "The Internet creates an opportunity to change the way people manage information technology," says Gary Bloom, an executive with Oracle Corp. "You will buy software as a service, just the way you buy telephone service today."[35]

CONCEPT CHECK

What is an application software provider?

Explain the concept of thin clients versus fat clients.

What is enterprise resource software?

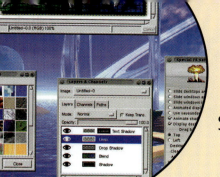

Experience Box
Student Use of Computers: Some Controversies

Information technology is very much a part of college. Elsewhere we discuss such matters as distance learning, Web research, online plagiarism, and Internet addiction. Here we describe some other issues regarding students' computer use.

Using Computers in the Classroom

Although they're more expensive than desktop computers, laptops are useful because you can take them not only to libraries, to help with reading or term-paper notes, but also to class to use in taking lecture notes.[36] You might even try using a palmtop computer, which costs half as much as a laptop and comes with word processing and spreadsheet software.[37] Be aware, however, that the small palmtop keyboard requires some dexterity. With either laptop or palmtop, battery life may also be a factor if you have several classes back to back and no chance to recharge your machine. (Some palmtops use penlight batteries, so you could easily carry spares.)

The use of computers in classrooms is still controversial in certain quarters. Some campuses allow—even require—students to bring laptops to class but are imposing rules about what they are allowed to do. "More students are sending instant messages to one another (chatting and note passing, 21st-century style)," says one article about business-school campuses, "day trading (as opposed to daydreaming), and even starting their own companies, all in class."[38] Computer-generated voices that accompany the downloading of e-mail attachments are another classroom annoyance. You should be careful not to be a classroom disturbance when using your computer.

Notes Posted on the Web

Wouldn't it be nice, when you're out sick, to be able to go to a Web site and get the notes of lectures for the classes you missed?

This is possible on campuses served by such commercial firms as StudentU.com, Versity.com, Study24-7.com, and Gethrucollege.com. Some colleges even have their own operations, such as Black Lightning Notes at the University of California, Berkeley. Many commercial sites are free, since they try to generate revenue by selling online advertising, but some charge a fee. Versity.com, for example, is free on many campuses but charges a subscription ($19.95 a semester) on others.[39]

Such services can be a real help to students who learn best by reading rather than by hearing. They also provide additional reinforcement to students who feel they have not been able to grasp all of a professor's ideas during the lecture. However, they are no substitute for the classroom experience, with its spontaneous exchange of ideas. Moreover, as one writer points out, "the very act of taking notes—not reading somebody else's notes, no matter how stellar—is a way of engaging the material, wrestling with it, struggling to comprehend or take issue."[40] In other words, you'll be better able to remember the lecture if you've reinforced the ideas by writing them down yourself.

Some faculty have no problem with note-taking operations. Indeed, for some time, some professors have been posting their lecture notes, practice exams, and reading lists on Web pages linked to the World Lecture Hall (*www.texas.edu*). Others disapprove, however. Among the criticisms: (1) Note-taking services don't always ask permission. (2) Instructors are reluctant to share their unpublished research in class if they think their ideas might end up posted in a public place and ripped off. (3) They might not wish to share controversial opinions with students if their views might be criticized in a world-wide forum. (4) Students might not come to class, especially if they think the lectures are boring or if they are given to chronic oversleeping or hangovers. (To deter falloffs in attendance, an instructor can take roll or pass around a sign-up sheet.) (5) The notes may be sloppy, inaccurate, or incomplete.

Sloppy notes are a serious concern. "The notes were very, very inaccurate and included gross errors," said one Yale University professor of environmental studies who checked out a Web site with material on her course. "For example, there was a review of an educational film that had obviously been shown in another course—or perhaps it was shown by a visiting professor last year."[41] A newspaper reporter investigating lecture-note Web sites took 118 pages of notes during her seven class visits, whereas the notes posted on the Web of those same events "were, at most, a few paragraphs per lecture."[42]

Bottom line: These Web sites may be helpful, but they are certainly no substitute for going to class.

Online Student Evaluations

Student evaluations of courses and professors have moved to the Internet. These can be useful, but since such evaluations are often expressed anonymously, they can also be inaccurate and unfair—even vicious, if students receiving poor grades take revenge by vilifying their instructors online. According to a spokesperson for the American Federation of Teachers, there has been a rise in false-accusation cases, which can have a severe impact on instructors' careers.[43]

Justine Heinze Giardello, 18, whose father is a professor, says students don't understand how hard instructors work. And she feels that students should not be able to bash teachers in a public forum. "I think it is horrible," she said, commenting on postings to a student-run teacher-evaluation Web site in which an openly gay community college instructor was anonymously called "homomanic," "racist," and "mentally ill." "How would we feel," she asks, "if there was a student review judging us on a personal basis?"[44]

Is there something about the instantaneity and anonymity of the online medium that invites such brutal expression? A composition instructor once used her class to examine the nature of e-mail. "After printing some of the class's mail with the names taken off," says an account of her experiment, "she asked her students to speculate about the character of the writer of each sample. Mostly they were horrified by how un-civil they came off in print; they claimed that their tone was merely a byproduct of the haste with which the writing was produced."[45]

In reading teacher evaluations, it's useful to pay close attention to how civil and fair-minded the reports seem. And, of course, you should try to be as considered and considerate as possible when writing them.

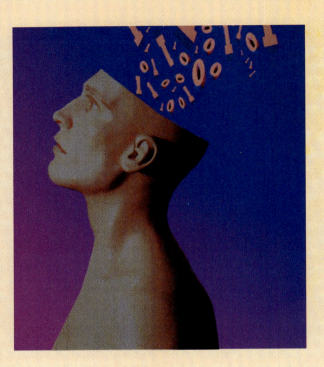

Visual Summary

antivirus software (p. 150, KQ 4.4) Utility program that scans hard disks, floppy disks, and memory to detect viruses. Why it's important: *The software helps avoid viruses, hidden programming instructions that can destroy programs and data.*

ASP (application service provider) (p. 156, KQ 4.6) Firm that leases software over the Internet. Why it's important: *You no longer have to buy software in a store, in shrink-wrapped packages. Instead, you can simply download a particular program when you need it, for as long as you need it.*

backup utility (p. 150, KQ 4.4) Utility program used to make a backup, or duplicate, copy of the information on your hard disk. Why it's important: *The program allows you to retrieve data in case your hard drive fails, and you have no more programs or files.*

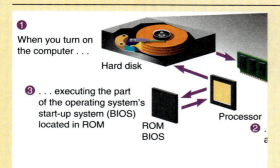

① When you turn on the computer . . .

Hard disk

③ . . . executing the part of the operating system's start-up system (BIOS) located in ROM

ROM BIOS

Processor

②

booting (p. 138, KQ 4.2) Loading an operating system into a computer's main memory. Why it's important: *Without booting, computers could not operate. The programs responsible for booting are stored permanently in the computer's electronic circuitry. When you turn on the machine, programs called diagnostic routines test the main memory, the central processing unit, and other parts of the system to make sure they are running properly. Next, BIOS (for basic input/output system) programs are copied to main memory and help the computer interpret keyboard characters or transmit characters to the display screen or to a diskette. Then the boot program obtains the operating system, usually from hard disk, and loads it into the computer's main memory, where it remains until you turn the computer off.*

BSD (p. 147, KQ 4.3) Free system software derived from Unix. Why it's important: *Three variations—FreeBSD Project, NetBSD, and OpenBSD—are distributed online and on CD-ROM for free and are widely used.*

data compression utility (p. 152, KQ 4.4) Utility that removes redundant elements, gaps, and unnecessary data from a computer's storage space so that less space (fewer bits) is required to store or transmit data. Why it's important: *With the increasing use of large graphic, sound, and video files, data compression is necessary to reduce hard-drive storage space required and to reduce time required to transmit such large files over a network.*

data-recovery utility (p. 150, KQ 4.4) Utility used to restore data that has been physically damaged or corrupted. Why it's important: *Data can be damaged by viruses, bad software, hardware failure, and power fluctuations that occur while data is being written/recorded. This utility may be able to restore it.*

defragmenter (p. 152, KQ 4.4) Commonly called a "defragger." This utility program will find all the scattered files on your hard disk and reorganize them as contiguous files. Why it's important: *Defragmenting the file will speed up the hard drive's operation.*

device drivers (p. 149, KQ 4.4) Specialized software programs—usually components of system software—that allow input and output devices to communicate with the rest of the computer system. Why it's important: *Drivers are needed so that the computer's operating system can recognize and run peripheral hardware.*

distributed system (p. 155, KQ 4.5) A noncentralized network consisting of several computers and other devices that can communicate with one another. Why it's important: *If the Internet becomes decentralized, with information distributed among millions of computers, efforts by Microsoft and others to create Internet "operating systems" could become less practical.*

DOS (Disk Operating System) (p. 142, KQ 4.3) Original operating system produced by Microsoft, with a hard-to-use command-driven user interface. Its initial 1982 version was designed to run on the IBM PC as PC-DOS. Later Microsoft licensed the same system to other computer makers as MS-DOS. *Why it's important: DOS used to be the most common microcomputer operating system, and it is still used on many microcomputers. Today the most popular operating systems use GUIs.*

emulation software (p. 143, KQ 4.3) Program that allows software designed for one computer (as for a Windows PC) to run on another kind of computer (as on a Macintosh). *Why it's important: Especially with Macintosh computers, you are no longer restricted from running PC-compatible software. (But you can't run Macintosh programs on Windows personal computers.)*

ERP (enterprise resource planning) software (p. 158, KQ 4.6) Large client/server software applications that help companies organize and operate their businesses. *Why it's important: ERP software coordinates a company's entire business and moves data speedily from one department to another.*

e-speak (p. 154, KQ. 4.5) Hewlett-Packard's version of an Internet operating system, or "universal language," that allows different Web sites to communicate with one another. *Why it's important: As with Microsoft.Net, the purpose of e-speak is to make the Internet easier and more useful. Also similarly, e-speak uses the programming standard XML, which helps computers communicate and sort through the millions of offerings on the Web.*

formatting (p. 142, KQ 4.2) Also called *initializing*, formatting is the process of preparing a floppy disk so it can store data or programs. *Why it's important: Different computers take disks with different formats; thus, you can't run a "Formatted IBM" disk on a Macintosh or a "Formatted Macintosh" disk on an IBM-compatible PC.*

fragmentation (p. 152, KQ 4.4) Over time, as old files are deleted from a hard disk and new ones are added, portions of files about the hard disk are scattered, or fragmented, in nonadjacent areas. *Why it's important: Fragmentation greatly slows access to the files.*

Jini (p. 154, KQ 4.5) Technology from Sun Microsystems; consists of a small layer of software designed to let all types of electronic gadgets on a wired or wireless network communicate with each other. Builds on another Sun technology called Java, which allows separate devices on a network to run different applications. *Why it's important: As with Microsoft.Net and Hewlett-Packard's e-speak, the purpose of Jini is to make the Internet easier and more useful.*

Linux (p. 147, KQ 4.3) Free version of Unix, supported by efforts of thousands of volunteer programmers. *Why it's important: Linux is an inexpensive, open-source operating system useful for online applications and to PC users who have to maintain a Web server or a network server.*

Macintosh operating system (Mac OS) (p. 142, KQ 4.3) System software that runs only on Apple Macintosh computers. *Why it's important: Although Macs are not as common as PCs, many people believe they are easier to use. Macs are often used for graphics and desktop publishing.*

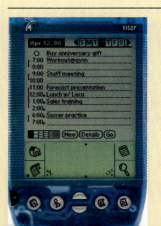

Microsoft Pocket PC (p. 149, KQ 4.3) Operating system for handhelds that is simpler and less cluttered than CE and looks and feels a lot less like desktop Windows. *Why it's important: Pocket PC offers pocket versions of Word and Excel that let users read standard word processing and spreadsheet files sent as e-mail attachments from their PCs.*

Microsoft Whistler (p. 144, KQ 4.3) Code name for Microsoft system software scheduled for release in late 2001. Whistler targets home users but is based on Windows NT. *Why it's important:* With this new version, Microsoft will finally be giving up the last of the Windows software carried forward from the aging DOS programming technology.

Microsoft Windows CE (p. 149, KQ 4.3) Greatly slimmed-down version of Windows 95 for handheld computing devices, such as those made by Casio, Compaq, and Hewlett-Packard. Windows CE had some of the familiar Windows look and feel and included rudimentary word processing, spreadsheet, e-mail, Web browsing, and other software. *Why it's important: Windows CE was Microsoft's first attempt to modify its Windows desktop operating system for use with handhelds. It has been succeeded by the Pocket PC system.*

Microsoft Windows Millennium Edition (WinMe) (p. 144, KQ 4.3) Successor to Windows 95 and 98, operating system designed to support desktop and portable computers. *Why it's important: Considered a boon to multimedia users because of its ability to handle still pictures, digital video, and audio files. Also claims to have reduced the problem of frequent "crashes."*

Microsoft Windows 95/98 (p. 144, KQ 4.3) Operating system for desktop and portable microcomputers, supporting the most hardware and the most application software. *Why it's important: Windows has become the most common system software used on microcomputers.*

Microsoft Windows NT/2000 (p. 146, KQ 4.3) Multitasking operating system designed to run on network servers. *Why it's important: It allows multiple users to share resources such as data and programs.*

Microsoft.Net (p. 153, KQ 4.5) Pronounced "dot-net," *.Net* is Microsoft's platform for an operating system for the entire Internet. Designed to link unrelated Web sites so that people can organize all the information in their lives, using PCs and smaller devices, such as cellphones, handheld computers, and set-top boxes. It will incorporate new technology being explored by Microsoft, including speech recognition, handwriting recognition, real-time video, and intelligent browsers. *Why it's important: As with Sun's Jini and Hewlett-Packard's e-speak, the purpose of .Net is to make the Internet easier and more useful.*

multiprocessing (p. 140, KQ 4.2) Feature of OS software allowing two or more computers or processors to be linked together to perform work simultaneously. *Why it's important: The processing is so fast that, by spending a little bit of time working on each program in turn, the computer can run several programs at the same time. An OS capable of multiprocessing is much more sophisticated than that required for multitasking.*

multiprogramming (p. 140, KQ 4.2) Feature of OS software in which two or more programs are executed concurrently on a multiuser operating system. *Why it's important: As with multitasking, the processor spends a certain amount of time executing each user's program. Because the processor works so quickly, it seems as though all the programs are being run at the same time.*

multitasking (p. 140, KQ 4.2) Feature of OS that allows the execution of two or more programs by one user concurrently on the same computer with one CPU. For instance, you might write a report on your computer with one program while another plays a music CD. *Why it's important: Multitasking allows the computer to switch rapidly back and forth among different tasks. The user is generally unaware of the switching process.*

NetWare (p. 146, KQ 4.3) Long-popular network operating system for coordinating microcomputer-based local area networks (LANs) throughout an organization. *Why it's important: LANs allow PCs to share programs, data files, and printers and other devices.*

network computer (p. 157, KQ 4.6) Stripped-down computer that connects people to networks and runs applications tied to servers. *Why it's important: It's less expensive than a personal computer, and, since the software is stored on a server, the user theoretically has fewer software headaches.*

open-source software (p. 147, KQ 4.3) Software that any programmer can download from the Internet for free and modify with suggested improvements. The only qualification is that changes can't be copyrighted; they must be made available to all and remain in the public domain. Why it's important: *Because this software is not proprietary, any programmer can make improvements, which can result in better-quality software.*

operating system (OS) (p. 137, KQ 4.2) Master system of programs that manage the basic operations of the computer. Why it's important: *These programs provide resource management services of many kinds. In particular, they handle the control and use of hardware resources, including disk space, memory, CPU time allocation, and peripheral devices. The operating system allows users to concentrate on their own tasks or applications rather than on the complexities of managing the computer.*

Palm OS (p. 148, KQ 4.3) Operating system for the Palm and Visor; the dominant operating system for handhelds. Why it's important: *Because it is not a Windows derivative but was specifically designed for handhelds, Palm OS is a smoother-running operating system.*

platform (p. 142, KQ 4.3) Particular processor model and operating system on which a computer system is based. Why it's important: *Generally, software written for one platform will not run on any other. Users should be aware that there are Mac platforms (Apple Macintosh) and Windows platforms, or "PC platforms" (for personal computers such as Dell, Compaq, Gateway, Hewlett-Packard, or IBM that run Microsoft Windows). Sometimes the latter are called "Wintel platforms," for "Windows + Intel," because they often combine the Windows operating system with the Intel processor chip.*

plug and play (p. 144, KQ 4.3) The ability of a computer to automatically configure a new hardware component that is added to it. When capitalized or when abbreviated PnP, Plug and Play refers to the specifications developed by Microsoft to allow users of Windows PCs to "plug" in peripherals such as monitors, modems, and printers and have them "play" or operate automatically. Why it's important: *Once a new component is plugged in, the operating system automatically recognizes it; the user does not have to manually reconfigure the system.*

Solaris (p. 147, KQ 4.3) A version of Unix made by Sun Microsystems that seems to be most popular for handling large e-commerce servers and large Web sites. Why it's important: *Popular because it is considered super-reliable. Also, Solaris 8 can handle servers with as many as 64 microprocessors, compared with 32 for Windows 2000.*

supervisor (p. 138, KQ 4.2) Also called *kernel;* the central component of the operating system that manages the CPU. Why it's important: *The supervisor remains in main memory while the computer is running. As well as managing the CPU, it directs other nonresident programs to perform tasks that support application programs.*

time-sharing (p. 140, KQ 4.2) Feature of OS software allowing a single computer to process the tasks of several users at different stations in round-robin fashion. Why it's important: *The computer first works on one user's task for a fraction of a second, then goes on to the next user's task, and so on. This is accomplished through time slicing; the computer rapidly switches back and forth among different tasks.*

Unix (p. 147, KQ 4.3) Multitasking operating system for multiple users that has built-in networking capability and versions that can run on all kinds of computers. Why it's important: *Government agencies, universities, research institutions, large corporations, and banks all use Unix for everything from designing airplane parts to currency trading. Unix is also used for Web-site management. The developers of the Internet built their communications system around Unix because it has the ability to keep large systems (with hundreds of processors) churning out transactions day in and day out for years without fail.*

utility programs (p. 150, KQ 4.4) Also known as *service programs;* system software component that performs tasks related to the control and allocation of computer resources. Why it's important: *Utility programs enhance existing functions or provide services not supplied by other system software programs. Most computers come with built-in utilities as part of the system software.*

off on

How many representations of 0s and 1s can be held in a computer or a storage device such as a hard disk? Capacity is denoted by *bits* and *bytes* and multiples thereof:

- **Bit:** In the binary system, **each 0 or 1 is called a _bit_, which is short for "binary digit."**

- **Byte:** To represent letters, numbers, or special characters (such as ! or *), bits are combined into groups. **A group of 8 bits is called a _byte_, and a byte represents one character, digit, or other value.** (As we mentioned, in one scheme, 01000111 represents the letter *G.*) The capacity of a computer's memory or of a floppy disk is expressed in numbers of bytes or multiples such as kilobytes and megabytes.

- **Kilobyte:** A _kilobyte_ **(_K, KB_) is about 1000 bytes.** (Actually, it's precisely 1024 bytes, but the figure is commonly rounded.) The kilobyte was a common unit of measure for memory or secondary-storage capacity on older computers.

- **Megabyte:** A _megabyte_ **(_M, MB_) is about 1 million bytes** (1,048,576 bytes). Measures of microcomputer primary-storage capacity today are expressed in megabytes.

- **Gigabyte:** A _gigabyte_ **(_G, GB_) is about 1 billion bytes** (1,073,741,824 bytes). This measure was formerly used mainly with "big iron" (mainframe) computers, but is typical of the secondary storage (hard disk) capacity of today's microcomputers.

- **Terabyte:** A _terabyte_ **(_T, TB_) represents about 1 trillion bytes** (1,009,511,627,776 bytes).

- **Petabyte:** A _petabyte_ **(_P, PB_) represents about 1 quadrillion bytes** (1,048,576 gigabytes).

Letters, numbers, and special characters are represented within a computer system by means of *binary coding schemes.* (See ● *Panel 5.5, next page.*) That is, the off/on 0s and 1s are arranged in such a way that they can be made to represent characters, digits, or other values.

- **ASCII:** Pronounced "*Ask-ee,*" **_ASCII (American Standard Code for Information Interchange)_ is the binary code most widely used with microcomputers.** Besides more conventional characters, ASCII includes such characters as math symbols and Greek letters.

Character	EBCDIC	ASCII-8	Character	EBCDIC	ASCII-8
A	1100 0001	0100 0001	N	1101 0101	0100 1110
B	1100 0010	0100 0010	O	1101 0110	0100 1111
C	1100 0011	0100 0011	P	1101 0111	0101 0000
D	1100 0100	0100 0100	Q	1101 1000	0101 0001
E	1100 0101	0100 0101	R	1101 1001	0101 0010
F	1100 0110	0100 0110	S	1110 0010	0101 0011
G	1100 0111	0100 0111	T	1110 0011	0101 0100
H	1100 1000	0100 1000	U	1110 0100	0101 0101
I	1100 1001	0100 1001	V	1110 0101	0101 0110
J	1101 0001	0100 1010	W	1110 0110	0101 0111
K	1101 0010	0100 1011	X	1110 0111	0101 1000
L	1101 0011	0100 1100	Y	1110 1000	0101 1001
M	1101 0100	0100 1101	Z	1110 1001	0101 1010
0	1111 0000	0011 0000	5	1111 0101	0011 0101
1	1111 0001	0011 0001	6	1111 0110	0011 0110
2	1111 0010	0011 0010	7	1111 0111	0011 0111
3	1111 0011	0011 0011	8	1111 1000	0011 1000
4	1111 0100	0011 0100	9	1111 1001	0011 1001
!	0101 1010	0010 0001	;	0101 1110	0011 1011

- **EBCDIC:** Pronounced "*Eb*-see-dick," **_EBCDIC (Extended Binary Coded Decimal Interchange Code)_ is a binary code used with large computers,** such as mainframes.
- **Unicode:** Unlike ASCII, **_Unicode_ uses two bytes (16 bits) for each character,** rather than one byte (8 bits). Instead of the 256 character combinations of ASCII, Unicode can handle 65,536 character combinations. Thus, it allows almost all the written languages of the world to be represented using a single character set.

The Parity Bit

Dust, electrical disturbance, weather conditions, and other factors can cause interference in a circuit or communications line that is transmitting a byte. How does the computer know if an error has occurred? Detection is accomplished by use of a parity bit. **A _parity bit_, also called a check bit, is an extra bit attached to the end of a byte for purposes of checking for accuracy.**

Parity schemes may be *even parity* or *odd parity*. In an even-parity scheme, for example, the ASCII letter *H* (01001000) contains two 1s. Thus, the ninth bit, the parity bit, would be 0 in order to make the sum of the bits come out even. With the letter *O* (01001111), which has five 1s, the ninth bit would be 1 to make the byte come out even. (See ● *Panel 5.6.*) The systems software in the computer automatically and continually checks the parity scheme for accuracy. (If the message "Parity Error" appears on your screen, you need a technician to look at the computer to see what is causing the problem.)

Machine Language

Why won't word processing software that runs on an Apple Macintosh run (without special arrangements) on an IBM microcomputer? It's because each computer has its own machine language. **_Machine language_ is a binary-type programming language that the computer can run directly.** To most people, an instruction written in machine language, consisting only of 0s and 1s, is incomprehensible. To the computer, however, the 0s and 1s represent precise storage locations and operations.

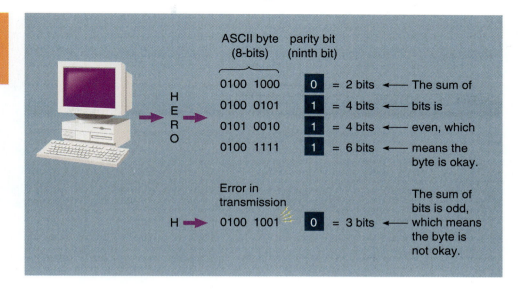

PANEL 5.6
Parity bit
This example uses an even-parity scheme.

How do people-comprehensible program instructions become computer-comprehensible machine language? Special systems programs called *language translators* rapidly convert the instructions into machine language. This translating occurs virtually instantaneously, so that you are not aware it is happening. Machine language is discussed in more detail in the Appendix.

Because the type of computer you will most likely be working with is the microcomputer, we'll now take a look at what's inside the microcomputer's system unit.

CONCEPT CHECK

What is the binary system?

Define bits, bytes, kilobytes, megabytes, gigabytes, terabytes, and petabytes.

Distinguish among ASCII, EBCDIC, and Unicode.

What is a parity bit?

What is machine language?

A line from the PC ad on page 176

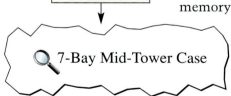

7-Bay Mid-Tower Case

The Computer Case: Bays, Buttons, & Boards

The *system unit* houses the motherboard (including the processor chip and memory chips), the power supply, and storage devices. *(See ● Panel 5.7, next page.)* In computer ads, the part of the system unit that is the empty box with just the power supply is called the *case* or *system cabinet*.

For today's desktop PC, the system unit may be advertised as something like a "4-bay micro-tower case" or a "7-bay mid-tower case." **A _bay_ is a shelf or opening used for the installation of electronic equipment,** generally storage devices such as a hard drive or DVD drive. A computer may come equipped with four or seven bays. (Empty bays are covered by a panel.) A *tower* is a cabinet that is tall, narrow, and deep (so that it can sit on the floor beside or under a table) rather than short, wide, and deep. Originally a tower was considered to be 24 inches high. Micro- and mid-towers may be less than half that size.

The number of buttons on the outside of the case will vary, but the on/off power switch will appear somewhere, either front or back. There may also be a "sleep" switch; this allows you to suspend operations without terminating them, so that you can conserve electrical power without the need for subsequent "rebooting," or restarting, the computer.

Hardware: The CPU & Storage

179

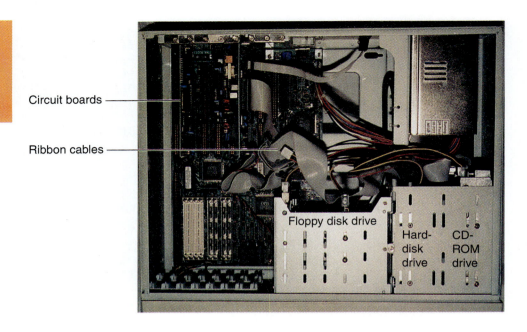

Circuit boards

Ribbon cables

Floppy disk drive

Hard-disk drive

CD-ROM drive

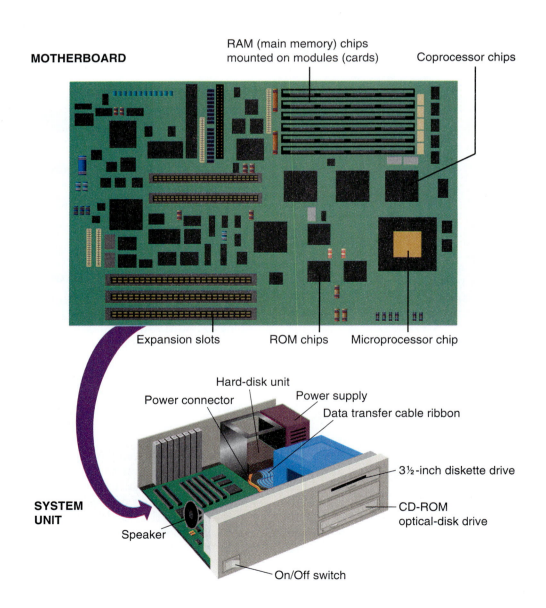

MOTHERBOARD

RAM (main memory) chips mounted on modules (cards)

Coprocessor chips

Expansion slots

ROM chips

Microprocessor chip

Hard-disk unit

Power connector

Power supply

Data transfer cable ribbon

3½-inch diskette drive

CD-ROM optical-disk drive

SYSTEM UNIT

Speaker

On/Off switch

Inside the case—not visible unless you remove the cabinet—are various electrical circuit boards, chief of which is the motherboard, as we'll discuss.

Power Supply

The electricity available from a standard wall outlet is alternating current (AC), but a microcomputer runs on direct current (DC). **The _power supply_ is a device that converts AC to DC to run the computer.** The on/off switch in your computer turns on or shuts off the electricity to the power supply. Because electricity can generate a lot of heat, a fan inside the computer keeps the power supply and other components from becoming too hot.

Electrical power drawn from a standard AC outlet can be quite uneven. For example, a sudden surge, or "spike," in AC voltage can burn out the low-voltage DC circuitry in your computer ("fry the motherboard"). Instead of plugging your computer directly into the wall electrical outlet, it's a good idea to plug it into a power protection device. The three principal types are surge protectors, voltage regulators, and UPS units.

- **Surge protector:** **A _surge protector_, or surge suppressor, is a device that protects a computer from being damaged by surges (spikes) of high voltage.** The computer is plugged into the surge protector, which in turn is plugged into a standard electrical outlet.
- **Voltage regulator:** **A _voltage regulator_, or line conditioner, is a device that protects a computer from being damaged by insufficient power—"brownouts" or "sags" in voltage.** Brownouts can occur when a large machine such as a power tool starts up and causes the lights in your house to dim. They also may occur on very hot summer days when the power company has to lower the voltage in an area because too many people are running their air conditioners all at once.
- **UPS:** **A _UPS (uninterruptible power supply)_ is a battery-operated device that provides a computer with electricity if there is a power failure.** The UPS will keep a computer going 5–30 minutes or more. It goes into operation as soon as the power to your computer fails.

The Motherboard & Microprocessor Chip

The _motherboard_, or _system board_, is the main circuit board in the system unit. The motherboard consists of a flat board that fills one side of the case. It contains both soldered, nonremovable components and sockets or slots for components that can be removed—microprocessor chip, RAM chips, and various expansion cards, as we explain later. (See ● Panel 5.8.) Making some components removable allows you to expand or upgrade your system. **_Expansion_ is a way of increasing a computer's capabilities by adding hardware to perform tasks that are beyond the scope of the basic system.** For example, you might want to add video and sound cards. **_Upgrading_ means changing to newer, usually more powerful or sophisticated versions,** such as a more powerful microprocessor or more memory chips.

The most fundamental part of the motherboard is the microprocessor chip. As mentioned, a _microprocessor_ is the miniaturized circuitry of a computer processor. It stores program instructions that process, or manipulate, data into information. The key parts of the microprocessor are transistors. _Transistors_, as we stated, are tiny electronic devices that act as on/off switches, which process the on/off 1/0 bits used to represent data. According to _Moore's law_, named for legendary Intel cofounder Gordon Moore, the number of transistors that can be packed onto a chip doubles every 18 months, while the price stays the same. In 1961 a chip had only 4 transistors; in 1971 it had 2300; in 1979 it had 30,000; and in 1997 it had 7.5 million. Moore's law actually accelerated with the debut of the 1-billion-transistor (1-gigahertz) chip in 2000.[11] (And in that year IBM announced a manufacturing breakthrough

RAM (main memory) chips mounted on modules (cards)

Microprocessor chip (with CPU)

that would result in even faster chips, permitting laptop batteries, for instance, to last twice as long.[12])

Two principal "architectures" or designs for microprocessors are CISC and RISC. **_CISC (complex instruction set computing) chips_, which are used mostly in PCs and in conventional mainframes, can support a large number of instructions,** but at relatively low processing speeds. **In _RISC (reduced instruction set computing) chips_, which are used mostly in workstations, a great many seldom-used instructions are eliminated.** As a result, workstations can work up to 10 times faster then most PCs. RISC chips have been used in many Macintosh computers since 1993.

Most personal computers today use microprocessors of two kinds—those based on the model made by Intel and those based on the model made by Motorola.

- **Intel-type chips—for PCs:** About 90% of microcomputers use Intel-type microprocessors. Indeed, the Microsoft Windows operating system is designed to run on Intel chips. As a result, people in the computer industry tend to refer to the Windows/Intel joint powerhouse as *Wintel.*

  Intel-type chips for PCs are made principally by Intel Corp.—but also by Advanced Micro Devices (AMD), Cyrix, DEC, and others. They are used by manufacturers such as Compaq, Dell, Gateway, Hewlett-Packard, and IBM. Since 1993, Intel has marketed its chips under the names *Pentium, Pentium Pro, Pentium MMX, Pentium II, Pentium III, Celeron,* and *Pentium IV.* Many ads for PCs contain the logo "Intel inside" to show that the systems run an Intel microprocessor.

- **Motorola-type chips—for Macintoshes:** **_Motorola-type chips_ are made by Motorola for Apple Macintosh computers.** Since 1993, Motorola has joined forces with IBM and Apple to produce the PowerPC family of chips. With certain hardware or software add-ons, a PowerPC can run PC as well as Mac applications software.

Processing Speeds: From Megahertz to Picoseconds

Often a PC ad will carry a line that says something like "Intel Celeron processor 500 MHz," "Intel Pentium III processor 866 MHz," or "AMD Athlon processor 1-GHZ." MHz stands for *megahertz* and GHz for *gigahertz*. These figures indicate how fast the microprocessor can process data and execute program instructions.

Every microprocessor contains a <u>**system clock**</u>, **which controls how fast all the operations within a computer take place.** The system clock uses fixed vibrations from a quartz crystal to deliver a steady stream of digital pulses or "ticks" to the CPU. These ticks are called *cycles.* Faster clock speeds will result in faster processing of data and execution of program instructions, as long as the computer's internal circuits can handle the increased speed.

There are four main ways in which processing speeds are measured:

- **For microcomputers—megahertz and gigahertz:** Microcomputer microprocessor speeds are usually expressed in <u>**megahertz (MHz)**</u>, **a measure of frequency equivalent to 1 million cycles (ticks of the system clock) per second.** The original IBM PC had a clock speed of 4.77 MHz. Today a 550-MHz Pentium III–based microcomputer processes 550

PRACTICAL ACTION BOX
Preventing Problems from Too Much or Too Little Electrical Power to Your Computer

"When the power disappears, so can your data," writes *San Jose Mercury News* computer columnist Phillip Robinson. "I say this with authority, sitting here in the dark in the wake of severe storms that have hit my part of California."[a] (Deprived of use of his computer, Robinson dictated his column by phone.)

Too little electricity can devastate your data. Too much electricity can devastate your computer hardware.

Here are a few things you can do to keep both safe:

- **Back up data regularly:** You should faithfully make backup (duplicate) copies of your data every few minutes as you're working. Then, if your computer has power problems you'll be able to get back in business fairly quickly once the machine is running again.

- **Use a surge protector to protect against too much electricity:** Plug all your hardware into a surge protector (suppressor), which will prevent damage to your equipment if there is a power surge. (You'll know you've experienced a power surge when the lights in the room suddenly get very bright.)

- **Use a voltage regulator to protect against too little electricity:** Plug your computer into a voltage regulator (also called a line conditioner) to adjust for power sags or brownouts. If power is too low for too long, it's as though the computer were turned off.

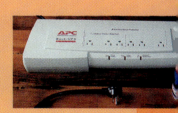

- **Consider using a UPS to protect against complete absence of electricity:** Consider plugging your computer into a UPS, or uninterruptible power supply. (A low-cost one, available at electronics stores, sells for about $150.) The UPS is kind of a short-term battery that, when the power fails, will keep your computer running long enough for you to save your data before you turn off the machine.

- **Turn ON highest-power-consuming hardware first:** When you turn on your computer system, you should turn on the devices that use the most power first. This will avoid causing a power drain on smaller devices. The most common advice is to turn on (1) external peripherals, (2) system unit, (3) monitor, (4) printer—in that order.

- **Turn OFF lowest-power-consuming hardware first:** When you turn off your system, follow the reverse order—first printer, then monitor, and so on. This avoids a power surge to the smaller devices.

- **Unplug your computer system during lightning storms:** Unplug all your system's components—including phone lines—during thunder and lightning storms. If lightning strikes your house or the power lines, it can ruin your equipment.

Hardware: The CPU & Storage

183

million cycles per second. The latest generation of processors (from AMD and Intel) operate in **gigahertz (GHz)—a billion cycles per second.** Intel's latest chip, the Pentium 4, operates at 1.4 gigahertz (it has 42 million transistors, up from 28 million in the Pentium III).[13] Since a new high-speed processor can cost many hundred dollars more than the previous generation of chip, experts often recommend that buyers fret less about the speed of the processor (since the work most people do on their PCs doesn't even tax the limits of the current hardware) and more about spending money on extra memory.[14]

- **For workstations, minicomputers, and mainframes—MIPS:** Processing speed can also be measured according to the number of instructions per second that a computer can process. **MIPS stands for millions of instructions per second.** A high-end microcomputer or workstation might perform at 100 MIPS or more, a mainframe at 200–1200 MIPS.

- **For supercomputers—flops:** The abbreviation **flops stands for floating-point operations per second.** A floating-point operation is a special kind of mathematical calculation. This measure, used mainly with supercomputers, is expressed as *megaflops* (mflops, or millions of floating-point operations per second), *gigaflops* (gflops, or billions), and *teraflops* (tflops, or trillions). The U.S. supercomputer known as Option Red cranks out 1.34 teraflops. (To put this in perspective, a person able to complete one arithmetic calculation every second would take about 31,000 years to do what Option Red does in a single second.) When IBM's new supercomputer, dubbed "Blue Gene," is completed, it will be 500 times faster than the world's swiftest computer, able to handle 1 *petaflop* (1 quadrillion operations per second).[15]

- **For all computers—fractions of a second:** Another way to measure cycle times is in fractions of a second. A microcomputer operates in microseconds, a supercomputer in nanoseconds or picoseconds—thousands or millions of times faster. A *millisecond* is one-thousandth of a second. A *microsecond* is one-millionth of a second. A *nanosecond* is one-billionth of a second. A *picosecond* is one-trillionth of a second.

CONCEPT CHECK

Distinguish expansion from upgrading.

What are three types of power protection devices?

Discuss the most fundamental part of the motherboard, its features, its two principal architectures, and the two principal kinds used in personal computers.

What is the system clock, and what are the various measures of processing speed?

How the Processor or CPU Works: Control Unit, ALU, & Registers

Once upon a time, the processor in a computer was measured in feet. A processing unit in the 1946 ENIAC (which had 20 such processors) was about 2 feet wide and 8 feet high. Today, computers are based on *micro*processors, less than 1 centimeter square. It may be difficult to visualize components so tiny. Yet it is necessary to understand how microprocessors work if you are to grasp what PC advertisers mean when they throw out terms such as "64 MB 133 MHz SDRAM" or "256K Advanced Transfer Cache."

Computer professionals often discuss a computer's word size. **_Word size_ is the number of bits that the processor may process at any one time.** The more bits in a word, the faster the computer. A 32-bit computer—that is, one with a 32-bit-word processor—will transfer data within each microprocessor chip in 32-bit chunks or 4 bytes at a time. (Recall there are 8 bits in a byte.) A 64-bit-word computer is faster; it transfers data in 64-bit chunks or 8 bytes at a time.

A processor is also called the *CPU,* and it works hand in hand with other circuits known as *main memory* to carry out processing. **The _CPU (central processing unit)_ is the "brain" of the computer; it follows the instructions of the software (program) to manipulate data into information. The CPU consists of two parts—(1) the control unit and (2) the arithmetic/logic unit (ALU), both of which contain registers, or high-speed storage areas** (as we discuss shortly). All are linked by a kind of electronic "roadway" called a *bus.* (See ● *Panel 5.9.*)

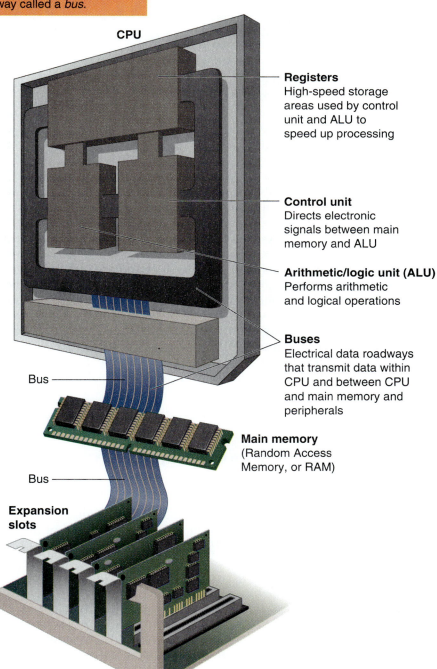

CPU

Registers
High-speed storage areas used by control unit and ALU to speed up processing

Control unit
Directs electronic signals between main memory and ALU

Arithmetic/logic unit (ALU)
Performs arithmetic and logical operations

Buses
Electrical data roadways that transmit data within CPU and between CPU and main memory and peripherals

Bus

Bus

Expansion slots

Main memory
(Random Access Memory, or RAM)

- **The control unit—for directing electronic signals:** The _control unit_ deciphers each instruction stored in it and then carries out the instruction. It directs the movement of electronic signals between main memory and the arithmetic/logic unit. It also directs these electronic signals between main memory and the input and output devices.

  For every instruction, the control unit carries out four basic operations, known as the machine cycle. In the _machine cycle,_ the CPU (1) fetches an instruction, (2) decodes the instruction, (3) executes the instruction, and (4) stores the result. *(See ● Panel 5.10.)*

- **The arithmetic/logic unit—for arithmetic and logical operations:** The _arithmetic/logic unit (ALU)_ performs arithmetic operations and logical operations and controls the speed of those operations.

  As you might guess, *arithmetic operations* are the fundamental math operations: addition, subtraction, multiplication, and division.

  Logical operations are comparisons. That is, the ALU compares two pieces of data to see whether one is equal to (=), greater

Hardware: The CPU & Storage

185

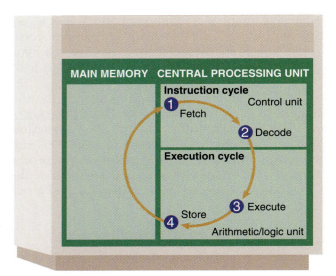

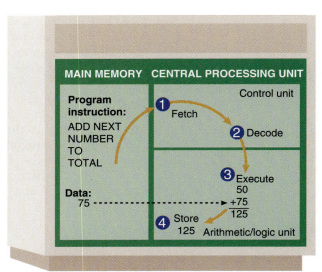

than (>), or less than (<) the other. (The comparisons can also be combined, as in "greater than or equal to" and "less than or equal to.")

- **Registers—special high-speed storage areas:** The control unit and the ALU also use registers, special areas that enhance the computer's performance. **<u>Registers</u> are high-speed storage areas that temporarily store data during processing.** They may store a program instruction while it is being decoded, store data while it is being processed by the ALU, or store the results of a calculation. (There are several types of registers, including *instruction register, address register, storage register,* and *accumulator register.*)

- **Buses—data roadways: <u>Buses</u>, or bus lines, are electrical data roadways through which bits are transmitted within the CPU and between the CPU and other components of the motherboard.** A bus resembles a multilane highway: The more lanes it has, the faster the bits can be transferred. The old-fashioned 8-bit-word bus of early microprocessors had only eight pathways. Data is transmitted four times faster in a computer with a 32-bit bus, which has 32 pathways, than in a computer with an 8-bit bus. Intel's Pentium chip is a 64-bit processor. Macintosh G4 microcomputers contain buses that are 128 bits, as do some supercomputers. Today there are several principal expansion bus standards, or "architectures," for microcomputers.

  We return to a discussion of buses in a few pages.

How Memory Works: RAM, ROM, CMOS, & Flash

So far we have described only the kinds of chips known as microprocessors. But other silicon chips called *memory chips* are attached to the motherboard. The four principal types of memory chips are *RAM, ROM, CMOS,* and *flash.*

- **RAM chips—to temporarily store program instructions and data:** Recall from Chapter 1 that there are two types of storage, primary and secondary. Primary storage is temporary or working storage and is often called *memory* or *main memory;* secondary storage is relatively permanent storage (for example, on floppy disk). **<u>RAM (random access memory) chips</u> are for primary storage; they temporarily hold (1) software instructions and (2) data before and after it is processed by the CPU.** Because its contents are temporary, RAM is said to be **<u>volatile</u>**—

the contents are lost when the power goes off or is turned off. This is why you should *frequently*—every 5–10 minutes, say—transfer (save) your work to a secondary-storage medium such as your hard disk, in case the electricity goes off while you're working. (However, there is one kind of RAM, called flash RAM, that is not temporary, as we'll discuss shortly.)

Three types of RAM chips are used in personal computers: *DRAM, SDRAM,* and *SRAM.* Pronounced *"dee-ram," DRAM (dynamic RAM)* must be constantly refreshed by the CPU or it will lose its contents. The type of dynamic RAM used in most PCs today is *SDRAM (synchronous dynamic RAM),* which is synchronized by the system clock and is much faster than DRAM. Often in computer ads, the speed of SDRAM is expressed in megahertz. The third type, pronounced *"ess-ram," SRAM (static RAM)* is faster than any DRAM and will retain its contents without having to be refreshed by the CPU.

SIMM
Single inline memory module

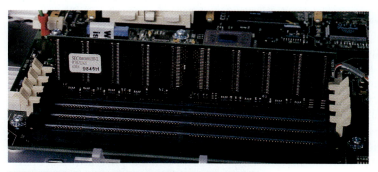

Microcomputers come with different amounts of RAM, which is usually measured in megabytes. An ad may list "64 MB SDRAM," but you can also get 128 megabytes of RAM. The more RAM you have, the faster the computer operates, and the better your software performs. *Having enough RAM is a critical matter.* Before you buy a software package, look at the outside of the box or check the manufacturer's Web site to see how much RAM is required. Microsoft Office 2000, for instance, states that a minimum of 16 megabytes of RAM is required.

If you're short on memory capacity, you can usually add more RAM chips by plugging them into the motherboard. Chips can be bought single or in so-called *memory modules,* circuit boards that can be plugged into expansion slots on the motherboard. There are two types of such modules: SIMMs and DIMMS, both of which use DRAM chips. A *SIMM (single inline memory module)* has RAM chips on only one side. A *DIMM (dual inline memory module)* has RAM chips on both sides.

DIMM
Dual inline memory module

● **ROM chips—to store fixed start-up instructions:** Unlike RAM, to which data is constantly being added and removed, **<u>ROM (read-only memory)</u> cannot be written on or erased by the computer user without special equipment. ROM chips contain fixed start-up instructions.** That is, ROM chips are loaded, at the factory, with programs containing special instructions for basic computer operations, such as those that start the computer or put characters on the screen. These chips are nonvolatile; their contents are not lost when power to the computer is turned off.

In computer terminology, **<u>read</u> means to transfer data from an input source into the computer's memory or CPU.** The opposite is **<u>write</u>—to transfer data from the computer's CPU or memory to an output device.** Thus, with a ROM chip, "read-only" means that the CPU can retrieve programs from the ROM chip but cannot modify or add to those programs. A variation is *PROM (programmable read-only memory),* which is a ROM chip that allows you, the user, to load read-only programs and data. However, this can be done only once.

● **CMOS chips—to store flexible start-up instructions:** Pronounced *"see-moss," <u>CMOS (complementary metal-oxide semiconductor) chips</u> are powered by a battery and thus don't lose their contents when the**

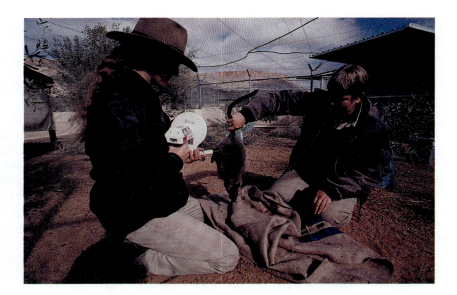

Intel inside?

Right: Nature conservancy workers in Alice Springs, Australia, check up on a wallaby by scanning the chip embedded under its skin. *Left:* Once implanted in a patient, this micro-processor and its subordi-nate valves can precisely administer intravenous drugs to a level of accuracy unparalleled by today's standards.

power is turned off. CMOS chips contain flexible start-up instructions, such as time, date, and calendar, that must be kept current even when the computer is turned off. Unlike ROM chips, CMOS chips can be reprogrammed, as when you need to change the time for daylight savings time.

- **Flash memory chips—to store flexible programs:** Also a nonvolatile form of memory, __flash memory chips__ **can be erased and reprogrammed more than once** (unlike PROM chips, which can be programmed only once). Flash memory, which doesn't require a battery and which can range from 1 to 64 megabytes in capacity, is used to store programs not only in personal computers but also in pagers, cellphones, MP3 players, Palm organizers, printers, and digital cameras. "Because flash chips retain information after the power is turned off," says a *Business Week* report, "they are going into new gadgets as quickly as they can be made."[16]

How Cache Works: Level 1 (Internal) & Level 2 (External)

Pronounced "cash," __cache__ **temporarily stores instructions and data that the processor is likely to use frequently. Thus, cache speeds up processing.** *(Refer back to* ● *Panel 5.9.)*

There are two kinds of cache—Level 1 and Level 2:

- **Level 1 (L1) cache—part of the microprocessor chip:** *Level 1 (L1) cache,* also called internal cache, is built into the processor chip. Ranging from 8 to 256 kilobytes, its capacity is less than that of Level 2 cache, although it operates faster.
- **Level 2 (L2) cache—not part of the microprocessor chip:** This is the kind of cache usually referred to in computer ads. *Level 2 (L2) cache,* also called external cache, resides outside the processor chip and consists of SRAM chips. Capacities range from 64 kilobytes to 2 megabytes. (In Intel ads, L2 is called Advanced Transfer Cache.)

In addition, most current computer operating systems allow for the use of __virtual memory__—**that is, some free hard-disk space is used to extend the capacity of RAM.** The processor searches for data or program instructions in the following order: first L1, then L2, then RAM, then hard disk (or CD-ROM). In this progression, each kind of memory or storage is slower than its predecessor.

Describe the following: word size, CPU, the control unit and the machine cycle, the ALU and arithmetic and logical operations, registers, and buses.

What is RAM and what are three variants?

What is ROM?

Discuss CMOS chips and flash memory chips.

How does L1 cache differ from L2 cache?

Ports & Cables

A _port_ is a connecting socket or jack on the outside of the system unit _(See_ ● _Panel 5.11.)_ **into which are plugged different kinds of cables.** A port allows

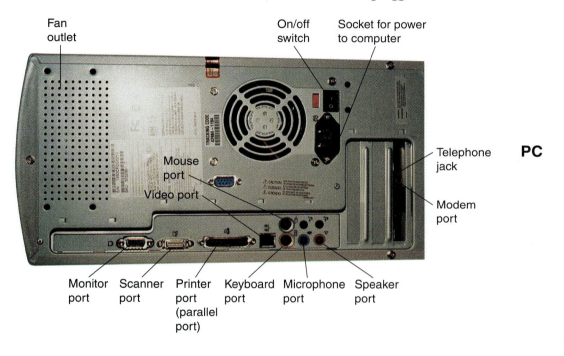

Fan outlet

On/off switch

Socket for power to computer

Telephone jack

PC

Mouse port

Video port

Modem port

Monitor port · Scanner port · Printer port (parallel port) · Keyboard port · Microphone port · Speaker port

Macintosh

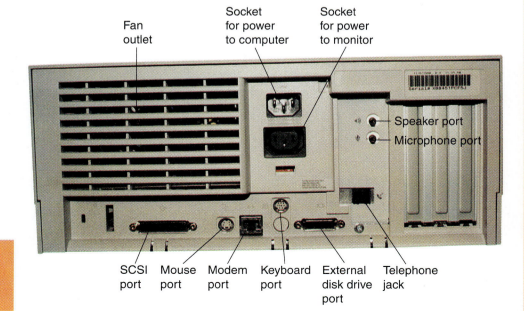

Fan outlet

Socket for power to computer

Socket for power to monitor

Speaker port

Microphone port

SCSI port · Mouse port · Modem port · Keyboard port · External disk drive port · Telephone jack

● **PANEL 5.11**

Ports

The backs of a PC and a Macintosh.

Hardware: The CPU & Storage

189

you to plug in a cable to connect a peripheral device, such as a monitor, printer, or modem, so that it can communicate with the computer system. Ports are of several types.

- **Serial ports—for transmitting slow data over long distances: A line connected to a _serial port_ will send bits one after another,** like cars on a one-lane highway. Because individual bits must follow each other, a serial port is usually used to connect devices that do not require fast transmission of data, such as keyboard, mouse, monitors, and modems. It is also useful for sending data over a long distance. The standard for PC serial ports is the 9-pin or 25-pin RS-232C connector.

- **Parallel ports—for transmitting fast data over short distances: A line connected to a _parallel port_ allows 8 bits (1 byte) to be transmitted simultaneously,** like cars on an eight-lane highway. Parallel lines move information faster than serial lines do, but they can transmit information efficiently only up to 15 feet. Thus, parallel ports are used principally for connecting printers or external disk or magnetic-tape backup storage devices.

- **SCSI ports—for transmitting fast data to up to seven devices in a daisy chain:** Pronounced "scuzzy," a **_SCSI (small computer system interface) port_ allows data to be transmitted in a "daisy chain" to up to 7 devices at speeds (32 bits at a time) higher than those possible with serial and parallel ports.** Among the devices that may be connected are external hard-disk drives, CD-ROM drives, scanners, and magnetic-tape backup units. The term _daisy chain_ means that several devices are connected in series to each other, so that data for the seventh device, for example, has to go through the other six devices first. Sometimes the equipment on the chain is inside the computer, an internal daisy chain; sometimes it is outside the computer, an external daisy chain. _(See ● Panel 5.12.)_

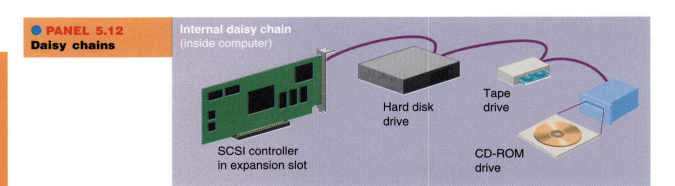

● **PANEL 5.12**
Daisy chains

Internal daisy chain (inside computer)

SCSI controller in expansion slot

Hard disk drive

Tape drive

CD-ROM drive

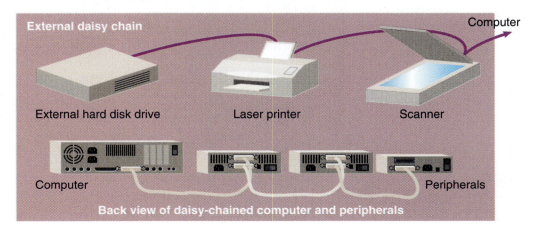

External daisy chain

Computer

External hard disk drive

Laser printer

Scanner

Computer

Peripherals

Back view of daisy-chained computer and peripherals

2 USB Ports

- **USB ports—for transmitting data to up to 127 devices in a daisy chain:** A _**USB (universal serial bus) port**_ **can theoretically connect up to 127 peripheral devices daisy-chained to one general-purpose port.** USB ports are useful for peripherals such as digital cameras, digital speakers, scanners, high-speed modems, and joysticks. The so-called USB *hot plug* or *hot swappable* allows such devices to be connected or disconnected even while the PC is running. In addition, USB permits _**Plug and Play**_**, which allows peripheral devices and expansion cards to be automatically configured while they are being installed.** This avoids the hassle of setting switching and creating special files that plagued earlier users. In 2000, computer makers were planning to introduce USB 2.0, expected to run 40 times faster than the earlier version.[17]

    Can you really connect up to 127 devices on a single chain? An Intel engineer did set a world record at an industry trade show before a live audience by connecting 111 peripheral devices to a single USB port on a PC. But respected technology writer Walter Mossberg says that "almost none of the USB peripherals I've seen support this [daisy-chaining] feature."[18] Thus, though most PCs contain only two USB ports, it's worth shopping around to find a model with extra USB connectors.

- **Dedicated ports—for keyboard, mouse, phone, and so on:** So far, we have been considering general-purpose ports, but the back of a computer also has other, *dedicated ports*—ports for special purposes. Among these are the round ports for connecting the keyboard and the mouse. There are also jacks for speakers and microphones and modem-to-telephone jacks. Finally, there is one connector that is not a port at all—the power plug socket, into which you insert the power cord that brings electricity from a wall plug.

- **Infrared ports—for cableless connections over a few feet:** When you use a handheld remote unit to change channels on a TV set, you're using invisible radio waves of the type known as infrared waves. **An** _**infrared port**_ **allows a computer to make a cableless connection with infrared-capable devices,** such as some printers. This type of connection requires an unobstructed line of sight between transmitting and receiving ports, and they can be only a few feet apart.

Expandability: Buses & Cards

Today many new microcomputer systems can be expanded. As mentioned earlier, *expansion* is a way of increasing a computer's capabilities by adding hardware to perform tasks that are not part of the basic system. *Upgrading* means changing to a newer, usually more powerful or sophisticated versions. (Computer ads often make no distinction between "expansion" and "upgrading." Their main interest is simply to sell you more hardware or software.)

Whether or not a computer can be expanded depends on its "architecture"—closed or open. *Closed architecture* means a computer has no expansion slots; *open architecture* means it does have expansion slots. (An alternative definition is that closed architecture is a computer design whose specifications are not made freely available by the manufacturer. Thus, other companies cannot create ancillary devices to work with it. With open architecture, the manufacturer shares specifications with outsiders.) _**Expansion slots**_ **are sockets on the motherboard into which you can plug expansion cards.** _**Expansion cards**_**—also known as expansion boards, adapter cards, interface cards, plug-in boards, controller cards, add-ins, or add-ons—are circuit boards that provide more memory or that control peripheral devices.** *(See*
● *Panel 5.13.)*

Expansion card Expansion slot

Type of Card (Board)	What It Does
Accelerator board	Speeds up processing; also known as turbo board or upgrade board
Cache card	Improves disk performance
Coprocessor board	Contains specialized processor chips that increase processing speed of computer system
Disk controller card	Allows certain type of disk drive to be connected to computer system
Emulator board	Permits microcomputer to be used as a terminal for a larger computer system
Fax modem board	Enables computer to transmit and receive fax messages and data over telephone lines
Graphics (Video) adapter board	Permits computer to have a particular graphics standard
Memory expansion board	Enables additional RAM to be added to computer system
Sound board	Enables certain types of systems to produce sound output

Common expansion cards connect to the monitor (graphics card), speakers and microphones (sound card), and network (network card), as we'll discuss. Most computers have four to eight expansion slots, some of which may already contain expansion cards included in your initial PC purchase.

Expansion cards are made to connect with different types of buses on the motherboard. (As we mentioned, *buses* are electrical data roadways through which bits are transmitted.) The bus that connects the CPU within itself and to main memory is the *system bus.* The bus that connects the CPU with expansion slots on the motherboard and thus with peripheral devices is the *expansion bus.* We already alluded to the universal serial bus (USB), whose purpose, in fact, is to *eliminate* the need for expansion slots and expansion cards, since you can just connect USB devices in a daisy chain outside the system unit. Three expansion buses to be aware of are *ISA, PCI,* and *AGP.*

- **ISA bus—for ordinary low-speed uses:** The _**ISA (industry standard architecture) bus**_ **is the most widely used expansion bus.** It is also the oldest and, at 8 or 16 bits, the slowest at transmitting data, though it is still used for mice, modem cards, and low-speed network cards.

- **PCI bus—for higher-speed uses:** The _**PCI (peripheral component interconnect) bus**_ **is a higher-speed bus,** and at 32 or 64 bits wide it is over four times faster than ISA buses. PCI is widely used to connect graphics cards, sound cards, modems, and high-speed network cards.

- **AGP bus—for even higher speeds and 3D graphics:** The _**AGP (accelerated graphics port) bus**_ **transmits data at even higher speeds and was designed to support video and three-dimensional (3D) graphics.** An AGP bus is twice as fast as a PCI bus.

Among the types of expansion cards are graphics, sound, modem, and network interface cards. A special kind of card is the PC card.

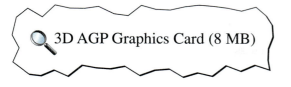

3D AGP Graphics Card (8 MB)

PCI Wavetable Sound Card

- **Graphics cards—for monitors:** Graphics cards are included in all PCs. **Also called a video card or video adapter, a _graphics card_ converts signals from the computer into video signals that can be displayed as images on a monitor.** A three-dimensional AGP card is an example of a graphics card. The power of an AGP graphics card is often expressed in megabytes, as in 8, 16, or 32 MB.

- **Sound cards—for speakers and audio output:** **A _sound card_ is used to transmit digital sounds through speakers, microphones, and headsets.** Sound cards come installed on most new PCs. Cards such as PCI wavetable sound cards are used to add music and sound effects to computer video games. _Wavetable synthesis_ is a method of creating music based on a wave table, which is a collection of digitized sound samples taken from recordings of actual instruments. The sound samples are then stored on a sound card and are edited and mixed together to produce music. Wavetable synthesis produces higher quality audio output than other sound techniques.

- **Modem cards—for remote communication via phone lines:** Very occasionally you may see a modem that is outside the computer. Most new PCs, however, come with internal modems—modems installed inside as circuit cards. The modem not only sends and receives digital data over telephone lines to and from other computers but can also transmit voice and fax signals.

- **Network interface cards—for remote communication via cable:** **A _network interface card_ allows the transmission of data over a cable network,** which connects various computers and other devices such as printers.

- **PC Cards—for laptop computers:** Originally called _PCMCIA cards_ (for the Personal Computer Memory Card International Association), **_PC cards_ are thin, credit-card size (2.1 by 3.4 inches) devices used principally on laptop computers to expand capabilities.** _(See_ ● _Panel 5.14.)_ Examples are extra memory (flash RAM—discussed later in the chapter), sound cards, modem, hard disks, and even pagers and cellular communicators. At present there are three sizes for PC cards—I (thin), II (thick), and III (thickest). Type I is used primarily for flash memory cards. Type II, the kind you'll find most often, is used for fax modems and network-interface cards. Type III is for rotating disk devices, such as hard-disk drives, and for wireless communication devices.

● **PANEL 5.14**
PC card
An example of a PC card used in a laptop

Distinguish among the following ports: serial, parallel, SCSI, USB, dedicated, and infrared.

Distinguish closed architecture from open architecture.

Define the following three expansion buses: ISA, PCI, AGP.

Why would you need the following cards: graphics, sound, modem, network interface, and PC?

5.3 Secondary Storage

KEY QUESTIONS

What are the features of floppy disks, hard disks, optical disks, magnetic tape, smart cards, and online secondary storage?

You're on a trip with your notebook, or maybe just a cell phone or a personal digital assistant, and you don't have a crucial file of data. Or maybe you need to look up a phone number that you can't get through the phone company's directory assistance. Fortunately, you backed up your data online, using any one of several services (Desktop.com, Driveway, FreeDesk.com, I-drive.com, MagicalDesk.com, MyInternetDesktop.com, MyWebOS.com, Visto.com, or X:Drive, for example), and are able to access it through your modem.[19]

Here is yet another example of how the World Wide Web is offering alternatives to traditional computer functions that once resided within stand-alone machines. We are not, however, fully into the all-online era just yet. Let us consider more traditional forms of **_secondary storage hardware_, devices that permanently hold data and information as well as programs.** We will look at the following types of secondary-storage devices:

- Floppy disks
- Hard disks
- Optical disks
- Magnetic tape
- Smart cards
- Flash memory cards

Finally, we return to online secondary storage, and look at how it actually works.

Floppy Disks

3.5" Floppy Drive

A **_floppy disk_, often called a diskette or simply a disk, is a removable flat piece of mylar plastic packaged in a 3.5-inch plastic case.** Data and programs are stored on the disk's coating by means of magnetized spots, following standard on/off patterns of data representation (such as ASCII). The plastic case protects the mylar disk from being touched by human hands. Originally, when most disks were larger (5.25 inches), the disks actually were "floppy," not rigid; now only the plastic disk inside is flexible or floppy.

Floppy disks are inserted into a floppy-disk drive, a device that holds, spins, reads data from, and writes data to a floppy disk. *Read* means that the data in secondary storage is converted to electronic signals and a copy of that data is transmitted to the computer's memory (RAM). *Write* means that a copy of the electronic information processed by the computer is transferred to secondary storage. Floppy disks have a **_write-protect notch_, which allows you to prevent a diskette from being written to.** In other words, it allows you to protect the data already on the disk. To write-protect, use your thumbnail or the tip of a pen to move the small sliding tab on the lower right side of the disk (viewed from the back), thereby uncovering the square hole. *(See* ● *Panel 5.15.)*

On the diskette, **data is recorded in concentric circles called _tracks_.** Unlike on a vinyl phonograph record, these tracks are neither visible grooves nor a

Front

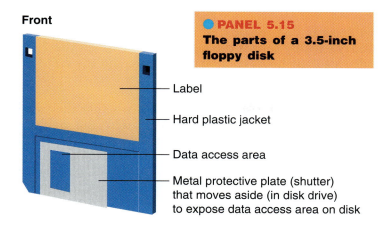

Label

Hard plastic jacket

Data access area

Metal protective plate (shutter) that moves aside (in disk drive) to expose data access area on disk

● PANEL 5.15

The parts of a 3.5-inch floppy disk

Back

Write-protect notch

Hub

Tracks and sectors

1 sector

track

Bits on 1 track

● PANEL 5.16

Floppy disk and drive

single spiral. Rather, they are closed concentric rings. On a formatted disk each track is divided into <u>**sectors**</u>, **invisible wedge-shaped sections used for storage reference purposes.** When you save data from your computer to a diskette, the data is distributed by tracks and sectors on the disk. That is, the system software uses the point at which a sector intersects a track to reference the data location.

When you insert a floppy disk into the slot (the *drive gate* or *drive door*) in the front of the disk drive, the disk is fixed in place over the spindle of the drive mechanism. The <u>***read/write head***</u> **is used to transfer data between the computer and the disk.** When the disk spins inside its case, the read/write head moves back and forth over the *data access area* on the disk. When the disk is not in the drive, a metal or plastic shutter covers this access area. An access light goes on when the disk is in use. After using the disk, you can retrieve it by pressing an eject button beside the drive. *(See ● Panel 5.16.)* Note: *Do not remove the disk when the access light is on.*

Let's compare the 3.5-inch floppy disk with some 3.5-inch <u>***floppy-disk cartridges***</u>**, or higher-capacity removable disks**—Zip disks, SuperDisks, and HiFD disks:

Floppy disk drive

Floppy disk

- **3.5-inch floppy disks—1.44 megabytes:** The present-day standard for traditional floppy disks is 1.44 megabytes, the equivalent of 400 typewritten pages. Today's floppy carries the label *2HD*, in which the *2* stands for "double-sided" (it holds data on both sides) and the *HD* stands for "high density" (which means it stores more data than the previous standard—*DD*, for "double density").

  When you buy a box of floppies, be sure to check whether they are "IBM formatted" (for PCs only) or "Macintosh formatted" (for Apple machines). You can also buy "unformatted" disks, which means you have to *format* or *initialize* them yourself—that is, prepare the disks for use so that the operating system can write information on them. The software commands for formatting are described in your computer's user manual.

Iomega 100 MB Zip Drive

- **Zip disks—100 megabytes:** Produced by Iomega Corp., **_Zip disks_ are special disks with a capacity of 100 or even 250 megabytes.** At 100 megabytes, this is 70 times the storage capacity of the standard floppy. Among other uses, Zip disks are used to store large spreadsheet files, database files, image files, multimedia presentation files, and Web sites. Zip disks require their own Zip disk drives, which may come installed on new computers, although external Zip drives are also available.

Zip drive
External Zip drive and disks

- **SuperDisks—120 megabytes:** Produced by Imation, **_SuperDisks_ are disks with a capacity of 120 megabytes; the SuperDisk drive can also read standard 1.44-megabyte floppy disks,** which Zip drives cannot do.

- **HiFD disks—200 megabytes:** Made by Sony Corp., **_HiFD disks_ have a capacity of 200 megabytes; the disk drive can also read standard 1.44-megabyte floppies.** HiFD disks have 140 times the capacity of today's standard floppy disks.

SuperDisk

Hard Disks

Floppy disks use flexible plastic, but hard disks use metal. **_Hard disks_ are thin but rigid metal platters covered with a substance that allows data to be held in the form of magnetized spots.** Hard disks are tightly sealed within an enclosed hard-disk-drive unit to prevent any foreign matter from getting inside. Data may be recorded on both sides of the disk platters. *(See ● Panel 5.17.)*

Hard disks are quite sensitive devices. The read/write head does not actually touch the disk but rather rides on a cushion of air about 0.000001 inch thick. *(See ● Panel 5.18.)* The disk is sealed from impurities within a container, and the whole apparatus is manufactured under sterile conditions. Otherwise, all it would take is a human hair, a dust particle, a fingerprint smudge, or a smoke particle to cause what is called a head crash. **A _head crash_ happens when the surface of the read/write head or particles on its surface come into contact with the surface of the hard-disk platter, causing the loss of some or all of the data on the disk.** A head crash can also happen when you bump a computer too hard or drop something heavy on the system cabinet. An incident of this sort could, of course, be a disaster if the data has not been backed up. There are firms that specialize in trying to

retrieve data from crashed hard disks (for a hefty price), though this cannot always be done.

There are two types of hard disks—nonremovable and removable.

- **Nonremovable hard disks: A _nonremovable hard disk_, also known as a fixed disk, is housed in a microcomputer system unit and is used to store nearly all programs and most data files.** Usually it consists of four 3.5-inch metallic platters sealed inside a drive case the size of a small sandwich, which contains disk platters on a drive spindle, read/write heads mounted on an access arm that moves back and forth, and power connections and circuitry. *(See ● Panel 5.19, next page.)* Operation is much the same as for a diskette drive: The read/write heads locate specific instructions or data files according to track or sector.

Microcomputer hard drives with capacities measured in tens of gigabytes—up to 40 gigabytes, according to current ads—are becoming essential because today's programs are so huge. Microsoft Office alone is 500 megabytes. As for speed, hard disks allow faster access to data than floppy disks do, because a hard disk spins many times faster. Computer ads frequently specify speeds in revolutions per minute. A floppy disk drive rotates at only 360 rpm; a 7200-rpm hard drive is going about 300 miles per hour.

40 GB Ultra ATA 7200 RPM Hard Drive

In addition, ads may specify the type of **_hard-disk controller_, a special-purpose circuit board that positions the disk and read/write heads and manages the flow of data and instructions to and from the disk.** Popular hard-disk controllers are *Ultra ATA* (or *EIDE*) and *SCSI*. Commonly found on new PCs, *Ultra ATA (advanced technology attachment)* allows fast data transfer and high storage capacity; it is also known as *EIDE (enhanced integrated drive electronics)*. Ultra ATA can support only one or two hard disks. By contrast,

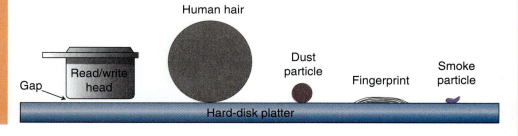

Human hair

Read/write head

Gap

Dust particle

Fingerprint

Smoke particle

Hard-disk platter

Hard disks

Drive spindle

Read/write heads

Read/write heads

Actuator arm

Platters (disks)

Power connection

Spindle

Power connection

Inside a microcomputer's nonremovable hard disk

These platters are installed inside a drive case to prevent contaminants from affecting the read/write heads.

SCSI (small computer system interface), pronounced "scuzzy," supports several disk drives as well as other peripheral devices by linking them in a daisy chain of up to seven devices. SCSI controllers are faster and have more store capacity than EIDE controllers; they are typically found in servers and workstations.

● **Removable hard disks:** *Removable hard disks*, **or hard-disk cartridges, consist of one or two platters enclosed along with read/write heads in a hard plastic case, which is inserted into a microcomputer's cartridge drive.** Typical capacity is 2 gigabytes. Two popular systems are the Iomega's Jaz and SyQuest's SparQ. *(See ● Panel 5.20.)* These cartridges

Removable hard disk

The Jaz disk from Iomega is a hard-disk cartridge system with a capacity of 2 gigabytes.

are often used to transport huge files, such as desktop-publishing files with color and graphics and large spreadsheets. They are also frequently used to back up data.

Hard-Disk Technology for Large Computer Systems

The large databases offered by such organizations as America Online and through the Internet and World Wide Web depend on secondary-storage technology for large computers.

Three types of secondary-storage devices are available for large computers:

RAID storage system
The RAID system not only holds more data than a fixed-disk drive within the same amount of space but is also more reliable. If one drive fails, others can take over.

- **Removable packs:** A _removable-pack hard-disk system_ **contains 6–20 hard disks, of 10.5- or 14-inch diameter, aligned one above the other in a sealed unit.** Capacity varies; some packs range into the terabytes.

  These removable hard-disk packs resemble a stack of phonograph records, except that there is space between disks to allow access arms to move in and out. Each access arm has two read/write heads—one reading the disk surface below, the other the disk surface above. However, only _one_ of the read/write heads is activated at any given moment.

- **Fixed-disk drives:** _Fixed-disk drives_ **are high-speed, high-capacity disk drives that are housed (sealed) in their own cabinets.** Although not removable or portable, they are now more common than removable packs, because of their greater storage capacity and reliability. A single mainframe computer might have 20–100 such fixed-disk drives attached to it.

- **RAID storage system:** A fixed-disk drive sends data to the computer along a single path. **A _RAID (redundant array of inexpensive disks) storage system_, which consists of two or more disk drives within a single cabinet or connected along a SCSI chain, sends data to the computer along several parallel paths simultaneously.** Response time is thereby significantly improved.

Optical Disks: CDs & DVDs

Everyone who has ever played an audio CD is familiar with optical disks. **An _optical disk_ is a removable disk, usually 4.75 inches in diameter and less than one-twentieth of an inch thick, on which data is written and read through the use of laser beams.** An audio CD holds up to 74 minutes (2 billion bits' worth) of high-fidelity stereo sound. Some optical disks are used strictly for digital data storage, but many are used to distribute multimedia programs that combine text, visuals, and sound.

With an optical disk, there is no mechanical arm, as with floppy disks and hard disks. Instead, a high-power laser beam is used to write data by burning tiny pits or indentations into the surface of a hard plastic disk. To read the data, a low-power laser light scans the disk surface: Pitted areas are not reflected and are interpreted as 0 bits; smooth areas are reflected and are interpreted as 1 bits. (See ● Panel 5.21, next page.) Because the pits are so tiny, a great deal more data can be represented than is possible in the same amount of space on a diskette and many hard disks. An optical disk can hold over 4.7 gigabytes of data, the equivalent of 1 million typewritten pages.

Nearly every PC marketed today contains a CD-ROM or DVD-ROM drive, which can also read audio CDs. These, along with their recordable and rewritable variations, are the two principal types of optical-disk technology used with computers. (See ● Panel 5.22, next page.)

- **CD-ROM—for reading only:** For microcomputer users, the best-known type of optical disk is the CD-ROM. **_CD-ROM (compact disk read-only memory)_ is an optical-disk format that is used to hold**

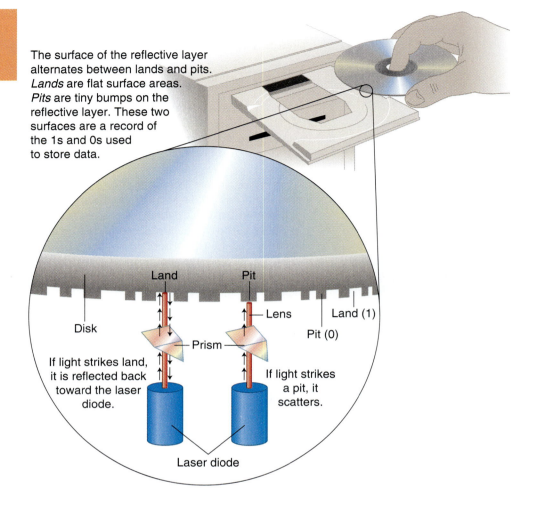

The surface of the reflective layer alternates between lands and pits. *Lands* are flat surface areas. *Pits* are tiny bumps on the reflective layer. These two surfaces are a record of the 1s and 0s used to store data.

Land · Pit

Disk

Lens

Land (1)

Pit (0)

Prism

If light strikes land, it is reflected back toward the laser diode.

If light strikes a pit, it scatters.

Laser diode

prerecorded text, graphics, and sound. Like music CDs, a CD-ROM is a read-only disk. *Read-only* means the disk's content is recorded at the time of manufacture and cannot be written on or erased by the user. As the user, you have access only to the data imprinted by the disk's manufacturer. A CD-ROM disk can hold up to 650 megabytes of data, equal to over 300,000 pages of text.

A CD-ROM drive's speed is important because with slower drives, images and sounds may appear choppy. In computer ads, drive speeds are indicated by the symbol "X," as in "44X," which is a high speed. "X" denotes the original data transfer rate of 150 kilobytes per second.

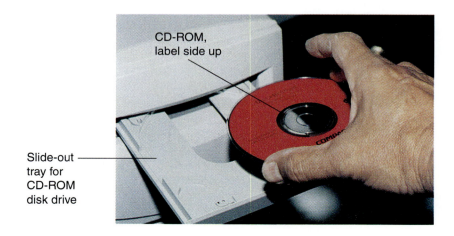

CD-ROM, label side up

Slide-out tray for CD-ROM disk drive

44X Max CD-ROM Drive

(The data transfer rate is the time the drive takes to transmit data to another device. A 44X drive runs at 44 times 150, or 6600 kilobytes (6.6 megabytes) per second. If an ad carries the word "Max," as in "44X Max," this indicates the device's maximum speed. Drives range in speed from 16X to 48X; the faster ones are more expensive.

• **CD-R—for recording on once:** ***CD-R (compact disk–recordable) disks*** **can be written to only once but can be read many times.** This allows consumers to make their own CD disks, though it's a slow process. (Recording a full disk takes 20–60 minutes.) Also, once recorded, the information cannot be erased. CD-R is often used by companies for archiving—that is, to store vast amounts of information. A variant is the Photo CD, an optical disk developed by Kodak that can digitally store photographs taken with an ordinary 35-millimeter camera. Once you've shot a roll of color photographs, you take it for processing to a photo shop, which produces a disk containing your images. You can view the disk on any personal computer with a CD-ROM drive and the right software.

CD-R/RW Drive

• **CD-RW—for rewriting many times:** A ***CD-RW (compact disk– rewritable) disk***, **also known as an erasable optical disk, allows users to record and erase data so that the disk can be used over and over again.** Special CD-RW drives and software are required. CD-RW disks are useful for archiving and backing up large amounts of data or work in multimedia production or desktop publishing. Because they are relatively slow, they are no substitute for a hard disk. But rewritable CDs may be poised to end the era of the floppy disk, particularly since users are now used to saving data-heavy multimedia music and videos.[20]

• **DVD-ROM—the versatile video disk:** A ***DVD-ROM (digital versatile disk or digital video disk, with read-only memory)*** **is a CD-style disk with extremely high capacity, able to store 4.7–17 gigabytes.** How is this done? Like a CD or CD-ROM, the surface of a DVD contains microscopic pits, which represent the 0s and 1s of digital code that can be read by a laser. The pits on the DVD, however, are much smaller and grouped more closely together than those on a CD, allowing far more information to be represented. Also, the laser beam used focuses on pits roughly half the size of those on current audio CDs. In addition, the DVD format allows for two layers of data-defining pits, not just one. Finally, engineers have succeeded in squeezing more data into fewer pits, principally through data compression.

Many new computer systems now come with a DVD drive as standard equipment. A great advantage is that these drives can also take standard CD-ROM disks, so that now you can watch DVD movies and play CD-ROMs using just one drive.[21] DVDs have enormous potential to replace CDs for archival storage, mass distribution of software, and entertainment. They not only store far more data but are different in quality from CDs. As one writer points out, "DVDs encompass much more: multiple dialogue tracks and screen formats, and best of all, smashing sound and video."[22] The theater-quality video and sound, of course, is what makes DVD a challenger to videotape as a vehicle for movie rentals. Since its 1997 introduction, DVD technology has gained steadily on videotape players.[23] Recent technology enables users to display Web pages on their television screens.[24]

As with CDs, DVDs have their recordable and rewritable variants. ***DVD-R (DVD–recordable) disks*** **allow one-time recording by the consumer.** Two types of reusable disks are *DVD-RW (DVD–rewritable)* and *DVD-RAM (DVD–random access memory)*, both of which can be recorded on and erased more than once.

● PANEL 5.23
Magnetic tape cartridge and three types of tapes
Three types of magnetic-tape cartridges and magnetic-tape cartridge drive

Magnetic-tape cartridge drive

Magnetic Tape

Similar to the tape used on an audio tape recorder (but of higher density), *magnetic tape* **is thin plastic tape coated with a substance that can be magnetized. Data is represented by magnetized spots (representing 1s) or nonmagnetized spots (representing 0s).** Today, "mag tape" is used mainly for backup and archiving—that is, for maintaining historical records—where there is no need for quick access.

On large computers, tapes are used on magnetic-tape units or reels, and in cartridges. On microcomputers, tape is used in the form of *tape cartridges*, **modules resembling audio cassettes that contain tape in rectangular, plastic housings.** There are three common types of drive tapes—QIC, DAT, and (most expensive) DLT. *(See* ● *Panel 5.23.)*

Tapes fell out of favor for a while, supplanted by such products as Iomega's Jaz and Zip drives. However, as hard drives have swelled to multigigabyte size, using Zip disks for backup has become less convenient. Since a single-tape cassette can hold up to 66 gigabytes, tape is looking like a better alternative.

Smart Cards

Today in the United States, most credit cards are old-fashioned magnetic-strip cards. A magnetic-strip card has a strip of magnetically encoded data on its back. The encoded data might include your name, account number, and PIN (personal identification number). Two other kinds of cards, smart cards and optical cards, which hold far more information, are already popular in Europe. Manufacturers are betting they will soon be popular in the U.S., which presently has less than 2% of the global $7-billion smart-card industry.[25]

- ● **Smart cards: A** *smart card* **looks like a credit card but contains a microprocessor and memory chip.** When inserted into a reader, it transfers data to and from a central computer, and it can store some basic financial records. Smart cards can be used as telephone debit

Smart card

Yifat Mabary uses a smart card to make a purchase at the Royale Kosher Bakery in New York City.

Flash memory card

cards. You insert the card into a slot in the phone, wait for a tone, and dial the number. The length of your call is automatically calculated on the chip inside the card, and the corresponding charge is deducted from the balance. Many colleges and universities issue student cards as smart cards.

- **Optical cards:** The conventional magnetic-stripe credit card holds the equivalent of a half page of data. The smart card with a microprocessor and memory chip holds the equivalent of 250 pages. The optical card presently holds about 2000 pages of data. Optical cards use the same type of technology as music compact disks but look like silvery credit cards. <u>**Optical cards** are plastic, laser-recordable, wallet-type cards used with an optical-card reader.</u> Because they can cram so much data (6.6 megabytes, as opposed to 16 kilobytes for smart cards) into so little space, they may become popular in the future. For instance, a health card based on an optical card would have room not only for the individual's medical history and health-insurance information but also for digital images, such as electrocardiograms.

Flash Memory Cards

Disk drives, whether for diskettes, hard disks, or CD-ROMs, all involve moving parts—and moving parts can break. Flash memory cards, by contrast, are variations on conventional computer-memory chips, which have no moving parts. <u>**Flash memory cards**, or flash RAM cards, consist of circuitry on credit-card-size cards that can be inserted into slots connecting to the motherboard.</u> The Memory Stick from Sony and the Secure Digital (SD) card from Panasonic presently hold up to 64 megabytes and are projected to expand to 1 gigabyte of capacity. What makes both the Memory Stick and SD particularly interesting is that they are interoperable between a wide range of computer and electronics devices.[26]

A promotional videotape demonstrates their advantage, as this description makes clear:

> In it, engineers strap a memory card onto one electric paint shaker and a disk drive onto another. Each storage device is linked to a personal computer, running identical graphics programs. Then the engineers switch on the paint shakers. Immediately, the disk drive fails, its delicate recording heads smashed against its spinning metal platters. The flash memory card takes the shaking and keeps on going.[27]

Flash memory cards are not infallible. Their circuits wear out after repeated use, limiting their lifespan. Still, unlike conventional computer memory (RAM or primary storage), flash memory is *nonvolatile*. That is, it retains data even when the power is turned off.

CONCEPT CHECK

What are four types of floppy disks and what are their features?

What are characteristics of hard disks, both nonremovable and removable?

Describe hard-disk technology for large computer systems.

Explain the various types of optical disks—CD-ROM, CD-R, CD-RW, and DVD-ROM.

How is magnetic tape used?

Describe smart cards and optical cards.

Describe flash memory cards.

Online Secondary Storage

If the network computer or thin-client computer actually becomes as popular as its promoters hope, the Internet itself will become, in effect, your hard disk. We described the concept of online storage at the start of this section. The services we mentioned included some applications software, such as calendar, address book, and even word processors. Other services, however, simply offer online storage for backup purposes. Examples are @Backup *(www.atbackup.com)*, Connected Online Backup *(www.connected.com)*, Network Associates Quick Backup to Personal Vault *(www.mcafee.com)*, and SafeGuard Interactive *(www.sgii.com)*. Monthly prices are generally in the $10–$15 range. When you sign up with the service, you usually download from a Web site free software that lets you upload whatever files you wish to the company's server. For security, you are given a password, and the files are supposedly encrypted to guard against anyone giving them an unwanted look.

From a practical standpoint, online backup should be used only for vital files. Removable hard-disk cartridges are the best medium for backing up entire hard disks, including files and programs.

5.4 Future Developments in Processing & Storage

KEY QUESTION

What are some forthcoming developments that could affect processing power and storage capacity?

Computer developers are obsessed with speed and power, constantly seeking ways to promote faster processing and more main memory in a smaller area. IBM, for instance, came up with a new manufacturing process (called silicon-on-insular, or SOI) that has the effect of increasing a chip's speed and reducing its power consumption. These chips, to be released in 2001, are expected to be up to 30% faster.[28] This increasing power, says physicist Michio Kaku, is the reason why you can get "a musical birthday card that contains more processing power than the combined computers of the Allied Forces in World War II."[29]

Does this mean that "Moore's law"—Intel co-founder Gordon Moore's 1965 prediction that the number of circuits on a silicon chip would keep doubling every 18 months—will never be repealed? After all, the smaller circuits get, the more chip manufacturers bump up against material limits. "The fact that chips are made of atoms has increasingly become a problem for us," said Moore in 2000. "In the next two or three generations, it may slow down to doubling every five years. People are predicting we will run out of gas in about 2020, and I don't see how we get around that."[30] *(See ● Panel 5.24.)*

What are the possible directions in which processors can go? Let us consider some of them.

The Ruputer

What may well be the world's first wristwatch personal computer, the Ruputer, has been produced by Japan's Seiko Instruments. It can download data, including text and pictures, from other PCs.

DSP Chips: Processors for the Post-PC Era

Millions of people may be familiar with the "Intel inside" slogan calling attention to the principal brand of microprocessor used in microcomputers. But they probably are unaware that they are more apt to go through the day using another kind of chip—*digital signal processors (DSPs)*, integrated circuits designed for high-speed data manipulation and used in audio, communications, and image manipulation. Made mainly by Texas Instruments but also by Lucent, Motorola, and Analog Devices, DSPs are designed to manipulate digital signals in speech, music, and video, and so they are found in pagers, cellphones, cars, hearing aids, and even washing machines.

Digital signal processing is presently only one-fifth the size of the $21 billion microprocessor business. But in the post-PC era, communications- and Internet-driven devices—which need

Moore's law: How much longer?

Miraculously, Gordon Moore's prediction that the number of transistors on a silicon chip will double every 18 months has held up since the 1960s. Eventually transistors will become so tiny that their components will approach the size of molecules, and the laws of physics will no longer allow this kind of doubling.

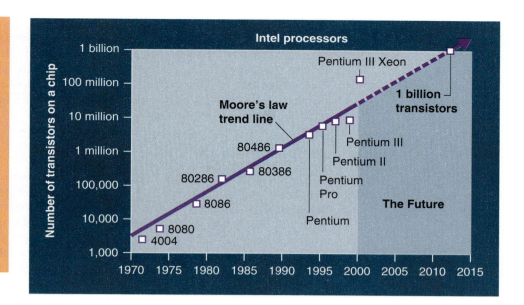

to handle enormous streams of real-world information, such as sounds and images—are expected to supplant the personal computer. Thus, in 10 years, it's possible that DSPs could outsell microprocessors.[31]

Nanotechnology

Nanotechnology, nanoelectronics, nanostructures, nanofabrication—all start with a measurement known as a nanometer. A *nanometer* is a billionth of a meter, which means we are operating at the level of atoms and molecules. A human hair is approximately 100,000 nanometers in diameter.

In *nanotechnology*, molecules are used to create tiny machines for holding data or performing tasks. Experts attempt to do nanofabrication by building tiny nanostructures one atom or molecule at a time. When applied to chips and other electronic devices, the field is called *nanoelectronics*.

Today scientists are trying to simulate the on/off of traditional transistors by creating transistor switches that manipulate a single *electron*, the subatomic particle that is the fundamental unit of electricity. In theory, a trillion of these electrons could be put on a "chip" the size of a fingernail. Scientists have already forged layers of individual molecules into tiny computer switches. The next step is to assemble these switches and other diminutive components into devices called *chemically assembled electronic devices*, or *CAENs*. These machines would be billions of times more powerful than today's personal computers.

CAEN components are supposed to be up and running within 10 years. But computer makers are already getting some payoffs from nanotechnology, which is being used to build read/write heads for hard-disk drives, improving the speed with which computers can access data.[32]

Optical Computing

Today's computers are electronic; tomorrow's might be optical, or opto-electronic—using light, not electricity. With optical technology, a machine using lasers, lenses, and mirrors would represent the on/off codes of data with pulses of light.

Light is much faster than electricity. Indeed, fiber-optic networks, which consist of hair-thin glass fibers instead of copper wire, can move information at speeds 3000 times faster than conventional networks. However, the signals get bogged down when they have to be processed by silicon chips. Optical chips would remove that bottleneck. (Some day, theoretically, it's conceivable that computers could operate even *faster* than the speed of light. For generations, physicists thought nothing was faster than light moving through a vacuum—186,000 miles per second. But in an experiment in Princeton, New Jersey, scientists succeeded in sending a pulse of light through cesium vapor so quickly that it left the chamber before it had even finished entering—310 times the distance it would have covered if the chamber had contained a vacuum.[33])

DNA Computing

Potentially, biotechnology could be used to grow cultures of bacteria that, when exposed to light, emit a small electrical charge, for example. The properties of this "biochip" could be used to represent the on/off digital signals used in computing. Or a strand of synthetic DNA might represent information as a pattern of molecules, and the information might be manipulated by subjecting it to precisely designed chemical reactions that could mark or lengthen the strand. For instance, instead of using binary, it could manipulate the four nucleic acids (represented by A, T, C, G), which holds the promise of processing big numbers. This is entirely *nondigital* way of thinking about computing.[34]

Imagine millions of nanomachines grown from microorganisms processing information at the speed of light and sending it over far-reaching pathways. What kind of changes could we expect with computers like these?

Quantum Computing

Sometimes called the "ultimate computer," the *quantum computer* is based on quantum mechanics, the theory of physics that explains the erratic world of the atom. Whereas an ordinary computer stores information as 0s and 1s represented by electrical currents or voltages that are either high or low, a quantum computer stores information by using states of elementary particles. Scientists envision using the energized and relaxed states of individual atoms to represent data. For example, hydrogen atoms could be made to switch off and on like a conventional computer's transistors by moving from low energy states (off) to high energy states (on).[35]

Other Possibilities: Molecular & Dot Computers

In the molecular computer, the silicon transistor is replaced with a single molecule. In the dot computer, the transistor is replaced by a single electron. These approaches, points out physicist Kaku, "face formidable technical problems, such as mass-producing atomic wires and insulators. No viable prototypes yet exist."[36]

Bits on disk
Magnetic bits on a disk surface, caught by a magnetic force microscope. The dark stripes are 0 bits, the bright stripes are 1 bits.

Greater Secondary Storage: Higher-Density Disks

As for developments in secondary storage, when IBM introduced the world's first disk drive in 1956, it was capable of storing 2000 bits per square inch. Today the company is shipping hard disks with densities of 14.3 billion bits, or gigabits, per square inch. In 2000, Hitachi Ltd. announced a hard drive capable of holding a record *56 billion bits* (56 gigabytes) per square inch—a nearly fourfold increase. "At this rate, says one report, "the terabyte hard drive—1000 GB—isn't very far off."[37]

Higher densities allow disks to be packaged in smaller sizes.[38] In 2000, IBM tripled the capacity of its silver-dollar-size removable hard drive from 340 megabytes to 1 gigabyte. The Microdrive, as it's called, is designed for digital cameras, MP3 players, and handheld PCs. This means a camera can now hold up to 1000 high-resolution digital photos. A handheld PC can store the of equivalent of 1000 200-page novels.

Molecular Electronics: Storage at the Subatomic Level

An emerging field, molecular electronics, may push secondary storage into another dimension entirely. Here are some possibilities:

- **Holograms as storage:** A *hologram* is a three-dimensional picture created by two lasers. You sometimes see holograms used as logos on bank credit cards. Dark and light areas of the hologram in a crystal could be used to code binary information.

  In the future, holograms could replace not only hard-disk drives but also memory chips. SRI International discovered that it could store different colors of light on the same crystal, thereby improving storage density up to 100 times and speeding the process of writing data into the crystal 1000-fold.[39]

- **Molecular magnets as storage:** Researchers have succeeded in creating a microscopic magnet, one molecule in size, derived from a special combination of materials (manganese, oxygen, carbon, and hydrogen). A data storage system based on such magnets could pack data thousands or millions of times more densely than is possible in today's computer systems. As one account puts it, "Using magnets the size of . . . molecules, it might some day be possible to store hundreds of gigabytes of data in an area no larger than the head of a pin."[40]

- **Subtomic lines as storage:** In what has been called "the world's smallest Etch-a-Sketch," physicists at NEC in Tokyo used a sophisticated probe—a tool called a scanning tunneling microscope (STM)—to paint and erase tiny lines roughly 20 atoms thick. This development could someday lead to ultra-high-capacity storage devices for computer data.

- **Bacteria as storage:** Scientists have reported research involving use of bacteria to store data in three dimensions. Said Robert Birge, who fashioned a 1-centimeter cube made of protein molecules that could store data in three dimensions, "Six of these cubes can store the entire Library of Congress."[41]

The Age of "Storewidth"

The arrival of broadband communication is sure to exacerbate the demand for storage, "with billions of bytes of digital video and graphical data begging for storage space," suggests one account. "At that rate, it won't be long before we enter the age of petabytes (2 petabytes is enough to store all of the information in all U.S. academic libraries)."[42]

Technology guru George Gilder already has a name for the future kind of storage: *storewidth.* Combining storage and bandwidth, he says, storewidth will enable swift and reliable access "to the ever expanding troves of content on the proliferating disks and cells and pages and repositories of the Internet."[43]

CONCEPT CHECK

Briefly describe some possible developments in processing.

Describe some possible future directions in secondary storage.

Experience Box
How to Buy a Notebook

"Selecting a notebook computer is much more complicated than buying a desktop PC," observes *Wall Street Journal* technology writer Walter Mossberg.[44] The reason: notebooks can vary a lot more than the desktop "generic boxes," which tend to be similar, at least within a price class.[45] Trying to choose among the many Windows-based notebooks is a particularly brow-wrinkling experience. (Macintosh notebooks tend to be more straightforward since there are only two models.)

Nevertheless, here are some suggestions:[46]

Purpose What are you going to use your notebook for? You can get a notebook that's essentially a desktop replacement and won't be moved much. If you expect to use the machine a lot in class, in libraries, or on airplanes, however, weight and battery life are important.

Budget & Weight Notebooks range from $1000 to $4000, with high-end brands aimed mainly at businesspeople. You may wish to concentrate on those in the range $1000–$1500 (for an eMachines eSlate or Toshiba), or $1500–$2200 (for an Apple, Dell or Gateway), or $2300–$2400 (for a Compaq, Hewlett Packard, or Sony).

In the $2000 range are the light machines, 3–4 pounds. Designed for mobility, they tend to lack internal disk drives and all the standard ports. The heavy machines (7 pounds and up) in the $3000-plus range generally include all the features, including DVD drives and big screens.

Batteries: The Life–Weight Trade-off A rechargeable lithium-ion battery lasts longer than the nickel–metal hydride battery. Even so, a battery in the less expensive machines will usually run continuously for only about 2¼ hours. (DVD players are particularly voracious consumers of battery power, so that it's the rare notebook that will allow you to finish watching a 2-hour movie.)

The trade-off is that heavier machines usually have longer battery life. (The Apple iBook, which weighs 6.6 pounds, is the champ at 4.5–6 hours.) The lightweight machines tend to get less than 2 hours, and toting extra batteries offsets the weight savings. (A battery can weigh a pound or so.)

Software Many notebooks come with less software than you would get with a typical desktop, though what you get will probably be adequate for most student purposes. On notebook PCs, count on getting Microsoft Works rather than the more powerful Microsoft Office 2000 to handle word processing, spreadsheets, databases, and the like.

Keyboards & Pointing Devices The keys on a notebook keyboard are usually the same size as those on a desktop machine, although they may be smaller. However, the up-and-down action feels different, and the keys may feel wobbly. In addition, some keys may be omitted altogether or keys may do double duty or appear in unaccustomed arrangements.

Most notebooks have a small touch-sensitive pad in lieu of a mouse—you drag your finger across the touchpad to move the cursor. Others use a pencil-eraser-size pointing stick in the middle of the keyboard.

Screens If you're not going to carry the notebook around much, go for a big, bright screen. Most people find they are comfortable with a 12- to 14-inch display, measured diagonally, though screens can be as small as 10.4 inches and as large as 15 inches.

The best screens are active-matrix display (TFT). However, some low-priced models have the cheaper passive-matrix screens (HPA, STN, or DSTN), which are harder to read, though you may find you can live with them. XGA screens (1024 x 768 pixels) have a higher resolution than SVGA screens (800 x 600 pixels), but fine detail may not be important to you.

Memory, Speed, & Storage Capacity If you're buying a notebook to complement your desktop, you may be able to get along with reduced memory, slow processor, small hard disk, and no CD-ROM. Otherwise, all these matters become important.

Memory (RAM) is the most important factor in computer performance, even though processor speed is more heavily hyped. Most notebooks have at least 64 megabytes (MB) of memory. Cheaper models have only 32 megabytes. In that case, the hard drive has to compensate, with corresponding loss of speed and battery life. A microprocessor running 350–500 megahertz (MHz) or higher is adequate, and more recent models are faster than this.

Sometimes notebooks are referred to as "three-spindle" or "two-spindle" machines. In a three-spindle machine, a hard drive, a floppy-disk drive, and a CD-ROM drive all reside internally (not as external peripherals). A two-spindle machine has a hard drive and space for either a floppy-disk drive or a CD-ROM drive (or a second battery). A hard drive of 6 gigabytes or more is sufficient for most people.

For more information about buying computers, go to *www.mhhe.com/cit4e*, *www.zdnet.com/computershopper*, and *micro.uoregon.edu/buyersguide*.

Visual Summary

AGP (accelerated graphics port) bus (p. 192, KQ 5.2) Bus that transmits data at high speeds; designed to support video and three-dimensional (3-D) graphics. Why it's important: *An AGP bus is twice as fast as a PCI bus.*

arithmetic/logic unit (ALU) (p. 185, KQ 5.2) Part of the CPU that performs arithmetic operations and logical operations and controls the speed of those operations. Why it's important: *Arithmetic operations are the fundamental math operations: addition, subtraction, multiplication, and division. Logical operations are comparisons such as is "equal to," "greater than," or "less than."*

ASCII-8
0100 1110
0100 1111
0101 0000

ASCII (American Standard Code for Information Interchange) (p. 177, KQ 5.2) Binary code used with microcomputers. Besides more conventional characters, ASCII includes such characters as math symbols and Greek letters. Why it's important: *ASCII is the binary code most widely used in microcomputers.*

bay (p. 179, KQ 5.2) Shelf or opening in the computer case used for the installation of electronic equipment, generally storage devices such as a hard drive or DVD drive. Why it's important: *Bays permit the expansion of system capabilities. A computer may come equipped with four or seven bays.*

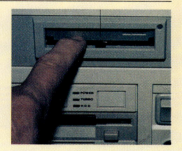

binary system (p. 176, KQ 5.2) A two-state system used for data representation in computers; has only two digits—0 and 1. Why it's important: *In the computer, 0 can be represented by electrical current being off and 1 by the current being on. All data and program instructions that go into the computer are represented in terms of these binary numbers.*

bit (p. 177, KQ 5.2) Short for "binary digit," which is either a 0 or a 1 in the binary system of data representation in computer systems. Why it's important: *The bit is the fundamental element of all data and information processed and stored in a computer system.*

bus (p. 186, KQ 5.2) Also called *bus line;* electrical data roadway through which bits are transmitted within the CPU and between the CPU and other components of the motherboard. Why it's important: *A bus resembles a multilane highway: The more lanes it has, the faster the bits can be transferred.*

byte (p. 177, KQ 5.2) Group of 8 bits. Why it's important: *A byte represents one character, digit, or other value. It is the basic unit used to measure the storage capacity of main memory and secondary storage devices (kilobytes and megabytes).*

cache (p. 188, KQ 5.2) Special high-speed memory area on a chip that the CPU can access quickly. It temporarily stores instructions and data that the processor is likely to use frequently. Why it's important: *Cache speeds up processing.*

CD-R (compact disk recordable) disks (p. 201, KQ 5.3) Optical-disk form of secondary storage that can be written to only once but can be read many times. Why it's important: *This format allows consumers to make their own CD disks, though it's a slow process. Once recorded, the information cannot be erased. CD-R is often used by companies for archiving—that is, to store vast amounts of information. A variant is the Photo CD, an optical disk developed by Kodak that can digitally store photographs taken with an ordinary 35-millimeter camera.*

CD-ROM (compact disk read-only memory) (p. 199, KQ 5.3) Optical-disk form of secondary storage that is used to hold prerecorded text, graphics, and sound. Why it's important: *Like music CDs, a CD-ROM is a read-only disk. Read-only means the disk's content is recorded at the time of manufacture and cannot be written on or erased by the user. A CD-ROM disk can hold up to 650 megabytes of data, equal to over 300,000 pages of text.*

CD-RW (compact disk–rewritable) disk (p. 201, KQ 5.3) Also known as *erasable optical disk;* optical-disk form of secondary storage. Users can record and erase data, so that the disk can be used over and over again. Special CD-RW drives and software are required. Why it's important: *CD-RW disks are useful for archiving and backing up large amounts of data or work in multimedia production or desktop publishing; however, they are relatively slow.*

chip (p. 173, KQ 5.1) Also called a *microchip,* or *integrated circuit;* consists of millions of electronic circuits printed on a tiny piece of silicon. Silicon is an element widely found in sand that has desirable electrical (or "semiconducting") properties. Why it's important: *Chips have made possible the development of small computers.*

CISC (complex instruction set computing) chips (p. 182, KQ 5.2) Design that allows a microprocessor to support a large number of instructions. Why it's important: *CISC chips are used mostly in PCs and in conventional mainframes. CISC chips are generally slower then RISC chips.*

CMOS (complementary metal-oxide semiconductor) chips (p. 187, KQ 5.2) Battery-powered chips that don't lose their contents when the power is turned off. Why it's important: *CMOS chips contain flexible start-up instructions, such as time, date, and calendar, that must be kept current even when the computer is turned off. Unlike ROM chips, CMOS chips can be reprogrammed—for example, when you need to change the time for daylight savings time.*

control unit (p. 185, KQ 5.2) Part of the CPU that deciphers each instruction stored in it and then carries out the instruction. Why it's important: *The control unit directs the movement of electronic signals between main memory and the arithmetic/logic unit. It also directs these electronic signals between main memory and the input and output devices.*

Microprocessor chip
(with CPU)

CPU (central processing unit) (p. 185, KQ 5.2) The processor; it follows the instructions of the software (program) to manipulate data into information. The CPU consists of two parts— (1) the control unit and (2) the arithmetic/logic unit (ALU), which both contain registers, or high-speed storage areas. All are linked by a kind of electronic "roadway" called a bus. Why it's important: *The CPU is the "brain" of the computer.*

DVD-R (DVD recordable) disks (p. 201, KQ 5.3) DVD disks that allow one-time recording by the consumer. Two types of reusable disks are DVD-RW (DVD rewritable) and DVD-RAM (DVD random access memory), both of which can be recorded on and erased more than once. Why it's important: *Recordable DVD disks offer the user yet another option for storing large amounts of data.*

DVD-ROM (digital versatile disk or digital video disk, with read-only memory) (p. 201, KQ 5.3) CD-type disk with extremely high capacity, able to store 4.7–17 gigabytes. Why it's important: *A powerful and versatile secondary storage medium.*

EBCDIC
1100 0001
1100 0010
1100 0011

EBCDIC (Extended Binary Coded Decimal Interchange Code) (p. 178, KQ 5.2) Binary code used with large computers. Why it's important: *EBCDIC is commonly used in mainframes.*

expansion (p. 181, KQ 5.2) Way of increasing a computer's capabilities by adding hardware to perform tasks that are beyond the scope of the basic system. Why it's important: *Expansion allows users to customize and/or upgrade their computer systems.*

expansion card (p. 191, KQ 5.2) Also known as *expansion board, adapter card, interface card, plug-in board, controller card, add-in,* or *add-on;* circuit board that provides more memory or that controls peripheral devices. Why it's important: *Common expansion cards connect to the monitor (graphics card), speakers and microphones (sound card), and network (network card). Most computers have four to eight expansion slots, some of which may already contain expansion cards included in your initial PC purchase.*

expansion slot (p. 191, KQ 5.2) Socket on the motherboard into which you can plug an expansion card. Why it's important: *See expansion card.*

fixed-disk drive (p. 199, KQ 5.3) High-speed, high-capacity disk drive housed in its own cabinet. Why it's important: More common than removable packs because of their greater storage capacity and reliability.

flash memory cards (p. 203, KQ 5.3) Also known as *flash RAM cards;* form of secondary storage consisting of circuitry on credit-card-size cards that can be inserted into slots connecting to the motherboard. Why it's important: *Flash memory is nonvolatile, so it retains data even when the power is turned off.*

flash memory chips (p. 188, KQ 5.2) Chips that can be erased and reprogrammed more than once (unlike PROM chips, which can be programmed only once). Why it's important: *Flash memory, which can range from 1 to 64 megabytes in capacity, is used to store programs not only in personal computers but also in pagers, cellphones, printers, and digital cameras. Unlike standard RAM chips, flash memory is nonvolatile—data is retained when the power is turned off.*

floppy disk (p. 194, KQ 5.3) Often called a *diskette* or simply a *disk;* removable flat piece of mylar plastic packaged in a 3.5-inch plastic case. Data and programs are stored on the disk's coating by means of magnetized spots, following standard on/off patterns of data representation (such as ASCII). The plastic case protects the mylar disk from being touched by human hands. Why it's important: *Floppy disks are used on all microcomputers.*

floppy-disk cartridges (p. 195, KQ 5.3) High-capacity removable 3.5-inch disks—Zip disks, SuperDisks, and HiFD disks. Why it's important: *These cartridges store more data than regular floppy disks and are just as portable.*

flops (p. 184, KQ 5.2) Stands for *floating-point operations per second.* A floating-point operation is a special kind of mathematical calculation. This measure, used mainly with supercomputers, is expressed as *megaflops* (mflops, or millions of floating-point operations per second), *gigaflops* (gflops, or billions), and *teraflops* (tflops, or trillions). Why it's important: *The measure is used to express the processing speed of supercomputers.*

gigabyte (G, GB) (p. 177, KQ 5.2) Approximately 1 billion bytes (1,073,741,824 bytes); a measure of storage capacity. Why it's important: *This measure was formerly used mainly with "big iron" (mainframe) computers but is typical of the secondary storage (hard disk) capacity of today's microcomputers.*

gigahertz (GHz) (p. 184, KQ 5.2) Measure of speed used for the latest generation of processors: a billion cycles per second. Why it's important: *Since a new high-speed processor can cost many hundred dollars more than the previous generation of chip, experts often recommend that buyers fret less about the speed of the processor (since the work most people do on their PCs doesn't even tax the limits of the current hardware) and more about spending money on extra memory.*

graphics cards (p. 193, KQ 5.2) Also called a *video card* or *video adapter;* a graphics card converts signals from the computer into video signals that can be displayed as images on a monitor. Why it's important: *The power of a graphics card, often expressed in megabytes, as in 8, 16, or 32 MB, determine the clarity of the images on the monitor.*

hard disk (p. 196, KQ 5.3) Secondary storage medium; thin but rigid metal platter covered with a substance that allows data to be stored in the form of magnetized spots. Hard disks are tightly sealed within an enclosed hard-disk-drive unit to prevent any foreign matter from getting inside. Data may be recorded on both sides of the disk platters. Why it's important: *Hard disks hold much more data than do floppy disks. All microcomputers use hard disks as their principal storage medium.*

hard-disk controller (p. 197, KQ 5.3) Special-purpose circuit board that positions the disk and read/write heads and manages the flow of data and instructions to and from the disk. Why it's important: *Common PC hard-disk controllers are Ultra ATA (or EIDE) and SCSI.*

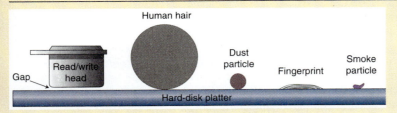

head crash (p. 196, KQ 5.3) Name for occurrence in which the surface of the read/write head or particles on its surface come into contact with the surface of the hard-disk platter, causing the loss of some or all of the data on the disk. Why it's important: *Because head crashes are always a possibility, users should always back up data from their hard disks on another storage medium, such as floppy disks.*

HiFD disk (p. 196, KQ 5.3) Disk with a capacity of 200 megabytes, made by Sony Corp. The disk drive can also read standard 1.44-megabyte floppies. Why it's important: *HiFD disks have 140 times the capacity of today's standard floppy disks.*

infrared port (p. 191, KQ 5.2) Port that allows a computer to make a cableless connection with infrared-capable devices, such as some printers. Why it's important: Infrared ports eliminate the need for cabling.

Intel-type chip (p. 182, KQ 5.2) Processor chip for PCs; made principally by Intel Corp., but also by Advanced Micro Devices (AMD), Cyrix, DEC, and others. Why it's important: *These chips are used by manufacturers such as Compaq, Dell, Gateway 2000, Hewlett-Packard, and IBM. Since 1993, Intel has marketed its chips under the names Pentium, Pentium Pro, Pentium MMX, Pentium II, Pentium III, and Celeron. Many ads for PCs contain the logo "Intel inside" to show that the systems run an Intel microprocessor.*

Integrated circuit (p. 173, KQ 5.1). An entire electronic circuit, including wires, formed on a single "chip," or piece, of special material, usually silicon. Why it's important: *In the old days, transistors were made individually and then formed into an electronic circuit with the use of wires and solder. An integrated circuit is formed as part of a single manufacturing process.*

ISA (industry standard architecture) bus (p. 192, KQ 5.2) The most widely used expansion bus. Why it's important: *ISA is also the oldest and, at 8 or 16 bits, the slowest at transmitting data, though it is still used for mouses, modem cards, and low-speed network cards.*

kilobyte (K, KB) (p. 177, KQ 5.2) Approximately 1000 bytes (1024 bytes); a measure of storage capacity. Why it's important: *The kilobyte was a common unit of measure for memory or secondary-storage capacity on older computers.*

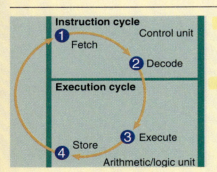

machine cycle (p. 185, KQ 5.2) Series of operations performed by the control unit to execute a single program instruction. It (1) fetches an instruction, (2) decodes the instruction, (3) executes the instruction, and (4) stores the result. Why it's important: *The machine cycle is the essence of computer-based processing.*

machine language (p. 178, KQ 5.2) Binary code (language) that the computer uses directly. The 0s and 1s represent precise storage locations and operations. Why it's important: *For a program to run, it must be in the machine language of the computer that is executing it.*

magnetic tape (p. 202, KQ 5.3) Thin plastic tape coated with a substance that can be magnetized. Data is represented by magnetized spots (representing 1s) or nonmagnetized spots (representing 0s). Why it's important: *Today, "mag tape" is used mainly for backup and archiving—that is, for maintaining historical records—where there is no need for quick access.*

megabyte (M, MB) (p. 177, KQ 5.2) Approximately 1 million bytes (1,048,576 bytes); measure of storage capacity. Why it's important: *Microcomputer primary-storage capacity is expressed in megabytes.*

megahertz (MHz) (p. 183, KQ 5.2) Measure of microcomputer processing speed, controlled by the system clock. Why it's important: *Generally, the higher the megahertz rate, the faster the computer can process data. A 550-MHz Pentium III–based microcomputer, for example, processes 550 million cycles per second, a 2-gigahertz chip processes 2 billion cycles per second.*

microprocessor (p. 175, KQ 5.1) Miniaturized circuitry of a computer processor. It stores program instructions that process, or manipulate, data into information. The key parts of the microprocessor are transistors. Why it's important: *Microprocessors enabled the development of microcomputers.*

MIPS (p. 184, KQ 5.2) Stands for *millions of instructions per second;* a measure of processing speed. Why it's important: *MIPS are used to measure processing speeds of mainframes, minicomputers, and workstations. A workstation might perform at 100 MIPS or more, a mainframe at 200–1200 MIPS.*

Motorola-type chips (p. 182, KQ 5.2) Microprocessors made by Motorola for Apple Macintosh computers. Why it's important: *Since 1993, Motorola has provided an alternative to the Intel-style chips made for PC microcomputers.*

network interface cards (p. 193, KQ 5.2) Expansion cards that allow the transmission of data over networks. Why it's important: *Installation of a network interface card in one's computer enables one to connect with various computers and other devices such as printers.*

nonremovable hard disk (p. 197, KQ 5.3) Also known as a *fixed disk;* hard disk housed in a microcomputer system unit and used to store nearly all programs and most data files. Usually it consists of four 3.5-inch metallic platters sealed inside a drive case the size of a small sandwich, which contains disk platters on a drive spindle, read/write heads mounted on an access arm that moves back and forth, and power connections and circuitry. Operation is much the same as for a diskette drive: The read/write heads locate specific instructions or data files according to track or sector. Hard disks can also come in removable cartridges. Why it's important: *See* hard disk.

optical card (p. 203, KQ 5.3) Plastic, laser-recordable, wallet-type cards used with an optical-card reader. Why it's important: *Because they can cram so much data (6.6 megabytes) into so little space, they may become popular in the future. For instance, a health card based on an optical card would have room not only for the individual's medical history and health-insurance information but also for digital images, such as electrocardiograms.*

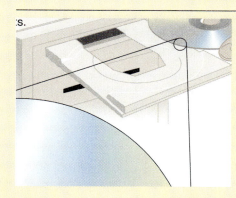

optical disk (p. 199, KQ 5.3) Removable disk, usually 4.75 inches in diameter and less than one-twentieth of an inch thick, on which data is written and read through the use of laser beams. Why it's important: *An audio CD holds up to 74 minutes (2 billion bits' worth) of high-fidelity stereo sound. Some optical disks are used strictly for digital data storage, but many are used to distribute multimedia programs that combine text, visuals, and sound.*

parallel port (p. 190, KQ 5.2) A line connected to a parallel port allows 8 bits (1 byte) to be transmitted simultaneously, like cars on an eight-lane highway. Why it's important: *Parallel lines move information faster than serial lines do. However, because they can transmit information efficiently only up to 15 feet, they are used principally for connecting printers or external disk or magnetic-tape backup storage devices.*

parity bit (p. 178, KQ 5.2) Also called a *check bit;* an extra bit attached to the end of a byte. Why it's important: *Enables a computer system to check for errors during transmission (the check bits are organized according to a particular coding scheme designed into the computer).*

PC card (p. 193, KQ 5.2) Thin, credit-card size (2.1 by 3.4 inches) hardware device. Why it's important: *PC cards are used principally on laptop computers to expand capabilities.*

PCI (peripheral component interconnect) bus (p. 192, KQ 5.2) High-speed bus; at 32 or 64 bits wide, it is more than four times faster than ISA buses. Why it's important: *PCI is widely used in microcomputers to connect graphics cards, sound cards, modems, and high-speed network cards.*

petabyte (P, PB) (p. 177, KQ 5.2) Approximately 1 quadrillion bytes (1,048,576 gigabytes); measure of storage capacity. Why it's important: *The huge storage capacities of modern databases are now expressed in petabytes.*

Plug and Play (p. 191, KQ 5.2) USB Peripheral connection standard that allows peripheral devices and expansion cards to be automatically configured while they are being installed. Why it's important: *Plug and Play avoids the hassle of setting switching and creating special files that plagued earlier users.*

port (p. 189, KQ 5.2) A connecting socket or jack on the outside of the system unit into which are plugged different kinds of cables. *Why it's important: Allows you to plug in a cable to connect a peripheral device, such as a monitor, printer, or modem, so that it can communicate with the computer system.*

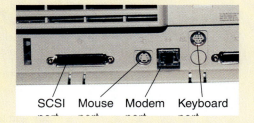

SCSI   Mouse   Modem   Keyboard

power supply (p. 181, KQ 5.2) Device that converts AC to DC to run the computer. *Why it's important: The electricity available from a standard wall outlet is alternating current (AC), but a microcomputer runs on direct current (DC).*

RAID (redundant array of independent disks) storage system (p. 199, KQ 5.3). Secondary-storage system consisting of two or more disk drives within a single cabinet or connected along a SCSI chain, sends data to the computer along parallel paths simultaneously. *Why it's important: The system not only holds more data than a fixed-disk drive within the same amount of space but is also more reliable.*

RAM (random access memory) chips (p. 186, KQ 5.2) Also called *primary storage* and *main memory;* these chips temporarily hold software instructions and data before and after it is processed by the CPU. RAM is a volatile form of storage. *Why it's important: RAM is the working memory of the computer. Having enough RAM is critical to your ability to run many software programs.*

read (p. 187, KQ 5.2) To transfer data from an input source into the computer's memory or CPU. *Why it's important: Reading, along with writing, is an essential computer activity.*

read/write head (p. 195, KQ 5.3) Mechanism used to transfer data between the computer and the disk. When the disk spins inside its case, the read/write head moves back and forth over the data access area on the disk. *Why it's important: The read/write head enables the essential activities of reading and writing data.*

registers (p. 186, KQ 5.2) High-speed storage areas that temporarily store data during processing. *Why it's important: Registers may store a program instruction while it is being decoded, store data while it is being processed by the ALU, or store the results of a calculation.*

removable hard disk (p. 198, KQ 5.3) Also called a *hard-disk cartridge;* one or two platters enclosed along with read/write heads in a hard plastic case, which is inserted into a microcomputer's cartridge drive. Typical capacity is 2 gigabytes. Two popular systems are Iomega's Jaz and SyQuest's SparQ. *Why it's important: These cartridges offer users greater storage capacity than do floppy disks but with the same portability.*

removable-pack hard-disk system (p. 199, KQ 5.3) Secondary storage with 6–20 hard disks, of 10.5- or 14-inch diameter, aligned one above the other in a sealed unit. *Why it's important: Such secondary-storage systems enable a large computer system to store massive amounts of data.*

RISC (reduced instruction set computing) chips (p. 182, KQ 5.2) Type of chip in which the complexity of the microprocessor is reduced by eliminating many seldom-used instructions, thereby increasing the processing speed. *Why it's important: RISC chips are used mostly in workstations. As a result, workstations can work up to 10 times faster then most PCs. RISC chips have been used in many Macintosh computers since 1993.*

ROM (read-only memory) (p. 187, KQ 5.2) Memory chip that cannot be written on or erased by the computer user without special equipment. *Why it's important: ROM chips contain fixed start-up instructions. They are loaded, at the factory, with programs containing special instructions for basic computer operations, such as starting the computer or putting characters on the screen. These chips are nonvolatile; their contents are not lost when power to the computer is turned off.*

SCSI (small computer system interface) port (p. 190, KQ 5.2) Pronounced "scuzzy," a SCSI allows data to be transmitted in a "daisy chain" to up to 7 devices at speeds (32 bits at a time) higher than those possible with serial and parallel ports. The term *daisy chain* means that several devices are connected in series to each other, so that data for the seventh device, for example, has to go through the other six devices first. *Why it's important: Enables users to connect external hard-disk drives, CD-ROM drives, scanners, and magnetic-tape backup units.*

secondary storage hardware (p. 194, KQ 5.3) Devices that permanently hold data and information as well as programs. *Why it's important: Secondary storage—as opposed to primary storage—is nonvolatile; that is, saved data and programs are permanent, or remain intact, when the power is turned off.*

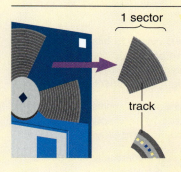

1 sector

track

sectors (p. 195, KQ 5.3) Wedge-shaped sections on a formatted diskette used for storage reference purposes. *Why it's important: When you save data from your computer to a diskette, the data is distributed by tracks and sectors on the disk. That is, the system software uses the point at which a sector intersects a track to reference the data location.*

semiconductor (p. 173, KQ 5.1) Material, such as silicon (in combination with other elements), whose electrical properties are intermediate between a good conductor and a nonconductor of electricity. When highly conducting materials are laid on the semiconducting material, an electronic circuit can be created. *Why it's important: Semiconductors are the materials from which integrated circuits (chips) are made.*

serial port (p. 190, KQ 5.2) A line connected to a serial port will send bits one after another, like cars on a one-lane highway. The standard for PC serial ports is the 9-pin or 25-pin RS-232C connector. *Why it's important: Because individual bits must follow each other, a serial port is usually used to connect devices that do not require fast transmission of data, such as keyboard, mouse, monitors, and modems. It is also useful for sending data over a long distance.*

silicon (p. 173, KQ 5.1) An element that is widely found in clay and sand used in the making of solid-state integrated circuits. *Why it's important: It is used not only because its abundance makes it cheap but also because it is a good semiconductor. As a result, highly conducting materials can be overlaid on the silicon to create the electronic circuitry of the integrated circuit.*

smart card (p. 202, KQ 5.3). Looks like a credit card but contains a microprocessor and memory chip. When inserted into a reader, it transfers data to and from a central computer. *Why it's important: Unlike conventional credit cards, smart cards can hold a fair amount of data and can store some basic financial records. Thus, they are used as telephone debit cards, health cards, and student cards.*

solid-state device (p. 173, KQ 5.1) Electronic component made of solid materials with no moving parts, such as an integrated circuit. *Why it's important: Solid-state integrated circuits are far more reliable, smaller, and less expensive than electronic circuits made from several components.*

sound card (p. 193, KQ 5.2) Used to transmit digital sounds through speakers, microphones, and headsets. *Why it's important: Cards such as PCI wavetable sound cards are used to add music and sound effects to computer video games.*

SuperDisk (p. 196, KQ 5.3) Disks with a capacity of 120 megabytes; produced by Imation. The SuperDisk drive can also read standard 1.44-megabyte floppy disks, which Zip drives cannot do. *Why it's important: See floppy-disk cartridges.*

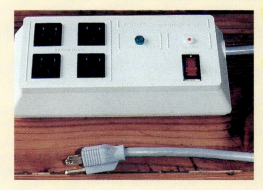

surge protector (p. 181, KQ 5.2) Also called *surge suppressor;* device keeps surges (spikes) of high voltage from entering the computer from its power source. *Why it's important: It protects the computer from damage.*

system clock (p. 183, KQ 5.2) Internal timing device that uses fixed vibrations from a quartz crystal to deliver a steady stream of digital pulses or "ticks" to the CPU. These ticks are called *cycles. Why it's important: Faster clock speeds will result in faster processing of data and execution of program instructions, as long as the computer's internal circuits can handle the increased speed.*

tape cartridge (p. 202, KQ 5.3) Module resembling an audio cassette that contains tape in a rectangular plastic housing. There are three common types of drive tapes—QIC, DAT, and (most expensive) DLT. *Why it's important: Tape cartridges are used for secondary storage on microcomputers and also on some large computers. Tape is used mainly for archiving purposes and backup.*

terabyte (T, TB) (p. 177, KQ 5.2) Approximately 1 trillion bytes (1,009,511,627,776 bytes); measure of storage capacity. *Why it's important: The storage capacities of some mainframes and supercomputers are expressed in terabytes.*

1 sector

track

tracks (p. 194, KQ 5.3) The rings on a diskette along which data is recorded. Why it's important: *See* sectors.

transistor (p. 172, KQ 5.1) Tiny electronic device that acts as an on/off switch, switching between on and off millions of times per second. Why it's important: *Transistors are part of the microprocessor.*

Unicode (p. 178, KQ 5.2) Binary coding scheme that uses two bytes (16 bits) for each character, rather than one byte (8 bits). Why it's important: *Instead of the 256 character combinations of ASCII, Unicode can handle 65,536 character combinations. Thus, it allows almost all the written languages of the world to be represented using a single character set.*

upgrading (p. 181, KQ 5.2) Changing to newer, usually more powerful or sophisticated versions, such as a more powerful microprocessor or more memory chips. Why it's important: *Through upgrading, users can improve their computer systems without buying completely new ones.*

USB (universal serial bus) port (p. 191, KQ 5.2) Port that can theoretically connect up to 127 peripheral devices daisy-chained to one general-purpose port. Why it's important: *USB ports are useful for peripherals such as digital cameras, digital speakers, scanners, high-speed modems, and joysticks. The so-called USB hot plug or hot swappable allows such devices to be connected or disconnected even while the PC is running.*

UPS (uninterruptible power supply) device (p. 181, KQ 5.2) Battery-operated device that provides a computer with electricity if there is a power failure. Why it's important: *A UPS will keep a computer going 5–30 minutes enabling users to take time to save their work before shutting down.*

virtual memory (p. 188, KQ 5.2) Type of hard disk space that mimics primary storage (RAM). Why it's important: *When RAM space is limited, virtual memory allows users to run more software at once, provided the computer's CPU and operating system are equipped to use it. The system allocates some free disk space as an extension of RAM; that is, the computer swaps parts of the software program between the hard disk and RAM as needed.*

volatile (p. 186, KQ 5.2) Temporary; the contents of volatile storage media, such as RAM, are lost when the power is turned off. Why it's important: *Save your work to a secondary-storage medium, such as a hard disk, in case the electricity goes off while you're working.*

voltage regulator (p. 181, KQ 5.2) Also called *line conditioner;* device that screens a computer from insufficient power—"brownouts" or "sags" in voltage. Why it's important: *Protects the computer from damage.*

word size (p. 184, KQ 5.2) Number of bits that the processor may process at any one time. Why it's important: *The more bits in a word, the faster the computer. A 32-bit computer— that is, one with a 32-bit-word processor—will transfer data within each microprocessor chip in 32-bit chunks, or 4 bytes at a time. A 64-bit computer transfers data in 64-bit chunks, or 8 bytes at a time.*

write (p. 187, KQ 5.2) To transfer data from the computer's CPU or memory to an output device. Why it's important: *See* read.

write-protect notch (p. 194, KQ 5.3) Floppy disk feature that prevents a diskette from being written to. Why it's important: *This feature allows you to protect the data already on the disk. To write-protect, use your thumbnail or the tip of a pen to move the small sliding tab on the lower right side of the disk (viewed from the back), thereby uncovering the square hole.*

Zip disk (p. 196, KQ 5.3) Floppy-disk cartridge with a capacity of 100 or 250 megabytes. At 100 megabytes, this is 70 times the storage capacity of the standard floppy. Why it's important: *Among other uses, Zip disks are used to store large spreadsheet files, database files, image files, multimedia presentation files, and Web sites. Zip disks require their own Zip disk drives, which may come installed on new computers, although external Zip drives are also available. See also floppy-disk cartridges.*

Chapter Review

Self-Test Questions

1. A(n) _____ is about 1000 bytes; a(n) _____ is about 1 million bytes.

2. The _____ is the part of the microprocessor that tells the rest of the computer how to carry out a program's instructions.

3. The process of retrieving data from a storage device is referred to as _____; the process of copying data to a storage device is called _____.

4. To avoid losing data, users should always _____ their files.

5. Formatted diskettes have _____ and _____ that the system software uses to reference data locations.

6. The _____ is often referred to as the "brain" of a computer.

Multiple-Choice Questions

1. Which of the following is another term for *primary storage*?
   a. ROM
   b. ALU
   c. CPU
   d. RAM
   e. CD-R

2. Which of the following is not included on a computer's motherboard?
   a. RAM chips
   b. ROM chips
   c. keyboard
   d. microprocessor
   e. expansion slots

3. Which of the following is used to hold data and instructions that will be used shortly by the CPU?
   a. ROM chips
   b. peripheral devices
   c. RAM chips
   d. CD-R
   e. hard disk

4. Which of the following coding schemes is widely used on microcomputers?
   a. EBCDIC
   b. Unicode
   c. ASCII
   d. Microcode
   e. Unix

5. Which of the following is used to measure processing speed in microcomputers?
   a. MIPS
   b. flops
   c. picoseconds
   d. megahertz
   e. millihertz

True/False Questions

T F   1. A bus connects a computer's control unit and ALU.

T F   2. The machine cycle comprises the instruction cycle and the execution cycle.

T F   3. Magnetic tape is the most common secondary storage medium used with microcomputers.

T F   4. Main memory is nonvolatile.

Short-Answer Questions

1. What is ASCII, and what do the letters stand for?

2. Why should measures of capacity matter to computer users?

3. What's the difference between RAM and ROM?

4. What is the significance of the term *megahertz*?

5. What is a motherboard? Name at least four components of a motherboard.

6. What advantage does a floppy-disk cartridge have over a regular floppy disk?

7. Why is it important for your computer to be expandable?

Concept Mapping

On a separate sheet of paper, draw a concept map, or visual diagram, linking concepts. Show how the following terms are related.

ALU	floppy disk
ASCII	gigabyte
binary system	hard disk
bit	ISA
bus	machine language
byte	megabyte
cache	microprocessor
CD-ROM	optical disk
chip	RAM
control unit	read/write
CPU	ROM
expansion card	secondary storage
expansion slot	volatile

Knowledge in Action

1. If you're using Windows 95 or 98, you can easily determine what microprocessor is in your computer and how much RAM it has. To begin, click the Start button in the Windows desktop pull-up menu bar and then choose Settings, Control Panel. Then locate the System icon in the Control Panel window and double-click on the icon.

   The System Properties dialog box will open. It contains four tabs: General, Device Manager, Hardware Profiles, and Performance. The name of your computer's microprocessor will display on the General tab. To see how much RAM is in your computer, click the Performance tab.

2. The Blue Mountain supercomputer is one of the most powerful computers in the world. Use an Internet search engine such as *www.yahoo.com* or *www.dogpile.com* to locate information on this machine. Determine why the computer was developed, how much RAM and how many processors the computer has, and what kind of secondary storage it uses.

3. The objective of this project is to introduce you to an online encyclopedia that's dedicated to computer technology. The *www.webopaedia* Web site is a good resource for deciphering computer ads and clearing up difficult concepts. For practice, visit the site and type "main memory" into the Search text box and then press the enter key. Print out the page that displays. Then locate information on other topics of interest to you.

4. Visit a local computer store and note the system requirements listed on five software packages. What are the requirements for: processor? RAM? operating system? available hard disk space? CD-ROM speed? Are there any output hardware requirements?

5. If you're using Windows 95 or 98, you can easily determine the storage capacity of your computer. Double-click the My Computer icon on the Windows desktop. Then right-click the hard disk icon ("C:"). From the mouse's right-click menu, choose Properties. Note your hard disk's capacity, amount of used space, and amount of free space.

6. Develop a binary system of your own. Use any two objects, states, or conditions, and encode the following statement: *I am a rocket scientist.*

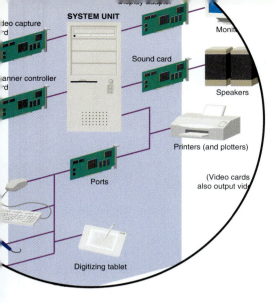

Hardware: Input & Output

Taking Charge of Computing & Communications

Key Questions

You should be able to answer the following questions.

6.1   Input & Output How is input and output hardware used by a computer system?

6.2   Input Hardware What are the three categories of input hardware, what devices do they include, and what are their features?

6.3   Output Hardware What are the two categories of output hardware, what devices do they include, and what are their features?

6.4   The Future of Input & Output What are some examples of the future of input and output technology?

6.5   Input & Output Technology & Quality of Life: Health & Ergonomics What are the principal health and ergonomic issues relating to computer use?

A Visual Overview of This Chapter

1 **Input & Output. Input hardware** consists of devices that translate data into a form the computer can process. The people-readable form of the data may be words, but the computer-readable form consists of binary 0s and 1s, or off and on electrical signals. **Output hardware** consists of devices that translate information processed by the computer into a form that humans can understand. The computer-processed information consists of 0s and 1s, which need to be translated into words, numbers, sounds, and pictures.

2 **Input Hardware.** *Input hardware*—devices that translate data into a form the computer can process—may be divided into three categories: *keyboards, pointing devices,* and *source-data entry devices.*

Keyboards: There are two categories of **keyboards,** devices that convert characters into electrical signals readable by the processor. The first is the *traditional computer keyboard,* which has all the keys of a typewriter plus some that are unique. The second category, *specialty keyboards and terminals,* includes three types of terminals: (1) A **dumb terminal** has screen and keyboard and can input and output but not process data. (2) An **intelligent terminal** has screen, keyboard, and its own processor and memory. One example is the *automated teller machine (ATM),* the self-service banking machine. Another is the *point-of-sale (POS) terminal,* used to record purchases in a store. (3) An **Internet terminal** provides access to the Internet. Examples are *set-top boxes* or *Web terminals,* stripped-down *network computers, online game players, PC/TVs,* and handheld *wireless pocket PCs* or *personal digital assistants (PDAs).*

Pointing Devices: Devices that control the cursor or pointer on a screen are **pointing devices.** They include the mouse and its variants, the touch screen, and various forms of pen input. (1) The *mouse,* which directs a pointer on the display screen, is moved on the desktop. Variants are the **trackball,** a movable ball mounted on a stationary device; the **pointing stick,** which protrudes from the keyboard; and the **touchpad,** a surface over which you move your finger. (2) The **touch screen** is a display screen that is sensitive to touch. (3) Devices for pen input include **pen-based computer systems,** in which users write with a penlike stylus on a screen; **light pens,** light-sensitive penlike devices; and **digitizers,** which convert drawings to digital data—one example is the **digitizing tablet.**

Source Data-Entry Devices: **Source-data entry devices** create machine-readable data on magnetic media or paper or feed it directly into the computer's processor. As well as various scanning devices—imaging systems, bar-code readers, mark- and character-recognition devices, and fax machines—they include audio-input devices, video input, photographic input (digital cameras), voice-recognition systems, sensors, radio-frequency identification devices, and human-biology input devices. (1) **Scanners** use laser beams and reflected light to translate images of text, drawings, and photographs into digital form. One type is an **imaging system,** which converts text, drawings, and photos into digital form that can be processed or stored in a computer system. This has led to the new industry of **electronic imaging,** the integration of separate images using scanners. Another scanning device is the **bar-code reader,** which reads the zebra-striped **bar codes** on products to translate them into digital code. **Magnetic-ink character recognition (MICR)** reads check numbers; **optical mark recognition (OMR)** reads pencil marks; **optical character recognition (OCR)** reads preprinted characters, such as those on store price tags. The **fax machine,** the last type of scanner, reads images and sends them over phone lines. **Dedicated fax machines** only send and receive fax documents; **fax modems** are modems with fax capabilities. (2) **Audio-input devices** translate analog sounds (those with continuously variable waves) into digital 0s and 1s, either through audio boards or MIDI boards. (3) **Video-input cards** translate analog film and videotape signals into digital form, using either frame-grabber video cards or full-motion video cards. (4) **Digital cameras** use light-sensitive processor chips to capture photographic images in digital form. (5) **Voice-recognition systems** convert speech into digital signals by comparing electrical patterns produced by voices with prerecorded patterns stored in a computer. (6) **Sensors** collect data directly from the environment and transmit it to a computer. (7) **Radio-frequency identification** (or RF-ID tagging) is based on an identifying tag bearing microchip that contains code numbers; these numbers are read by radio waves of a scanner linked to a database. (8) **Human-biology input devices** include *biometric systems,* which use **biometrics,** the study of body characteristics, to identify people through biological characteristics, and *line-of-site systems,* in which people point their eyes at a screen.

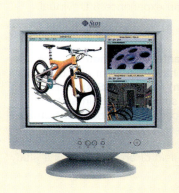

3 Output Hardware. *Output hardware* converts machine-readable information into people-readable form. Three types of output are softcopy, hardcopy, and other.

Softcopy: Softcopy refers to nonprinted data, such as that shown on a display screen. A **display screen (monitor, screen)** shows programming instructions and data as they are being input and information after it is processed. Screen clarity is affected by **dot pitch,** or space between **pixels** (the small units on screen that can be turned on or off); by **resolution,** which involves the number of pixels per square inch; and by **refresh rate,** the number of times per second pixels are recharged. Two types of monitors are CRT and flat-panel. A **CRT** (cathode-ray tube) is a vacuum tube. A **flat-panel display** consists of two plates of glass separated by a layer of a substance in which light is manipulated; one technology is **liquid crystal display (LCD),** in which molecules of liquid crystal create images by transmitting or blocking light. Flat-panel screens are either **active-matrix display,** in which each pixel on screen is controlled by its own transistor and so the image is brighter and sharper, or **passive-matrix display,** in which a transistor controls a row or column of pixels. Two common color and resolution standards for monitors are **SVGA** (the most common), which can produce 16 million possible colors, and **XGA,** which can produce 65,536 possible colors.

Hardcopy: Hardcopy refers to printed output. A **printer** prints characters or images on paper or another medium. Resolution of the image is measured by **dpi (dots per inch),** with more dots producing greater sharpness. Two types of printers are impact printers and nonimpact printers. **Impact printers** form images by striking a print hammer or wheel against an inked ribbon, leaving an image on paper; one type is the **dot-matrix printer,** which contains a print head of small pins. **Nonimpact printers** form characters or images without direct physical contact between printing mechanism and paper. Three types of nonimpact printers are laser, ink-jet, and thermal. A **laser printer** creates images with dots like a photocopying machine; the printer uses a **page description language,** software that describes the images to the printer. An **ink-jet printer** sprays electrically charged droplets of ink at high speed onto paper. A **thermal printer** uses colored waxes and heat to burn dots onto special paper. A special

Dumb terminals used by airline reservations clerks

A *__dumb terminal__* has a display screen and a keyboard and can input and output but not process data. For instance, airline reservations clerks use these terminals to access a mainframe computer containing flight information. Dumb terminals can perform no functions independent of the mainframe to which they are linked.

An *__intelligent terminal__* has its own memory and processor, as well as a display screen and keyboard. Such a terminal can perform some functions independent of any mainframe to which it is linked. One example is the familiar *automated teller machine (ATM),* a self-service banking machine that is connected through a telephone network to a central computer. Another example is the *point-of-sale (POS) terminal,* used to record purchases at a store's checkout counter. Recently, many intelligent terminals have been replaced by personal computers.

An *__Internet terminal__* provides access to the Internet. There are several variants: (1) the *set-top box* or *Web terminal,* which displays Web pages on a TV set; (2) the *network computer,* a cheap (less than $500), stripped-down computer that connects people to networks; (3) the *online game player,* which not only lets you play games but also connects to the Internet; (4) the full-blown *PC/TV* (or TV/PC), which merges the personal computer with the television set; and (5) the *wireless pocket PC* or *personal digital assistant (PDA),* a handheld computer with a tiny keyboard that can do two-way wireless messaging.

Palm
This Palm has a portable, collapsible keyboard.

Pointing Devices

One of the most natural of all human gestures, the act of pointing, is incorporated in several kinds of input devices. *__Pointing devices__* control the position of the cursor or pointer on the screen. Pointing devices include the *mouse* and its variants, the *touch screen,* and various forms of *pen input.*

• **The mouse and its variants—trackball, pointing stick, and touchpad:** The principal pointing tool used with microcomputers is the *mouse,* a device that is rolled about on a desktop and directs a pointer on the computer's display screen. The *mouse pointer*—an arrow, a rectangle, a pointing finger—is the symbol that indicates the position of the mouse on the display screen. When the mouse pointer changes to the shape of an I-beam, it shows the place where text may be selected for special treatment or to activate icons.

On the bottom side of the mouse is a ball that translates the mouse movement into digital signals. On the top side are one to five buttons. The first button is used for common functions, such as clicking and dragging. The functions of the other buttons are determined by whatever software you're using. Microsoft's IntelliMouse Optical has five programmable buttons, a scroll wheel, and no rolling ball to get gummed up, which allows the mouse to work on almost any surface.[7]

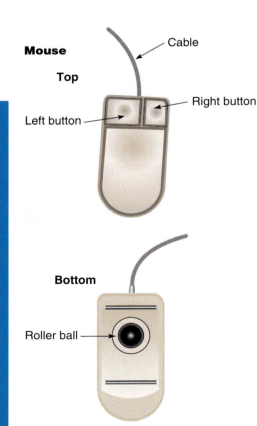

Mouse

Top

Cable

Right button

Left button

Bottom

Roller ball

Microsoft IntelliMouse

Mouse variants: (top) trackball, (middle) pointing stick, and (bottom) touchpad

There are three main variations on the mouse:

The **_trackball_ is a movable ball, mounted on top of a stationary device, that can be rotated using your fingers or palm.** In fact, the trackball looks like the mouse turned upside down. Instead of moving the mouse around on the desktop, you move the trackball with the tips of your fingers.

A **_pointing stick_ looks like a pencil eraser protruding from the keyboard between the G, H, and B keys. You move the pointing stick with your forefinger while using your thumb to press buttons located in front of the space bar.** (A forerunner of the pointing stick is the _joystick,_ which consists of a vertical handle like a gearshift lever mounted on a base with one or two buttons.)

A **_touchpad_ is a small, flat surface over which you slide your finger, using the same movements as you would with a mouse.** The cursor follows the movement of your finger. You "click" by tapping your finger on the pad's surface or by pressing buttons positioned close by the pad.

- **Touch screen:** A **_touch screen_ is a video display screen that has been sensitized to receive input from the touch of a finger.** _(See_ ● _Panel 6.3.)_ The screen is covered with a plastic layer, behind which are invisible beams of infrared light. You can input requests for information by pressing on buttons or menus displayed. The answers to your requests are displayed as output in words or pictures on the screen. (There may also be sound.) You find touch screens in kiosks, ATMs, airport tourist directories, hotel TV screens (for guest checkout), and campus information kiosks making available everything from lists of coming events to (with proper ID and personal code) student financial-aid records and grades.

- **Pen input:** Some input devices use variations on an electronic pen. Examples are _pen-based systems, light pens,_ and _digitizers._

● **PANEL 6.3**

Touch screens

Top left: In many restaurants and bars, the staff place orders using touch screens. _Top right:_ Elizabeth Young uses a touch screen at Duke University Medical Center to obtain information about her medical history. _Bottom:_ A hotel touch screen provides information about hotel services.

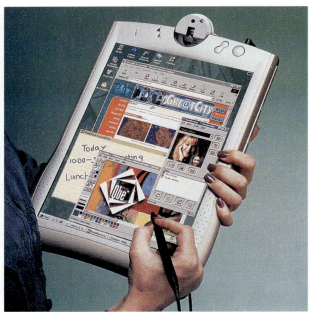

● PANEL 6.4
Pen-based computer systems
Handheld computer and *(left)* digital notebook

Pen-based computer systems **allow users to enter handwriting and marks onto a computer screen by means of a penlike stylus rather than by typing on a keyboard.** *(See* ● *Panel 6.4.)* Pen computers use handwriting recognition software that translates handwritten characters made by the stylus into data that is usable by the computer. Many handheld computers and PDAs have pen input, as do digital notebooks.

The *light pen* **is a light-sensitive penlike device connected by a wire to the computer terminal.** The user brings the pen to a desired point on the display screen and presses the pen button, which identifies that screen location to the computer. *(See* ● *Panel 6.5.)* Light pens are used by engineers, graphic designers, and illustrators.

A *digitizer* **uses a mouselike copying device called a puck, which can convert drawings and photos to digital data.** One form of digitizer is the *digitizing tablet*, **an electronic plastic board on which each specific location corresponds to a location on the screen.** When you use a puck, the tablet converts your movements into digital signals that are input to the computer. Digitizing tablets are often used to make maps and engineering drawings. *(See* ● *Panel 6.6.)*

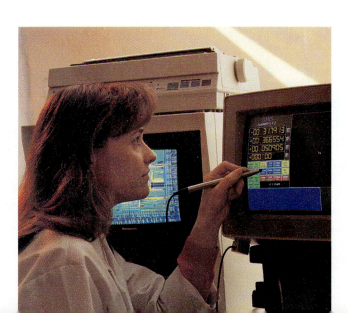

Source Data-Entry Devices

In old-fashioned grocery stores, checkout clerks read the price on every can and box, then enter those prices on the keyboard—a wasteful, duplicated effort. In new stores, of course, the clerks merely wave the products over a scanner, which automatically enters the price (from the bar code) into digital form. This is the difference between *keyboard entry* and *source data entry*.

Source data-input devices do not require keystrokes (or require only a few keystrokes) to input data to the computer. Rather data is entered directly from the *source*, without human intervention. **Source data-entry devices create machine-readable data on magnetic media or paper or feed it directly into the computer's processor.** In this section, we cover the following:

1. Scanning devices—imaging systems, bar-code readers, mark- and character-recognition devices, and fax machines
2. Audio-input devices, Web cameras and video input, and photographic input (digital cameras)
3. Voice-recognition systems, sensors, radio-frequency iden- tification devices, and human-biology input devices

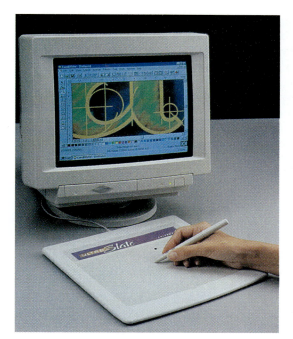

- **Scanning devices—imaging systems:** Anthony J. Scalise, 80, of Utica, New York, found a 1922 picture of his father and other immigrants from the Italian city of Scandale. "It was wonderful, all those people with walrus mustaches," he said. He immediately had prints made for friends and relatives. It's easy to make such duplicates using the self-service Kodak scanners called imaging systems now found in many photo stores.[8]

  Scanners **use light-sensing equipment to translate images of text, drawings, photos, and the like into digital form.** The images can then be processed by a computer, displayed on a monitor, stored on a stor- age device, or communicated to another computer. One type of scan- ner is the *imaging system*—**or image scanner, or graphics scanner— which converts text, drawings, and photographs into digital form that can be stored in a computer system and then manipulated, output, or sent via modem to another computer.** *(See ● Panel 6.7.)* The

Source data
Keiko, the "killer whale" known to moviegoers as the star of *Free Willy*, wears a transmitter that records heart and respiration rate. The data is stored on a computer chip and later downloaded by researchers.

● PANEL 6.8
Bar-codes and bar-code reader

system scans each image—color or black and white—with light and breaks the image into light and dark dots or color dots, which are then converted to digital code. For example, the imaging system used in desktop publishing scans in artwork or photos, which can then be positioned within a page of text, using desktop publishing software.

Imaging-system technology has led to a whole new art or industry called *electronic imaging*. **_Electronic imaging_ is the software-controlled integration of separate images, using scanners, digital cameras, and advanced graphic computers.** This technology has become an important part of multimedia.

- **Scanning devices—bar-code readers:** Another scanning device reads **_bar codes_, the vertical zebra-striped marks you see on most manufactured retail products**—everything from candy to cosmetics to comic books. *(See ● Panel 6.8.)* In North America, supermarkets, food manufacturers, and others have agreed to use a bar-code system called the *Universal Product Code.* Other kinds of bar-code systems are used on everything from FedEx packages, to railroad cars, to the jerseys of long-distance runners. **_Bar-code readers_ are photoelectric scanners that translate the symbols in the bar code into digital code.** In this system, the price of a particular item is set within the store's computer. Once the bar code has been scanned, the corresponding price appears on the salesclerk's point-of-sale terminal and on your receipt. Records of sales from the bar-code readers are input to the store's computer and used for accounting, restocking store inventory, and weeding out products that don't sell well.

- **Scanning devices—mark-recognition and character-recognition devices:** There are three types of scanning devices that sense marks or characters. They are usually referred to by their abbreviations *MICR*, *OMR*, and *OCR*.

  Magnetic-ink character recognition (MICR) reads the strange-looking numbers printed at the bottom of checks. MICR characters, which are printed with magnetized ink, are read by MICR equipment, producing a digitized signal. The bank's reader/sorter machine employs this signal to sort checks.

```
OCR-A
NUMERIC     0123456789
ALPHA       ABCDEFGHIJKLMNOPQRSTUVWXYZ
SYMBOLS     >$/-+-#"

OCR-B
NUMERIC     00123456789
ALPHA       ACENPSTVX
SYMBOLS     <+>-¥
```

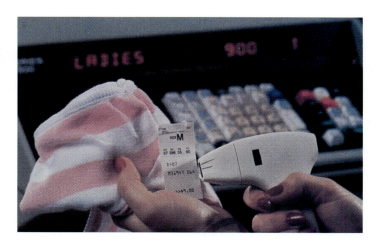

● PANEL 6.9

Optical character recognition

Special typefaces can be read by a scanning device called a wand reader.

Optical mark recognition (OMR) **uses a device that reads pencil marks and converts them into computer-usable form.** The best-known example is the OMR technology used to read the College Board Scholastic Aptitude Test (SAT) and the Graduate Record Examination (GRE).

Optical character recognition (OCR) **uses a device that reads preprinted characters in a particular font (typeface design) and converts them to digital code.** OCR characters appear on utility bills and price tags on department-store merchandise; for example, the wand reader is a common OCR scanning device. *(See ● Panel 6.9.)*

● **Scanning devices—fax machines:** A *fax machine*—or facsimile transmission machine—scans an image and sends it as electronic signals over telephone lines to a receiving fax machine, which prints out the image on paper.

There are two types of fax machines—*dedicated fax machines* and *fax modems*. *Dedicated fax machines* **are specialized devices that do nothing except send and receive fax documents.** These are what we usually think of as fax machines. They are found not only in offices and homes but also alongside regular phones in public places such as airports.

A *fax modem* **is installed as a circuit board inside the computer's system cabinet. It is a modem with fax capability that enables you to send signals directly from your computer to someone else's fax machine or computer fax modem.** With this device, you don't have to print out the material from your printer and then turn around and run it through the scanner on a fax machine. The fax modem allows you to send information more quickly than if you had to feed it page by page into a machine.

Dedicated fax machine

Fax modem

Wireless fax

A field worker views a fax message received via a wireless computer display and a cell phone.

The fax modem is another feature of mobile computing; it's especially powerful as a receiving device. Fax modems are installed inside portable computers, including pocket PCs and PDAs. If you link up a cellular phone to a fax modem in your portable computer, you can send and receive wireless fax messages no matter where you are in the world.

- **Audio-input devices:** An **_audio-input device_ records analog sound and translates it for digital storage and processing.** An analog sound signal is a continuously variable wave within a certain frequency range. For the computer to process them, these variable waves must be converted to digital 0s and 1s. The principal use of audio-input devices is to produce digital input for multimedia computers.

  An audio signal can be digitized in two ways—by an *audio board* or a *MIDI board.* Analog sound from a cassette player or a microphone goes through a special circuit board called an audio board (or card). An *audio board* is an add-on circuit board in a computer that converts analog sound to digital sound and stores it for further processing and/or plays it back, providing output directly to speakers or an external amplifier. A *MIDI board*—MIDI, pronounced "middie," stands for *Musical Instrument Digital Interface*—provides a standard for the interchange of musical information between musical instruments, synthesizers, and computers.

- **Webcams and video-input cards:** Are you the type who likes to show off for the camera? Maybe, then, you'd like to acquire a *Webcam*—a camera that attaches to a computer to record moving images that can then be posted on a Web site in real time. You could join the 10,000 other Web camera users out there who are hosting such riveting material as a 24-hour view of the aquarium of a turtle named Pixel. Or, like some of those so-called reality-TV programs, you could show your living quarters or messy desk for all of computerland to see.[9]

  As with sound, most film and videotape is in analog form; the signal is a continuously variable wave. For computer use, the signals that come from a VCR or a camcorder must be converted to digital form through a special digitizing card—a *video-capture card*—that is installed in the computer. Two types of video cards are frame-grabber video and full-motion video. *Frame-grabber video cards* can capture and digitize only a single frame at a time. Full-motion video cards can convert analog to digital signals at rates of up to 30 frames per second, giving the effect of a continuously flowing motion picture. *(See ● Panel 6.10.)*

- **Digital cameras:** Digital cameras are particularly interesting because they foreshadow a major change for the entire industry of photography. Instead of using traditional (chemical) film, a **_digital camera_ uses a light-sensitive processor chip to capture photographic images in digital form on a small diskette inserted in the camera or on flash-memory chips.** *(See ● Panel 6.11.)* The bits of digital information can then be copied right into a computer's hard disk for manipulation and printing out.

Analog camera

Analog videotape

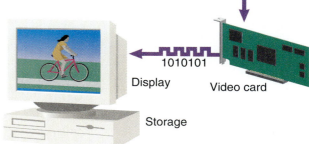

1010101

Display

Storage

Video card

Digital video

VIDEO INPUT

Analog to Digital

Full-motion video is accomplished by taking multiple pictures in sequence. Movie theater film uses 24 frames per second, which is the minimum frequency required to eliminate the perception of moving frames and make the images appear visually fluid to the eye.

TV video generates 30 interlaced frames per second, which is actually transmitted as 60 half frames ("fields" in TV lingo) per second.

Video that has been digitized and stored in the computer can be displayed at varying frame rates, depending on the speed of the computer. The slower the computer, the jerkier the movement.

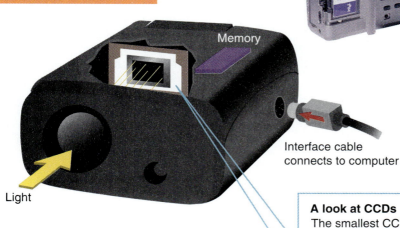

Memory

Interface cable connects to computer

Light

3. The digital information is stored in the camera's electronic memory, either built-in or removable.

4. Using an interface cable, the digital photo can be downloaded onto a computer, where it can be manipulated, printed, placed on a Web page, or e-mailed.

1. Light enters the camera through the lens.

2. The light is focused on the charge-coupled device (CCD), a solid-state chip made up of tiny, light-sensitive photosites. When light hits the CCD, it records the image electronically, just like film records images in a standard camera. The photosites convert light into electrons, which are then converted into digital information.

A look at CCDs

The smallest CCDs are 1/8 the size of a frame of 35mm film. The largest are the same size as a 35mm frame.

Smallest CCD

- Lower-end cameras start with 180,000 photosites.
- Professional cameras can have up to 6 million photosites.

CCD detail

Light-sensitive photosite

● PANEL 6.12

How a voice-recognition system works

Speech
A person who's going to use speech recognition software must first go through an *enrollment*. This consists of the person dictating text that is already known to the software for 10 minutes to an hour. From this sampling, the software creates a table of *vocal references*, which are the ways in which the speaker's pronunciation of phonemes varies from models of speech based on a sampling of hundreds to thousands of people. *Phonemes* are the smallest sound units that combine into words, such as "duh," "aw" and "guh" in "dog." There are 48 phonemes in English. After enrollment, the speaker dictates the text he wants the software to transcribe into a microphone, preferably one that uses *noise-cancellation* to eliminate background sounds. The quality of the microphone and the computer's processing power are the most important hardware factors in speech recognition. The speaker can use continuous speech, which is normal speech without pauses between words.

Signal Processing
The sound wave is transformed into a sequence of codes that represent speech sounds.

Output
Computer recognizes word string and prints it on the screen.

Recognition Search
Using the data from 1, 2, and 3, the computer tries to find the best matching sequence of words as learned from a variety of examples.

1. Phonetic Models
Describe what codes may occur for a given speech sound.
    In the word *how*, what is the probability of the "ow" sound appearing between an H and a D?

2. Dictionary
Defines the phonetic pronunciation (sequence of sounds) of each word.
    How does it work
    haw daz it werk

3. Grammar
Defines what words may follow each other, using part of speech.
    How does <it> [work]
    adv vt pron v

• **Voice-recognition systems:** Can your computer tell whether you want it to "recognize speech" or "wreck a nice beach"? A **_voice-recognition system_**, **using a microphone (or a telephone) as an input device, converts a person's speech into digital signals by comparing the electrical patterns produced by the speaker's voice with a set of prerecorded patterns stored in the computer.** *(See ● Panel 6.12.)* Programs let you accomplish two tasks: turn spoken dictation into typed text, and issue oral commands (such as "Print file") to control your computer.[10]

Voice-recognition systems have had to overcome many difficulties, such as different voices, pronunciations, and accents. Recently, however, the systems have measurably improved. Today's programs reach above 98% accuracy, with speeds as fast as you can talk, but you can't just do this without training. "High levels [of accuracy] are reached," says one report, "after you have trained the program on how you speak and the program has trained you on how to speak. The training is essential. You can't just go up to a computer, start talking to it, and expect to have your words accurately transcribed into text."[11]

Speech-recognition systems are finding many uses. Warehouse workers are able to speed inventory-taking by recording inventory counts verbally. Traders on stock exchanges can communicate their trades by speaking to computers. Radiologists can dictate their interpretations of X-rays directly into transcription machines. Nurses can fill out patient charts by talking to a computer. Drivers can talk to their car radios to change stations. Speakers of Chinese can speak to machines that will print out Chinese characters. Indeed, for many individuals with disabilities, a computer isn't so much a luxury or a

PRACTICAL ACTION BOX
Photo Opportunities: Working with Digitized Photographs

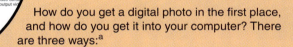

How do you get a digital photo in the first place, and how do you get it into your computer? There are three ways:[a]

1. You can ask your film developer to develop your standard (chemical-based) film not only into prints or slides but also onto a diskette or CD-ROM that you can then put in your computer's disk drive.

2. You can use an image scanner attached to your computer to scan in a traditional photographic print (or its negative).

3. You can take a picture with a digital camera, which will store the image on a diskette or removable memory card. You can then use a cable supplied with the camera to transfer the stored image to your computer.

Once the digitized images are on the hard drive in your computer, they are ready for viewing, printing, editing, or sending via the World Wide Web. Now the real fun begins.

Viewing. One of the major advantages of digital cameras is that you can view your shots immediately, on your computer's monitor. But, no matter how you get the image into the computer, on the screen it will probably look as good as, or better than, any print. Once you've viewed the image, you can decide whether to delete it, store it, have it professionally printed by a photo shop, print it out on your own printer, edit it, or send it via the Web to someone else.

Printing. Ink-jet printers can now produce output that is almost indistinguishable from a photograph. Special paper is available that gives the same look and feel as a photographic print.

Editing. Before printing the image, you might want to manipulate or, as the professionals say, "edit" it. Image-editing software—such as PhotoDeluxe (from Adobe, maker of PhotoShop, which serves professionals), Picture It, PictureWorks, Photo Soap, PhotoImpact, and Live Picture—enables you to make changes from the minor to the outrageous. (Erase scratches on the photo, the wrinkles from your face, or the teenager's nose ring in the special photo for Grandma. Put yourself in a picture of Hong Kong, which you've never visited. Even change Santa's clothes from red to blue and give him two heads.)

You can stretch, shrink, and distort images; remove the flash-photo "red eye"; sharpen the focus of fuzzy pictures; crop awkwardly shot photos; turn pictures around; cut and paste segments of photos; apply special-effects filters; retouch documents; and merge and otherwise alter images. Says one report, "Not only can you place pictures against different backgrounds—there's you grinning sheepishly on the cover of a major newsweekly, your smitten visage smack in the middle of a big, red valentine heart!—but you can morph, swirl, stretch, and squish photos any which way you please."[b]

Transmitting via the Web. You can use the World Wide Web to transmit pictures to people wherever they are, as long as they are connected to an online computer. For instance, a program called Pictra Album allows you to compile a digital album of your photos and send it to Pictra's Website for display.[c] (The pictures can be viewed publicly or privately through use of a password you select.) Kodak also offers the Kodak Picture Network, which allows photographers to store their photos on the Web and make them available by e-mail to friends and family.[d]

Sensor

Pummeling a "smart" punching bag, Olympic boxer Kerry Deshawn Jackson works out at the U.S. Olympic training center. The bag is outfitted with a sensor designed to gauge the force of the punches. The data is charted on a computer.

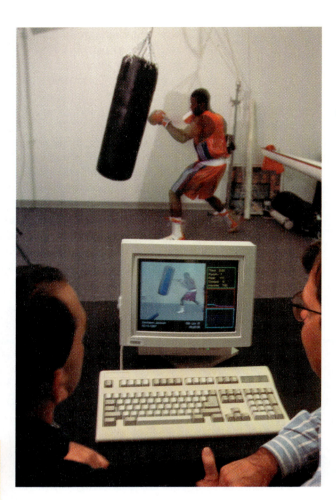

● PANEL 6.13
Earthquake sensor
Southern California's advanced seismic-monitoring system uses sensors, computers, and networks to capture data and analyze it. Within minutes of a quake, the system, known as TriNet, produces a map showing regions of intensity and areas that were probably hardest hit, enabling more efficient dispatch of rescue crews.

productivity tool as a necessity. It provides freedom of expression, independence, and empowerment.

- **Sensors:** A _**sensor**_ is **an input device that collects specific data directly from the environment and transmits it to a computer.** Although you are unlikely to see such input devices connected to a PC in an office, they exist all around us, often in nearly invisible form. Sensors can be used to detect all kinds of things: speed, movement, weight, pressure, temperature, humidity, wind, current, fog, gas, smoke, light, shapes, images, and so on.

  Sensors are used to detect the speed and volume of traffic and adjust traffic lights. They are used on mountain highways in wintertime in the Sierra Nevada as weather-sensing devices to tell workers when to roll out snowplows. In California, sensors have been planted along major earthquake fault lines in an experiment to see whether scientists can predict major earth movements. *(See ● Panel 6.13.)* In aviation, sensors are used to detect ice buildup on airplane wings or to alert pilots to sudden changes in wind direction.

- **Radio-frequency identification devices:** **Also known as RF-ID tagging,** _**radio-frequency identification technology**_ **is based on an identifying tag bearing a microchip that contains specific code numbers. These code numbers are read by the radio waves of a scanner linked to a database.** Drivers with RF-ID tags can breeze through the tollbooths without having to even roll down their windows; the toll is automatically charged to their accounts. Radio-readable ID tags are also used by the Postal Service to monitor the flow of mail; by stores for inventory control and warehousing; and in the railroad industry to keep track of rail cars. They are even injected into dogs and cats, so that veterinarians with the right scanning equipment can identify them if they become separated from their owners.

- **Human-biology input devices:** Characteristics and movements of the human body, when interpreted by sensors, optical scanners, voice recognition, and other technologies, can become forms of input. Two examples are *biometric systems* and *line-of-sight systems.*

  **Biometrics** **is the science of measuring individual body characteristics.** Biometric security devices identify a person through a fingerprint, voice intonation, or other biological characteristic. For example, retinal-identification devices use a ray of light to identify the distinctive network of blood vessels at the back of the eyeball.

Line-of-sight input

After receiving a diagnosis of Lou Gehrig's disease, Intel physicist and engineer Mike Ward began drawing blueprints for a computer system that would allow him to continue working as the disease progressed. "Typing" an e-mail message using eye-gaze technology, Ward says, "I cannot move or speak, but I can still be productive."

Line-of-sight systems enable you to use your eyes to point at the screen. This technology allows some physically disabled users to direct a computer. For example, the Eyegaze System from LC Technologies allows you to operate a computer by focusing on particular areas of a display screen. A camera mounted on the computer analyzes the point of focus of the eye to determine where you are looking. You operate the computer by looking at icons on the screen and "press a key" by looking at one spot for a specified period of time.

CONCEPT CHECK

Describe the various types of keyboards.

Discuss the mouse and its variants.

Identify the other types of pointing devices.

Name and characterize the various types of scanning devices.

Describe all the other types of source-data entry devices.

6.3 Output Hardware

KEY QUESTIONS

What are the two categories of output hardware, what devices do they include, and what are their features?

Are we back to old-time radio? Almost. Except that you can call up local programs by downloading them from the Internet. Want to listen to comedian Robin Williams, with all his dopey foreign accents and cackling laughs? You can get his weekly audio show on audible.com.[12] The sound quality isn't even as good as that of AM radio, but no doubt that will improve eventually. Computer output is taking more and more innovative forms and getting better and better.

As mentioned, *output hardware* consists of devices that convert machine-readable information, obtained as the result of processing, into people-readable form. The principal kinds of output are *softcopy* and *hardcopy*. (See ● *Panel 6.14, next page.*)

- Softcopy: _Softcopy_ **is data that is shown on a display screen or is in audio or voice form.** This kind of output is not tangible; it cannot be touched.
- Hardcopy: _Hardcopy_ **is printed output.** The principal examples are printouts, whether text or graphics, from printers. Film, including microfilm and microfiche, is also considered hardcopy output.

Softcopy Devices	Hardcopy Devices	Other Devices
CRT display screens	Impact printers: dot-matrix printer	Sound output
Flat-panel display screen (e.g., liquid-crystal display)	Nonimpact printers: laser, ink-jet, thermal	Voice output
		Video output

There are several types of output devices. In the following three sections, we discuss, first, *softcopy output—display screens*; second, *hardcopy output—printers*; and, third, *other output—sound, voice, animation, and video.*

Softcopy Output: Display Screens

<u>*Display screens*</u>**—also variously called monitors, CRTs, or simply screens—are output devices that show programming instructions and data as they are being input and information after it is processed.**

As for television screens, the size of a computer screen is measured diagonally from corner to corner in inches. For desktop microcomputers, the most common sizes are 13, 15, 17, 19, and 21 inches; for laptop computers, 12.1, 13.3, and 14.1 inches. Increasingly, computer ads state the actual display area, called the *viewable image size (vis)*, which may be an inch or so less. A 15-inch monitor may have a 13.8-inch vis; a 17-inch monitor may have a 16-inch vis.

In deciding which display screen to buy, you will need to consider issues of screen clarity (dot pitch, resolution, and refresh rate), types of monitor (CRT versus flat panel, active-matrix flat panel versus passive-matrix flat panel), and color and resolution standards (SVGA and XGA).

Monitor screen size	Viewable image area
15 inches	14 inches
17 inches	16 inches
21 inches	20 inches

- **Screen clarity—dot pitch, resolution, and refresh rate:** Among the factors affecting screen clarity (often mentioned in ads) are *dot pitch, resolution,* and *refresh rate.* These relate to the individual dots known as pixels, which represent the images on the screen. A <u>***pixel***</u>, for "picture element," is the smallest unit on the screen that can be turned on and off or made different shades.

<u>*Dot pitch (dp)*</u> **is the amount of space between the centers of adjacent pixels; the closer the dots, the crisper the image.** For a .28dp monitor, for instance, the dots are 28/100ths of a millimeter apart. Generally, a dot pitch of .28dp will provide clear images.

<u>*Resolution*</u> **is the image sharpness of a display screen; the more pixels there are per square inch, the finer the level of detail attained.** Resolution is expressed in terms of the formula *horizontal pixels* × *vertical pixels.* Each pixel can be assigned a color or a particular shade of gray. Standard resolutions are 640 × 480, 800 × 600, 1024 × 768, 1280 × 1024, and 1600 × 1200 pixels.

<u>*Refresh rate*</u> **is the number of times per second that the pixels are recharged so that their glow remains bright.** In general, displays are refreshed 45–100 times per second. The higher the refresh rate, the more solid the image looks on the screen—that is, the less it flickers.

● **PANEL 6.15**
CRT *(left)* **versus flat-panel displays**

- **Two types of monitors—CRT and flat-panel:** Display screens are of two types: *CRT* and *flat-panel.* (See ● *Panel 6.15.*)

  A ___CRT___, **for cathode-ray tube, is a vacuum tube used as a display screen in a computer or video display terminal.** The same kind of technology is found not only in the screens of desktop computers but also in television sets and flight-information monitors in airports. *Note:* Advertisements for desktop computers often *do not* include a monitor as part of the price of the system. You need to be prepared to spend a few hundred dollars extra for the monitor.

  Compared to CRTs, flat-panel displays are much thinner, weigh less, and consume less power. Thus, they are better for portable computers, although they are available for desktop computers as well. ___Flat-panel displays___ **are made up of two plates of glass separated by a layer of a substance in which light is manipulated.** One technology used is ___liquid crystal display (LCD)___, **in which molecules of liquid crystal line up in a way that alters their optical properties, creating images on the screen by transmitting or blocking out light.**

  Flat-panel monitors are available for desktop computers. Because they are smaller than CRTs, they fit more easily onto a crowded desk. However, CRTs are considerably cheaper. A new 15-inch CRT costs as little as $100. A flat-panel costs 5–10 times as much. (You can get flat-panel TVs, too—at even scarier prices.)[13]

- **Active-matrix versus passive-matrix flat-panel displays:** Flat-panel screens are either active-matrix or passive-matrix displays, according to where their transistors are located.

  In an ___active-matrix display___, **also known as TFT (thin-film transistor) display, each pixel on the screen is controlled by its own transistor.** Active-matrix screens are much brighter and sharper than passive-matrix screens, but they are more complicated and thus more expensive. They also require more power, affecting the battery life in laptop computers.

  In a ___passive-matrix display___, **a transistor controls a whole row or column of pixels.** Passive matrix provides a sharp image for one-color (monochrome) screens but is more subdued for color. The advantage is that passive-matrix displays are less expensive and use less power than active-matrix displays, but they aren't as clear and bright and can leave "ghosts" when the display changes quickly. Passive-matrix displays go by the abbreviations HPA, STN, or DSTN.

- **Color and resolution standards for monitors—SVGA and XGA:** As we mentioned earlier in the chapter, PCs come with *graphics cards* (also known as *video cards* or *video adapters*) that convert signals from the computer into video signals that can be displayed as images on a monitor. The monitor then separates the video signal into three colors: red, green, and blue signals. Inside the monitor, these three colors combine to make up each individual pixel.

  The common color and resolution standards for monitors are *SVGA, XGA, SXGA,* and *UXGA. (See ● Panel 6.16.)*

  SVGA (super video graphics array) supports a resolution of 800 × 600 pixels, or variations, producing 16 million possible simultaneous colors. SVGA is the most common standard used today with 15-inch monitors.

  XGA (extended graphics array) has a resolution of up to 1024 × 768 pixels, with 65,536 possible colors. It is used mainly with 17- and 19-inch monitors.

  SXGA (super extended graphics array) has a resolution of 1280 × 1024 pixels. It is often used with 19- and 21-inch monitors.

  UXGA (ultra extended graphics array) has a resolution of 1600 × 1200 pixels. It is expected to become more popular with graphic artists, engineering designers, and others using 21-inch monitors.

Hardcopy Output: Printers

The prices for computer systems in ads often do not include a printer. Thus, you will need to budget an additional $150 to $1000 or more for a printer.

A **_printer_ is an output device that prints characters, symbols, and perhaps graphics on paper or another hardcopy medium.** The resolution, or quality of sharpness of the image, is indicated by **_dpi (dots per inch),_ which is a measure of the number of dots that are printed in a linear inch.** For microcomputer printers, the resolution is in the range 60–1500 dpi.

Printers can be separated into two categories, according to whether or not the image produced is formed by physical contact of the print mechanism with the paper. *Impact printers* do have contact with paper; *nonimpact printers* do not. We will also consider plotters and multifunction printers.

HP DeskJet 970Cse Printer

- **Impact printers:** An **_impact printer_ forms characters or images by striking a mechanism such as a print hammer or wheel against an inked ribbon, leaving an image on paper.** A **_dot-matrix printer_** contains a print head of small pins, which strike an inked ribbon against

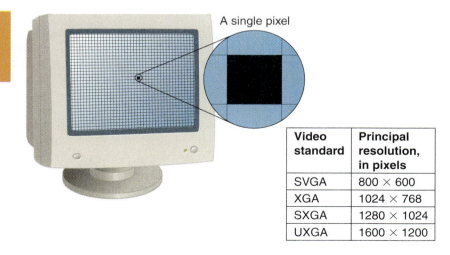

● **PANEL 6.16**
Videographics standards compared for pixels

A single pixel

Video standard	Principal resolution, in pixels
SVGA	800 × 600
XGA	1024 × 768
SXGA	1280 × 1024
UXGA	1600 × 1200

paper, to form characters or images. Print heads are available with 9, 18, or 24 pins; the 24-pin head offers the best quality. Dot-matrix printers can print *draft quality,* a coarser-looking 72 dpi, or *near-letter-quality (NLQ),* a crisper-looking 144 dpi. The machines print 40–300 characters per second and can handle graphics as well as text. A disadvantage is the noise they produce. Nowadays impact printers are more commonly used with mainframes than with personal computers.

- **Nonimpact printers:** Nonimpact printers are faster and quieter than impact printers because they have fewer moving parts. **_Nonimpact printers_ form characters and images without direct physical contact between the printing mechanism and paper.** Two types of nonimpact printers often used with microcomputers are *laser printers* and *ink-jet printers.* A third kind, the *thermal printer,* is seen less frequently.

  Like a dot-matrix printer, a **_laser printer_ creates images with dots. However, as in a photocopying machine, these images are produced on a drum, treated with a magnetically charged ink-like toner (powder), and then transferred from drum to paper.** *(See ● Panel 6.17.)*

  There are good reasons that laser printers are among the most common types of nonimpact printer. They produce sharp, crisp images of both text and graphics. They are quiet and fast—able to print 4–32 text-only pages per minute for individual microcomputers and up to 200 pages per minute for mainframes. They can print in different *fonts*—that is, type styles and sizes. The more expensive models can print in different colors.

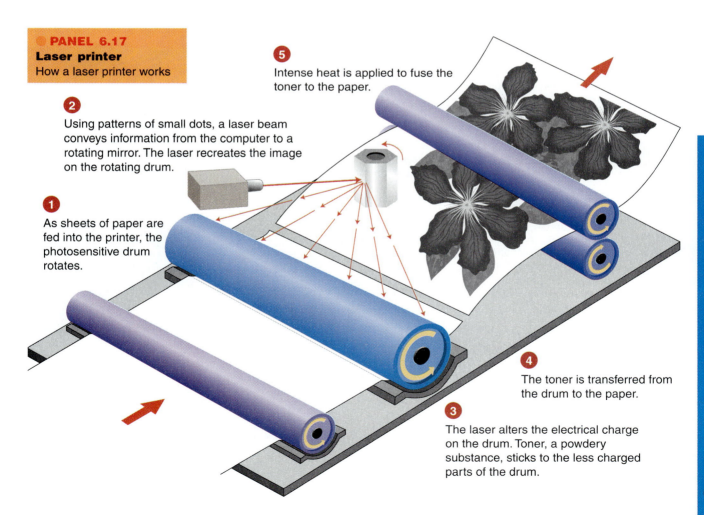

● **PANEL 6.17**
Laser printer
How a laser printer works

5 Intense heat is applied to fuse the toner to the paper.

2 Using patterns of small dots, a laser beam conveys information from the computer to a rotating mirror. The laser recreates the image on the rotating drum.

1 As sheets of paper are fed into the printer, the photosensitive drum rotates.

4 The toner is transferred from the drum to the paper.

3 The laser alters the electrical charge on the drum. Toner, a powdery substance, sticks to the less charged parts of the drum.

Printer toner cartridge

Replacing a toner cartridge

To be able to manage graphics and complex page design, a laser printer works with a page description language. A **_page description language_ is software that describes the shape and position of characters and graphics to the printer.** PostScript (from Adobe Systems) is one common type of page description language; Hewlett-Packard Graphic Language (HPGL) is another.

Ink-jet printers spray small, electrically charged droplets of ink from four nozzles through holes in a matrix at high speed onto paper. *(See* ● *Panel 6.18.)* Like laser and dot-matrix printers, ink-jet printers form images with little dots.

The advantages of ink-jet printers are that they can print in color, are quieter, and are much less expensive than color laser printers. The disadvantages are that they print in a somewhat lower resolution than laser printers and they are slower. Printing a document with high-resolution color graphics may take 10 minutes or more for a single page. Ink-jets, which spray ink onto the page a line at a time, can produce both high-quality black-and-white text and high-quality color graphics. However, if you print a lot of color, you'll find ink-jets much slower and more expensive to operate than laser printers. Moreover, a freshly printed page is apt to smear unless handled carefully. Still, a color ink-jet printer's cost is considerably less than color laser printers. The rock-bottom price is only about $150.

1 Four removable ink cartridges are attached to print heads with 64 firing chambers and nozzles apiece.

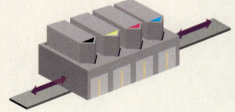

2 As the print heads move back and forth across the page, software instructs them where to apply dots of ink, what colors to use, and in what quantity.

3 To follow those instructions, the printer sends electrical pulses to thin resistors at the base of the firing chambers behind each nozzle.

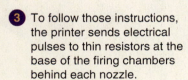

Resistor
Vapor bubble
Ink

4 The resistor heats a thin layer of ink, which in turn forms a vapor bubble. That expansion forces ink through the nozzle and onto the paper at a rate of about 6,000 dots per second.

5 A matrix of dots forms characters and pictures. Colors are created by layering multiple color dots in varying densities.

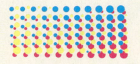

● PANEL 6.18
Ink-jet printer

Chapter 6

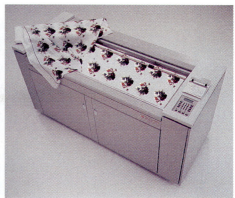

Thermal printers **use colored waxes and heat to produce images by burning dots onto special paper.** The colored wax sheets are not required for black-and-white output. However, thermal printers are expensive, and they require expensive paper. For people who want the highest-quality color printing available with a desktop printer, thermal printers are the answer.

- **Plotters:** A *plotter* **is a specialized output device designed to produce high-quality graphics in a variety of colors.** *(See ● Panel 6.19.)* Plotters are used to create hardcopy items such as maps, architectural drawings, and three-dimensional illustrations, which are usually too large for regular printers.

  The two principal kinds of plotters are ink-jet and electrostatic. An *ink-jet plotter* employs the same principle as an ink-jet printer; the paper is output over a drum, enabling continuous output. In an *electrostatic plotter,* paper lies partially flat on a table-like surface, and toner is used in a photocopier-like manner. Though more expensive, it is faster than the ink-jet plotter.

- **Multifunction printers—printers that do more than print:** *Multifunction printers* **combine several capabilities, such as printing, scanning, copying, and faxing.** *(See ● Panel 6.20.)* Xerox and Hewlett-Packard make machines that combine a photocopier, fax machine, scanner, and laser printer. Multifunction printers take up less space and cost

Hardware: Input & Output

Do I need color, or will black-only do? Are you mainly printing text or will you need to produce color charts and illustrations (and, if so, how often)? If you print lots of black text, consider getting a laser printer. If you might occasionally print color, get an ink-jet that will accept cartridges for both black and color.

Do I have other special output requirements? Do you need to print envelopes or labels? special fonts (type styles)? multiple copies? transparencies or on heavy stock? Find out if the printer comes with envelope feeders, sheet feeders holding at least 100 sheets, or whatever will meet your requirements.

Is the printer easy to set up? Can you easily put the unit together, plug in the hardware, and adjust the software (the "driver" programs) to make the printer work with your computer?

Is the printer easy to operate? Can you add paper, replace ink/toner cartridges or ribbons, and otherwise operate the printer without much difficulty?

Does the printer provide the speed and quality I want? Will the machine print at least three pages a minute of black text and two pages a minute of color? Are the blacks dark enough and the colors vivid enough?

Will I get a reasonable cost per page? Special paper, ink or toner cartridges (especially color), and ribbons are all ongoing costs. Ink-jet color cartridges, for example, may last 100–500 pages and cost $25–$30 new. Laser toner cartridges are cheaper. Ribbons for dot-matrix printers are cheaper still. Ask the seller what the cost per page works out to.

Does the manufacturer offer a good warranty and good telephone technical support? Find out if the warranty lasts at least 2 years. See if the printer's manufacturer offers telephone support in case you have technical problems. The best support systems offer toll-free numbers and operate evenings and weekends as well as weekdays.

less than the four separate office machines that they replace, but if one component breaks, nothing works.

The accompanying box gives some questions to consider when you're buying a printer. *(See ● Panel 6.21.)*

Other Output: Sound, Voice, Animation, & Video

Most PCs are now multimedia computers, capable of outputting not only text and graphics but also sound, voice, and video, as we consider next.

- Sound output: *Sound-output devices* **produce digitized sounds, ranging from beeps and chirps to music.** To use sound output, you need appropriate software and a sound card. The sound card could be Sound Blaster or, since that brand has become a de facto standard, one that is "Sound Blaster compatible." Well-known brands include Creative Labs, Diamond, and Turtle Beach. The sound card plugs into an expansion slot in your computer; on newer computers, it is integrated with the motherboard.

- Voice output: *Voice-output devices* **convert digital data into speech-like sounds.** You hear such forms of voice output on telephones ("Please hang up and dial your call again"), in soft-drink machines, in cars, in toys and games, and recently in mapping software for vehicle-navigation devices. Voice portals now read news and other information to users on the go. [14]

- Some uses of speech output are simply frivolous or amusing. You can replace your computer start-up beep with the sound of James Brown screaming "I feel gooooooood!" But some uses are quite serious. For people with physical challenges, computers with voice output help to level the playing field.

Voice output will probably become more important in the future, and it will probably get better. Text-to-speech programs have been developed that produce computer voices that sound human rather than tinny and robotic.[15]

- **Video output:** *Video* **consists of photographic images, which are played at 15–29 frames per second to give the appearance of full motion.** Video is input into a multimedia system using a video camera or VCR and, after editing, is output on a computer's display screen. Because video files can require a great deal of storage—a 3-minute video may require 1 gigabyte of storage—video is often compressed (a topic we discuss in Chapter 10). Good video output requires a powerful processor as well as a video card.

  Another form of video output is *videoconferencing*, **in which people in different geographical locations can have a meeting—can see and hear one another—using computers and communications.** Videoconferencing systems range from videophones to group conference rooms with cameras and multimedia equipment to desktop systems with small video cameras, microphones, and speakers. (We discuss this topic further in Chapter 7.)

CONCEPT CHECK

What is the difference between softcopy and hardcopy output?

What are the different characteristics of display screens?

What is the difference between impact and nonimpact printers?

Identify the characteristics of dot-matrix, laser, ink-jet, thermal, and multifunction printers.

Describe sound, voice, and video output.

6.4 The Future of Input & Output

KEY QUESTION

What are some examples of future input and output technology?

Singer/songwriter Stevie Wonder has been blind all his life. But he said he's considered surgery to restore his vision with an experimental procedure that would implant chips in his eyes, giving him "bionic eyesight."[16] Although Wonder wasn't accepted as a candidate, electronic "eyes" for the blind—printed-circuit boards about the size of a jumbo postage stamp implanted in the brain, with signals processed by a computer worn in a backpack or on the belt—are one of the high-tech input devices that have the capability of revolutionizing disabled people's lives.[17] Here let us consider some other scenarios for input/output devices in the future, as we discuss in a few pages.

Input technology seems headed in two directions: (1) toward more input devices in remote locations and (2) toward more refinements in source data automation. Output is distinguished by (1) more output in remote locations and (2) increasingly realistic—even lifelike—forms, as we discuss in a few pages.

Toward More Input from Remote Locations

When management consultant Steve Kaye of Santa Ana, California, wants to change a brochure or company letterhead, he doesn't have to drop everything and drive over to a printer. He simply enters his requests through the phone line to an electronic bulletin board at the Sir Speedy print shop that he deals with. "What this does is free me up to focus on my business," Kaye says.[18]

The linkage of computers and telecommunications means that data may be input from nearly anywhere. For instance, X-ray machines are now going digital, which means that a medical technician in the jungles of South America can take an X-ray of a patient and then transmit a perfect copy of it by satellite uplink to a hospital in Boston. Visa and MasterCard are moving closer to using "smart cards," or stored-value cards, for Internet transactions.

Toward More Source Data Automation

Keyboards and pointing devices will no doubt always be with us but increasingly, input technology is being designed to capture data at its source. This will reduce the costs and mistakes that come with copying or otherwise preparing data in a form suitable for processing.

Some reports from the input-technology front:

- **High-capacity bar codes:** Traditional bar codes read only horizontally. A new generation of bar codes reads vertically as well, which enables them to store more than 100 times as much data. With that added capacity, bar codes can now include digitized photos, along with a person's date of birth, eye color, blood type, and other personal data.

  Some grocery stores are experimenting with personal scanning guns. Customers speed through the store, waving the scanners over the bar code of any product they put in the basket, and then pay at an express checkout stand. (Random rescans make sure customers haven't "forgotten" to charge themselves for any items.)

  Experiments are also being tried with Web-enabled cellphones equipped with scanners that could read bar codes of nonelectronic things stamped with bar codes. Thus, for example, if you see someone across the library or airport reading a book that looks interesting, you could aim your bar-code–reading cellphone at the book, which would instantly produce an online bookseller (such as Amazon.com) with information about the book.[19]

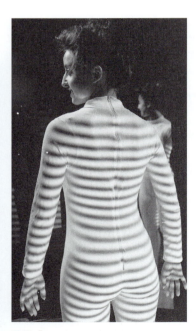

3-D jeans scanner

- **3-D scanners:** Do you have trouble finding the perfect fit in blue jeans? Clothes are designed to fit mannequins, which is why it's difficult to get that "sprayed-on" look. However, clothing makers (including Levi Strauss, the world's largest jeans maker) have inaugurated a body scanner that enables people to buy clothes that fit precisely.

  This is how it works. You enter a store, put on a body suit, and the scanner (which doesn't use lasers, to alleviate possible customer health concerns) measures you three-dimensionally. You can then select the clothes you're interested in, superimpose them on your body-scanned image on a screen, and order them custom-manufactured.

  Another experimental device allows you to laser-scan yourself in three dimensions, then digitally insert your replica (a so-called avatar) into a favorite videogame.

- **More sophisticated touch devices:** Touch screens are becoming commonplace. Sometime in the near future, futurists have suggested, your car may have a dashboard touch screen linked to mobile electronic "yellow pages," which would enable you to reserve a motel room for the night or find the nearest Chinese restaurant.

  More interesting is the invention of touch-sensitive "smart paper." TV Interactive Inc. has devised a way to coat plain paper with a crisscross grid of carbon-based ink that can conduct electricity. When you press a spot on the paper, an electrical contact sends a signal to a cheap, wafer-thin microchip embedded in the page. The technology is already finding a use in Japan, where karaoke fans are using "touch and view" song catalogs to initiate sing-along videos on their microcomputers.[20]

Implanted microchips
(Top) Sue Smith, of the Marin, California, Humane Society, scans a pet cat to identify its owner. The society has implanted more than 15,000 pets with microchips *(bottom)* containing owners' names, addresses, and phone numbers. The data is then logged into a nationwide database to help pet owners retrieve lost or stolen pets.

Finally, researchers in what is known as *hepatic*—active touch—systems are exploring how to create devices that will allow people to feel what isn't there. With this kind of "virtual touch," a dental student could train in drilling down into the decay of a simulated tooth without fear of destroying a real healthy tooth.[21]

• **Smarter smart cards:** Over the next few years, stored-value smart cards with microchips, acting as "electronic purses," will no doubt begin to displace cash in many transactions. Targets for smart cards are not only convenience stores and toll booths but also battery-powered card readers in newspaper racks and similar devices.

Already, as we mentioned earlier, microchips with identification numbers are being injected into dogs and cats so that, with the help of a scanner, stolen or lost pets can be identified. Although it's doubtful chip implantation for identification purposes would be extended to people (though it could), smart cards and optical cards could evolve into all-purpose cards including biotechnological identifiers. These could contain medical records, driver's license data, insurance information, security codes for the office, and frequent flier program information. Already, some health insurance companies are taking steps to give enrollees smart cards that contain their coverage information; when doctor's offices swipe the cards through card readers, patients can see what their insurance will pay for.

• **More sensors:** The long-elusive sense of smell is coming to machines. One "digital nose" has 32 sensors wired to it, each of which reacts with a wide variety of chemicals. This technology could extend the range of human smell in the same way that microscopes and binoculars now extend human vision. Such e-noses might come in handy, for instance in diagnosing disease or detecting whether Kona coffee beans are the real thing.[22] Military researchers have devised a "canary on a chip," a sensor that detects and identifies threatening chemical and biological agents—a modern twist on the dying canaries that alerted miners to the presence of dangerous gases.[23]

Sensors are beginning to find all kinds of other new uses as well. Luxury cars, for instance, employ sensors that automatically inflate or deflate air cells in the seat to provide the best support. In the Disklavier, a grand piano without strings, hammers pass across optical sensors that are used to generate a piano-like sound.[24] Miami Beach has tested use of infrared sensors that detect when a car leaves a parking place, then reset the time on the meter to zero—so that, alas, the next person wanting to park is deprived of a few free minutes on the dial.[25] Coca-Cola has experimented with a vending machine that has a temperature sensor that can automatically raise prices for cold drinks in hot weather.[26]

A special "smart needle" with embedded sensors can be inserted into a tumor to provide more information about the cancer, such as its response to chemotherapy.[27] Wheelchair-bound quadriplegics whose injuries are low enough on the spinal cord to preserve some shoulder and arm movement are undergoing experiments with shoulder-position sensors; these devices translate small movements in the shoulder into electrical impulses that can direct the muscles of the hand and arm.[28]

• **Better voice recognition:** "Typing is an aberration of our time," suggests vice president Roger Matus of Dragon Systems, one of the speech-recognition software makers. "In 10 years, it will be very rare that you'll get a device with a keyboard," he predicts. "And in 100 years, people will barely remember keyboards."[29]

While we're willing to bet that keyboards will always be popular, no doubt voice recognition is getting better. It's possible, for instance, that voice recognition may some day fulfill world travelers' fondest dream: You'll be able to speak in English, and a voice-recognition device will instantly translate your remarks into another language, whether French, Swahili, or Japanese. At the moment, translation programs such as Easy Translator can translate text on Web pages (English to Spanish, French, and German and the reverse), although they do so imperfectly.

- **Smaller electronic cameras:** Digital still cameras and video cameras are fast becoming commonplace. The next development may be the camera-on-a-chip, which will contain all the components necessary to take a photograph or make a movie. Such a device, called an *active pixel sensor*, based on NASA space technology, is now being made by a company called Photobit. Because it can be made on standard semiconductor production lines, the camera-on-a-chip can be made incredibly cheaply, perhaps for $20 apiece.

  Such micro-cameras could be put anywhere as security devices or for other purposes. "They might even make possible the camera in the glasses frame used by Tom Cruise in *Mission: Impossible*," says one report. "And instead of highlighting a passage in a book with a colored marker, a student of the future might make a snapshot of the passage with a camera built into a pen."[30]

- **Pattern-recognition and biometric devices:** Would you believe a computer could read people's emotions from changes in their facial patterns, like surprise and sadness? Such devices are being worked on at Georgia Institute of Technology and elsewhere.[31] *(See ● Panel 6.22.)*

● **PANEL 6.22**

Computers tuned to emotions

At Georgia Institute of Technology, researchers have devised a computer system that can read people's emotions from changes in their faces. Expressions of happiness, surprise, and other movements of facial muscles are converted by a special camera into digitized renderings of energy patterns.

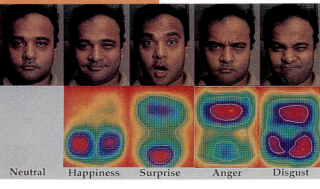

Neutral Happiness Surprise Anger Disgust

Indeed, you can buy a face-recognition program so that you can have your computer respond only to *your* smile.

In another development, IBM has demonstrated something it calls Emotion Mouse, which can measure your heart rate and body temperature and ultimately sense whether you are happy, sad, fearful, angry, or perplexed. IBM is also working on another intelligent system called Suitor (for Simple User Interest Tracker), which uses a miniature camera to pay attention to a user's behavior and actions, following one's gaze to determine the user's interests and information needs.[32]

- **Brainwave devices:** Perhaps the ultimate input device analyzes the electrical signals of the brain and translates them into computer commands. In experiments, users have successfully moved a cursor on the screen through sheer power of thought.[33] Individuals have even typed a letter by slowly spelling out the words in their heads.[34]

  Although there is a very long way to go before brainwave input technology becomes practical, the consequences not only for people with disabilities but for all of us could be tremendous.

Toward More Output in Remote Locations

What will happen to paper-and-ink newspapers in the Digital Age? Roger Fidler, former head of a research lab for the Knight-Ridder newspaper chain and now at Kent State University, has conceptualized a device called the *electronic news tablet,* which could be given or rented to newspaper readers.[35] This gadget would be about the same size as this textbook and perhaps half the thickness, and most of its surface area would be a screen. Readers would stick the tablet in a slot in an electronic rack and download a digital newspaper into its memory. The racks, of course, would be connected by some sort of electronic network transmitting information from a central source.

Clearly, output in remote locations is the wave of the future. As TV and the personal computer further converge, we can expect more scenarios like this: Your PC continually receives any Web sites covering topics of interest to you—CNN's home page for news on Yugoslavia, for example—which are stored on your hard disk for later viewing. The information, on up to 5000 Web sites a day, is transmitted from a television broadcast satellite to an 18-inch satellite dish on your roof. Interested? The technology is available now with direct broadcast services such as DirecTV Inc. You will need to install a $400 circuit board that permits Web surfing at speeds up to 30 megabits per second—more than 100 times as fast as current PC modems.[36]

Toward More Realistic Output

Once upon a time, having a "home theater" probably meant you were a wealthy film-industry figure with a large room containing a wall-size screen and movie-house seats. Now, says technology writer Phillip Robinson, "home theater has become a middle-income commodity with new audio and video gear promising to improve your television viewing so much you'll practically think you're in a theater instead of your living room or den."[37]

The enhanced qualities of home theater won't be limited to television viewing, of course. As analog and digital technologies merge, we will no doubt see this type of increased realism appearing in all forms of output. Let's consider what's coming into view.

- **Display screens—better and cheaper:** Computer screens will probably become crisper, brighter, bigger, and cheaper. Newer LCD monitors, for instance, have as high a resolution as previous larger CRT monitors. Although prices of flat-panel screens have not dropped as predicted, they will probably do so in the future.

  New gas-plasma technology is being employed to build flat-panel hang-on-the-wall screens as large as 50 inches from corner to corner. Using a technique known as microreplication, researchers have constructed a thin transparent sheet of plastic prisms that allows builders of portable computer screens to halve the amount of battery power required. Another technique, field-emission display (FED), also promises to lower power requirements and to simplify manufacturing of computer screens, thereby lowering the cost. All this is good news in an industry in which screens make up about 40% of a portable computer's cost.

  If the electronic information tablet designed to replace traditional newspapers seems exotic, consider another futuristic display technology: newspapers made of special electronic cloth—paper that is actually a digital display device. "It would look like newspaper, feel like newspaper," says technology writer Kevin Maney. "But each day (or hour, or minute!), new news and pictures would be downloaded, overriding the old."[38] E Ink, one company working on it, calls it "radio paper."

- **Audio—higher fidelity:** Not long ago, pointed out one writer, "the sound wafting from the PC [had] the crackly ring of grandpa's radio—the one he retired to the attic long ago."[39] Then came an audio card delivering stereo sound. In 1992 came wavetable synthesizers, with a wide library of sound samples recorded from musical instruments, from which digitally synthesized musical sounds can be drawn and blended. Then, in 1997, audio in personal computers began to shift to three-dimensional sound. For instance, mathematicians and musicians have devised an audio chip that can make a PC with two speakers "sound more like a 'surround sound' movie house, immersing the listener in waves of high-fidelity sound," according to one description.[40]

Unlike conventional stereo sound, 3-D audio describes an expanded field of sound—a broad arc starting at the right ear and curving around to the left. Thus, in a videogame, you might hear a rocket approach, go by, and explode off your right shoulder. The effect is achieved by boosting certain frequencies that provide clues to a sound's location in the room or by varying the timing of sounds from different speakers.

To test their technology, researchers use blind people who have acute hearing. Some researchers think that recent advances will put PC sound on the same par as home-theater audio systems.

- **Video—movie quality for PCs:** Today the movement of most video images displayed on a microcomputer comes across as fuzzy and jerky, with a person's lip movements out of sync with his or her voice. This is because currently available equipment is capable of running only about eight frames a second.

New technology based on digital wavelet theory, a complicated mathematical theory, has led to software that can compress digitized pictures into fewer bytes and do it more quickly than current standards. Indeed, the technology can display 30–38 frames a second—"real-time video." Images have the look and feel of a movie. Although this advance has more to do with software than hardware, it will clearly affect future video output. For example, it will allow two people to easily converse over microcomputer connections in a way that cannot be accomplished easily now. (We discuss compression in more detail in Chapter 7.)

- **Three-dimensional display:** In the 1930s, radiologists tried to create three-dimensional images by holding up two slightly offset X-rays of the same object and crossing their eyes. Now the same effects can be achieved by computers. With 3-D technology, flat, cartoon-like images give way to rounded objects with shadows and textures. Artists can even add "radiosity," so that a dog standing next to a red car, for instance, will pick up a red glow.

During the past several years, many people have come to experience 3-D images not only through random-dot stereograms like the Magic Eye books but also through videogames using virtual-reality head-mounted displays. Now companies are moving beyond VR goggles to design 3-D displays that produce two different eye views without glasses. In one design, for instance, the display remotely senses the viewer's head movements and moves lenses to change the scene presented to each eye.

In the 1990s, the Japanese manufacturer Sanyo demonstrated two types of 3-D systems on television screens. One was an experimental system that projected a 120-inch picture on a theater-type screen and required the use of special glasses with polarized lenses. The other, which did not involve special glasses, used what appeared to be a normal 40-inch TV set, with a screen that employed hundreds of tiny prisms/lenses. "The 3-D effect was stunning," wrote one reporter who saw it. "In one scene, water was sprayed from a hose directly at the camera. When watching the replay, I had to control the urge to jump aside."[41]

This kind of display technology has many uses beyond moving videogames to the next stage. It could give PCs the kind of 3-D graphics capability previously available only on expensive workstations and supercomputers.[42] It could enable online shoppers to "try on" clothes using 3-D models of their own bodies. Plastic surgeons could give patients a preview of their work. Images of organs inside bodies could help doctors plan and perform surgeries. Engineering design firms could exchange 3-D graphic designs over the Internet.

6.5 Input & Output Technology & Quality of Life: Health & Ergonomics

KEY QUESTION

What are the principal health and ergonomic issues relating to computer use?

Susan Harrigan, a financial reporter for *Newsday*, a daily newspaper based on New York's Long Island, had to learn to write her stories using a voice-activated computer. She did not do so by choice, nor was she as efficient as she had been. She did it because she was too disabled to type at all.

After more than two decades of writing articles with deadline-driven fingers at the keyboard, she had developed a crippling hand disorder. At first the pain was so severe she couldn't even hold a subway token. "Also, I couldn't open doors," she said, "so I'd have to stand in front of doors and ask someone to open them for me."[43]

Health Matters

Harrigan suffers from one of the computer-induced disorders classified as repetitive stress injuries (RSIs). The computer is supposed to make us efficient. Unfortunately, it has made some users—journalists, postal workers, data-entry clerks—anything but. The reasons are repetitive strain injuries, eyestrain and headache, and back and neck pains. In this section, we consider these health matters along with the effects of electromagnetic fields and noise.

- **Repetitive stress injuries:** *Repetitive stress (or strain) injuries (RSIs)* **are several wrist, hand, arm, and neck injuries resulting when muscle groups are forced through fast, repetitive motions.** The Bureau of Labor Statistics says 25% of all injuries that result in lost work time are due to RSI problems.[44] Most victims of RSI are in meat-packing, automobile manufacturing, poultry slaughtering, and clothing manufacturing. Musicians, too, are often troubled by RSI (because of long hours of practice).

  People who use computer keyboards—some of whom make as many as 21,600 keystrokes *an hour*—account for about 12% of RSI cases that result in lost work time.[45] Before computers came along, typists would stop to make corrections or change paper. These motions had the effect of providing many small rest breaks. Today keyboard users must devise their own mini-breaks to prevent excessive use of hands and wrists. People who use a mouse for more than a few hours a day—graphic designers, desktop publishing professionals, and the like—are also showing up in increased RSI injuries. The best advice is to find a mouse that fits your hand size and that supports the whole hand from the fingers to the palm.[46]

  Among the various RSIs, some, such as muscle strain and tendinitis, are painful but usually not crippling. These injuries, often caused by hitting the keys too hard, may be cured by rest, anti-inflammatory medication, and change in typing technique. However, carpal tunnel syndrome is disabling and often requires surgery. *Carpal tunnel syndrome (CTS)* **consists of a debilitating condition caused by pressure on the median nerve in the wrist, producing damage and pain to nerves and tendons in the hands.** *(See ● Panel 6.23.)*

  It's important to point out, however, that scientists still don't know what causes RSIs. They don't know why some people operating keyboards develop upper body and wrist pains and others don't. The working list of possible explanations for RSI includes "wrist size, stress level, relationship with supervisors, job pace, posture, length of workday, exercise routine, workplace furniture, [and] job security." Other possible contributors are diabetes, weight, and menopause.[47]

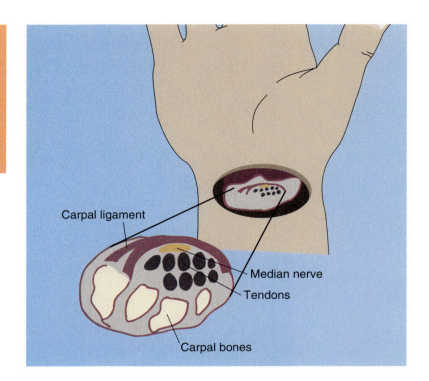

PANEL 6.23
Carpal tunnel syndrome
The carpal ligament creates a tunnel across the bones of the wrist. When the tendons passing through the carpal tunnel become swollen, they press against the median nerve, which runs to the thumb and the first three fingers.

Carpal ligament

Median nerve

Tendons

Carpal bones

- **Eyestrain and headaches:** Vision problems are actually more common than RSI problems among computer users.[48] Computers compel people to use their eyes at close range for a long time. However, our eyes were made to see most efficiently at a distance. It's not surprising, then, that people develop what's called computer vision syndrome.

     **Computer vision syndrome (CVS)** **consists of eyestrain, headaches, double vision, and other problems caused by improper use of computer display screens.** By "improper use," we mean not only staring at the screen for too long, but also failing to correct faulty lighting and screen glare and using screens with poor resolution.

- **Back and neck pains:** Improper chairs or improper positioning of keyboards and display screens can lead to back and neck pains. All kinds of adjustable, special-purpose furniture and equipment is available to avoid or diminish such maladies.

- **Electromagnetic fields:** Like kitchen appliances, hairdryers, and television sets, many devices related to computers and communications generate low-level electromagnetic field emissions. _**Electromagnetic fields (EMFs)**_ **are waves of electrical energy and magnetic energy.**

     In recent years, stories have appeared in the mass media reflecting concerns that high-voltage power lines, cellular phones, and CRT-type computer monitors might be harmful. There have been worries that monitors might be linked to miscarriages and birth defects, and that cellular phones and power lines might lead to some types of cancers.

     Is there anything to this? The answer is: So far no one is sure. The evidence seems scant that weak electromagnetic fields, such as those used for cellular phones and found near high-voltage lines, cause cancer. Still, handheld cellular phones do put the radio transmitter next to the user's head. This causes some health professionals concern about the effects of radio waves entering the brain as they seek out the nearest cellular transmitter. In 2000, the Independent Expert Group on Mobile Phones recommended that children should use cellphones only when necessary.[49] A researcher who reviewed 75 studies found none that suggested cellphones cause brain cancer, but he nevertheless recommended that users keep cellphone antennas at least 2½ inches from the head.[50] Meanwhile, the research continues.

As for CRT monitors, monitors made since the early 1980s produce very low emissions. Even so, users are advised to work no closer than arm's length to a CRT monitor. The strongest fields are emitted from the sides and backs of terminals. Alternatively, you can use laptop computers, because their liquid crystal display (LCD) screens emit negligible radiation.

The current advice from the Environmental Protection Agency is to exercise *prudent avoidance.* That is, we should take precautions that are relatively easy. However, we should not feel compelled to change our whole lives or to spend a fortune minimizing exposure to electromagnetic fields. Thus, we can take steps to put some distance between ourselves and a CRT monitor. However, it is probably not necessary to change residences because we happen to be living near some high-tension power lines.

- **Noise:** The chatter of impact printers or hum of fans in computer power units can be psychologically stressful to many people. Sound-muffling covers are available for impact printers. Some system units may be placed on the floor under the desk to minimize noise from fans.

Ergonomics: Design with People in Mind

Previously, workers had to fit themselves to the job environment. However, health and productivity issues have spurred the development of a relatively new field, called *ergonomics,* that is concerned with fitting the job environment to the worker.

The purpose of **_ergonomics_ is to make working conditions and equipment safer and more efficient.** It is concerned with designing hardware and software that is less stressful and more comfortable to use, that blends more smoothly with a person's body or actions. Examples of ergonomic hardware are tilting display screens, detachable keyboards, and keyboards hinged in the middle to allow the users' wrists to rest in a more natural position. *(See* ● *Panel 6.24.)*

We address some further ergonomic issues in the Experience Box immediately following.

● **PANEL 6.24**

Three ergonomic keyboards

Top: Keyboard by Kinesis. *Middle:* ComKey ergonomic keyboard. *Bottom:* Keyboard for one hand, used at rehabilitation center.

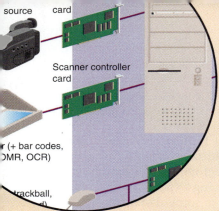
source card

Scanner controller card

r (+ bar codes, OMR, OCR)

trackball,

Experience Box

Good Habits: Protecting Your Computer System, Your Data, & Your Health

Whether you set up a desktop computer and never move it or tote a portable PC from place to place, you need to be concerned about protecting not only your computer but yourself. You don't want your computer to get stolen or zapped by a power surge. You don't want to lose your data. And you certainly don't want to lose your health for computer-related reasons. Here are some tips for taking care of these vital areas.

Guarding Against Hardware Theft & Loss

Portable computers are easy targets for thieves. Obviously, anything conveniently small enough to be slipped into your briefcase or backpack can be slipped into someone else's. Never leave a portable computer unattended in a public place.

It's also possible to simply lose a portable, as in forgetting it's in the overhead-luggage bin in an airplane. To help in its return, use a wide piece of clear tape to tape a card with your name and address to the outside of the machine. You should tape a similar card to the inside also. In addition, scatter a few such cards in the pockets of the carrying case.

Desktop computers are also easily stolen. However, for under $25, you can buy a cable and lock, like those used for bicycles, and secure the computer, monitor, and printer to a work area. If your hardware does get stolen, its recovery may be helped if you have inscribed your driver's license number, Social Security number, or home address on each piece. Some campus and city police departments lend inscribing tools for such purposes. Finally, insurance to cover computer theft or damage is surprisingly cheap. Look for advertisements in computer magazines. (If you have standard tenants' or homeowners' insurance, it may not cover your computer. Ask your insurance agent.)

Guarding Against Heat, Cold, Spills, & Drops

"We dropped 'em, baked 'em, we even froze 'em," proclaimed the *PC Computing* cover, ballyhooing a story about its notebook "torture test."[51]

The magazine put eight notebook computers through durability trials. One approximated putting these machines in a car trunk in the desert heat, another with leaving them outdoors in a Buffalo, New York, winter. A third test simulated sloshing coffee on a keyboard, and a fourth dropped them the equivalent of from desktop height to a carpeted floor. All passed the bake test, but one failed the freeze test. Three completely flunked the coffee-spill test, one other revived, and the rest passed. One that was dropped lost the right side of its display; the others were unharmed. Of the eight, half passed all tests unscathed. In a more recent torture test, nine notebooks survived the heat, cold, and spill tests, but three failed the drop test.[52]

This gives you an idea of how durable computers are. Designed for portability, notebooks may be hardier than desktop machines. Even so, you really don't want to tempt fate by dropping your computer, which could cause your hard-disk drive to fail.

Guarding Against Damage to Software

Systems software and applications software generally come on CD-ROM disks or flexible diskettes. The unbreakable rule is simply this: Copy the original disk, either onto your hard-disk drive or onto another diskette. Then store the original disk in a safe place. If your computer gets stolen or your software destroyed, you can retrieve the original and make another copy.

Protecting Your Data

Computer hardware and commercial software are nearly always replaceable, although perhaps with some expense and difficulty. Data, however, may be major trouble to replace or even irreplaceable. If your hard-disk drive crashes, do you have the same data on a back-up disk? Almost every microcomputer user sooner or later has the experience of accidentally wiping out or losing material and having no copy. This is what makes people true believers in backing up their data—making a duplicate in some form. If you're working on a research paper, for example, it's fairly easy to copy your work onto a floppy disk at the end of your work session.

Floppy disks can be harmed by any number of enemies. These include spills, dirt, heat, moisture, weights, and magnetic fields and magnetized objects. Here are some diskette-maintenance tips:

- Insert the floppy disk *carefully* into the disk drive.
- Don't manipulate the metal shutter on the floppy; it protects the surface of the magnetic material inside.
- Do not place heavy objects on the diskette.
- Do not expose the floppy to excessive heat or light.
- Do not use or place diskettes near a magnetic field, such as a telephone or paper clips stored in magnetic holders. Data can be lost if exposed.
- Do not use alcohol, thinners, or freon to clean the diskette.

Protecting Your Health

More important than any computer system and (probably) any data is your health. What adverse effects might computers cause? As we discussed earlier in the chapter, the most serious are painful hand and wrist injuries, eyestrain and headache, and back and neck pains.

Many people set up their computers in the same way as they would a typewriter. However, the two machines are ergonomically different for various reasons. With a computer, it's important to sit with both feet on the floor, thighs at right angles to your body. The chair should be adjustable and support your lower back. Your forearms should be parallel to the floor. You should look down slightly at the screen. *(See ● Panel 6.25.)* This setup is particularly important if you are going to be sitting at a computer for hours at a stretch.

To avoid wrist and forearm injuries, you should keep your wrists straight and hands relaxed as you type. Instead of putting the keyboard on top of a desk, therefore, you should put it on a low table or in a keyboard drawer under the desk. Otherwise the nerves in your wrists will rub against the sheaths surrounding them, possibly leading to RSI pains. Some experts also suggest using a padded, adjustable wrist rest, attached to the keyboard.

Eyestrain and headaches usually arise because of improper lighting, screen glare, and long shifts staring at the screen. Make sure your windows and lights don't throw a glare on the screen, and that your computer is not framed by an uncovered window. Headaches may also result from too much noise, such as listening for hours to an impact printer printing out.

Back and neck pains occur because furniture is not adjusted correctly or because of heavy computer use. Adjustable furniture and frequent breaks should provide relief here.

Some people worry about emissions of electromagnetic waves and whether they could cause problems in pregnancy or even cause cancer. The best approach is to simply work at an arm's length from computers with CRT-type monitors.

● PANEL 6.25
How to set up your computer work area

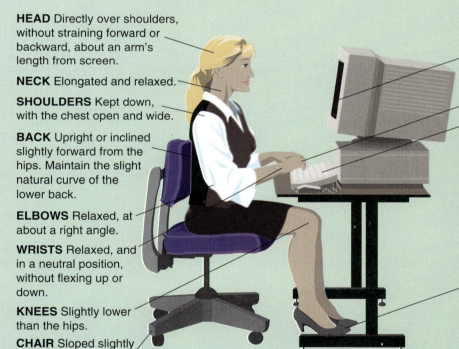

HEAD Directly over shoulders, without straining forward or backward, about an arm's length from screen.

NECK Elongated and relaxed.

SHOULDERS Kept down, with the chest open and wide.

BACK Upright or inclined slightly forward from the hips. Maintain the slight natural curve of the lower back.

ELBOWS Relaxed, at about a right angle.

WRISTS Relaxed, and in a neutral position, without flexing up or down.

KNEES Slightly lower than the hips.

CHAIR Sloped slightly forward to facilitate proper knee position.

LIGHT SOURCE Should come from behind the head.

SCREEN At eye level or slightly lower. Use an anti-glare screen.

FINGERS Gently curved.

KEYBOARD Best when kept flat (for proper wrist positioning) and at or just below elbow level. Computer keys that are far away should be reached by moving the entire arm, starting from the shoulders, rather than by twisting the wrists or straining the fingers. Take frequent rest breaks.

FEET Firmly planted on the floor. Shorter people may need a footrest.

Dig

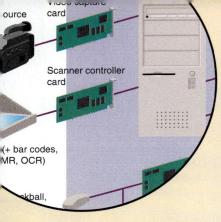

Video capture card

Scanner controller card

(+ bar codes, MR, OCR)

ckball,

Visual Summary

active-matrix display (p. 239, KQ 6.3) Also known as *TFT (thin-film transistor) display;* flat-panel display in which each pixel on the screen is controlled by its own transistor. Why it's important: *Active-matrix screens are much brighter and sharper than passive-matrix screens, but they are more complicated and thus more expensive. They also require more power, affecting the battery life in laptop computers.*

audio-input device (p. 232, KQ 6.2) Hardware that records analog sound and translates it for digital storage and processing. Why it's important: *Analog sound signals are continuous variable waves within a certain frequency range. For the computer to process them, these variable waves must be converted to digital 0s and 1s. The principal use of audio-input devices is to produce digital input for multimedia computers. An audio signal can be digitized in two ways—by an audio board or a MIDI board.*

bar-code reader (p. 230, KQ 6.2) Photoelectric scanner that translates bar codes into digital codes. Why it's important: *With bar-code readers and the appropriate software system, store clerks can total purchases and produce invoices with increased speed and accuracy; and stores and other businesses can monitor inventory and services with increased efficiency.*

bar codes (p. 230, KQ 6.2) Vertical zebra-striped marks imprinted on most manufactured retail products. Why it's important: *Bar codes provide a convenient means of identifying and tracking items. In North America, supermarkets, food manufacturers, and others have agreed to use a bar-code system called the Universal Product Code. Other kinds of bar-code systems are used on everything from FedEx packages, to railroad cars, to the jerseys of long-distance runners.*

biometrics (p. 236, KQ 6.2) Science of measuring individual body characteristics. Why it's important: *Biometric security devices identify a person through a fingerprint, voice intonation, or other biological characteristic. For example, retinal-identification devices use a ray of light to identify the distinctive network of blood vessels at the back of the eyeball.*

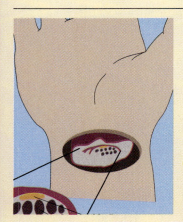

carpal tunnel syndrome (CTS) (p. 251, KQ 6.5) Debilitating condition caused by pressure on the median nerve in the wrist, producing damage and pain to nerves and tendons in the hands. Why it's important: *CTS can be caused by overuse or misuse of computer keyboards.*

computer vision syndrome (CVS) (p. 251, KQ 6.5) Eyestrain, headaches, double vision, and other problems caused by improper use of computer display screens. Why it's important: *CVS can be prevented by not staring at the display screen for too long, correcting faulty lighting, avoiding screen glare and not using screens with poor resolution.*

CRT (cathode-ray tube) (p. 239, KQ 6.3) Vacuum tube used as a display screen in a computer or video display terminal. Why it's important: *This technology is found not only in the screens of desktop computers but also in television sets and flight-information monitors in airports.*

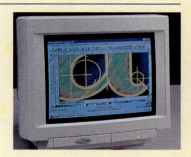

dedicated fax machine (p. 231, KQ 6.2) Specialized device that does nothing except send and receive fax documents. Why it's important: *Fax machines permit the transmission of text and graphic data over telephone lines quickly and inexpensively. They are found not only in offices and homes but also alongside regular phones in public places such as airports. See also* fax modem.

digital camera (p. 232, KQ 6.2) Electronic camera that uses a light-sensitive processor chip to capture photographic images in digital form on a small diskette inserted in the camera or on flash-memory chips. Why it's important: *The bits of digital information—the snapshots you have taken, say—can be copied right onto a computer's hard disk for manipulation and printing out. The environmentally undesirable stage of chemical development required for conventional film is completely eliminated.*

digitizer (p. 228, KQ 6.2) Input unit based on a mouselike copying device called a *puck,* which converts drawings and photos to digital data. Why it's important: *See* digitizing tablet.

digitizing tablet (p. 228, KQ 6.2) One form of digitizer; an electronic plastic board on which each specific location corresponds to a location on the screen. When you use a puck, the tablet converts your movements into digital signals that are input to the computer. Why it's important: *Digitizing tablets are often used to make maps and engineering drawings, as well as to trace drawings.*

display screen (p. 238, KQ 6.3) Also called *monitor, CRT,* or simply *screen*; output device that shows programming instructions and data as they are being input and information after it is processed. Why it's important: *Screens are needed to display softcopy output.*

dot-matrix printer (p. 240, KQ 6.3) Impact printer with a print head of small pins that strike an inked ribbon against paper to form characters or images. Print heads are available with 9, 18, or 24 pins; the 24-pin head offers the best quality. Why it's important: *Dot-matrix printers are employed much less frequently than laser printers and ink-jet printers but are still useful for some purposes. Dot-matrix printers can print draft quality, a coarser-looking 72 dpi, or near-letter-quality (NLQ), a crisper-looking 144 dpi. The machines print 40–300 characters per second and can handle graphics as well as text. Dot-matrix printers are used much less frequently than laser printers and ink-jet printers.*

dot pitch (dp) (p. 238, KQ 6.3) Amount of space between the centers of adjacent pixels; the closer the dots, the crisper the image. Why it's important: *Dot pitch is one of the measures of display-screen crispness. For a .28dp monitor, for instance, the dots are 28/100ths of a millimeter apart. Generally, a dot pitch of .28dp will provide clear images.*

dpi (dots per inch) (p. 240, KQ 6.3) Measure of the number of dots that are printed in a linear inch. For microcomputer printers, the resolution is in the range 60–1500 dpi. Why it's important: *The higher the dpi, the better the resolution.*

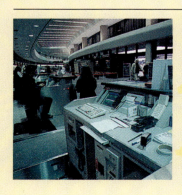

dumb terminal (p. 225, KQ 6.2) Display screen and a keyboard hooked up to a computer system; it can input and output but not process data. Why it's important: *Dumb terminals are used, for example, by airline reservations clerks to access a mainframe computer containing flight information.*

electromagnetic fields (EMFs) (p. 252, KQ 6.5) Waves of electrical energy and magnetic energy. Why it's important: *Some people have worried that CRT monitors might be linked to miscarriages and birth defects and that cellular phones and power lines might lead to some types of cancers. However, the evidence is unclear.*

electronic imaging (p. 230, KQ 6.2) Software-controlled integration of separate images, using scanners, digital cameras, and advanced graphic computers. Why it's important: *Electronic imaging is a whole new industry based on imaging-system technology. It has become an important part of multimedia.*

Digi

ergonomics (p. 253, KQ 6.5) Study, or science, of working conditions and equipment with the goal of improving worker safety and efficiency. Why it's important: *On the basis of ergonomic principles, stress, illness, and injuries associated with computer use may be minimized.*

fax machine (p. 231, KQ 6.2). Also called a *facsimile transmission machine;* input device that scans an image and sends it as electronic signals over telephone lines to a receiving fax machine, which prints the image on paper. Why it's important: *See* dedicated fax machine *and* fax modem.

fax modem (p. 231, KQ 6.2) Input device installed as a circuit board inside the computer's system cabinet; a modem with fax capability that enables you to send signals directly from your computer to someone else's fax machine or computer fax modem. Why it's important: *With this device, you don't have to print out the material from your printer and then turn around and run it through the scanner on a fax machine. The fax modem allows you to send information more quickly than if you had to feed it page by page into a machine. Fax modems are installed inside portable computers, including pocket PCs and PDAs. If you can also link up a cellular phone to a fax modem in your portable computer, you can send and receive wireless fax messages no matter where you are in the world.*

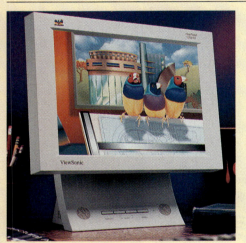

flat-panel display (p. 239, KQ 6.3) Display screens that are much thinner, weigh less, and consume less power than CRTs. Flat-panel displays are made up of two plates of glass separated by a layer of a substance in which light is manipulated. Why it's important: *Flat-panel displays are essential to portable computers, although they are available for desktop computers as well.*

hardcopy (p. 237, KQ 6.3) Printed output. Why it's important: *The principal examples are printouts, whether text or graphics, from printers. Film, including microfilm and microfiche, is also considered hardcopy output.*

imaging system (p. 229, KQ 6.2) Also called *image scanner* or *graphics scanner;* type of scanner that converts text, drawings, and photographs into digital form that can be stored in a computer system and then manipulated, output, or sent via modem to another computer. Why it's important: *An important part of multimedia. For example, the imaging system used in desktop publishing scans in artwork or photos, which can then be positioned within a page of text, using desktop publishing software.*

impact printer (p. 240, KQ 6.3) Printer that forms characters or images by striking a mechanism such as a print hammer or wheel against an inked ribbon, leaving an image on paper. Why it's important: *Nonimpact printers are more commonly used than impact printers, but dot-matrix printers are still used in some businesses.*

ink-jet printer (p. 242, KQ 6.3) Printer that sprays small, electrically charged droplets of ink from four nozzles through holes in a matrix at high speed onto paper. Like laser and dot-matrix printers, ink-jet printers form images with little dots. Why it's important: *Because they produce high-quality images on special paper, ink-jet printers are often used in graphic design and desktop publishing. However, ink-jet printers are slower than laser printers and print at a lower resolution on regular paper.*

input hardware (p. 224, KQ 6.1) Devices that translate data into a form the computer can process. Why it's important: *Without input hardware, computers could not function. The computer-readable form consists of 0s and 1s, represented as off and on electrical signals. Input hardware devices are categorized as three types: keyboards, pointing devices, and source data-entry devices.*

intelligent terminal (p. 226, KQ 6.2) Hardware unit with its own memory and processor, as well as a display screen and keyboard, hooked to a larger computer system. Why it's important: *Such a terminal can perform some functions independent of any mainframe to which it is linked. Examples include the automated teller machine (ATM), a self-service banking machine connected through a telephone network to a central computer, and the point-of-sale (POS) terminal, used to record purchases at a store's customer checkout counter. Recently, many intelligent terminals have been replaced by personal computers.*

Internet terminal (p. 226, KQ 6.2) Terminal that provides access to the Internet. There are several variants of Internet terminal: (1) the set-top box or Web terminal, which displays Web pages on a TV set; (2) the network computer, a cheap, stripped-down computer that connects people to networks; (3) the online game player, which not only lets you play games but also connects to the Internet; (4) the full-blown PC/TV (or TV/PC), which merges the personal computer with the television set; and (5) the wireless pocket PC or personal digital assistant (PDA), a handheld computer with a tiny keyboard that can do two-way wireless messaging. Why it's important: *In the near future, most likely, Internet terminals will be everywhere.*

keyboard (p. 225, KQ 6.2) Input device that converts letters, numbers, and other characters into electrical signals that can be read by the computer's processor. Why it's important: *Keyboards are the most popular kind of input device.*

laser printer (p. 241, KQ 6.3) Nonimpact printer that creates images with dots. As in a photocopying machine, images are produced on a drum, treated with a magnetically charged ink-like toner (powder), and then transferred from drum to paper. Why it's important: *Laser printers produce much better image quality than do dot-matrix printers and can print in many more colors; they are also quieter. Laser printers, along with page description languages, enabled the development of desktop publishing.*

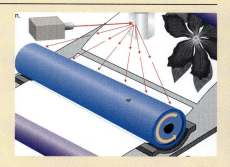

light pen (p. 228, KQ 6.2) Light-sensitive penlike device connected by a wire to the computer terminal. The user brings the pen to a desired point on the display screen and presses the pen button, which identifies that screen location to the computer. Why it's important: *Light pens are used by engineers, graphic designers, and illustrators.*

liquid crystal display (LCD) (p. 239, KQ 6.3) Flat-panel display in which molecules of liquid crystal line up in a way that alters their optical properties, creating images on the screen by transmitting or blocking out light. Why it's important: *LCD is useful not only for portable computers but also as a display for various electronic devices, such as watches and radios.*

magnetic-ink character recognition (MICR) (p. 230, KQ 6.2) Scanning technology that reads magnetized-ink characters printed at the bottom of checks and converts them to digital form. Why it's important: *MICR technology is used by banks to sort checks.*

multifunction printer (p. 243, KQ 6.3) Hardware device that combines several capabilities, such as printing, scanning, copying, and faxing. Why it's important: *Multifunction printers take up less space and cost less than the four separate office machines that they replace. The downside, however, is that, if one component breaks, nothing works.*

nonimpact printer (p. 241, KQ 6.3) Printer that forms characters and images without direct physical contact between the printing mechanism and paper. Two types of nonimpact printers often used with microcomputers are laser printers and ink-jet printers. A third kind, the thermal printer, is seen less frequently. Why it's important: *Nonimpact printers are faster and quieter than impact printers.*

```
0123456789
ABCDEFGHIJKLMNOPQR
>$/-+-#ⁿ
```

optical character recognition (OCR) (p. 231, KQ 6.2) Scanning technology that reads special preprinted characters in a particular font (typeface design) and converts them to digital code. Why it's important: *OCR characters appear on utility bills and price tags on department-store merchandise.*

```
00123456789
ACENPSTVX
<+>-¥
```

optical mark recognition (OMR) (p. 231, KQ 6.2) Scanning technology that reads pencil marks and converts them into computer-usable form. Why it's important: *OMR technology is used to read the College Board Scholastic Aptitude Test (SAT) and the Graduate Record Examination (GRE).*

Dig

output hardware (p. 224 KQ 6.1) Hardware devices that convert machine-readable information, obtained as the result of processing, into people-readable form. The principal kinds of output are softcopy and hardcopy. Why it's important: *Without output devices, people would have no access to processed data and information.*

page description language (p. 242, KQ 6.3) Software that describes the shape and position of characters and graphics to the printer. PostScript (from Adobe Systems) and Hewlett-Packard Graphic Language (HPGL) are common page description languages. Why it's important: *Page description languages are essential to desktop publishing.*

passive-matrix display (p. 239, KQ 6.3) Flat-panel display in which a transistor controls a whole row or column of pixels. Passive matrix provides a sharp image for one-color (monochrome) screens but is more subdued for color. Why it's important: *Passive-matrix displays are less expensive and use less power than active-matrix displays, but they aren't as clear and bright and can leave "ghosts" when the display changes quickly. Passive-matrix displays go by the abbreviations HPA, STN, or DSTN.*

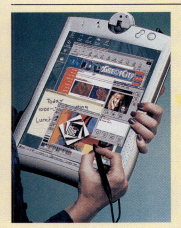

pen-based computer system (p. 228, KQ 6.2) Input system that allows users to enter handwriting and marks onto a computer screen by means of a penlike stylus rather than by typing on a keyboard. Pen computers use handwriting recognition software that translates handwritten characters made by the stylus into data that is usable by the computer. Why it's important: *Many handheld computers and PDAs have pen input, as do digital notebooks.*

pixel (p. 238, KQ 6.3) Short for "picture element"; the smallest unit on the screen that can be turned on and off or made different shades. Why it's important: *Pixels are the building blocks that allow text and graphical images to be displayed on a screen.*

plotters (p. 243, KQ 6.3) Specialized output device designed to produce high-quality graphics in a variety of colors. The *ink-jet plotter* employs the same principle as an ink-jet printer; the paper is output over a drum, enabling continuous output. In an *electrostatic plotter,* paper lies partially flat on a table-like surface, and toner is used in a photocopier-like manner. Why it's important: *Plotters are used to create hardcopy items such as maps, architectural drawings, and three-dimensional illustrations, which are usually too large for regular printers.*

pointing devices (p. 226, KQ 6.2) Hardware that controls the position of the cursor or pointer on the screen. It includes the mouse and its variants, the touch screen, and various forms of pen input. Why it's important: *In many contexts, pointing devices permit quick and convenient data input.*

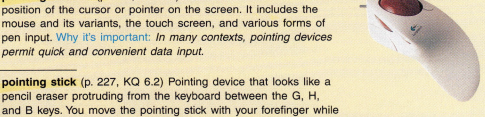

pointing stick (p. 227, KQ 6.2) Pointing device that looks like a pencil eraser protruding from the keyboard between the G, H, and B keys. You move the pointing stick with your forefinger while using your thumb to press buttons located in front of the space bar. Why it's important: *Pointing sticks are used principally in videogames, in computer-aided design systems, and in robots.*

printer (p. 240, KQ 6.3) Output device that prints characters, symbols, and perhaps graphics on paper or another hardcopy medium. Why it's important: *Printers provide one of the principal forms of computer output.*

radio-frequency identification technology (p. 236, KQ 6.2) Also known as RF-ID tagging; a source data-entry technology based on an identifying tag bearing a microchip that contains specific code numbers. These code numbers are read by the radio waves of a scanner linked to a database. Why it's important: *Drivers with RF-ID tags can breeze through tollbooths without having to even roll down their windows; the toll is automatically charged to their accounts. Radio-readable ID tags are also used by the Postal Service to monitor the flow of mail, by stores for inventory control and warehousing, and in the railroad industry to keep track of rail cars.*

refresh rate (p. 238, KQ 6.3) Number of times per second that screen pixels are recharged so that their glow remains bright. In general, displays are refreshed 45–100 times per second. Why it's important: *The higher the refresh rate, the more solid the image looks on the screen—that is, the less it flickers.*

repetitive stress injuries (RSIs) (p. 251, KQ 6.5) Several wrist, hand, arm, and neck injuries resulting when muscle groups are forced through fast, repetitive motions. They include muscle strain and tendinitis, which are painful but usually not crippling, and carpal tunnel syndrome, which is disabling and often requires surgery. Why it's important: *People who use computer keyboards account for about 12% of RSI cases that result in lost work time.*

resolution (p. 238, KQ 6.3) Clarity or sharpness of display-screen images; the more pixels there are per square inch, the finer the level of detail attained. Resolution is expressed in terms of the formula *horizontal pixels* × *vertical pixels.* Each pixel can be assigned a color or a particular shade of gray. Standard resolutions are 640 × 480, 800 × 600, 1024 × 768, 1280 × 1024, and 1600 × 1200 pixels. Why it's important: *Users need to know what screen resolution is appropriate for their purposes.*

scanner (p. 229, KQ 6.2) Source-data input device that uses light-sensing equipment to translate images of text, drawings, photos, and the like into digital form. Why it's important: *Scanners simplify the input of complex data. The images can be processed by a computer, displayed on a monitor, stored on a storage device, or communicated to another computer.*

sensor (p. 236, KQ 6.2) Input device that collects specific data directly from the environment and transmits it to a computer. Why it's important: *Although you are unlikely to see such input devices connected to a PC in an office, they exist all around us, often in nearly invisible form. Sensors can be used to detect all kinds of things: speed, movement, weight, pressure, temperature, humidity, wind, current, fog, gas, smoke, light, shapes, images, and so on. In aviation, for example, sensors are used to detect ice buildup on airplane wings and to alert pilots to sudden changes in wind direction.*

softcopy (p. 237, KQ 6.3) Data on a display screen or in audio or voice form. This kind of output is not tangible; it cannot be touched. Why it's important: *This term is used to distinguish nonprinted output from printed (hardcopy) output.*

sound-output devices (p. 244, KQ 6.3) Hardware that produces digitized sounds, ranging from beeps and chirps to music. Why it's important: *To use sound output, you need appropriate software and a sound card. Such devices are used to produce the sound effects when you play a CD-ROM, for example.*

source data-entry devices (p. 229, KQ 6.2) Data-entry devices that create machine-readable data on magnetic media or paper or feed it directly into the computer's processor, without the use of a keyboard. Categories include scanning devices (imaging systems, bar-code readers, mark- and character-recognition devices, and fax machines), audio-input devices, video input, and photographic input (digital cameras), voice-recognition systems, sensors, radio-frequency identification devices, and human-biology input devices. Why it's important: *Source-data entry devices lessen reliance on keyboards for data entry and can make data entry more accurate.*

SVGA (super video graphics array) (p. 240, KQ 6.3) Graphics board standard that supports a resolution of 800 × 600 pixels, or variations, producing 16 million possible simultaneous colors. Why it's important: *SVGA is the most common standard used today.*

SXGA (super extended graphics array) (p. 240, KQ 6.3) Graphics board standard that supports a resolution of 1280 × 1024 pixels. Why it's important: *SXGA is often used with 19- and 21-inch monitors.*

thermal printer (p. 243, KQ 6.3) Printer that uses colored waxes and heat to produce images by burning dots onto special paper. The colored wax sheets are not required for black-and-white output. Thermal printers are expensive, and they require expensive paper. Why it's important: *For people who want the highest-quality color printing available with a desktop printer, thermal printers are the answer.*

touchpad (p. 227, KQ 6.2) Input device; a small, flat surface over which you slide your finger, using the same movements as you would with a mouse. The cursor follows the movement of your finger. You "click" by tapping your finger on the pad's surface or by pressing buttons positioned close by the pad. Why it's important: *Touchpads let you control the cursor/pointer with your finger, and they require very little space to use. Most laptops have touchpads.*

touch screen (p. 227, KQ 6.2) Video display screen that has been sensitized to receive input from the touch of a finger. The screen is covered with a plastic layer, behind which are invisible beams of infrared light. Why it's important: *You can input requests for information by pressing on buttons or menus displayed. The answers to your requests are displayed as output in words or pictures on the screen. (There may also be sound.) You find touch screens in kiosks, ATMs, airport tourist directories, hotel TV screens (for guest checkout), and campus information kiosks making available everything from lists of coming events to (with proper ID and personal code) student financial-aid records and grades.*

trackball (p. 227, KQ 6.2) Movable ball, mounted on top of a stationary device, that can be rotated using your fingers or palm. It looks like the mouse turned upside down. Instead of moving the mouse around on the desktop, you move the trackball with the tips of your fingers. Why it's important: *Trackballs require less space to use than does a mouse.*

UXGA (ultra extended graphics array) (p. 240, KQ 6.3) Graphics board standard that supports a resolution of 1600 × 1200 pixels. Why it's important: *UXGA is expected to become more popular with graphic artists, engineering designers, and others using 21-inch monitors.*

video (p. 245, KQ 6.3) Output consisting of photographic images played at 15–29 frames per second to give the appearance of full motion. Why it's important: *Video is input into a multimedia system using a video camera or VCR and, after editing, is output on a computer's display screen. Because video files can require a great deal of storage—a 3-minute video may require 1 gigabyte of storage—video is often compressed. Digital video has revolutionized the movie industry, as in the use of special effects.*

videoconferencing (p. 245, KQ 6.3) Form of video output in which people in different geographical locations can have a meeting—can see and hear one another—using computers and communications. Why it's important: *Many organizations use videoconferencing to take the place of face-to-face meetings. Videoconferencing systems range from videophones to group conference rooms with cameras and multimedia equipment to desktop systems with small video cameras, microphones, and speakers.*

voice-output device (p. 244, KQ 6.3) Hardware that converts digital data into speech-like sounds. Why it's important: *You hear such voice output on telephones ("Please hang up and dial your call again"), in soft-drink machines, in cars, in toys and games, and recently in mapping software for vehicle-navigation devices. For people with physical challenges, computers with voice output help to level the playing field.*

voice-recognition system (p. 234, KQ 6.2) Input system that uses a microphone (or a telephone) as an input device and converts a person's speech into digital signals by comparing the electrical patterns produced by the speaker's voice with a set of prerecorded patterns stored in the computer. Why it's important: *Voice-recognition technology is useful in situations where people are unable to use their hands to input data or need their hands free for other purposes.*

XGA (extended graphics array) (p. 240, KQ 6.3) Graphics board display standard with a resolution of up to 1024 × 768 pixels, corresponding to 65,536 possible colors. Why it's important: *XGA offers the most sophisticated standard for color and resolution. It is used mainly on workstation systems, such as those employed by engineering designers.*

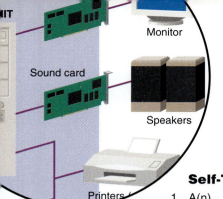

Chapter Review

Self-Test Questions

1. A(n) _____ terminal is entirely dependent for all its processing activities on the computer system to which it is connected.

2. The two main categories of printer are _____ and _____.

3. _____ is the study of the physical relationships between people and their work environment.

4. A(n) _____ is an input device that is rolled about on a desktop and directs a pointer on the computer's display screen.

5. _____ consists of devices that translate information processed by the computer into a form that humans can understand.

6. _____ is the science of measuring individual body characteristics.

7. CRT is short for _____.

8. LCD is short for _____.

9. A _____ is software that describes the shape and position of characters and graphics to the printer.

10. When people in different geographical locations can have a meeting using computers and communications, it is called _____.

Multiple-Choice Questions

1. Which of the following is not a pointing device?
   a. mouse
   b. touchpad
   c. keyboard
   d. joystick

2. Which of the following is not a source data-entry device?
   a. bar-code reader
   b. sensor
   c. digital camera
   d. scanner
   e. mouse

True/False Questions

T F   1. On a computer screen, the more pixels that appear per square inch, the higher the resolution.

T F   2. Photos taken with a digital camera can be downloaded to a computer's hard disk.

T F   3. *Resolution* is the amount of space between the centers of adjacent pixels.

T F   4. The abbreviation *dpi* stands for *dense pixel intervals.*

T F   5. Pointing devices control the position of the cursor on the screen.

Short-Answer Questions

1. What determines how a keyboard's function keys work?

2. What characteristics determine the clarity of a computer screen?

3. Describe two situations in which scanning is useful.

4. What is source-data entry?

5. Why is it important for your computer to be expandable?

6. What is *pixel* short for? What is a pixel?

7. Briefly describe *RSI* and *CTS.* Why are they a problem?

Concept Mapping

On a separate sheet of paper, draw a concept map, or visual diagram, linking concepts. Show how the following terms are related.

bar-code reader	laser printer
CRT	LCD
digitizer	nonimpact printer
display screen	OCR
dot-matrix printer	output hardware
dot pitch	PDL
dpi	pixel
fax machine	plotter
flat-panel display	pointing device
hardcopy	resolution
input hardware	scanner
keyboard	source-data entry

Knowledge in Action

1. Visit a local electronics store or an online shopping site such as *www.beyond.com/hardware/printers.htm* and investigate five different types of printers for sale. Note the (a) type of printer, (b) price, and (c) resolution, as well as whether the printer is PC- or Mac-compatible. Click on "Specs" for additional information on each printer. Which printer would you choose? Why?

2. Cut out an advertisement from a newspaper or a magazine that features a new microcomputer system. Circle all the terms that are familiar to you now that you have read the first four chapters of this text. Define these terms on a separate sheet of paper. Is this computer expandable?

How much does it cost? Is the monitor included in the price? A printer?

3. *Paperless office* is a term that has been around for some time. However, the paperless office has not yet been achieved. Do you think the paperless office is a good idea? Do you think it's possible? Why do you think it has not yet been achieved?

Telecommunications

Networks & Communications: The "New Story" in Computing

Key Questions

You should be able to answer the following questions.

7.1   **From the Analog to the Digital Age** How do digital and analog data differ, and what does a modem do?

7.2   **The Practical Uses of Communications** What are some offerings of new telecommunications technology?

7.3   **Communications Channels: The Conduits of Communications** What are types of wired and wireless channels and some types of wireless communications?

7.4   **Networks** What are the benefits of networks, and what are their types, components, and variations?

7.5   **The Future of Communications** What are the characteristics of the next generation of wireless communications, satellite systems, gigabit Ethernets, Bluetooth, photonics, and power lines?

7.6   **Cyberethics: Controversial Material, Privacy, & Intellectual Property** What are important issues in cyberethics?

A Visual Overview of This Chapter

Graphical Interface

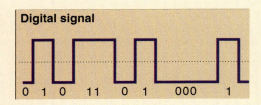

Digital signal

0   1   0   11   0   1   000   1

1 From the Analog to the Digital Age. Computers use **digital** signals, which present information in a binary way. Most other systems, such as telephones and TV, use **analog** signals, which continuously vary in strength or quality. A **modem** converts digital signals into analog signals, so that computer signals can be sent over phone lines.

2 The Practical Uses of Communications. There are many forms of connectivity, or communications connections. (1) **Videoconferencing,** the linking of people through TV video and sound plus computers, is useful for long-distance meetings. (2) **Workgroup computing,** in which microcomputer networks enable workers to cooperate on projects, allows people to work on the same information at the same time. (3) **Telecommuting,** working at home with telecommunications, can increase productivity. (4) **Virtual offices,** nonpermanent, mobile offices run with telecommunications and computers, add workplace flexibility. (5) Home networks enable households to link and share all kinds of peripheral devices. (6) Information appliances deliver all types of data anywhere at any time. (7) Smart television consists of **digital television (DTV),** which uses a digital signal, or series of 0s and 1s. The high-resolution type of DTV is **high-definition television (HDTV),** which comes in 720- or 1080-line mode (compared to 525-line mode for analog TV). However, an alternate DTV is **standard-definition television (SDTV),** which has 480 lines, allowing broadcasters to transmit more information within the HDTV bandwidth so that they can broadcast six channels instead of one.

3 Communications Channels: The Conduits of Communication. The following factors affect how data is transmitted.

The **electromagnetic spectrum** consists of fields of electrical energy and magnetic energy, which travel in waves. In the middle is the **radio frequency spectrum,** fields of electrical and magnetic energy that carry communications signals, which vary according to frequency, or repeating waves. A range of frequencies is called a **band** or **bandwidth.** The wider the band, the faster data can be transmitted. **Broadband connections** are very high speed.

A **communications channel** is the path over which information travels in a telecommunications system. Channels may be wired or wireless.

Three types of *wired channels* are the following. (1) **Twisted-pair wire,** or standard telephone wire,

Radar

dio frequencies

HF VHF UHF SHF EHF

High Frequency 3 MHz–30 MHz: Shortwave broadcast; amateur radio; CB (citizen band) radio

Very High Frequency 30 MHz–300 MHz: Private radio land mobile services such as police, fire, and taxi dispatch; TV channels (2-13); FM broadcasting; cordless phones; baby

Ultrahigh Frequency 300 MHz–3000 MHz (3 GHz): UHF TV channels; cellular phones; common carrier point-to-point microwave transmission used by long-distance

consists of two strands of insulated copper wire twisted around each other; it is used for both voice and data transmission. (2) **Coaxial cable** consists of insulated copper wire wrapped in other materials; it is better than twisted-pair for resisting noise. (3) **Fiber-optic cable** consists of thin strands of glass or plastic that transmit beams of light rather than electricity; it is very fast and noise-resistant.

Four types of *wireless channels* are the following. (1) **Infrared transmission** sends data via infrared-light waves, as in some wireless mouses. (2) **Broadcast radio** sends data over long distances, as between states. (3) **Microwave radio** transmits voice and data via superhigh-frequency radio waves, as between hilltops. (4) **Communications satellites** are microwave relay stations that orbit the earth, occupying low, medium, or high (geostationary) earth orbits.

Types of long-distance wireless communications may be one-way or two-way. One-way is exemplified by (1) the **Global Positioning System (GPS),** timed radio signals sent by satellites that can be used to identify earth locations, and by (2) **pagers,** radio receivers that receive data from special radio transmitters. Two-way is exemplified by (1) two-way pagers; (2) **analog cellular phones,** designed for communicating by voice through a system of **cells,** each 8 miles or less in diameter and served by a transmitter-receiving tower; (3) **packet-radio-based communications,** which use a nationwide system of radio towers to send data to handheld computers; and (4) **CDPD,** for Cellular Digital Packet Data, which places messages in digital envelopes, or packets, and sends them through underused radio channels between pauses in cellular phone conversations.

Compression is a method of removing repetitive elements from a file so that it requires less time to transmit; at the receiving end, the file is *decompressed*—the repeated patterns are restored. Two methods of compression are lossless and lossy. *Lossless compression* uses mathematical techniques to replace repetitive patterns of bits with a kind of coded summary; on decompression, the bits are restored, so that the data is the same as what went in—important in database and similar computer data. *Lossy compression* permanently discards some data during compression; it is often used for graphics files and sound files. Two compression standards are **JPEG,** for still images, and **MPEG,** for moving images.

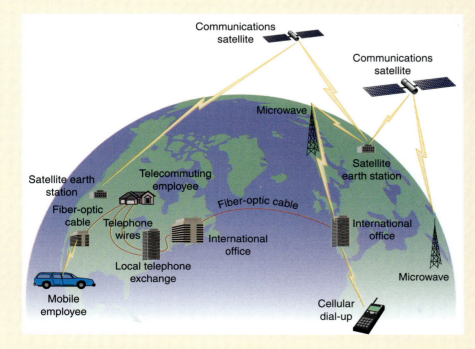

Communications satellite

Communications satellite

Microwave

Satellite earth station

Satellite earth station

Telecommuting employee

Fiber-optic cable

Fiber-optic cable

Telephone wires

Local telephone exchange

International office

International office

Microwave

Mobile employee

Cellular dial-up

4 **Networks.** A communications **network** is a system of interconnected computers, phones, or other communications devices that can share applications and data. Among the benefits: Networks enable sharing of peripheral devices, programs, and data; better communications; improved security of information; and access to numerous databases.

Types of networks are as follows. A **wide area network (WAN)** covers a wide geographical area, such as a country. A **metropolitan area network (MAN)** covers a city or suburb. A **local area network (LAN)** covers a limited area such as an office or a building. Most large networks have a **host computer,** a mainframe or midsize central computer to control the network. Any device attached to a network is called a **node.** MANs and LANs may be connected to the Internet by a high-speed network called a **backbone.**

Two types of LANs are client/server and peer-to-peer. A **client/server LAN** consists of microcomputers requesting data (*clients*) and powerful computers supplying data (*servers*). A **file server,** for example, stores programs and data files; other servers are database server, printer server, Web server, and mail server. In a **peer-to-peer LAN,** there is no server; microcomputers on a network communicate with each other directly. Several standard components of a LAN are the connection or cabling system, microcomputers with network interface cards, network operating system (Novell NetWare, Microsoft Windows NT/2000, Unix, or Linux), and other shared devices (printers, storage devices). Other components are a **router,** a special computer that directs communicating messages when networks are tied together; a **bridge,** an interface used to connect the same types of networks; and a **gateway,** an interface permitting communication between dissimilar networks.

Networks can be laid out in three different **topologies,** or shapes: star, ring, and bus. In a **star network,** all devices are connected to a central server. In a **ring network,** all devices are connected in a continuous loop. In a **bus network,** all devices are connected to a common channel.

Organizations now use two variant networks that use the Internet's infrastructure and standards. One is an **intranet,** an organization's internal private network for employee use. The other is an **extranet,** for selected suppliers and other strategic parties as well as employees. Security for such networks is maintained through a **firewall,** a system of hardware and software that blocks unauthorized users inside and outside the organization.

5 The Future of Communications. Among new developments: (1) **Digital cellular phone** networks are replacing analog networks and are turning voice messages into digital bits. (2) **Personal communications services (PCS)** are digital wireless services that use a new band of frequencies and transmitter-receivers in thousands of microcells. (3) Satellite-based systems consisting of *global high-speed satellite networks* are going up, which will permit users to exchange a broader range of data, including Internet pages and videophone calls, anywhere in the world. (4) The Ethernet, a bus network devised in the 1970s, could be replaced by *Gigabit Ethernet,* which can move data at 1 gigabit per second. (5) **Bluetooth** is a wireless technology that uses simple shortwave radio links to allow as many as eight electronic devices to communicate within 33 feet of one another. (6) *Photonics* is the science of sending data bits by means of light pulses carried on glass fibers; it would enable more light signals to be carried on fiber-optic lines.

6 Cyberethics: Controversial Material, Privacy, & Intellectual Property. Three important issues of cyberethics are as follows.

To protect children against access to controversial material, parents may employ blocking software that screens objectionable material based on keywords;

browsers that contain built-in ratings for Internet and World Wide Web use; and the V-chip, which allows the screening of TV programs high in violence, sex, and the like.

Privacy, the right of people not to reveal information about themselves, is under pressure from information technology. Web **cookies,** files stored on a user's hard drive when he or she visits a Web site, allow Web site operators to track user movements online. Medical records and office e-mail are other areas in which there may be privacy intrusions.

File-sharing programs have allowed users to get copyrighted materials without having to compensate the copyright owners.

The essence of all revolution, stated philosopher Hannah Arendt, is the start of a *new story* in human experience.

Before the 1950s, computing devices processed data into information, and communications devices communicated information over distances. The two streams of technology developed pretty much independently, like rails on a railroad track that never merge. Now we have a new story, a revolution.

For us, the new story has been *digital convergence*—the gradual merger of computing and communications into a new information environment, in which *the same information is exchanged among many kinds of equipment, using the language of computers.* (See ● Panel 7.1.) At the same time, there has been a convergence of several important industries—computers, telecommunications, consumer electronics, entertainment, mass media—producing new electronic products that perform multiple functions.

An example is what's been happening in television. *WebTV* consists of a set-top box powered by WebTV Networks, a subsidiary of Microsoft. The box makes a conventional television set function like three different devices: a television with satellite service, an Internet-linked computer, and an enhanced digital video recorder. The result of this convergence of technologies is that, besides watching TV programs, you can also interact with them in real time; for instance, you might participate in *Jeopardy!* along with the contestants you're viewing on the screen.[1] Or you can order a Domino's Pizza by pointing and clicking with a remote control.[2] Late arriving, as we discuss later in the chapter, is *digital television (DTV)* and the variant called *high-definition television (HDTV)*, which are supposed to deliver not only crisper images but also more channels and computer connections.[3]

● **PANEL 7.1**
Digital convergence—the fusion of computer and communications technologies
Today's new information environment came about gradually from the merger of two separate streams of technological development—computers and communications.

Computer Technology

1621 AD	1642	1833	1843	
Slide rule invented (Edmund Gunther)	First mechanical adding machine (Blaise Pascal)	Babbage's difference engine (automatic calculator)	World's first computer programmer, Ada Lovelace, publishes her notes	

Communications Technology

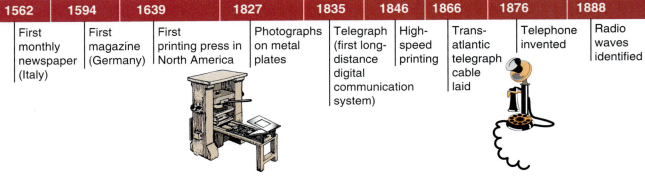

1562	1594	1639	1827	1835	1846	1866	1876	1888
First monthly newspaper (Italy)	First magazine (Germany)	First printing press in North America	Photographs on metal plates	Telegraph (first long-distance digital communication system)	High-speed printing	Trans-atlantic telegraph cable laid	Telephone invented	Radio waves identified

7.1 From the Analog to the Digital Age

Why have the worlds of computers and of telecommunications been so long in coming together? Because *computers are digital, but most of the world has been analog.* Let's take a look at what this means.

The Digital Basis of Computers: Electrical Signals as Discontinuous Bursts

Computers may seem like incredibly complicated devices but, as we saw in Chapter 1, their underlying principle is simple. Because they are based on on/off electrical states, they use the *binary system,* which consists of only two digits—0 and 1. Today **_digital_ specifically refers to communications signals or information represented in a two-stat (binary) way.** More generally, *digital* is usually synonymous with "computer-based."

Digital data consists of data (expressed as 0s and 1s) represented by on/off electrical pulses. These pulses are transmitted in discontinuous bursts rather than (as with analog devices) in continuous waves.

The Analog Basis of Life: Electrical Signals as Continuous Waves

"The shades of a sunset, the flight of a bird, or the voice of a singer would seem to defy the black or white simplicity of binary representation," points out one writer.[4] Indeed, these and most other phenomena of the world are **_analog_, continuously varying in strength and/or quality.** Sound, light, temperature, and pressure values, for instance, can fall anywhere on a continuum or range. The highs, lows, and in-between states have historically been represented with analog devices rather than in digital form. Examples of analog devices are a speedometer, a thermometer, and a tire-pressure gauge, all

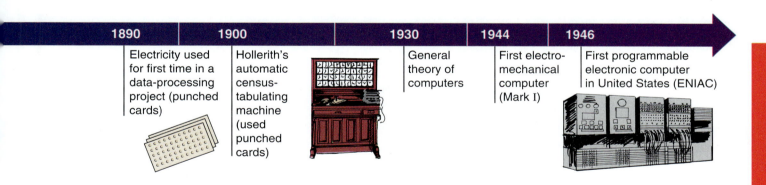

1890	1900		1930	1944	1946
Electricity used for first time in a data-processing project (punched cards)	Hollerith's automatic census-tabulating machine (used punched cards)		General theory of computers	First electro-mechanical computer (Mark I)	First programmable electronic computer in United States (ENIAC)

1894	1895	1912	1915	1928	1939	1946	1947	1948	1950
Edison makes a movie	Marconi develops radio; motion-picture camera invented	Motion pictures become a big business	AT&T long-distance service reaches San Francisco	First TV demonstrated; first sound movie	Commercial TV broad-casting	Color TV demon-strated	Transistor invented	Reel-to-reel tape recorder	Cable TV

of which can measure continuous fluctuations. The electrical signals on a telephone line, for instance, have traditionally been analog-data representations of the original voices.

Thus, *analog data* is transmitted in a continuous form—a continuous electrical signal in the shape of a wave (called a *carrier wave*). Telephone, radio, television, and cable-TV technologies have long been based on analog data.

Purpose of the Modem: Converting Digital Signals to Analog Signals & Back

To understand the differences between digital and analog transmission, look at a graphic representation of an on/off digital signal emitted from a computer. Like a regular light switch, this signal has only two states—on and off. Compare this with a graphic representation of a wavy analog signal emitted as a signal. The changes in this signal are gradual, as in a dimmer switch, which gradually increases or decreases brightness.

Because telephone lines have traditionally been analog, you need to have a *modem* if your computer is to send communications signals over a telephone line. The modem translates the computer's digital signals into the telephone line's analog signals. The receiving computer also needs a modem to translate the analog signals back into digital signals. *(See ● Panel 7.2.)*

How, in fact, does a modem convert the continuous analog wave to a discontinuous digital pulse that can represent 0s and 1s? The modem can make adjustments to the *frequency*—the number of cycles per second, or the number of times a wave repeats during a specific time interval (the fastness/slowness). Or it can make adjustments to the analog signal's *amplitude*—the

1952	1963	1964	1967	1969	1970	1971	1975	1977	1978
UNIVAC computer correctly predicts election of Eisenhower as U.S. President	BASIC developed at Dartmouth	IBM introduces 360 line of computers	Hand-held calculator	ARPA-Net established, led to Internet	Micro-processor chips come into use; floppy disk introduced for storing data	First pocket calculator	First micro-computer (MITs Altair 8800)	Apple II computer (first personal computer sold in assembled form)	5 ¼" floppy disk; Atari home videogame

Fusing of computer and communications lines of development

1952	1957	1961	1968	1975	1976	1977	1979	1982
Direct-distance dialing (no need to go through operator); transistor radio introduced	First satellite launched (Russia's Sputnik)	Push-button telephones	Portable video recorders; video cassettes	Flat-screen TV	First wide-scale marketing of TV computer games (Atari)	First inter-active cable TV	3-D TV demonstrated	Compact disks; European consortium launches multiple communications satellites

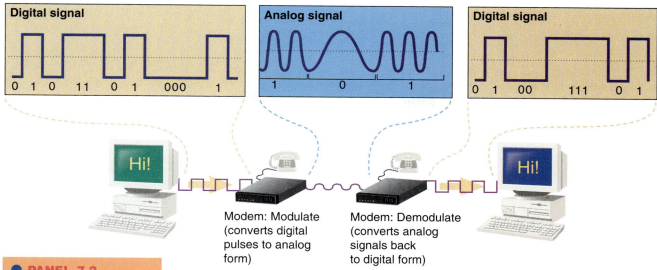

Digital signal

0 1 0 11 0 1 000 1

Analog signal

1 0 1

Digital signal

0 1 00 111 0 1

Modem: Modulate (converts digital pulses to analog form)

Modem: Demodulate (converts analog signals back to digital form)

● **PANEL 7.2**

Analog versus digital signals, and the modem

Note that an analog signal represents a continuous electrical signal in the form of a wave. A digital signal is discontinuous, expressed as discrete bursts in on/off electrical pulses.

height of the wave (the loudness/softness). Thus, in frequency, a slow wave might represent a 0 and a quick wave might represent a 1. In amplitude, a low wave might represent a 0 and a high wave might represent a 1. *(See* ● *Panel 7.3, next page.)*

Modem **is short for "*modul*ate/*dem*odulate"; a sending modem modulates digital signals into analog signals for transmission over phone lines. A receiving modem demodulates the analog signals back into digital signals.** The modem provides a means for computers to communicate with one another

Computer Technology

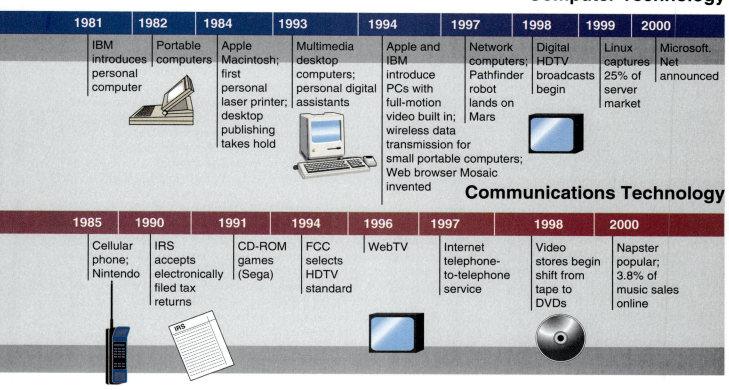

1981	1982	1984	1993	1994	1997	1998	1999	2000
IBM introduces personal computer	Portable computers	Apple Macintosh; first personal laser printer; desktop publishing takes hold	Multimedia desktop computers; personal digital assistants	Apple and IBM introduce PCs with full-motion video built in; wireless data transmission for small portable computers; Web browser Mosaic invented	Network computers; Pathfinder robot lands on Mars	Digital HDTV broadcasts begin	Linux captures 25% of server market	Microsoft. Net announced

Communications Technology

1985	1990	1991	1994	1996	1997	1998	2000
Cellular phone; Nintendo	IRS accepts electronically filed tax returns	CD-ROM games (Sega)	FCC selects HDTV standard	WebTV	Internet telephone-to-telephone service	Video stores begin shift from tape to DVDs	Napster popular; 3.8% of music sales online

● **PANEL 7.3**
How analog waves are modified to resemble digital pulses

The continuous, even cycle of an analog wave...

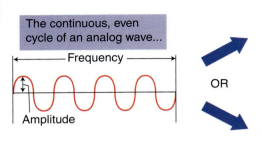

OR

... is converted to digital form through *frequency modulation*—the frequency of the cycle increases to represent a 1 and stays the same to represent a 0.

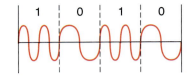

... or is converted to digital form through *amplitude modulation*—the height of the wave is increased to represent a 1 and stays the same to represent a 0.

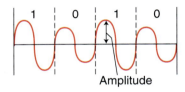

using the standard copper-wire telephone network, an analog system that was built to transmit the human voice but not computer signals.

Our concern, however, goes far beyond telephone transmission. How can the analog realities of the world be expressed in digital form? How can light, sounds, colors, temperatures, and other dynamic values be represented so that they can be manipulated by a computer? Let us consider this.

Converting Reality to Digital Form

Suppose you are using an analog tape recorder to record a singer during a performance. The analog process will produce a near duplicate of the sounds. This will include distortions, such as buzzings and clicks, or electronic hums if an amplified guitar is used.

The digital recording process is different. The way in which music is captured for audio CDs does not provide a duplicate of a musical performance. Rather, the digital process uses *representative selections (samples)* to record the sounds. The copy obtained is virtually exact and free from distortion and noise. Computer-based equipment takes samples of sounds at regular intervals—nearly 44,100 times a second. The samples are converted to numbers that the computer then uses to express the sounds. The sample rate of 44,100 times per second and the high precision fool our ears into hearing a smooth, continuous sound. Similarly, for visual material, a computer can take samples of values such as brightness and color. The same is true of other aspects of real-life experience, such as pressure, temperature, and motion.

Are we being cheated out of our experience of "reality" by allowing computers to sample sounds, images, and so on? Actually, people willingly made this compromise years ago, before computers were invented. Movies, for instance, carve up reality into 24 frames a second. Television frames are drawn at 30 lines per second. These processes happen so quickly that our eyes and brains easily jump the visual gaps. Digital processing of analog experience is just one more way of expressing or translating reality.

Turning analog reality into digital form provides tremendous opportunities. One of the most important is that all kinds of multimedia can now be changed into digital form and transmitted as data to all kinds of devices.

Now let us examine the digital world of telecommunications.

Distinguish between digital and analog signals.

Explain what a modem does.

7.2 The Practical Uses of Communications

Guyana is the lone English-speaking country in South America. In the village of Lethem (population 2000), a collective of women from two native tribes has revived the ancient art of hand-weaving large hammocks. Thanks to donated computer equipment and Internet training for one member, plus free satellite-assisted online access, the collective sold 17 hammocks around the world in 1999 through its Web site (*www.gol.net.gy/rweavers*). Though it might take 600 hours to make, a hammock could bring as much as $1000—a gigantic sum in this area.[5]

Such is the power of communications connections, or *connectivity*, which give us all instant, around-the-clock information from all over the globe. Let's consider some of the forms this connectivity takes:

- **Videoconferencing and videophones**
- **Workgroup computing and groupware**
- **Telecommuting and virtual offices**
- **Information or Internet appliances**
- **Smart television**

BOOKMARK IT!

PRACTICAL ACTION BOX

Web-Authoring Tools: How to Create Your Own Simple Web Site, Easily & for Free

"The full power of the Web will only be unleashed," says technology writer Stephen Wildstrom, "when it's as easy to post information online as it is to write a memo."[a]

We may be getting closer. The World Wide Web is a great way to get information about yourself and your work to co-workers or to potential customers. To do so, however, you need to create a Web page and put it online.

Fortunately, some free, easy-to-use Web-authoring tools for building simple Web sites are available. All of the following let you create Web pages using icons and menus; you don't need to know HTML to get the job done.[b]

- **Online services:** If you're a member of America Online, you can use AOL's Personal Publisher (keyword: *personal publisher*).
- **Browsers:** Netscape's Communicator comes with a Web-building program called Composer. The recent releases of Microsoft Explorer also offer Web authoring tools.

- **Word processing software:** Microsoft offers Internet Assistant, which can be used with Word, its word processing program. (You can download Internet Assistant at *www.microsoft.com/word/internet/ia/*)
- **Desktop publishing software:** If you've taken up desktop publishing, you should know that Adobe Page-Maker and Microsoft Publisher come with simple Web page editors.

Once you've created your Web site, you'll need to "publish" it—upload it to a Web server for viewing on the Internet. You can get upload instructions from your online service or Internet service provider (ISP). Some ISPs will give you free space on their servers.

Other specific packages for building Web sites are Adobe's PageMill, Soft Quad's Hot Metal Pro, and Microsoft's FrontPage Editor. If you want to create more sophisticated sites, you can try using tools such as Home Page from Claris or Trellix from Trellix Corp. (A trial version of Trellix is downloadable from *www.trellix.com*)

Telecommunications

Videoconferencing
Three online paticipants discuss an architectural drawing (shown in the right half of the screen).

Videoconferencing & Videophones: Video/Voice Communication

"I was a little nervous about going in front of the camera," said job applicant Mark Dillard, "but I calmed down pretty quickly after we got going, and it went well."[6]

Interviewing for a job can be uncomfortable for many people. However, Dillard had just undergone a high-tech interview. He had talked to a job recruiter in New York while sitting in front of a video camera in a booth at a local Kinko's store in Atlanta.

Videoconferencing, **also called teleconferencing, is the use of television video and sound technology as well as computers to enable people in different locations to see, hear, and talk with one another.** For a videoconference, people may go to conference rooms or booths with specially equipped television cameras. Alternatively, videoconferencing equipment can be set up on people's desks, with a camera and microphone to capture the person speaking and a monitor and speakers for the person being spoken to. The *videophone* is a telephone with a TV-like screen and a built-in camera that allows you to see the person you're calling, and vice versa.

The main difficulty with videoconferencing and videophones is that POTS ("plain old telephone service") equipment based on standard copper wire cannot transmit or receive images very rapidly. Thus, unless you can afford expensive high-speed communications lines, present-day screens will convey a series of jerky, stop-action images of the participants' faces.

Workgroup Computing & Groupware

When microcomputers were first brought into the workplace, they were used simply as another personal-productivity tool, like typewriters or calculators. Gradually, however, companies began to link microcomputers together on a network, usually to share an expensive piece of hardware, such as a laser printer. Then employees found that networks allowed them to share files and databases. Networking using common software also allowed users to buy equipment from different manufacturers—a mix of computers from both Sun

Microsystems and Hewlett-Packard, for example. Sharing resources has led to workgroup computing.

In **_workgroup computing_, also called collaborative computing, teams of co-workers use networks of microcomputers to share information and to cooperate on projects.** Workgroup computing is made possible not only by networks and microcomputers but also by *groupware*. As we stated in Chapter 3, *groupware* is software that allows two or more people on a network to work on the same information at the same time.

In general, groupware, such as Lotus Notes or MS NetMeeting, permits office workers to collaborate with colleagues and to tap into company information through computer networks. It also enables them to link up with crucial contacts outside their organization.

Telecommuting & Virtual Offices

Computers and communications tools have led to telecommuting and virtual offices.

Telecommuter
An at-home worker stays connected to his office.

- **Telecommuting:** **Working at home while in telecommunication with the office is called _telecommuting_.** The number of U.S. households in which someone telecommutes—works at home at least three days a month—rose to 10.7 million in 1999 and is expected to hit 13.9 million in 2002.[7] In one study, 41% of workers surveyed said they believe they could perform part of their work as telecommuters and would like to do so. Only 9% are doing so now.[8]

  Telecommuting can have many benefits. The advantages to society are reduced traffic congestion, energy consumption, and air pollution. What are the advantages to employers? Productivity can increase, because telecommuters may experience fewer distractions at home and can work flexible hours. Teamwork can improve, and the labor pool can be expanded, because hard-to-get employees don't have to uproot themselves from where they want to live. A disadvantage is that people may feel isolated and may not develop firm bonds with office employees.

  A related term is *telework*, which includes not only those who work at least part time from home but also those who work at remote or satellite offices (sometimes called *telework centers*).

- **Virtual offices:** The **_virtual office_ is an often nonpermanent and mobile office run with computer and communications technology.** Employees work from their homes, cars, and other new work sites. They use pocket pagers, portable computers, fax machines, and various phone and network services to conduct business.

  Could you stand not having a permanent office at all? Here's how one variant called "hoteling" works: You call ahead to book a room and speak to the concierge. However, the "hotel" isn't a Hilton but an organization such as advertising agency Chiat/Day. The concierge is an administrator who handles the scheduling of office cubicles. There are fewer cubicles than there are employees, who are often out in the field. When you check in, you pick up your personal effects and files from your locker and take them to the cubicle you will use for the next few days or weeks. This system depends on computers, which facilitate cubicle scheduling and the reprogramming of phones. Laptops allow employees to carry their work around with them, stored on their hard drives. Cellular phones, fax machines, and e-mail permit employees to stay in touch with supervisors and coworkers.

The rise in telecommuting and virtual offices is only part of a larger trend. "Powerful economic forces are turning the whole labor force into an army of freelancers—temps, contingents, and independent contractors and

consultants," says business strategy consultant David Kline. The result, he believes, is that "computers, the Net, and telecommuting systems will become as central to the conduct of 21st-century business as the automobile, freeways, and corporate parking lots were to the conduct of mid-20th-century business."[9] The transformation will really gather momentum, some observers feel, when homes have broadband Internet access (as explained shortly).[10]

Home Networks

As we shall see, computers linked by telephone lines, cable, or wireless systems are an established component of information technology. Today, however, many new buildings and even homes are built as "network enabled." The new superconnected home or small office is equipped with a *local area network (LAN)*, which allows all the personal computers under the same roof to share peripherals (such as a printer or a fax machine) and a single modem and Internet service.[11] The next development is supposed to be the networking of home appliances, linking stereos, lights, heating systems, phones, and TV sets. Once that has been accomplished, you could walk into your house and give a voice command to turn on the lights or bring up music, for example. Even kitchen appliances would be linked, so that your refrigerator, for instance, could alert a grocery store that you need more milk.[12]

The Information/Internet Appliance

An *information appliance* is a device merging computing capabilities with communications gadgets. Examples include TV set-top boxes, Internet phones, and personal digital assistants. Especially as cable and wireless channels become speedier, these devices will offer the ability to deliver all types of data—text, audio, video, film, still pictures—anywhere at any time. We discuss various types of information appliances in the Experience Box at the end of this chapter.

Smart Television: DTV, HDTV, & SDTV

Today experts differentiate between interactive TV, personalized TV, and Internet TV. *Interactive TV,* which is popular in Europe, lets you interact with the show you're watching, so you can request information about a product or play along with a game show. *Personalized TV* consists of hard-drive–equipped personal video recorders (PVRs), such as TiVo and ReplayTV, that let you not only record shows but also pause, rewind, and replay live TV programs. *Internet TV* lets you read Internet text and Web pages on your television set. The foremost example is WebTV; another is Liberate.[13] In the future, interactive, personal, and Internet TV will probably come together in a single box that goes under the umbrella name of digital television. Let's consider this subject.

- **Digital television (DTV):** When most of us tune in our TV sets, we get *analog television*, a system of varying signal amplitude and frequency that represents picture and sound elements. In 1996, however, broadcasters and their government regulator, the Federal Communications Commission (FCC), adopted a standard called **_digital television (DTV)_, which uses a digital signal, or series of 0s and 1s.** DTV is much clearer and less prone to interference than analog TV. (For instance, analog TV has a width-to-height ratio, or aspect ratio, of 4 to 3. One form of digital TV, HDTV, has a ratio of 16 to 9, which is similar to the wide-screen approach used in movies.)

- **High-definition television (HDTV):** If you acquire something advertised as a digital TV, you'll find that it may handle a greater number of channels, but the picture quality won't be much different from traditional analog sets. This is because the cable box often can't communicate with the set. Nevertheless, real DTV is here, in the form known as **_high-definition television (HDTV)_, the high-resolution type of DTV,** which comes in either a 720- or 1080-line mode (compared to 525-line resolution for analog TV). Why don't more people have HDTV sets? The biggest reason is expense—a set may cost $7000, although 95% of the regular televisions purchased in the U.S. retail for under $1500.[14] Another reason is that broadcasters and content producers haven't yet fully backed the standard. Indeed, they might well favor another DTV standard known as SDTV.

- **Standard-definition television (SDTV):** HDTV takes a lot of bandwidth that broadcasters could use instead for **_standard-definition television (SDTV)_, which has a minimum of 480 vertical lines, allowing broadcasters to transmit more information within the HDTV bandwidth.** SDTV would enable broadcasters to effectively multicast their products, on up to as many as six channels instead of one.[15] It also frees up bandwidth for data transmission. Thus, in the future there might be separate channels carrying video, audio, and data.

What might make HDTV finally arrive? In a word, movies. "While these expensive sets may not be the best choice for TV addicts," says one analysis, "HDTV lets film buffs create an impressive home theater."[16]

CONCEPT CHECK

Explain videoconferencing, workgroup computing, telecommuting and virtual offices, home networks, and information/Internet appliances.

Distinguish among DTV, HDTV, and SDTV.

If you are of a certain age, you may recall when two-way individual communications were accomplished mainly in two ways. They were carried by (1) a telephone wire or (2) a wireless method such as shortwave radio. Today there are many kinds of communications channels, although they are still wired or wireless. A **_communications channel_ is the path—the physical medium—over which information travels in a telecommunications system from its source to its destination.** (Channels are also called *links, lines,* or *media.*)

- **The electromagnetic spectrum**
- **Wired communications channels**
- **Wireless communications channels**
- **Types of long-distance wireless communications**
- **Compression and decompression**

The Electromagnetic Spectrum, the Radio Spectrum, & Bandwidth

The basis for all telecommunications channels, both wired and wireless, is the electromagnetic spectrum. Telephone signals, radar waves, and the invisible commands from a garage-door opener all represent different waves on what is called the electromagnetic spectrum. The **_electromagnetic spectrum_ consists of fields of electrical energy and magnetic energy, which travel in waves.** In the middle of the electromagnetic spectrum is the **_radio frequency spectrum,_ fields of electrical energy and magnetic energy that carry communications signals.** *(See ● Panel 7.4.)*

The waves vary according to *frequency*—the number of times a wave repeats, or makes a cycle, in a second. The radio spectrum ranges from low-frequency waves, such as those used for aeronautical and marine navigation equipment, through the medium frequencies for CB radios, cordless phones, and baby monitors, to ultrahigh frequency bands for cell phones and also microwave bands for communications satellites.

A range of frequencies is called a _band_ or _bandwidth._ Bandwidth is a measure of the amount of information that can be delivered within a given period of time. For analog signals, bandwidth is expressed in hertz (Hz), or cycles per second. For digital signals, bandwidth is expressed in bits per second (bps). In the United States, certain bands are assigned by the Federal Communications Commission for certain purposes—cell phones within one range, automated teller machines within another, broadcast television within yet another, and so on.

The bandwidth is the difference between the lowest and the highest frequencies transmitted. For example, cellular phones operate within the range 800–900 megahertz—that is, their bandwidth is 100 megahertz. *The wider the bandwidth, the faster data can be transmitted. The narrower the band, the greater the loss of transmission power.* This loss of power must be overcome by using relays or repeaters that rebroadcast the original signal. **_Broadband connections_ are characterized by very high speed.** For instance, the connections that carry broadcast video range in bandwidth from 10 megabits to 30 gigabits per second.

Let us now look more closely at the various types of channels:

- Wired channels—twisted-pair wire, coaxial cable, fiber-optic cable
- Wireless channels—infrared transmission, broadcast radio, microwave radio, and communications satellite
- Types of long-distance wireless communications—GPS, pagers, analog cellular, packet radio, CDPD

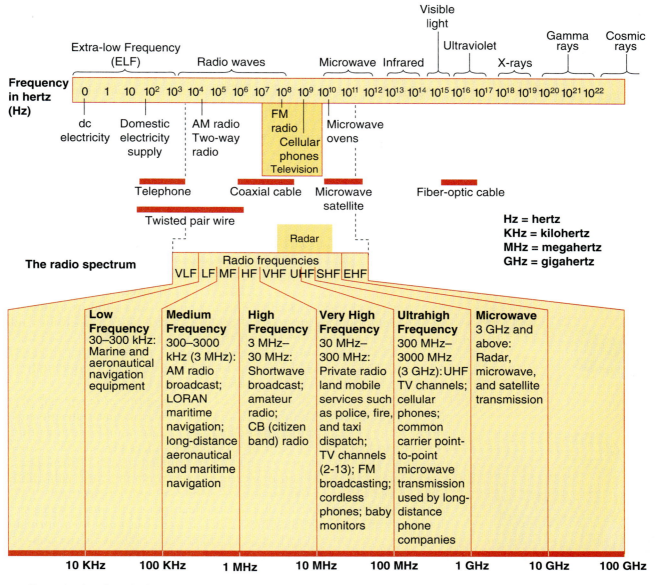

The electromagnetic spectrum

The radio spectrum

Frequencies for wireless data communications

Cellular	Private land mobile	Narrowband PCS	Industrial	Common carrier paging	Point-to-multipoint Point-to-point	PCS	Industrial
824–849 MHz 869–894 MHz	896–901 MHz 930–931 MHz Includes RF packet radio services	901–902 MHz 930–931 MHz	902–928 MHz Unlicensed commercial use such as cordless phones and LANs	931–932 MHz Includes national paging services	932–935 MHz 941–944 MHz	1850–1970 MHz 2130–2150 MHz 2180–2200 MHz	2400–2483.5 MHz Unlicensed commercial use such as LANs

Hz = hertz
KHz = kilohertz
MHz = megahertz
GHz = gigahertz

● PANEL 7.4

The electromagnetic spectrum
The radio frequency spectrum, which carries most communications signals, appears as part of the electromagnetic spectrum.

Wired Communications Channels: Transmitting Data by Physical Means

Three types of wired channels are twisted-pair wire (conventional telephone lines), coaxial cable, and fiber-optic cable.

Twisted-pair wire

- **Twisted-pair wire (1–128 Mbps):** The telephone line that runs from your house to the pole outside, or underground, is probably twisted-pair wire. _**Twisted-pair wire**_ **consists of two strands of insulated copper wire, twisted around each other. This twisted-pair configuration somewhat reduces interference from electrical fields.** Twisted-pair is relatively slow. Moreover, it does not protect well against electrical interference. However, because so much of the world is already served by twisted-pair wire, it will no doubt be used for years to come, both for voice messages and for modem-transmitted computer data.

    The prevalence of twisted-pair wire gives rise to what experts call the "final mile problem." That is, it is relatively easy for telecommunications companies to upgrade the physical connections between cities and even between neighborhoods. But it is expensive for them to replace the "final mile" of twisted-pair wire that connects to individual houses.

Coaxial cable

- **Coaxial cable (up to 200 Mbps):** _**Coaxial cable,**_ **commonly called "co-ax," consists of insulated copper wire wrapped in a solid or braided metal shield, then in an external cover.** Co-ax is widely used for cable television. Thanks to the extra insulation, coaxial cable is much better than twisted-pair wiring at resisting noise. Moreover, it can carry voice and data at a faster rate (up to 200 megabits per second). Often many coaxial cables will be bundled together.

Fiber-optic cable

- **Fiber-optic cable (100 Mbps to 2 Gbps):** A _**fiber-optic cable**_ **consists of dozens or hundreds of thin strands of glass or plastic that transmit pulsating beams of light rather than electricity.** These strands, each as thin as a human hair, can transmit up to 2 billion pulses per second (2 Gbps); each "on" pulse represents one bit. When bundled together, fiber-optic strands in a cable 0.12 inch thick can support a quarter- to a half-million voice conversations at the same time. Moreover, unlike electrical signals, light pulses are not affected by random electromagnetic interference in the environment. Thus, they have much lower error rates than normal telephone wire and cable. In addition, fiber-optic cable is lighter and more durable than twisted-pair and co-ax cable. A final advantage is that it cannot easily be wiretapped, so transmissions are more secure.

The various kinds of wired Internet connections discussed in Chapter 2—dial-up modem, DSL, ISDN, cable modem, T1 lines—are effected through these alternative wired channels.

Wireless Communications Channels: Transmitting Data through the Air

Four types of wireless channels are infrared transmission, broadcast radio, microwave radio, and communications satellite.

Infrared access device

- **Infrared transmission (1–4 Mbps):** _**Infrared wireless transmission**_ **sends data signals using infrared-light waves.** Infrared ports can be found on some laptop computers and printers, as well as wireless mouses. The drawbacks are that _line-of-sight_ communication is required—there must be an unobstructed view between transmitter and receiver—and transmission is confined to short range.

- **Broadcast radio (up to 2 Mbps):** When you tune in to an AM or FM radio station, you are using **_broadcast radio_, a wireless transmission medium that sends data over long distances—between regions, states, or countries.** A transmitter is required to send messages and a receiver to receive them; sometimes both sending and receiving functions are combined in a *transceiver*.

  In the lower frequencies of the radio spectrum, several broadcast radio bands are reserved not only for conventional AM/FM radio but also for broadcast television, CB (citizens band) radio, ham (amateur) radio, cellular phones, and private radio land mobile services (such as police, fire, and taxi dispatch). Some organizations use specific radio frequencies and networks to support wireless communications. For example, UPC (Universal Product Code) readers are used by grocery-store clerks restocking store shelves to communicate with a main computer so that the store can control inventory levels.

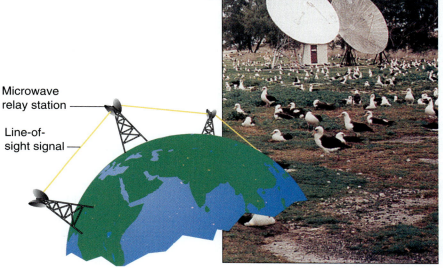

- **Microwave radio (45 Mbps):** **_Microwave radio_ transmits voice and data through the atmosphere as super-high-frequency radio waves called microwaves,** which vibrate at 1 gigahertz (1 billion hertz) per second or higher. These frequencies are used not only to operate microwave ovens but also to transmit messages between ground-based stations and satellite communications systems.

  Nowadays dish- or horn-shaped microwave reflective dishes, which contain transceivers and antennas, are nearly everywhere—on towers, buildings, and hilltops.

Microwave radio

Microwave relay station

Line-of-sight signal

Microwave radio
These dishes are on Midway Island, 1100 miles from Hawaii.

Why, you might wonder, do we have to interfere with nature by putting a microwave dish on top of a mountain? As with infrared waves, microwaves are line-of-sight; they cannot bend around corners or around the earth's curvature, so there must be an unobstructed view between transmitter and receiver. Thus, microwave stations need to be placed within 25–30 miles of each other, with no obstructions in between. The size of the dish varies with the distance (perhaps 2–4 feet in diameter for short distances, 10 feet or more for long distances). A string of microwave relay stations will each receive incoming messages, boost the signal strength, and relay the signal to the next station.

More than half of today's telephone system uses dish microwave transmission. However, the airwaves are becoming so saturated with microwave signals that future needs will have to be satisfied by other channels, such as satellite systems.

- **Communications satellites:** To avoid some of the limitations of microwave earth stations, communications companies have added microwave "sky stations"—communications satellites. **_Communications satellites_ are microwave relay stations in orbit around the earth.** Transmitting a signal from a ground station to a satellite is called *uplinking*; the reverse is called *downlinking*. The delivery process will be slowed if, as is often the case, more than one satellite is required to get the message delivered.

Communications satellite

GEO

Orbit:
22,300 miles
at the equator

MEO

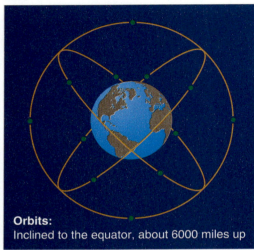

Orbits:
Inclined to the equator, about 6000 miles up

LEO

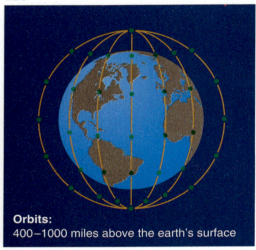

Orbits:
400–1000 miles above the earth's surface

Satellite systems may occupy one of three zones in space: *GEO, MEO,* and *LEO.*

The highest level, known as *geostationary earth orbit (GEO),* is 22,300 miles and up and is always directly above the equator. Because the satellites in this orbit travel at the same speed as the earth, they appear to an observer on the ground to be stationary in space—that is, they are *geostationary.* Consequently, microwave earth stations are always able to beam signals to a fixed location above. The orbiting satellite has solar-powered transceivers to receive the signals, amplify them, and retransmit them to another earth station. At this high orbit, fewer satellites are required for global coverage; however, their quarter-second delay makes two-way conversations difficult.

The *medium-earth orbit (MEO)* is 5000–10,000 miles up. It requires more satellites for global coverage than GEO.

The *low-earth orbit (LEO)* is 400–1000 miles up and has no signal delay. LEO satellites may be smaller and are cheaper to launch.

Types of Long-Distance Wireless Communications

His friends thought Hank Kahrs, an insurance auditor, was just being silly in taking along a cell phone on their hike up California's Mount Whitney, the highest mountain in the continental United States. One reason for getting outdoors, they chided him, was to get away from civilization and its gadgets; anyway, they said, the phone probably wouldn't work at 14,494 feet. Kahrs, however, didn't want to break his long-standing custom of calling his wife every day.

At the top of the peak, he got a pleasant surprise. "It took 30 seconds before the phone started ringing" at his wife's number, he said. After he made his call, his hiking companions' attitudes changed. "When they saw the signal was fine, my friends all wanted to use the phone," said Kahrs.[17]

Call this a glimpse of the future's "unwired planet." Very soon, it will be nearly impossible to *not* make a phone call or at least page someone from anywhere on earth (except from a cave, perhaps). Already this has produced headaches for forest rangers, who have had to rescue too many hikers who embarked on wilderness treks without being properly equipped—except for a cellular phone with which to call for help.

Mobile wireless communications have been around for some time. The Detroit Police Department started using two-way car radios in 1921. Mobile telephones were introduced in 1946. Today, however, we are witnessing an explosion in mobile wireless use that is making worldwide changes.

There are essentially two ways to move information through the air long-distance on radio frequencies. The first is via *one-way communications,* as typified by the satellite navigation system known as the Global Positioning System, and by most pagers. The second is via *two-way communications:* (1) two-way pagers, (2) analog cellular phones, (3) packet radio, and (4) Cellular Digital Packet Data. (Other wireless methods operate at short distances.)

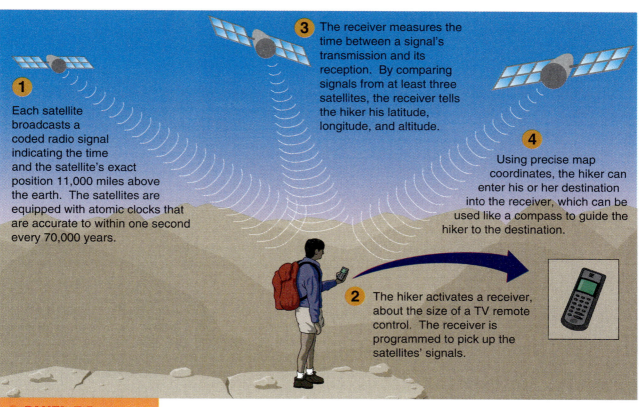

1 Each satellite broadcasts a coded radio signal indicating the time and the satellite's exact position 11,000 miles above the earth. The satellites are equipped with atomic clocks that are accurate to within one second every 70,000 years.

3 The receiver measures the time between a signal's transmission and its reception. By comparing signals from at least three satellites, the receiver tells the hiker his latitude, longitude, and altitude.

4 Using precise map coordinates, the hiker can enter his or her destination into the receiver, which can be used like a compass to guide the hiker to the destination.

2 The hiker activates a receiver, about the size of a TV remote control. The receiver is programmed to pick up the satellites' signals.

GPS receiver: hand-held compass

The Global Positioning System uses 24 satellites, developed for military use, to pinpoint a location on the earth's surface.

• **One-way communications—the Global Positioning System:** A $10 billion infrastructure developed by the military in the mid-1980s, the ***Global Positioning System (GPS)*** **consists of a series of earth-orbiting satellites continuously transmitting timed radio signals that can be used to identify earth locations.** A GPS receiver—handheld or mounted in a vehicle, plane, or boat—can pick up transmissions from any four satellites, interpret the information from each, and pinpoint the receiver's longitude, latitude, and altitude. *(See ● Panel 7.5.)* Some GPS receivers include map software for finding your way around, as with the navigation systems available with some rental cars.

The system, accurate within 3–100 feet, is used to tell military units carrying special receivers where they are. GPS is used for such civilian activities as tracking trucks and taxis, locating stolen cars, orienting hikers, and aiding in surveying. Some public-transportation systems have installed GPS receivers on buses, where they can tell drivers when they fall behind schedule. The makers of the film *Forrest Gump* used a GPS device to track the sun and time their sunrise and sunset shots so they didn't get shadows.[18] GPS has also been used by scientists to keep a satellite watch over a Hawaiian volcano, Mauna Loa, and to capture infinitesimal movements that may be used to predict eruptions.[19]

A GPS car-navigation system

GPS

A visually impaired man with a guide dog uses a special GPS computer device to help navigate his way.

Telecommunications

285

- **One-way communications—pagers:** In a Clearwater, Florida, child-care center, if one child bites another or otherwise misbehaves, the head of the center can instantly alert the parents by paging them. All parents are given a pager as part of the child-care service.[20]

  Once stereotyped as devices for doctors and drug dealers, pagers are now consumer items. Commonly known as beepers, for the sound they make when activated, _**pagers**_ **are simple radio receivers that receive data (but not voice messages) sent from a special radio transmitter.** Often the pager has its own telephone number. When the number is dialed from a phone, the call goes by way of the transmitter straight to the designated pager. Pagers are very efficient for transmitting one-way information—emergency messages, news, prices, stock quotations, delivery-route assignments, even sports news and scores—at low cost to single or multiple receivers.

  Pagers do more than beep, transmitting full-blown alphanumeric text (such as four-line, 80-character messages) and other data. Newer ones are mini-answering machines, capable of relaying digitized voice messages.

- **Two-way communications—pagers:** Recently advances have given us _two-way paging_ or _enhanced paging._ Service is provided by carriers like SkyTel, PageMart, PageNet, and Bell South Wireless Data. For instance, in one version, users can send a preprogrammed message or acknowledgment that they have received a message. Another version allows consumers to compose and send e-mail to anyone on the Internet and to other pagers. Typing a message on a tiny keyboard no larger than those on pocket calculators can pose a challenge, however.

- **Two-way communications—analog cellular:** _**Analog cellular phones**_ **are designed primarily for communicating by voice through a system of ground-area cells. Each** _**cell**_ **is hexagonal in shape, usually 8 miles or less in diameter, and is served by a transmitter-receiving tower.** Communications are handled in the bandwidth of 824–894 megahertz. Calls are directed between cells by a mobile telephone switching office. Movement between cells requires that calls be "handed off" by this switching office. (See ● _Panel 7.6._)

  Handing off voice calls between cells poses only minimal problems. However, handing off data transmission (where every bit counts), with the inevitable gaps and pauses on moving from one cell to another, is much more difficult. In the long run, data transmissions will probably have to be handled by the technology we discuss next, packet radio.

- **Two-way communications—packet radio:** _**Packet-radio-based communications**_ **use a nationwide system of radio towers that send data to handheld computers.** Packet radio is the basis for services such as RAM Mobile Data and Ardis. The advantage of packet-radio transmission is that the wireless computer identifies itself to the local base station, which can transmit over as many as 16 separate radio channels. Packet switching encapsulates the data in "envelopes," which ensures that the information arrives intact.

  Packet-radio data networks are useful for mobile workers who need to communicate frequently with a corporate database. For example, National Car Rental System sends workers with handheld terminals to prowl parking lots, recording the location of rental cars and noting the latest scratches and dents. They can thereby easily check a customer's claim that a car was already damaged or find out quickly when one is stolen.

- **Two-way communications—CDPD:** Short for _Cellular Digital Packet Data,_ _**CDPD**_ **places messages in packets, or digital electronic "envelopes," and sends them through underused radio channels or between pauses in cellular phone conversations.** CDPD is thus an enhancement

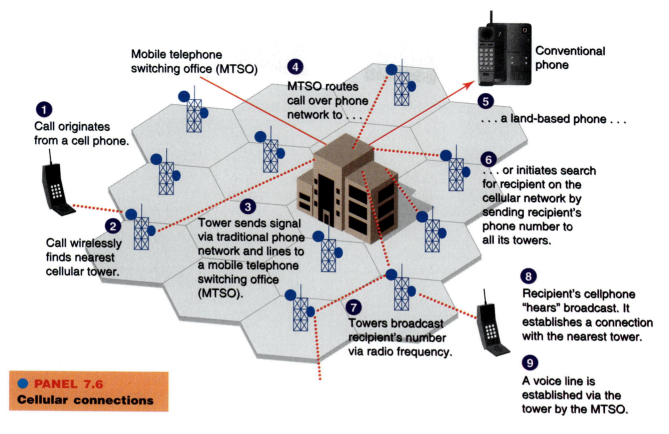

Mobile telephone switching office (MTSO)

① Call originates from a cell phone.

② Call wirelessly finds nearest cellular tower.

③ Tower sends signal via traditional phone network and lines to a mobile telephone switching office (MTSO).

④ MTSO routes call over phone network to . . .

Conventional phone

⑤ . . . a land-based phone . . .

⑥ . . . or initiates search for recipient on the cellular network by sending recipient's phone number to all its towers.

⑦ Towers broadcast recipient's number via radio frequency.

⑧ Recipient's cellphone "hears" broadcast. It establishes a connection with the nearest tower.

⑨ A voice line is established via the tower by the MTSO.

● PANEL 7.6
Cellular connections

to today's analog cellular phone systems, allowing packets of data to "hop" between temporarily free voice channels. As a result, a user carrying a CDPD device could have access to both voice and data. One problem with CDPD so far, however, is that it is not widely available.

Compression & Decompression: Putting More Data in Less Space

The vast streams of text, audio, and visual information threaten to overwhelm us. The file of a 2-hour movie, for instance, contains so much sound and visual information that, if stored without modification on a standard CD-ROM, it would require 360 disk changes during a single showing. A broadcast of *Oprah* that presently fits into one conventional, or analog, television channel would require 45 channels if sent in digital language.

To fit more data into less space, we use the mathematical process called compression. **_Compression_, or _digital-data compression_, is a method of removing repetitive elements from a file so that the file requires less storage space and therefore less time to transmit.** Before we use the data, it is decompressed—the repeated patterns are restored. These methods are sometimes referred to as *codec* (for *compression/decompression*) techniques.

- **Lossless versus lossy compression:** There are two principal methods of compressing data—lossless and lossy. In any situation, which of these two techniques is more appropriate will depend on whether data quality or storage space is more critical.

  Lossless compression uses mathematical techniques to replace repetitive patterns of bits with a kind of coded summary. During decompression, the coded summaries are replaced with the original patterns of bits. In this method, the data that comes out is exactly the same as what went in; it has merely been repackaged for purposes of storage or transmission. Lossless techniques are used when it's important that

Three MPEG standards

Standard	Mission
MPEG-1	For microcomputers and consumer gadgets. Provides full-screen video (VHS-like quality) of images similar to those on videocassette.
MPEG-2	For broadcast and cable television. Provides digital-TV-quality video for use with cable networks, satellite dishes, and new types of CD-ROMs. (MPEG-3 was incorporated into MPEG-4.)
MPEG-4	For wireless videoconferencing.

nothing be lost—for instance, for computer data, database records, spreadsheets, and word processing files.

Lossy compression techniques permanently discard some data during compression. Lossy data compression involves a certain loss of accuracy in exchange for a high degree of compression (to as little as 5% of the original file size). This method of compression is often used for graphics files and sound files. Thus, a lossy codec might discard subtle shades of color or very soft sounds. Most users wouldn't notice the absence of these details.

- **Compression standards—JPEG and MPEG:** Several standards exist for compression, particularly of visual data. Data recorded and compressed in one standard cannot be played back in another. The main reason for the lack of agreement is that different industries have different priorities. What will satisfy the users of still photographs, for instance, will not work for the users of movies.

As we have seen, lossless compression schemes are used for text and numeric data files, whereas lossy compression schemes are used with graphics and video files. The principal lossy compression schemes are *JPEG* and *MPEG*.

The leading compression standard for still images is <u>JPEG</u> (pronounced "jay-peg"), which stands for the Joint Photographic Experts Group of the International Standards Organization. The file extension that identifies graphic images files in the JPEG format is *.jpeg* or *.jpg*. In storing and transmitting still photographs, the data must remain of high quality. The JPEG codec looks for a way to squeeze a single image, mainly by eliminating repetitive pixels (picture-element dots) within the image. Higher or lower degrees of JPEG compression may be chosen; greater compression corresponds to greater image loss.

The leading compression standard for moving images is <u>MPEG</u> ("em-peg"), for Motion Picture Experts Group. *(See ● Panel 7.7.)* The file extension that identifies video (and sound) files compressed in this format is *.mpeg* or *.mpg*. People who work with videos are mainly interested in storing or transmitting an enormous amount of visual information in economical form; preserving details is a secondary consideration. The Motion Picture Experts Group sets standards for weeding out redundancies between neighboring images in a stream of video. Three MPEG standards have been developed for compressing visual information—MPEG-1, MPEG-2, and MPEG-4.

CONCEPT CHECK

What are the electromagnetic spectrum, the radio frequency spectrum, and bandwidth?

How video images are compressed
MPEG is a method of computerized compression/decompression that can reduce the size of a video signal by 95%. As a result the signal can be stored or transmitted more efficiently and economically.

1 COMPRESS EACH FRAME

MPEG divides a frame of video into many tiny blocks, each containing 64 picture elements (pixels). The patterns in each block are transformed into a set of numbers. A few of these numbers (bars) contain most of the important picture information and everything else is discarded.

I   B   B   P   B   B   I

2 COMPRESS BETWEEN FRAMES

MPEG divides the video signal into three types of frames. Every 1/3 of a second, an intraframe picture (I) captures all the information in the compressed signal. A predicted frame (P) based on the previous I frame, and bidirectional frames (B) interpolated between the two, contain less data but preserve video quality.

3 REVERSE THE PROCESS

On playback, the restored frames lack some information but the eye is fooled into seeing detail that doesn't exist.

Describe the three types of wired channels.

Distinguish among four types of wireless channels.

Describe how information is moved through the air long-distance—two one-way methods and four two-way methods.

Discuss compression and decompression and the leading compression standards.

7.4 Networks

KEY QUESTIONS

What are the benefits of networks, and what are their types, components, and variations?

Whether wired, wireless, or both, all the channels we've described can be used singly or in mix-and-match fashion to form networks. A **_network_, or communications network, is a system of interconnected computers, telephones, or other communications devices that can communicate with one another and share applications and data.** The tying together of so many communications devices in so many ways is changing the world we live in.

The Benefits of Networks

People and organizations use computers in networks for several reasons. These include the following:

- **Sharing of peripheral devices:** Peripheral devices such as laser printers, disk drives, and scanners are often quite expensive. Consequently, to justify their purchase, management wants to maximize their use. Usually the best way to do this is to connect the peripheral to a network serving several computer users.

- **Sharing of programs and data:** In most organizations, people use the same software and need access to the same information. It is less expensive for a company to buy a separate word processing program that will serve many employees than to buy a separate word processing program for each employee.

  Moreover, if all employees have access to the same data on a shared storage device, the organization can save money and avoid serious problems. If each employee has a separate machine, some employees may update customer addresses, while others remain ignorant of the changes. Updating information on a shared server is much easier than updating every user's individual system.

  Finally, network-linked employees can more easily work together online on shared projects.

- **Better communications:** One of the greatest features of networks is electronic mail. With e-mail, everyone on a network can easily keep others posted about important information.

- **Security of information:** Before networks became commonplace, an individual employee might be the only one with a particular piece of information, which was stored in his or her desktop computer. If the employee was dismissed—or if a fire or flood demolished the office—the company would lose that information. Today such data would be backed up or duplicated on a networked storage device shared by others.

- **Access to databases:** Networks enable users to tap into numerous databases, whether private company databases or public databases available online through the Internet.

Types of Networks: WANs, MANs, & LANs

Networks, which consist of various combinations of computers, storage devices, and communications devices, may be divided into three main categories, differing primarily in their geographical range.

- **Wide area network:** A **_wide area network (WAN)_** **is a communications network that covers a wide geographical area, such as a country or the world.** Most long-distance and regional Bell telephone companies are WANs. A WAN may use a combination of satellites, fiber-optic cable, microwave, and copper wire connections and link a variety of computers, from mainframes to terminals. *(See ● Panel 7.8.)*

- **Metropolitan area network:** A **_metropolitan area network (MAN)_** **is a communications network covering a city or a suburb.** The purpose of a MAN is often to bypass local telephone companies when accessing long-distance services. Many cellular phone systems are MANs.

- **Local area network:** A **_local area network (LAN)_** **connects computers and devices in a limited geographical area,** such as one office, one building, or a group of buildings close together (for instance, a college campus). A small LAN in a modest office, or even in a home, might link a file server with a few terminals or PCs and a printer or two. Such small LANs have been called *TANs,* for "tiny area networks."

Most large computer networks have at least one **_host computer_**, **a mainframe or midsize central computer that controls the network.** The other devices within the network are called nodes. A **_node_** **is any device that is attached to a network**—for example, a microcomputer, terminal, storage device, or printer.

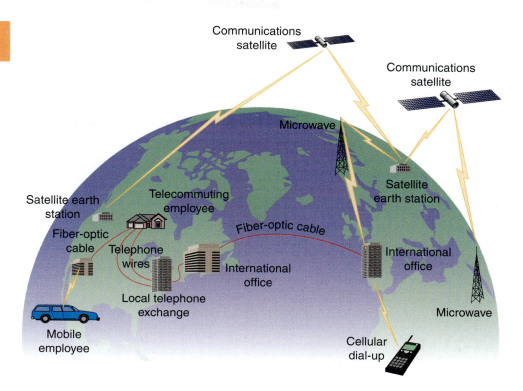

Networks may be connected together—LANs to MANs and MANs to WANs. A **_backbone_ is a high-speed network that connects LANs and MANs to the Internet.**

Types of LANs: Client-Server & Peer-to-Peer

Local area networks consist of two principal types: client/server and peer-to-peer. *(See ● Panel 7.9, next page.)*

- **Client/server LANs:** A *client/server LAN* consists of clients, which are microcomputers that request data, and servers, which are computers used to supply data. The server is a powerful microcomputer that manages shared devices, such as laser printers. It runs server software for applications such as e-mail and Web browsing.

  Different servers may be used to manage different tasks. A **_file server_ is a computer that acts like a disk drive, storing the programs and data files shared by users on a LAN.** A *database server* is a computer in a LAN that stores data but doesn't store programs. A *print server* controls one or more printers and stores the print-image output from all the microcomputers on the system. *Web servers* contain Web pages that can be viewed using a browser. *Mail servers* manage e-mail.

- **Peer-to-peer LANs:** The word *peer* denotes one who is equal in standing with another (as in the phrases "peer pressure" and "jury of one's peers"). In a **_peer-to-peer LAN_, all microcomputers on the network communicate directly with one another without relying on a server.** Peer-to-peer networks are less expensive than client-server networks and work effectively for up to 25 computers. Beyond that, they slow down under heavy use. They are appropriate for small networks.

Many LANs mix elements from both client-server and peer-to-peer models.

Components of a LAN

Local area networks are made up of several standard components.

Client/server LAN
In a client/server LAN, individual microcomputer users, or "clients," share the services of a centralized computer called a "server." In this case, the server is a file server, allowing users to share files of data and some programs.

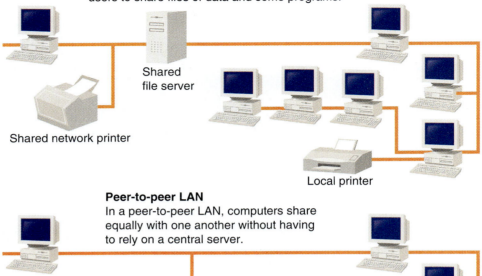

Shared file server

Shared network printer

Local printer

Peer-to-peer LAN
In a peer-to-peer LAN, computers share equally with one another without having to rely on a central server.

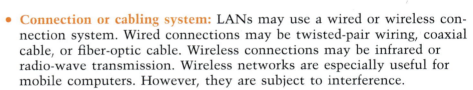

Shared network printer

Local printer

- **Connection or cabling system:** LANs may use a wired or wireless connection system. Wired connections may be twisted-pair wiring, coaxial cable, or fiber-optic cable. Wireless connections may be infrared or radio-wave transmission. Wireless networks are especially useful for mobile computers. However, they are subject to interference.

- **Microcomputers with network interface cards:** Two or more microcomputers are required, along with network interface cards. As we mentioned in Chapter 5, a *network interface card* enables the computer to send and receive messages over a cable network. The network card can be inserted into an expansion slot in a PC. Alternatively, a network card in a stand-alone box may serve a number of devices. Many new computers come with network cards already installed.

- **Network operating system:** The *network operating system (NOS)* is the system software that manages the activity of a network. The NOS supports access by multiple users and provides for recognition of users based on passwords and terminal identifications. Depending on whether the LAN is client/server or peer-to-peer, the operating system may be stored on the file server, on each microcomputer on the network, or a combination of both.

  Examples of popular NOS software are Novell NetWare, Microsoft Windows NT/2000, Unix, and Linux. Peer-to-peer networking can also be accomplished with Microsoft Windows 95/98/Me and Microsoft Windows for Workgroups.

- **Other shared devices:** Printers, scanners, storage devices, and other peripherals may be added to the network as necessary and shared by all users.

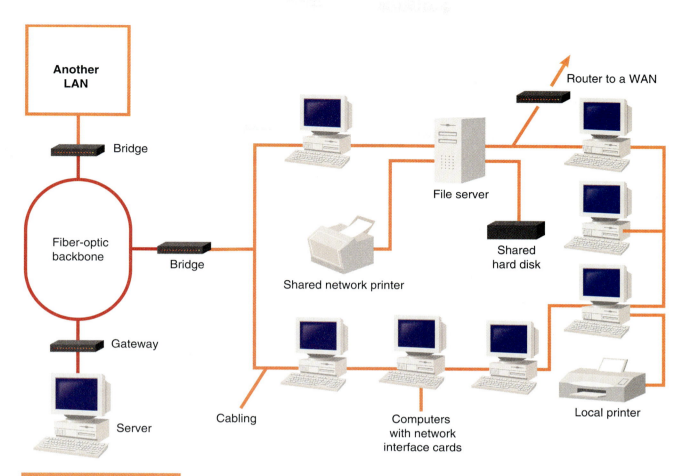

Another
LAN

Bridge

Fiber-optic
backbone

Bridge

Gateway

Server

Router to a WAN

File server

Shared
hard disk

Shared network printer

Cabling

Computers
with network
interface cards

Local printer

● **PANEL 7.10**
**Components of a
typical LAN**

LAN

At an early-morning flower auction in Alsmeer, Netherlands, buyers bid electronically via a LAN. Within three hours, 17 million flowers will have been snapped up. By noon, the flowers will be jetting toward shops around the world.

● **Routers, bridges, and gateways:** In principle, a LAN may stand alone. Today, however, it invariably connects to other networks, especially the Internet. Network designers determine the types of hardware and software necessary as interfaces to make these connections. Routers, bridges, and gateways are used for this purpose. *(See* ● *Panel 7.10.)*

A ***router*** **is a special computer that directs communicating messages when several networks are connected together.** High-speed routers can serve as part of the Internet backbone, or transmission path, handling the major data traffic.

A ***bridge*** **is an interface used to connect the same types of networks.**

A ***gateway*** **is an interface permitting communication between dissimilar networks**—for instance, between a LAN and a WAN or between two LANs based on different network operating systems or different layouts.

Topology of LANs

Networks can be laid out in different ways. **The logical layout, or shape, of a network is called a _topology_.** The three basic topologies are *star, ring,* and *bus.*

● **Star network:** A ***star network*** **is one in which all microcomputers and other communications devices are connected to a central server.** *(See* ● *Panel 7.11, next page.)*

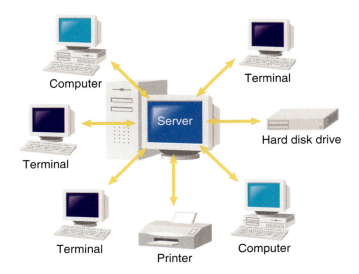

Electronic messages are routed through the central hub to their destinations. The central hub monitors the flow of traffic. A PBX system is an example of a star network.

The advantage of a star network is that the hub prevents collisions between messages. Moreover, if a connection is broken between any communications device and the hub, the rest of the devices on the network will continue operating. However, if the hub goes down, the entire network will stop.

● **Ring network:** A ***ring network*** **is one in which all microcomputers and other communications devices are connected in a continuous loop.** *(See* ● *Panel 7.12.)*

Electronic messages are passed around the ring until they reach the right destination. There is no central server. An example of a ring network is IBM's Token Ring Network, in which a bit pattern (called a "token") determines which user on the network can send information.

The advantage of a ring network is that messages flow in only one direction. Thus, there is no danger of collisions. The disadvantage is that, if a connection is broken, the entire network stops working.

● **Bus network:** The bus network works like a bus system at rush hour, with various buses pausing in different bus zones to pick up passen-

PANEL 7.12
Ring network
This arrangement connects the network's devices in a closed loop.

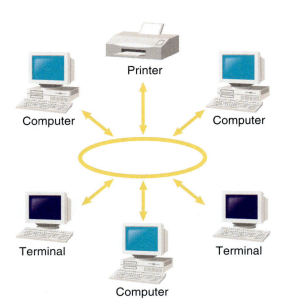

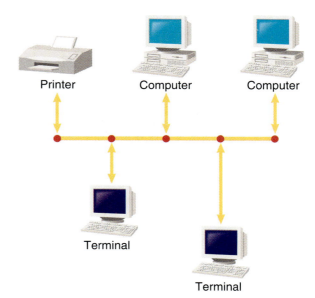

Printer   Computer   Computer

Terminal

Terminal

gers. In a ***bus network*, all communications devices are connected to a common channel.** (*See Panel* ● *7.13.*)

In a bus network, there is no central server. Each communications device transmits electronic messages to other devices. If some of those messages collide, the device waits and tries to transmit again. An example of a bus network is Xerox's Ethernet (which can also be configured in a star topology).

The advantage of a bus network is that it may be organized as a client-server or peer-to-peer network. The disadvantage is that extra circuitry and software are needed to avoid collisions between data. Also, if a connection in the bus is broken—as when someone moves a desk and knocks the connection out—the entire network may stop working.

Intranets, Extranets, & Firewalls: Private Internet Networks

Early in the Online Age, businesses discovered the benefits of using the World Wide Web to get information to customers, suppliers, or investors. For example, in the mid-1990s, Federal Express found it could save millions by allowing customers to click through Web pages to trace their parcels, instead of having FedEx customer-service agents do it. From there, it was a short step to the application of the same technology inside companies—in internal Internet networks called *intranets.*

● **Intranets—for internal use only:** An ***intranet* is an organization's internal private network that uses the infrastructure and standards of the Internet and the World Wide Web.** When a corporation develops a public Web site, it is making selected information available to consumers and other interested parties. When it creates an intranet, it enables employees to have quicker access to internal information and to share knowledge so that they can do their jobs better. Information exchanged on intranets may include employee e-mail addresses and telephone numbers, product information, sales data, employee benefit information, and lists of jobs available within the organization.

● **Extranets—for certain outsiders:** Taking intranet technology a few steps further, extranets offer security and controlled access. As we have seen, intranets are internal systems, designed to connect the members of a specific group or a single company. By contrast, ***extranets* are private** intranets that connect not only internal personnel but also selected suppliers and other strategic parties. Extranets have become popular for standard transactions such as purchasing. Ford Motor Company, for instance, has an extranet that connects more than 15,000 Ford dealers worldwide. Called FocalPt, the extranet supports sales and servicing of cars, with the aim of improving service to Ford customers.

- **Firewalls:** Security is essential to an intranet (or even an extranet). Sensitive company data, such as payroll information, must be kept private, by means of a *firewall*. A *__firewall__* **is a system of hardware and software that blocks unauthorized users inside and outside the organization from entering the intranet.**

  A firewall consists of two parts, a choke and a gate. The *choke* forces all data packets flowing between the Internet and the intranet to pass through a gate. The *gate* regulates the flow between the two networks. It identifies authorized users, searches for viruses, and implements other security measures. Thus, intranet users can gain access to the Internet (including key sites connected by hyperlinks), but outside Internet users cannot enter the intranet.

CONCEPT CHECK

What are the benefits of networks?

Characterize WANs, MANs, and LANs.

What are the differences between client-server and peer-to-peer LANs?

What are the components of a LAN?

Discuss intranets, extranets, and firewalls.

7.5 The Future of Communications

KEY QUESTION

What are the characteristics of the next generation of wireless communications, satellite systems, gigabit Ethernets, Bluetooth, photonics, and power lines?

Clearly, we are very near the time when voices, images, and information can be transmitted to any place on earth. Says C. Michael Armstrong, now president and CEO of AT&T, these advances already are revolutionizing "how we talk to each other and how we relate to each other."[21] Let's consider the outlook for the future. New kinds of wireless data services are rapidly being put in place, promising to offer us lots of choices. The following are a few such developments.

Digital Cellular Phones

Cellular telephone companies are trying to rectify the problem of faulty data transmission by switching from analog to digital. ***Digital cellular phone* networks turn your voice message into digital bits, which are sent through the airwaves, then decoded back into your voice by the cellular handset.**

Unlike analog cellular phones, digital phones can handle short e-mail messages, paging, and some headline news items in addition to voice transmission. Currently, however, these extra features won't work if the user is traveling outside the digital network service area. Moreover, to make and receive analog phone calls, your handset has to be able to work in both digital and analog modes.

A digital cell phone costs more than an analog one, but the monthly bill may be less, especially for heavy users. Digital phone networks promise clearer sound, although some consumers don't agree. They also offer more privacy.

Despite advances in wireless technology, American cell phones are useless outside North America. If you're an American traveling abroad, you'll need to rent a temporary unit. All this promises to change in the next six years or so, however. Under the umbrella of the World Trade Organization, 69 nations have agreed to certain standards in worldwide telecommunications systems.

Personal Communications Services

Special portable phones known as PCS phones have seen a surge in popularity. The lure is service that's noticeably better-sounding and less expensive than the standard analog cellular phones.

Like digital cellular, but lower-powered, ***personal communications services (PCS)*, or personal communications networks (PCN), are digital wireless services that use a new band of frequencies (1850–1990 megahertz) and transmitter-receivers in thousands of microcells.** PCS systems operate at superhigh frequencies, where the spectrum isn't crowded. The microcells are smaller than the cells of today's cellular phone systems.

At the moment, PCS phones have spotty availability. For instance, says one report, "a PCS subscriber driving east out of downtown Dallas on a surface street loses the signal but can hold onto it by driving on the highway."[22]

Satellite-Based Systems

The first communications satellite, AT&T's Telstar, went up in 1962. Now all of a sudden it looks like we will have a traffic jam in space. More than half the people in the world, mostly in underdeveloped countries, live more than 2 hours from the nearest telephone. (China has only four telephone lines for every 100 people.) These people, as well as business travelers and corporations needing speedy data transmission, are demanding more than wire-line or cellular service can deliver. At a World Trade Organization meeting in 1997, for instance, 67 nations agreed to open their communications markets to foreign satellite systems.

In the next few years, four kinds of satellite systems will dot the skies to provide a variety of consumer services. The first is the TV direct-broadcast system, and the second is the GPS system, both of which we described earlier. The third type is designed to handle cellular-phone and paging services, using satellite transmissions in place of tower-to-tower microwave transmissions.

Probably most interesting is the fourth, which consists of global high-speed satellite networks that will let users exchange a much broader range of data, including Internet pages and videophone calls, anywhere in the world. Five giant companies or consortiums of companies have been competing: GlobalStar, Iridium, Orbcomm, Teledesic, and Motorola. Iridium (a $5 billion international consortium led by Motorola and including Raytheon, Lockheed Martin, and Sprint) was planning to have 66 satellites operational by late 1998, but users found the phones clumsy and expensive and the firm has had to go through bankruptcy reorganization. Teledesic, known as the "Internet in the Sky" project, is perhaps the most ambitious scheme. By 2002 it aims to have no less than 220 satellites (scaled back from 840) orbiting at an altitude of 435 miles. Its purpose is to make broadband multimedia connections anywhere, just like fiber-optic cables, that will be 1000 times faster than most Internet access today. It will be used for broadband multimedia for the Internet, corporate intranets, and videoconferencing.

All these companies are taking huge risks, because the consensus of experts is that not all these ventures can survive. The spread of cellular

International phones
Many parts of the world adhere to a universal cell-phone standard, but not the United States.

phones, especially in developing countries, could make phone systems like Iridium unnecessary, says one commentator. "And technologies such as [DSL] for phones, cable modems, and a host of so-called gigabit ethernet technologies would make consumer and corporate networks faster—and satellite systems less attractive."[23]

Gigabit Ethernet: Sending Data at Supercomputer Speeds

Most corporate networks use Ethernet, a bus network devised in the 1970s. It moves data at the rate of 10 million bits (10 megabits) per second. In 1995, Fast Ethernet was invented, which can transmit data at 100 million bits (100 megabits) per second.

Now a hot new technology promises to make office networks 100 times faster than standard. "It will help end data roadblocks at major financial companies, Internet service providers, heavy-duty engineering firms, and other big enterprises," says one report.[24] Developed at government and defense laboratories to help supercomputers communicate, Gigabit Ethernet can move data at 1000 million bits (1 gigabit) per second. Recently, commercial firms have jumped in to make the technology more widely available.

Bluetooth: The Next Big Wireless Wave?

Bluetooth is a wireless technology that uses simple shortwave radio links to allow as many as eight electronic devices to communicate within about 33 feet of one another. This will allow individuals to form their own personal networks, using cell phones, laptops, and PDAs to exchange e-mail, schedules, and address information; compete at video games; or access the Internet. One of the first gadgets is a wireless headset that links wirelessly to a cellphone, so that users can talk hands-free while driving. Already more than 1600 companies are developing devices and applications for Bluetooth.[25]

Photonics: Optical Technologies at Warp Speed

"Moore's Law said that chip power would double every 18 months," writes journalist Howard Banks. "That's plodding. The new law of the photon says that bandwidth triples every year."[26]

Photonics is the science of sending data bits by means of light pulses carried on hair-thin glass fibers. For 15 years, the glass fibers of fiber-optic lines have been used to carry light pulses representing voice and data in long-distance telephone lines. Photonics has achieved breakthroughs that enable glass fibers to carry more light signals than ever before.

Older fiber-optic technologies were limited to only a few dozen miles. Then the light beams had to be converted to electrical signals, amplified, and converted back into light signals. This made the technique slow, unreliable, and expensive.

Then, in 1988, researcher David Payne at the University of Southampton in England developed an optical amplifier. This device boosts light signals without converting them first to electrical signals. The amplifier, in one description, "gives a huge push to the incoming signal, letting it carry on for dozens of more miles to the next amplifier."[27] Engineers also devised another technology (called wave-division multiplexing or dense wavelength division multiplexing) that allows laser pulses of different hues to be sent down the same tiny fiber.[28] This allows carriers to transmit at least 16 channels per fiber, which may grow to 100 channels per fiber in a few years. The upshot is that in the laboratory researchers have been able to produce glass fiber systems that carry 2 trillion bits (2 terabits) per second. That is six times the volume of all phone calls in the United States on an average day.[29]

Discuss the significance of digital cellular phones and PCS.

What could be the future of satellite systems?

Describe gigabit Ethernet and Bluetooth.

Discuss photonics.

7.6 Cyberethics: Controversial Material, Privacy, & Intellectual Property

KEY QUESTION
What are important issues in cyberethics?

Communications technology gives us more choices of nearly every sort. It provides us with different ways of working, thinking, and playing. It also presents us with some different moral choices—determining right actions in the digital and online universe. Let's consider some important aspects of "cyberethics"—controversial material and censorship, and matters of privacy.

Controversial Material & Censorship

Since computers are simply another way of communicating, there should be no surprise that many people use them to communicate about sex. Yahoo!, the Internet directory company, says that the word "sex" is the most popular search word on the Net.[30] All kinds of online X-rated message boards, chat rooms, and Usenet newsgroups exist. These raise serious issues for parents. Do we want children to have access to sexual conversations, to download hard-core pictures, or to encounter criminals who might try to meet them offline? "Parents should never use [a computer] as an electronic baby sitter," computer columnist Lawrence Magid says. People online are not always what they seem to be, he points out, and a message seemingly from a 12-year-old girl could really be from a 30-year-old man. "Children should be warned never to give out personal information," says Magid, "and to tell their parents if they encounter mail or messages that make them uncomfortable."[31]

What can be done about X-rated materials? Some possibilities:

Protecting children
Blocking software can help restrict objectionable material from children.

- **Blocking software:** Some software developers have discovered a golden opportunity in making programs like SurfWatch, Net Nanny, and CYBERsitter. These "blocking" programs screen out objectionable material, typically by identifying certain unapproved keywords in a user's request or comparing the user's request for information against a list of prohibited sites.

- **Browsers with ratings:** Another proposal in the works is browser software that contains built-in ratings for Internet, Usenet, and World Wide Web files. Parents could, for example, choose a browser that has been endorsed by the local school board or the online service provider.

- **The V-chip:** The 1996 Telecommunications Law officially launched the era of the V-chip, a device that will be required equipment in most new television sets. The *V-chip* allows parents to automatically block out programs that have been labeled as high in violence, sex, or other objectionable material.

However, any attempts at restricting the flow of information are hindered by the basic design of the Internet itself, with its strategy of offering different roads to the same place. "If access to information on a computer is blocked by one route," writes the *New York Times*'s Peter Lewis, "a moderately skilled computer user can simply tap into another computer by an alternative route." Lewis cites an Internet axiom attributed to an engineer named John Gilmore: "The Internet interprets censorship as damage and routes around it."[32]

Privacy

Privacy **is the right of people not to reveal information about themselves.** Technology, however, puts constant pressure on this right.

Consider Web cookies, little pieces of data left in your computer by some sites you visit.[33] A _cookie_ **is a file that the Web server stores on your hard-disk drive when you visit a Web site.** Thus, unknown to you, a Web site operator or companies advertising on the site can log your movements within the site. These records provide information that marketers can use to target customers for their products. Other Web sites can also get access to the cookies and acquire information about you.

There are other intrusions on your privacy. Think your medical records are inviolable? Actually, private medical information is bought and sold freely by various companies since there is no federal law prohibiting it. (And they simply ignore the patchwork of state laws.)

Think the boss can't snoop on your e-mail at work? The law allows employers to "intercept" employee communications if one of the parties involved agrees to the "interception." The party who agrees in this case is the employer. Indeed, employer snooping seems to be widespread.

A great many people are concerned about the loss of their right to privacy. Indeed, one survey found that 80% of the people contacted worried that they had lost "all control" of the personal information being collected and tracked by computers.[34] Although several laws restrain the government's ability to acquire and disseminate information and to listen in on private conversations, there are reasons to be alarmed.

Respecting Intellectual Property

The photocopier allows people to copy all kinds of print material without permission from the people who created it or own it. Two software programs, Napster and Gnutella, which allow file sharing among thousands of computers around the world, made it easy to get music over the Internet at no charge.[35]

Hopefully, you are the kind of person who wouldn't shoplift a CD in a store. Hopefully, you're also the sort who would return a backpack containing CDs to its owner. But is there any difference between getting easily available copyrighted materials for free off the Internet and stealing them? A CD's digital file available in some other student's computer may not have cost anyone anything. But it certainly cost a good deal of time and money to create the recording in the first place.

"Artists who create music for a living have a right to control what happens to their work," says Lawrence Magid.[36] This is why there are laws protecting intellectual property—music, art, words. If you've ever had someone steal something on which you spent a lot of time or take credit for an idea that was yours, you can appreciate this.

CONCEPT CHECK

Discuss some ways to protect children from X-rated material.

What is privacy?

Why should intellectual property copyrights be respected?

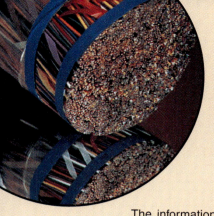

Experience Box

Information Appliances: All Kinds of Smart Gadgets for All Kinds of Digital Data

The information appliance is a specialized computer that does just a few things, such as traversing the Internet or exchanging e-mail. Here are some examples.

- **Internet appliances:** Oracle Corp. unveiled the concept of an *Internet appliance* in 1996. This device—also called a "network computer," or NC—is really just an Internet terminal, consisting of a monitor and a keyboard. The newest versions are the i-Opener (from Netpliance), the NetVista Internet Appliance (IBM), the MSN Web Companion (Compaq), the Qubit (Qubit Technology), and the i-Station (Acer America).[37] Some of these come with factory-set function keys that put weather, news, e-mail, and e-commerce services at your fingertips. For instance, there may be a button that allows you to order pizza from a local pizza parlor.

  Experts say Internet appliances represent an important transition for computers. After two decades of "faster and more powerful" PCs, Net appliances signal a major shift to "simpler and cheaper." "While it's gotten simpler, the typical PC is still complex for a lot of people," says the director of marketing for IBM's personal systems group. "The Internet appliance is all about connectivity. You plug it in, turn it on, and you're connected."[38]

- **Internet phones:** *Internet phones,* also called *Web phones,* offer wireless access to the Internet. Their primary drawback at present is the limited amount of information they can access and the limited number of Web sites they are able to log on to. On most of these devices, the screens display only 4–11 lines of information. Moreover, they can't show graphics. However, a number of companies (Yahoo!, ABCNews.com, and others) have adapted a format called *Wireless Application Protocol,* which requires sites to strip out graphics and shorten stories. Other drawbacks are the small area of the keypad and the slow procedure for typing in letters and numbers.[39]

  Among the portable Internet phones available are PDQ 800 Smart Phone (Qualcomm), Timeport (Motorola), and Touchpoint (Denso). A desktop phone, iPhone (Infogear), combines e-mail, voice mail, and Internet access in a touch-screen device.

Web phone

- **Set-top boxes:** The *set-top box* is a keypad that allows TV viewers to change channels or, in the case of interactive systems, to display Web pages on a TV set. For example, WebTV, offered by Microsoft, allows you to interact with the Internet and World Wide Web through your television set.

  Incidentally, experts differentiate between *Internet TV,* such as the services offered by WebTV and Liberate, which allow Internet access through your TV set, and *interactive TV,* such as that offered by OpenTV, which allows you to interact with the show you're watching (for example, *Jeopardy!*). These are also distinct from *personalized TV,* or *personal video recorders (PVRs),* such as the stand-alone boxes offered by TiVo and Replay TV, which allow viewers to pause, rewind, and replay live TV and to record shows.[40]

- **PC/TVs:** The full-blown *PC/TV* (or TV/PC) merges the personal computer with the television set. A circuit board in your computer allows you to receive TV programs. For instance, you can watch TV shows in a window in one corner of the screen while you're working on your computer. A different kind of device is the iCEBOX CounterTOP (from CMI), a cable-ready TV with Internet access that also features a built-in CD drive.

- **Online game players:** The *online game player* not only lets you play games but also connects to the Internet. The principal contenders on this battlefield are Sega, with its 200-megahertz Dreamcast, and Sony, with its 300-megahertz PlayStation 2. Microsoft is plan-

Sony PlayStation2

ning to enter the fray in late 2001 with its X-Box. Nintendo will offer Dolphin. The new game players far outstrip older models. The PlayStation 2, for instance, provides unparalleled animation details in video games, plays DVD movies, and offers high-speed Internet access and a hard drive to store data.[41]

- **Personal digital assistants:** A variety of devices known as electronic organizers, personal communicators, and handheld computers may be classified together as *personal digital assistants (PDAs)*. Examples are the Palm (from Palm Inc.), the Visor (Handspring), and the Pocket PC (Hewlett-Packard). Electronic organizers, such as the Palm, have a calendar, an address book, a calculator, and a stylus with which to write notes on-screen, along with a modem for e-mail connections. Handheld computers, such as the Pocket PC, have all these plus consumer versions of desktop-computer software. Some, such as the Visor, have a slot into which you can insert specially designed cartridges, turning the device into an e-book, MP3 player, camera, and game machine.[42]

- **Two-way pagers:** Motorola offers the PageWriter 2000X, a *two-way pager* with a miniature screen and a very small keyboard.[43] Besides two-way paging, the device can be used to send and receive e-mail.

- **E-mail appliances:** Various miniature *e-mail appliances* that allow you to access your e-mail from anywhere, with or without a PDA. One such device is the BackFlip (from PocketScience), a 5.5-ounce device that you connect to your Palm. If you hold the Palm and the BackFlip against a telephone's handset, you can dial a toll-free service that will forward your e-mail. The 5.4-ounce Minstrel (Novatel Wireless) connects to a Palm device and lets you browse the Internet and send and receive e-mail.[44] Using the TelMail E-Mail Organizer (Sharp), which has a built-in acoustic modem, you can send and receive e-mail without a PDA.[45]

- **Short-range wireless toys:** Some playthings marketed to children may well be harbingers of new technologies for adults.[46] These *wireless toys* are low-cost devices capable of beaming voice and text messages to fellow students carrying similar devices, if they are within close range. Examples are Lightning Mail (Tiger Electronics, operates up to 50 feet); Quik Writer (Tiger Electronics, up to 90 feet); V-Mail (Toy-Biz, up to 100 feet); and Cybiko (up to 300 feet). Some of these come with a miniature keyboard, some with a stylus for writing on a screen. Numerous devices are also being developed for Bluetooth short-wave radio technology.

Visual Summary

analog (p. 271, KQ 7.1) Continuous and varying in strength and/or quantity. An analog signal is a continuous electrical signal with such variation. *Why it's important: Sound, light, temperature, and pressure values, for instance, can fall anywhere on a continuum or range. The highs, lows, and in-between states have historically been represented with analog devices rather than in digital form. Examples of analog devices are a speedometer, a thermometer, and a tire-pressure gauge, all of which can measure continuous fluctuations. The electrical signals on a telephone line have traditionally been analog-data representations of the original voices. Telephone, radio, television, and cable-TV technologies have long been based on analog data.*

analog cellular phone (p. 286, KQ 7.3) Mobile telephone designed primarily for communicating by voice through a system of ground-area cells. Calls are directed to cells by a mobile telephone switching office (MTSO). Moving between cells requires that calls be "handed off" by the MTSO. *Why it's important: Cellular phone systems allow callers mobility.*

backbone (p. 291, KQ 7.4) High-speed network that connects LANs and MANs to the Internet. *Why it's important: The backbone is an essential part of the Internet.*

Radio frequencies
VLF LF MF HF VHF UHF SHF EH

Medium Frequency	High Frequency	Very H Freque
300–3000 kHz (3 MHz): AM radio broadcast; LORAN maritime navigation;	3 MHz– 30 MHz: Shortwave broadcast; amateur radio; CB (citizen	30 MHz 300 MH Private land mo service: as polic and tax

band (p. 280, KQ 7.3) Also called *bandwidth;* range of frequencies serving as a measure of the amount of information that can be delivered within a given period of time. The bandwidth is the difference between the lowest and the highest frequencies transmitted. For analog signals, bandwidth is expressed in hertz (Hz), or cycles per second. For digital signals, bandwidth is expressed in bits per second (bps). In the United States, certain bands are assigned by the Federal Communications Commission (FCC) for certain purposes. *Why it's important: The wider the bandwidth, the faster data can be transmitted. The narrower the band, the greater the loss of transmission power. This loss of power must be overcome by using relays or repeaters that rebroadcast the original signal.*

bridge (p. 293, KQ 7.4) Interface used to connect the same types of networks. *Why it's important: Similar networks (local area networks) can be joined together to create larger area networks.*

broadband connections (p. 280, KQ 7.3) Connections characterized by very high speed (wide bandwidth). *Why it's important: Broadband connections are necessary for reliable, high-speed Internet hook-ups.*

broadcast radio (p. 283, KQ 7.3) Wireless transmission medium that sends data over long distances—between regions, states, or countries. A transmitter is required to send messages and a receiver to receive them; sometimes both sending and receiving functions are combined in a transceiver. *Why it's important: In the lower frequencies of the radio spectrum, several broadcast radio bands are reserved not only for conventional AM/FM radio but also for broadcast television, CB (citizens band) radio, ham (amateur) radio, cellular phones, and private radio land mobile services (such as police, fire, and taxi dispatch). Some organizations use specific radio frequencies and networks to support wireless communications.*

bus network (p. 295, KQ 7.4) Type of network in which all communications devices are connected to a common channel, with no central server. Each device transmits electronic messages to other devices. If some of those messages collide, the device waits and then tries to retransmit. *Why it's important: The bus network is relatively inexpensive to install. However, if the bus itself fails, the entire network fails..*

Cellular Digital Packet Radio (CDPD) (p. 286, KQ 7.3) Wireless two-way communications system that places messages in packets—digital electronic "envelopes"—and sends them through underused radio channels or between pauses in cellular phone conversations. *Why it's important: CDPD devices allow users to send and receive voice and data messages, and the packet technique keeps data intact.*

coaxial cable (p. 282, KQ 7.3) Commonly called "co-ax"; insulated copper wire wrapped in a solid or braided metal shield, then in an external cover. *Why it's important: Co-ax is widely used for cable television. Because of the extra insulation, coaxial cable is much better than twisted-pair wiring at resisting noise. Moreover, it can carry voice and data at a faster rate.*

communications channel (p. 280, KQ 7.3) Path over which information travels in a telecommunications system from its source to its destination. *Why it's important: Channels may be wired or wireless. Three types of wired channels are twisted-pair wire (conventional telephone lines), coaxial cable, and fiber-optic cable.*

communications satellite (p. 283, KQ 7.3) Microwave relay station in orbit around the earth. *Why it's important: Transmitting a signal from a ground station to a satellite is called uplinking; the reverse is called downlinking. The delivery process will be slowed if, as is often the case, more than one satellite is required to get the message delivered.*

compression (p. 287, KQ 7.3) Also called *digital-data compression;* method of removing repetitive elements from a file so that the file requires less storage space and therefore less time to transmit. Before we use the data, it is decompressed—the repeated patterns are restored. These methods are sometimes referred to as *codec* (for *compression/decompression*) techniques. *Why it's important: Many of today's files, with graphics, sound, and video, require huge amounts of storage space; data compression makes the storage and transmission of these files more feasible.*

cookie (p. 300, KQ 7.6) A file that the Web server stores on a computer user's hard-disk drive when he or she visits a Web site. *Why it's important: Unknown to users, a Web site operator or a company advertising on the site can log their movements within the site. These records provide information that marketers can use to target customers for their products.*

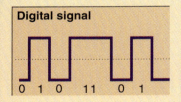

Digital signal

digital (p. 271, KQ 7.1) Represented in a two-state (binary) way. *Why it's important: Digital signals are the basis of computer-based communications. "Digital" is usually synonymous with "computer-based."*

digital cellular phone (p. 296, KQ 7.5) Mobile phone system that uses cells like an analog cellular phone system but transmits digital signals. Voice messages are decoded by the cellular handset, and short data items can also be transmitted. *Why it's important: Like an analog cellular phone, it offers the user mobility, but digital service also provides lower costs for heavy users, greater privacy, and perhaps clearer sound.*

digital television (DTV) (p. 279, KQ 7.2) A television standard that uses a digital signal, or series of 0s and 1s, rather than the customary analog standard, a system of varying signal amplitude and frequency that represents picture and sound elements. DTV was adopted as a standard in 1996 by television broadcasters and the Federal Communications Commission. *Why it's important: DTV is much clearer and less prone to interference than analog TV and is better suited to handling computer and Internet data.*

electromagnetic spectrum (p. 280, KQ 7.3) All the fields of electrical energy and magnetic energy, which travel in waves. This includes all radio signals, light rays, X-rays, and radioactivity. *Why it's important: The part of the electromagnetic spectrum of particular interest is the area in the middle, which is used for communications purposes. Various frequencies are assigned by the federal government for different purposes.*

extranet (p. 295, KQ 7.4) Private intranet that connects not only internal personnel but also selected suppliers and other strategic parties. Why it's important: *Extranets have become popular for standard transactions such as purchasing.*

fiber-optic cable (p. 282, KQ 7.3) Cable that consists of dozens or hundreds of thin strands of glass or plastic that transmit pulsating beams of light rather than electricity. Why it's important: *These strands, each as thin as a human hair, can transmit up to 2 billion pulses per second (2 Gbps); each "on" pulse represents one bit. When bundled together, fiber-optic strands in a cable 0.12 inch thick can support a quarter-million to a half-million voice conversations at the same time. Moreover, unlike electrical signals, light pulses are not affected by random electromagnetic interference in the environment. Thus, they have much lower error rates than normal telephone wire and cable. In addition, fiber-optic cable is lighter and more durable than twisted-pair and co-ax cable. A final advantage is that it cannot easily be wiretapped, so transmissions are more secure.*

file server (p. 291, KQ 7.4) Computer in a client/server network that acts like a disk drive, storing the programs and data files shared by users. Why it's important: *A file server enables all users of a LAN to have access to the same programs and data.*

firewall (p. 296, KQ 7.4) System of hardware and software that blocks unauthorized users inside and outside the organization from entering the intranet. Why it's important: *Security is essential to an intranet. A firewall consists of two parts, a choke and a gate. The choke forces all data packets flowing between the Internet and the intranet to pass through a gate. The gate regulates the flow between the two networks. It identifies authorized users, searches for viruses, and implements other security measures. Thus, intranet users can gain access to the Internet (including key sites connected by hyperlinks), but outside Internet users cannot enter the intranet.*

gateway (p. 293, KQ 7.4) Interface permitting communication between dissimilar networks. Why it's important: *Gateways permit communication between a LAN and a WAN or between two LANs based on different network operating systems or different layouts.*

Global Positioning System (GPS) (p. 285, KQ 7.3) A series of earth-orbiting satellites continuously transmitting timed radio signals that can be used to identify earth locations. Why it's important: *A GPS receiver—handheld or mounted in a vehicle, plane, or boat—can pick up transmissions from any four satellites, interpret the information from each, and calculate to within a few hundred feet or less the receiver's longitude, latitude, and altitude. Some GPS receivers include map software for finding your way around, as with the Guidestar system available with some rental cars.*

high-definition television (HDTV) (p. 279, KQ 7.2) A high-resolution type of digital television (DTV), which comes in either a 720- or 1080-line mode (compared to 525-line resolution for analog TV). Why it's important: *The Federal Communications Commission expects that eventually HDTV will supplant analog TV as the dominant digital standard. At present, most consumers consider the HDTV sets available to be too expensive.*

host computer (p. 290, KQ 7.4) Mainframe or midsize central computer that controls a network. Why it's important: *The host is responsible for managing the entire network.*

infrared wireless transmission (p. 282, KQ 7.3) Transmission of data signals using infrared-light waves. Why it's important: *Infrared ports can be found on some laptop computers and printers, as well as wireless mouses. The advantage is that no physical connection is required among devices. The drawbacks are that line-of-sight communication is required—there must be an unobstructed view between transmitter and receiver—and transmission is confined to short range.*

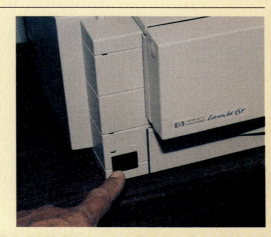

intranet (p. 295, KQ 7.4) An organization's internal private network that uses the infrastructure and standards of the Internet and the World Wide Web. Why it's important: *When an organization creates an intranet, it enables employees to have quicker access to internal information and to share knowledge so that they can do their jobs better. Information exchanged on intranets may include employee e-mail addresses and telephone numbers, product information, sales data, employee benefit information, and lists of jobs available within the organization.*

JPEG (p. 288, KQ 7.3) The leading compression standard for still images; it stands for Joint Photographic Experts Group of the International Standards Organization. The file extension that identifies graphic images files in the JPEG format is .jpeg or .jpg. Why it's important: *JPEG compression is commonly used to store and transmit still graphic images. Higher or lower degrees of JPEG compression may be chosen; greater compression corresponds to greater image loss.*

local area network (LAN) (p. 290, KQ 7.4) Network that connects computers and devices in a limited geographical area, such as one office, one building, or a group of buildings close together (for instance, a college campus). Why it's important: *LANS have replaced large computers for many functions and are considerably less expensive.*

metropolitan area network (MAN) (p. 290, KQ 7.4) Communications network covering a city or a suburb. Why it's important: *The purpose of a MAN is often to bypass local telephone companies when accessing long-distance services. Many cellular phone systems are MANs.*

microwave radio (p. 283, KQ 7.3) Transmission of voice and data through the atmosphere as superhigh-frequency radio waves called *microwaves*. These frequencies are used to transmit messages between ground-based stations and satellite communications systems. Why it's important: *Microwaves are line-of-sight; they cannot bend around corners or around the earth's curvature, so there must be an unobstructed view between transmitter and receiver. Thus, microwave stations need to be placed within 25–30 miles of each other, with no obstructions in between. A string of microwave relay stations will each receive incoming messages, boost the signal strength, and relay the signal to the next station. Nowadays dish- or horn-shaped microwave reflective dishes, which contain transceivers and antennas, are nearly everywhere.*

Modem: Demodulate (converts analog signals back to digital form)

modem (p. 273, KQ 7.1) Short for "**mo**dulate/**dem**odulate"; device that converts digital signals into a representation of analog form (modulation) to send over phone lines; a receiving modem then converts the analog signal back to a digital signal (demodulation). Why it's important: *The modem provides a means for computers to communicate with one another using the standard copper-wire telephone network, an analog system that was built to transmit the human voice but not computer signals.*

MPEG (p. 288, KQ 7.3) Stands for Motion Picture Experts Group. MPEG is the leading compression standard for video images. The file extension that identifies video (and sound) files compressed in this format is .mpeg or .mpg. Why it's important: *People who work with videos are mainly interested in storing or transmitting an enormous amount of visual information in economical form; preserving details is a secondary consideration. The Motion Picture Experts Group sets standards for weeding out redundancies between neighboring images in a stream of video. Three MPEG standards have been developed for compressing visual information—MPEG-1, MPEG-2, and MPEG-4.*

network (p. 289, KQ 7.4) Also called *communications network;* system of interconnected computers, telephones, or other communications devices that can communicate with one another and share applications and data. Why it's important: *The tying together of so many communications devices in so many ways is changing the world we live in.*

node (p. 290, KQ 7.4) Any device that is attached to a network. Why it's important: *A node may be a microcomputer, terminal, storage device, or peripheral device, any of which enhance the usefulness of the network.*

packet-radio-based communications (p. 286, KQ 7.3) Wireless two-way communications system that uses a nationwide system of radio towers to send data to handheld computers. Why it's important: *Uses packet switching to encapsulate the data in envelopes so that it arrives intact, which may not be the case with data transmission over analog cellular systems. Used by mobile workers who need to communicate frequently with a corporate database.*

Another LAN

Bridge

Fiber-optic backbone

Bridge

Gateway

pager (p. 286, KQ 7.3) Simple radio receiver that receives data, but not voice messages, sent from a special radio transmitter. The pager number is dialed from a phone and travels via the transmitter to the pager. Why it's important: *Pagers have become a common way of receiving notification of phone calls so that the user can return the calls immediately; some pagers can also display messages of up to 80 characters and send preprogrammed messages.*

peer-to-peer LAN (p. 291, KQ 7.4) Type of local area network in which all microcomputers on the network communicate directly with one another without relying on a server. Why it's important: *Peer-to-peer networks are less expensive than client-server networks and work effectively for up to 25 computers. Beyond that, they slow down under heavy use. They are appropriate for small networks.*

personal communications services (PCS) (p. 297, KQ 7.5) Also called *personal communications networks (PCN);* it is a lower-powered version of digital cellular phone service. PCS is a digital wireless phone service that uses a relatively new band of frequencies (1850–1990 megahertz) and transmitter-receivers in thousands of microcells. Why it's important: *PCS takes advantage of the uncrowded superhigh frequencies, but with its small cells users may find themselves out of range of the service in many areas.*

privacy (p. 300, KQ 7.6) The right of people not to reveal information about themselves. Why it's important: *Privacy is a basic democratic right. Information technology presents constant threats to this right.*

radio frequency spectrum (p. 280, KQ 7.3) The part of the electromagnetic spectrum that carries communications signals. Why it's important: *The radio spectrum ranges from low-frequency waves, such as those used for aeronautical and marine navigation equipment, through the medium frequencies for CB radios, cordless phones, and baby monitors, to ultrahigh frequency bands for cell phones and also microwave bands for communications satellites.*

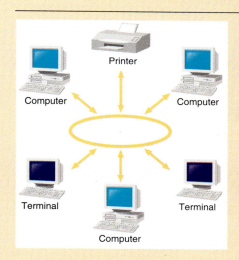

Printer

Computer Computer

Terminal Terminal

Computer

ring network (p. 294, KQ 7.4) Type of local area network (LAN) in which all communications devices are connected in a continuous loop and messages are passed around the ring until they reach the right destination. There is no central server. Why it's important: *The advantage of a ring network is that messages flow in only one direction and so there is no danger of collisions. The disadvantage is that if a connection is broken, the entire network stops working.*

router (p. 293, KQ 7.4) Special computer that directs communicating messages when several networks are connected together. Why it's important: *High-speed routers can serve as part of the Internet backbone, or transmission path, handling the major data traffic.*

standard-definition television (SDTV) (p. 279, KQ 7.2) A standard of digital television that has a minimum of 480 vertical lines, allowing broadcasters to transmit more information within the HDTV bandwidth. Why it's important: *SDTV would enable broadcasters to effectively multicast their products, on up to as many as six channels instead of one. It also frees up bandwidth for data transmission. Thus, in the future there might be separate channels carrying video, audio, and data.*

star network (p. 293, KQ 7.4) Type of local area network (LAN) in which all microcomputers and other communications devices are connected to a central hub, such as a file server. Electronic messages are routed through the central hub to their destinations. The central hub monitors the flow of traffic. Why it's important: *The advantage of a star network is that the hub prevents collisions between messages. Moreover, if a connection is broken between any communications device and the hub, the rest of the devices on the network will continue operating.*

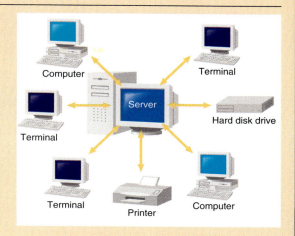

Computer

Server

Terminal

Terminal

Hard disk drive

Terminal

Printer

Computer

telecommuting (p. 277, KQ 7.2) Working at home while in telecommunication with the office. Why it's important: *Telecommuting has many benefits. Examples are reduced traffic congestion, energy consumption, and air pollution, increased productivity, and improved teamwork. A disadvantage is that people may feel isolated.*

topology (p. 293, KQ 7.4) The logical layout, or shape, of a local area network. The five basic topologies are star, ring, bus, hybrid, and FDDI. Why it's important: *Different topologies can be used to suit different office and equipment configurations.*

twisted-pair wire (p. 282, KQ 7.3) Two strands of insulated copper wire, twisted around each other. Why it's important: *Twisted-pair wire has been the most common channel or medium used for telephone systems. However, it is relatively slow and does not protect well against electrical interference.*

videoconferencing (p. 275, KQ 7.2) Also called *teleconferencing;* use of television video and sound technology as well as computers to enable people in different locations to see, hear, and talk with one another. Why it's important: *Videoconferencing may eliminate the need for some travel for the purpose of meetings and allow people who cannot travel to visit face to face.*

virtual office (p. 277, KQ 7.2) Nonpermanent and mobile office run with computer and communications technology. Why it's important: *Employees work from their homes, cars, and other new work sites, using pagers, portable computers, and cell phones to conduct business. This reduces the office expenses of their employer.*

wide area network (WAN) (p. 290, KQ 7.4) Communications network that covers a wide geographical area, such as a country or the world. Why it's important: *Most long-distance and regional Bell telephone companies are WANs. A WAN may use a combination of satellites, fiber-optic cable, microwave, and copper wire connections and link a variety of computers, from mainframes to terminals.*

workgroup computing (p. 277, KQ 7.2) Also called *collaborative computing;* technology by which teams of co-workers can use networks of microcomputers to share information and to cooperate on projects. Workgroup computing is made possible not only by networks and microcomputers but also by groupware. Why it's important: *Workgroup computing allows co-workers to collaborate with colleagues, suppliers, and customers and to tap into company information through computer networks.*

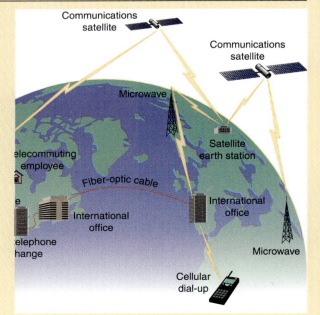

Chapter Review

Self-Test Questions

1. A(n)_____ converts digital signals into analog signals for transmission over phone lines.

2. A(n) _____ network covers a wide geographical area, such as a state or a country.

3. _____ cable transmits data as pulses of light rather than as electricity.

4. _____refers to waves continuously varying in strength and/or quantity; _____ refers to communications signals or information in a binary form.

5. _____ is a method of removing repetitive elements from a file so that the file requires less storage space.

6. A(n) _____is a computer that acts as a disk drive, storing programs and data files shared by users on a LAN.

7. The _____ is the system software that manages the activities of a network.

8. *Modem* is short for _____.

9. The leading compression standard for still images is _____.

10. _____ programs can screen out objectionable material on the Internet.

Multiple-Choice Questions

1. Which of the following best describes the telephone line that is used in most homes today?
   a. coaxial cable
   b. modem cable
   c. twisted-wire pair
   d. fiber-optic cable
   e. LAN

2. Which of the following do local area networks enable?
   a. sharing of peripheral devices
   b. sharing of programs and data
   c. better communications
   d. access to databases
   e. all of the above

3. Which of the following is not a data compression standard/method?
   a. lossless
   b. JPEG
   c. MPEG
   d. lossy
   e. NOS

4. Which of the following is not a type of server?
   a. file server
   b. print server
   c. mail server
   d. disk server
   e. database server

True/False Questions

T F 1. In a LAN, a bridge is used to connect the same type of networks, whereas a gateway is used to enable dissimilar networks to communicate.

T F 2. Frequency and amplitude are two characteristics of analog carrier waves.

T F 3. A range of frequencies is called a *spectrum*.

T F 4. Twisted-pair wire commonly connects residences to external telephone systems.

T F 5. A cookie is an ID number used to visit specific Web sites.

Short-Answer Questions

1. What is the difference between an intranet and an extranet?
2. What is workgroup computing?
3. What is the difference between a LAN and a WAN?
4. Why is bandwidth a factor in data transmission?
5. What is a firewall?

Concept Mapping

On a separate sheet of paper, draw a concept map, or visual diagram, linking concepts. Show how the following terms are related.

analog	digital
backbone	extranet
band	fiber-optic cable
bridge	file server
broadband connections	firewall
coaxial cable	gateway
communications channel	GPS
host computer	ring network
Intranet	router
JPEG	satellite
LAN	spectrum

6 **The Ethics of Using Databases: Concerns about Accuracy & Privacy.** In **morphing,** a film image is altered pixel by pixel, so that the image becomes something else. This manipulation of digitized images and sounds raises some ethical issues. Sound performances can be misrepresented, photos may be manipulated, and video and TV images may be altered in undetectable ways and all stored in a database.

Databases are also limited in accuracy and completeness, since not all facts can be found in a database, nor are all data items true. In addition, databases raise several concerns about privacy. Finally, those who own databases may be in a position to monopolize information.

If information exists in one place, it exists in more than one place."

So says Carole A. Lane, a database expert.[1] This is what she calls "Lane's First Law of Information." Perhaps it could stand as a summary of one of the most important developments of the Digital Age.

How does information in one place get to be in more than one place? The answer has to do with databases. A database is not just the computerization of what used to go into manila folders and a filing cabinet. A database is an organized collection of related files, a technology for pulling together facts that allows the slicing and dicing and mixing and matching of data in all kinds of ways. As a result, the arrival of databases—especially when linked to the Internet—has stood many of our business and social institutions on their heads.

Your name and some facts about you can probably be found in scores—if not hundreds—of far-flung databases. How is that data being used? That is a very interesting question. In this chapter, we will examine the importance of databases and how they work.

8.1 Databases & the New Economy: E-Commerce, Data Mining, & B2B Systems

KEY QUESTIONS

What are databases, and how are e-commerce, data mining, and business-to-business systems using them?

It used to be there was a difference between the Old Economy and the New Economy. The first consisted of traditional companies—car makers, pharmaceuticals, retailers, publishers. The second consisted of computer, telecommunications, and Internet companies (AOL, Amazon, eBay, and a raft of "dot-com" firms). Until recently, New Economy companies were fast growing. From 1996 to 1997, for example, the U.S. Internet-based economy nearly doubled, from $15.5 billion to $29 billion. By 2001, it was predicted to reach $350 billion.[2] Now, however, Old Economy companies have begun to absorb the new Internet-driven technologies, and the differences between the two sectors are dwindling.

One sign of growth is that Internet host computers has been almost doubling every year. But the mushrooming of computer networks and the booming popularity of the World Wide Web are only the most obvious signs of the digital economy. Behind them lies something equally important: the growth of vast stores of information in the form known as *databases*, organized collections of related files.

How are databases underpinning the New Economy? Let us consider three aspects: *e-commerce, data mining,* and *business-to-business (B2B) systems.*

E-Commerce

The Internet might have remained a text-based realm, the province of academicians and researchers, had it not been for the creative contributions of Tim Berners-Lee. He was the computer scientist who came up with the coding system (HyperText Markup Language), linkages, and addressing scheme (URLs) that debuted in 1991 as the graphics-laden and multimedia World Wide Web. "It's hard to overstate the impact of the global system he created," writes *Time* technology writer Joshua Quittner. "He took a powerful communications system [the Internet] that only the elite could use, and turned it into a mass medium."[3]

The arrival of the Web quickly led to ***e-commerce*, or electronic commerce, the buying and selling of products and services through computer networks.**

By 2003, total U.S. e-commerce sales to consumers were expected to reach $108 billion, or 6% of consumer retail spending.[4] Indeed, online shopping is growing even faster than the increase in computer use, which has been fueled by the falling price of personal computers. (Half of American households now have a PC.)[5] Among the best-known e-firms are bookseller Amazon.com; auction network eBay; and Priceline, which lets you name the price you're willing to pay for airline tickets and hotel rooms.

Probably the foremost example of e-commerce is Amazon.com.[6] In 1994, seeing the potential for electronic retailing on the World Wide Web, Jeffrey Bezos left a successful career on Wall Street to launch an online bookstore called Amazon.com. Why the name *Amazon*?

"Earth's biggest river, Earth's biggest bookstore," said Bezos in a 1996 interview. "The Amazon River is ten times as large as the next largest river, which is the Mississippi, in terms of volume of water. Twenty percent of the world's fresh water is in the Amazon River Basin, and we have six times as many titles as the world's largest physical bookstore."[7] A more hard-headed reason is that, according to consumer tests, words starting with "A" show up on search-engine lists first.[8]

Still, Bezos realized that no bookstore with four walls could possibly stock the more than 2.5 million titles that are now active and in print. Moreover, he saw that an online bookstore wouldn't

E-commerce
This Irish retailer of woolens and clothing goods sells not only in its store but also over the World Wide Web.

have to make the same investment in retail clerks, store real estate, or warehouse space (in the beginning, Amazon.com ordered books from the publisher *after* it took an order), so it could pass savings along to customers in the form of discounts. In addition, he appreciated that there would be opportunities to obtain demographic information about customers in order to offer personalized services. For example, Amazon could let customers know of books that might be of interest to them. Such personalized attention is difficult for traditional bookstores. Finally, Bezos saw that there could be a good deal of online interaction: Customers could post reviews of books they read and could reach authors by e-mail to provide feedback. All this was made possible on the Web by the recording of information on giant databases.

Amazon.com sold its first book in July 1995 and by early 2000 had 1.1 million customers and a market capitalization of $18.3 billion, with Bezos owning about a third of that.[9] The firm also had expanded into the online retailing of music CDs, toys, electronics, drugs, cosmetics, pet supplies, and other goods and also into online auctions.

In 2000, Amazon.com's stock plummeted, along with that of many other so-called "dot-com" companies, which had become overvalued in the stock-market gold rush earlier that year. But the lasting impact of Amazon's trail blazing is clear from the surge of Old Economy "brick and mortar" companies into the online sector. Indeed, proof that the two sectors probably need each other more than they thought came in the summer of 2000 when Amazon.com, the Internet superstore, and Toys R Us, the top brick-and-mortar toy seller, announced they would join their online toy forces.

Data Mining

A personal database, such as the address list of friends you have on your microcomputer, is generally small. But some databases are almost unimaginably vast, involving records for millions of households and trillions of bytes of data. Some of these activities require the use of so-called massively parallel database computers that cost $1 million or more. "These machines gang together scores or even hundreds of the fastest microprocessors around," says one description, "giving them the oomph to respond in minutes to complex database queries."[10]

These large-scale efforts go under the name *data mining*. **_Data mining (DM)_ is the computer-assisted process of sifting through and analyzing vast amounts of data in order to extract meaning and discover new knowledge.** The purpose of DM is to describe past trends and predict future trends. Thus, data-mining tools might sift through a company's immense collections of customer, marketing, production, and financial data and identify what's worth noting and what's not.

Data mining begins with acquiring data and preparing it for what is known as the "data warehouse," by the following steps.[11] *(See ● Panel 8.1.)*

● **PANEL 8.1**

The data-mining process

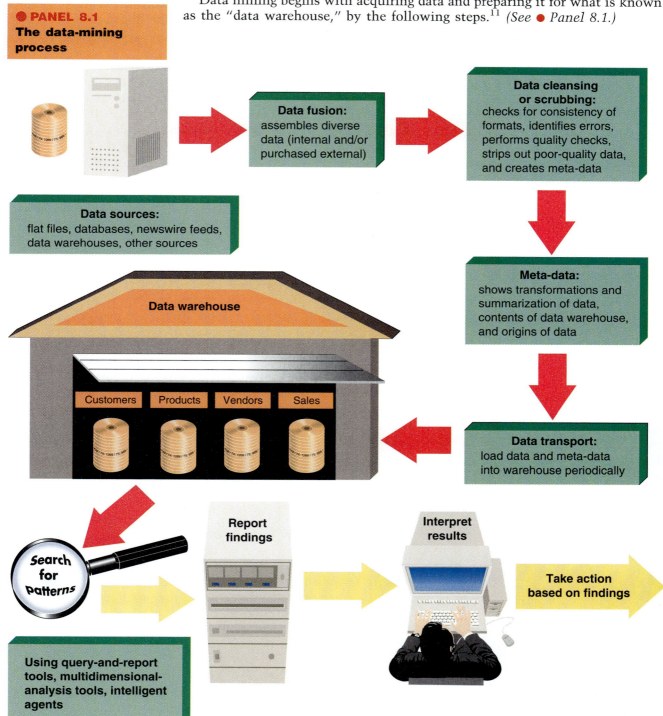

Data sources: flat files, databases, newswire feeds, data warehouses, other sources

Data fusion: assembles diverse data (internal and/or purchased external)

Data cleansing or scrubbing: checks for consistency of formats, identifies errors, performs quality checks, strips out poor-quality data, and creates meta-data

Meta-data: shows transformations and summarization of data, contents of data warehouse, and origins of data

Data transport: load data and meta-data into warehouse periodically

Data warehouse

Customers Products Vendors Sales

Search for patterns

Using query-and-report tools, multidimensional-analysis tools, intelligent agents

Report findings

Interpret results

Take action based on findings

1. **Data sources:** Data may come from a number of sources: (1) point-of-sale transactions in files (flat files) managed by file management systems on mainframes, (2) databases of all kinds, and (3) other—for example, news articles transmitted over newswires or online sources such as the Internet. To the mix may also be added (4) data from data warehouses, as we describe.

2. **Data fusion and cleansing:** Data from diverse sources, whether from inside the company (internal data) or purchased from outside the company (external data), must be fused together, then put through a process known as *data cleansing,* or *scrubbing.* Even if the data comes from just one source, such as one company's mainframe, the data may be of poor quality, full of errors and inconsistencies. Therefore, for data mining to produce accurate results, the source data has to be "scrubbed"—that is, cleaned of errors and checked for consistency of formats.

3. **Data and meta-data:** The cleansing process yields both the cleaned-up data and a variation of it called *meta-data.* Meta-data (or "data about data") shows the origins of the data, the transformations it has undergone, and summary information about it, which makes it more useful than the cleansed but unintegrated, unsummarized data. The meta-data also describes the contents of the data warehouse.

4. **The data warehouse:** Both the data and the meta-data are sent to the data warehouse. A ___data warehouse___ **is a special database of cleaned up data and meta-data.** It is a replica, or close reproduction, of a mainframe's data. The data warehouse is stored on disk using storage technology such as RAID (redundant arrays of independent disks). Small data warehouses may hold 100 gigabytes of data or less. Once 500 gigabytes are reached, massively parallel processing computers are needed. Projections call for large data warehouses holding hundreds of terabytes within the next few years.

Three kinds of software, or "siftware," tools are used to perform data mining—that is, to do finding and analyzing tasks.[12]

Databases for marketing
Harrah's casinos use data mining to identify gambling patron preferences for marketing and special promotions.

- **Query-and-reporting tools:** *Query-and-reporting tools* (examples are Focus Reporter and Esperant) require a database structure and work well with relational databases. Their best use is for specific questions to verify hypotheses. For example, if a company decides to mine its database to find customers most likely to respond to a mail-order promotion, it might use a query-and-reporting tool and construct a query: "How many credit-card customers who made purchases of over $100 on sporting goods in August have at least $2000 of available credit?"[13]

- **Multidimensional-analysis tools:** Multidimensional-analysis (MDA) tools (examples are Essbase and Lightship) can do "data surfing" to explore all dimensions of a particular subset of data. In one writer's example, "The idea [with MDA] is to load a multidimensional server with data that is likely to be combined. Imagine all the possible ways of analyzing clothing sales: by brand name, size, color, location, advertising, and so on."[14] Using MDA tools, you can analyze this multidimensional database from all points of view.

- **Intelligent agents:** An intelligent agent is a computer program that roams through networks performing complex work tasks for people. Intelligent agents are best used for turning up unsuspected relationships and patterns. "These patterns may be so nonobvious as to appear almost nonsensical," says one writer, "such as that people who have bought scuba gear are good candidates for taking Australian vacations."[15]

Data mining has come about because companies find that, in today's fierce competitive business environment, they need to turn the gazillions of bytes of raw data at their disposal to new uses for further profitability. However, nonprofit institutions have also found DM methods useful, as in the pursuit of scientific and medical discoveries.

Some applications of data mining:[16]

- **Marketing:** Marketers use DM tools (such as one called Spotlight) to mine point-of-sale databases of retail stores, which contain facts (such as prices, quantities sold, dates of sale) for thousands of products in hundreds of geographic areas. By understanding customer preferences and buying patterns, marketers hope to target consumers' individual needs.

- **Health:** A coach in the U.S. Gymnastics Federation is using a DM system (called IDIS) to discover what long-term factors contribute to an athlete's performance, so as to know what problems to treat early on. A Los Angeles hospital is using the same tool to see what subtle factors affect success and failure in back surgery. Another system helps health-care organizations pinpoint groups whose costs are likely to increase in the near future, so that medical interventions can be made.

- **Science:** DM techniques are being employed to find new patterns in genetic data, molecular structures, global climate changes, and more. For instance, one DM tool (called SKICAT) is being used to catalog more than 50 million galaxies, which will be reduced to a 3-terabyte galaxy catalog.

Clearly, short-term payoffs can be dramatic. One telephone company, for instance, mined its existing billing data to identify 10,000 supposedly "residential" customers who spent more than $1000 a month on their phone bills. When it looked more closely, the company found these customers were really small businesses trying to avoid paying the more expensive business rates for their telephone service.[17]

However, the payoffs in the long term could be truly astonishing. Sifting medical-research data or subatomic-particle information may reveal new treatments for diseases or new insights into the nature of the universe.[18]

Business-to-Business (B2B) Systems

In a **_business-to-business (B2B) system_, a business sells to other businesses, using the Internet or a private network to cut transaction costs and increase efficiencies.** Business-to-business activity is expected to balloon to a $1 trillion industry by 2002 and a $2.8 trillion industry by 2004.[19]

One of the most famous examples of B2B is the auto industry online exchange developed by the big three U.S. automakers—General Motors, Ford, and DaimlerChrysler. The companies are putting their entire system of purchasing, involving more than $250 billion in parts and materials and 60,000 suppliers, on the Internet. Already so-called "reverse auctions," in which suppliers bid to provide the lowest price, have driven down the cost of parts such as tires and window sealers. This system replaces the old-fashioned bureaucratic procurement process built on phone calls and fax machines, and provides substantial cost savings.[20]

Online B2B exchanges have been developed to serve a variety of businesses, from manufacturers of steel and airplanes to convenience stores to olive oil producers.[21] B2B exchanges are expected to revolutionize business by moving beyond pricing mechanisms and encompassing product quality, customer support, credit terms, and shipping reliability, which often count for more than price. The name given to this system is the *business web*, or *b-web*, in which suppliers, distributors, customers, and e-commerce service providers use the Internet for communications and transactions. In addition, b-webs are expected to provide extra revenue from ancillary services, such as financing and logistics.[22] *(See ● Panel 8.2.)*

None of these sectors of the New Economy is possible without databases and the communications lines connecting them. Let us start our discussion of databases by looking at basic concepts, beginning with files.

CONCEPT CHECK

Describe what e-commerce is.

What is data mining, and how is it used?

What is a B2B system?

KEY QUESTIONS

What are the data storage hierarchy, the key field, types of files, ways of keeping files, sequential versus direct access, and offline versus online storage?

When you create a letter or a spreadsheet or an address file on a computer, you often want to save it for future reference. Let's look at the mechanics of this process.

How Data Is Organized: The Data Storage Hierarchy

Data can be grouped into a hierarchy of categories, each increasingly more complex. The ***data storage hierarchy* consists of the levels of data stored in a computer: bits, bytes (characters), fields, records, files, and databases.** *(See* ● *Panel 8.3.)*

Computers, we have said, are based on the principle that electricity may be on or off. Thus, individual items of data are represented by the bits 0 for off and 1 for on. Bits and bytes are the building blocks for representing data, whether it is being processed, stored, or telecommunicated. The computer deals with the bits and bytes; you, however, will need to deal with characters, fields, records, files, and databases.

- **Characters:** A ***character*** **(byte) is a letter, number, or special character.** A, B, C, 1, 2, 3, #, $, % are all examples of single characters.
- **Field:** A ***field*** **is a unit of data consisting of one or more characters (bytes).** An example of a field is your name, your address, or your Social Security number.

● **PANEL 8.3**
How data is organized

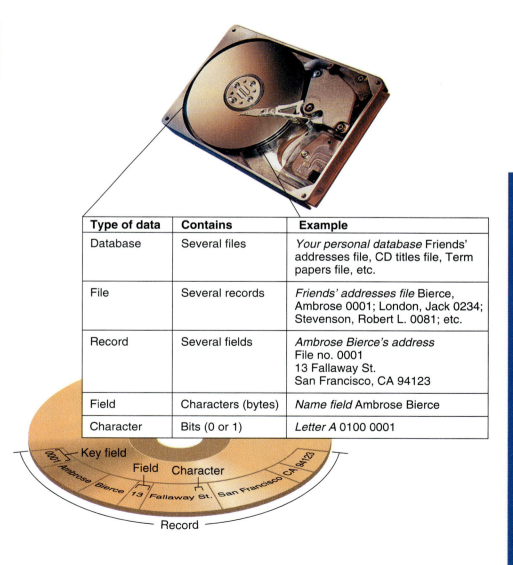

Type of data	Contains	Example
Database	Several files	*Your personal database* Friends' addresses file, CD titles file, Term papers file, etc.
File	Several records	*Friends' addresses file* Bierce, Ambrose 0001; London, Jack 0234; Stevenson, Robert L. 0081; etc.
Record	Several fields	*Ambrose Bierce's address* File no. 0001 13 Fallaway St. San Francisco, CA 94123
Field	Characters (bytes)	*Name field* Ambrose Bierce
Character	Bits (0 or 1)	*Letter A* 0100 0001

Key field

0001 Ambrose Bierce

Field   Character

13 Fallaway St.   San Francisco   CA 94123

Record

- **Record:** A _record_ **is a collection of related fields.** An example of a record would be your name _and_ address _and_ Social Security number.
- **File:** A _file_ **is a collection of related records.** An example of a file is data collected on everyone employed in the same department of a company, including all names, addresses, and Social Security numbers. You use files a lot because the file is the collection of data or information that is treated as a unit by the computer.
- **Database:** As we've stated, a _database_ **is an organized collection of integrated files.** A company database might include files on all past and current employees in all departments. There would be various files for each employee: payroll, retirement benefits, sales quotas and achievements (if in sales), and so on.

The Key Field

An important concept in data organization is that of the _key field_. A _key field_ **is a field that is chosen to uniquely identify a record so that it can be easily retrieved and processed.** The key field is often an identification number, Social Security number, customer account number, or the like. The primary characteristic of the key field is that it is _unique_. Thus, numbers are clearly preferable to names as key fields because there are many people with common names like James Johnson, Susan Williams, Ann Wong, or Roberto Sanchez, whose records might be confused.

Types of Files: Program Files, Data Files, & Others

As we said, the _file_ is the collection of data or information that is treated as a unit by the computer. **Files are given names—_filenames_.** If you're using a word processing program to write a psychology term paper, you might name it Psychreport.

Filenames also have _extension names_. These extensions of up to three letters are added after a period following the filename—for example, the _.doc_ in Psychreport.doc is recognized by Microsoft Word as a "document." Extensions are usually inserted automatically by the application software.

When you look up the filenames listed on your hard drive (on the directory, as we will explain), you will notice a number of extensions, such as _.doc, .exe,_ and _.com._ There are many kinds of files, but perhaps the two principal ones are _program files_ and _data files._

- **Program files:** _Program files_ **are files containing software instructions.** Examples are word processing or spreadsheet programs, which are made up of several different program files. The two most important are source program files and executable files.

  Source program files contain high-level computer instructions in the original form written by the programmer. Some source program files have the extension of the language in which they are written, such as _.bas_ for BASIC, _.pas_ for Pascal, or _.jav_ for Java.

  For the processor to use source program instructions, they must be translated into an _executable file,_ which contains the instructions that tell the computer how to perform a particular task. You can identify an executable file by its extension, _.exe._ You use an executable file by running it—as when you select Microsoft Excel from your onscreen menu and run it. (There are some executable files that you cannot run—another computer program causes them to execute. These are identified by such extensions as _.dll, .drv, ocx, .sys,_ and _.vbx._)

- **Data files:** _Data files_ **are files that contain data**—words, numbers, pictures, sounds, and so on. Unlike program files, data files don't instruct the computer to do anything. Rather, data files are there to be acted

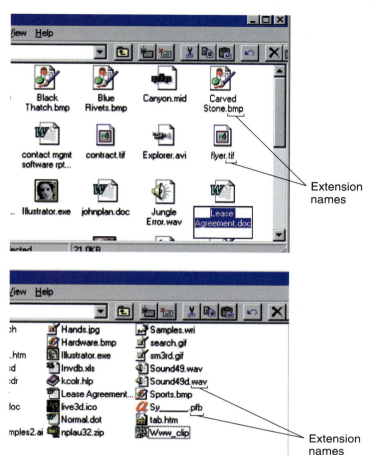

Extension names

Extension names

upon by program files. Examples of common extensions in data files are *.txt* (text) and *.xls* (spreadsheets). Certain proprietary software programs have their own extensions, such as *.ppt* for Power-Point and *.mdb* for Access.

Other common types of files are *ASCII files*, *image files*, *audio files*, *animation/video files*, and *Web files*.

- **ASCII files:** ASCII is a common binary coding scheme used to represent data in a computer. *ASCII ("as-key") files* are text-only files that contain no graphics and no formatting, such as boldface or italics. This format is used to transfer documents between incompatible computers, such as PC and Macintosh. Such files may use the *.txt* extension.

- **Image files:** If ASCII files are for text, *image files* are for digitized graphics, such as art or photographs. They are indicated by such extensions as *.bmp*, *.gif*, *.jpg*, *.pcx*, *.tif*, and *.wmf*.

- **Audio files:** *Audio files* contain digitized sound and are used for conveying sound in CD-ROM multimedia and over the Internet. They have extensions such as *.wav* and *.mid*.

- **Animation/video files:** *Video files*, used for such purposes as conveying moving images over the Internet, contain digitized video images. Common extensions are *.avi, flc, .fli*, and *.mpg*.

- **Web files:** *Web files* are files carried over the World Wide Web. Their extensions include *.html, .htm*, and *.xml*.

Two Types of Data Files: Master File & Transaction File

Among the several types of data files, two are commonly used to update data: a master file and a transaction file.

- **Master file:** The **_master file_** is a data file containing relatively perma-**nent records that are generally updated periodically.** An example of a master file would be the address-label file for all students currently enrolled at your college.

- **Transaction file:** The **_transaction file_** is a temporary holding file that **holds all changes to be made to the master file: additions, deletions, revisions.** For example, in the case of the address labels for your college, a transaction file would hold new names and addresses to be added (because over time new students enroll) and names and addresses to be deleted (because students leave). It would also hold revised names and addresses (because students change their names or move). Each month or so, the master file would be *updated* with the changes called for in the transaction file.

Data Access Methods: Sequential versus Direct Access

The way that a secondary-storage device allows access to the data stored on it affects its speed and its usefulness for certain applications. The two main types of data access are sequential and direct.

- Sequential storage: _Sequential storage_ **means that data is stored and retrieved in sequence,** such as alphabetically. Tape storage falls in the category of sequential storage. Thus, if you are looking for employee number 8888 on a tape, the computer will have to start with 0001, then go past 0002, 0003, and so on, until it finally comes to 8888. This data access method is less expensive than other methods because it uses magnetic tape, which is cheaper than disks. The disadvantage of sequential file organization is that searching for data is slow.

- Direct access storage: _Direct access storage_ **means that the computer can go directly to the information you want**—just as a CD player can go directly to a particular track on a music CD. The data is retrieved (accessed) according to a unique data identifier called a _key field,_ as we will discuss. It also uses a _file allocation table (FAT),_ a hidden on-disk table that records exactly where the parts of a given file are stored.

  This method of file organization is used with hard disks and other types of disks. It is ideal for applications where there is no fixed pattern to the requests for data—for example, in airline reservation systems or computer-based directory-assistance operations.

  If you need to find specific data, direct file access is much faster than sequential access. However, direct file access is also more expensive, for two reasons: (1) the complexity involved in maintaining a file allocation table and (2) the need to use hard-disk technology, rather than cheaper magnetic tape technology.

Offline versus Online Storage

Whether it's on magnetic tape or on some form of disk, data may be stored either offline or online.

- Offline: _Offline storage_ **means that data is not directly accessible for processing until the tape or disk it's on has been loaded onto an input device.** That is, the storage is not under the direct, immediate control of the central processing unit.

- Online: _Online storage_ **means that stored data is randomly (directly) accessible for processing.** That is, storage is under the direct, immediate control of the central processing unit. You need not wait for a tape or disk to be loaded onto an input device.

For processing to be online, the storage must be online and _fast._ This nearly always means storage on disk (direct access storage) rather than magnetic tape (sequential storage).

CONCEPT CHECK

Describe the data storage hierarchy and the concept of a key field.

Distinguish program files from data files.

List other types of files.

Explain how to keep track of files on your computer.

Distinguish sequential from direct access.

Discuss offline and online storage.

As we said earlier, a *database* is an organized collection of related (integrated) files. A database may be small, contained entirely within your own personal computer. Or, as we've seen, it may be massive, available through online connections. Such massive databases are of particular interest to us in this book because they offer phenomenal resources that until recently were unavailable to most ordinary computer users.

Database Management Systems

In the 1950s, when commercial use of computers was just beginning, a large organization would have different files for different purposes. For example, a university might have one file for course grades, another for student records, another for tuition billing, and so on. In a corporation, people in the accounting, order-entry, and customer-service departments all had their own separate files. Thus, if an address had to be changed, for example, each file would have to be updated separately. The database files were stored on magnetic tape and had to be accessed in sequence. Later magnetic disk technology came along, allowing any file to be accessed randomly. This permitted the development of new technology and new software: the database management system. A ***database management system (DBMS)*, or database manager, consists of programs that control the structure of a database and access to the data.** In a DBMS, an address change need be entered only once, and the updated information is then available in any relevant file.

The advantages of database management systems are as follows:

- **File sharing:** This is perhaps the biggest benefit. Now all authorized users can work with the same set of files.

- **Reduced data redundancy:** *Data redundancy* means that the same data fields (a person's address, say) appear in many different files and often in different formats. In the old file system, separate files would repeat the same data, wasting storage space. In a database, the information appears just once, freeing up more storage capacity. Moreover, the *same* information is available to *different* users.

- **Improved data integrity:** *Data integrity* means that data is accurate, consistent, and up to date. In the old system, when a change was made in one file, it might not get made in another. The result was

Fingerprint database
Law-enforcement officers are able to compare a fingerprint against a database of 8 million stored prints. If such a search were conducted manually, it would take more than 40 years. Today it takes 75 minutes.

that some reports were produced with erroneous information. In a DBMS, reduced redundancy increases the chances of data integrity—the chances that the data is accurate, consistent, and up to date—because each updating change is made in only one place.

- **Increased security:** Although various departments may share data, access to specific information can be limited to selected users. Thus, through the use of passwords, a student's financial, medical, and grade information in a university database is made available only to those who have a legitimate need to know.

Four Types of Databases

We may classify databases into four types: *individual, shared (company),* and *distributed databases* and what might be called *public databanks. (See* ● *Panel 8.4.)*

- **Individual databases:** *Individual databases* **are collections of integrated files used by one person.** As we discussed in Chapter 3, microcomputer users can set up their own individual databases using popular database management software; the information is stored on the hard drives of their personal computers. Today the principal database programs are Microsoft Access, Corel Paradox, and Lotus Approach. Such programs are used, for example, by graduate students to conduct research, by salespeople to keep track of clients, by purchasing agents to monitor orders, and by coaches to keep watch on other teams and players.

  In addition, types of individual databases known as *personal information managers (PIMs)* can help you keep track of and manage information you use on a daily basis, such as addresses, telephone numbers, appointments, to-do lists, and miscellaneous notes. Popular PIMs are Microsoft Outlook, Lotus Organizer, and Act.

- **Shared (company) databases:** A *shared database* **or company database is shared by users in one company or organization in one location.** The organization owns the database, which may be stored on a server such as a mainframe. Users are linked to the database via a local area or wide area network; the users access the network through terminals or microcomputers.

  Shared databases, such as those you find when surfing the Web, are the foundation for a great deal of electronic commerce, particularly B2B commerce, as we have seen.

- **Distributed databases:** A *distributed database* **is stored on different computers in different locations connected by a client/server network.**

PANEL 8.4
Four types of databases

Database	Description
Individual database	Collection of integrated files used by one person
Shared database	Database shared by users in one organization in one location
Distributed database	Database stored on different computers in different locations connected by a client/server network
Public databank	Compilation of data available to the public

For example, Cisco Systems, which calls itself the fastest-growing company in the history of the computing industry (it supplies the vast network that connects computers to the Internet), uses its worldwide intranet to connect its distributed databases, even those located overseas. As a result, it is able to execute what is called a "virtual close"—defined as the ability to close the financial books with a one-hour notice. "By connecting an entire company via intranet, even one with operations in dozens of countries," explains Cisco CEO John Chambers, "what was once done quarterly can now be done anytime."[23]

- **Public databanks:** If you're looking for very specific information, you can use your browser to do a Web search, investigating hundreds or thousands of Web sites. Many of these Web sites represent _**public databanks**_, **compilations of data that are available to the public.**

How long are those floppy disks or CD-ROMs on which you're storing your important documents going to last? Will you, or anyone else, be able to make use of them 15 or 25 or 50 years from now?

In 1982, software pioneer Jaron Lanier created a video game called Moondust for the then-popular Commodore 64 personal computer. Fifteen years later, when asked by a museum to display the game, he couldn't find a way to do it—until he had tracked down an old microcomputer of exactly that brand, type, and age, along with a joystick and video interface that would work with it.[a]

Would this have been a problem if Lanier had originally published a game in a _book_? Probably not. Books have been around since about 1453, when Johannes Gutenberg developed the printing press and used it to print 150 copies of the Bible in Latin. Some of these Gutenberg Bibles still exist—and are still readable (if you can read Latin).[b]

Digital storage has a serious problem: It isn't as long-lived as older forms of data storage. Today's books printed on "permanent" (low-acid, buffered) paper may last up to 500 years. Even books printed on cheap paper that crumbles will still be readable.

By contrast, data stored on diskettes, magnetic tape, and optical disks is subject to two hazards:[c]

- **Short life span of storage media:** The storage media themselves have a short life expectancy, and often the degradation is not apparent until it's too late. The maximum time seems to be 50 years, the longevity of a high-quality CD-ROM. Some average-quality CD-ROMs won't last 5 years, according to tests run at the National Media Laboratory.

  The magnetic tapes holding government records, which are stored in the National Archives in Washing-ton, D.C., need to be "refreshed"—copied onto more advanced tapes—every 10 years.

- **Hardware and software obsolescence:** As Jaron Lanier found out, even when tapes and disks remain intact, the hardware and software needed to read them may no longer be available. Without the programs and computers used to encode data, digital information may no longer be readable.

  "Eight-inch floppy disks and drives, popular as recently as a dozen years ago, are now virtually extinct," said a 1998 article, "and their 5¼-inch successors are rapidly disappearing. Optical and magnetic disks recorded under nonstandard storage schemes will be increasingly useless because of the lack of working equipment to read them."[d]

What about the personal records you would store on your own PC, such as financial records, inventories, genealogies, and photographs? _New York Times_ technology writer Stephen Manes has a number of suggestions:[e]

1. Choose your storage media carefully. CD-R disks are probably best for archiving—especially if you also keep a paper record.
2. Keep it simple. Store files in a standard format, such as text files and uncompressed bitmapped files.
3. Store data along with the software that created it.
4. Keep two copies, stored in separate places, preferably cool, dry environments.
5. Use high-quality media, not off brands.
6. When you upgrade to a new hardware or software product, have a strategy for migrating the old data.

Some public databanks are fee-based, such as Dialog Information Services, which offers scientific and technical information, and Dow Jones Interactive Publishing, which provides business information. Certain fee-based public databases are specialized, such as Lexis, which gives lawyers access to local, state, and federal laws, or Nexis, which gives journalists access to published articles in a range of newspapers.

Other public databanks are free. For instance, the U.S. government provides a great deal of free information, such as economic figures from the Bureau of Labor Statistics. Finally, many public databanks, such as Yahoo! or Amazon.com, are supported by advertising or online sales. The most popular revenue-producing Web sites—those supported by selling products or ads—are devoted to shopping, news and media, entertainment, online games, sweepstakes and lotteries, travel, finance, health and family, sports, and home and food.[25]

The last three databases should be managed by a specialist called a database administrator. The _database administrator (DBA)_ **coordinates all related activities and needs for an organization's database.** The DBA determines user access privileges; sets standards, guidelines, and control procedures; assists in establishing priorities for requests; prioritizes conflicting user needs; and develops user documentation and input procedures. He or she is also concerned with security—setting up and monitoring a system for preventing unauthorized access and making sure that the system is regularly backed up and that data can be recovered should a failure or disaster occur. Finally, the DBA establishes and enforces policies about privacy.

CONCEPT CHECK

What are the advantages of database management systems?

Distinguish among four types of databases.

8.4 The Ways Databases Are Organized

KEY QUESTION
What are four types of database organization?

Just as files can be organized in different ways (sequentially, for example), so databases can be organized in ways to best fit their use. The four most common arrangements are _hierarchical, network, relational,_ and _object-oriented_.

Hierarchical Database

In a _hierarchical database_, **fields or records are arranged in related groups resembling a family tree, with child (lower-level) records subordinate to parent (higher-level) records.** The parent record at the top of the database is called the _root record_. (See ● _Panel 8.5._)

Used principally on mainframes, hierarchical databases are the oldest of the four types. They are still used in some types of passenger reservation systems. In hierarchical databases, accessing or updating data is very fast, because the relationships have been predefined. However, because the structure must be defined in advance, it is quite rigid. There may be only one parent per child, and no relationships among the child records are possible. Moreover, adding new fields to database records requires that the entire database be redefined.

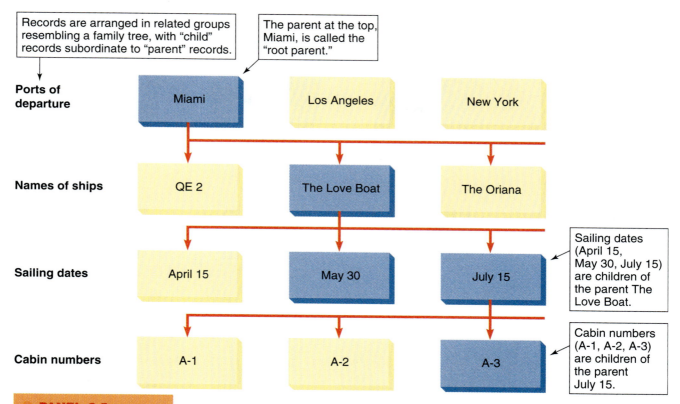

Records are arranged in related groups resembling a family tree, with "child" records subordinate to "parent" records.

The parent at the top, Miami, is called the "root parent."

Ports of departure — Miami · Los Angeles · New York

Names of ships — QE 2 · The Love Boat · The Oriana

Sailing dates — April 15 · May 30 · July 15

Sailing dates (April 15, May 30, July 15) are children of the parent The Love Boat.

Cabin numbers — A-1 · A-2 · A-3

Cabin numbers (A-1, A-2, A-3) are children of the parent July 15.

● **PANEL 8.5**
Hierarchical data-base
Example of a cruise ship reservation system

Network Database

A **_network database_** is similar to a hierarchical database, but each child record can have more than one parent record. (See ● Panel 8.6, next page.) Thus, a child record, which in network database terminology is called a *member*, may be reached through more than one parent, which is called an *owner*.

Also used principally with mainframes, the network database is more flexible than the hierarchical arrangement, because different relationships may be established between different branches of data. However, it still requires that the structure be defined in advance. Moreover, there are limits to the number of possible links among records.

Relational Database

More flexible than hierarchical and network database models, the **_relational database_** relates, or connects, data in different files through the use of a key field, or common data element. (See ● Panel 8.7, next page.) In this arrangement there are no access paths down through a hierarchy. Instead, data elements are stored in different tables made up of rows and columns. In database terminology, the tables are called *relations* (files), the rows are called *tuples* (records), and the columns are called *attributes* (fields). All related tables must have a key field that uniquely identifies each row; that is, the key field must be in *all* tables.

The advantage of relational databases is that the user does not have to be aware of any structure. Thus, they can be used with little training. Moreover, entries can easily be added, deleted, or modified. A disadvantage is that some searches can be time-consuming. Nevertheless, the relational model has become popular for microcomputer database management programs, such as Paradox and Access.

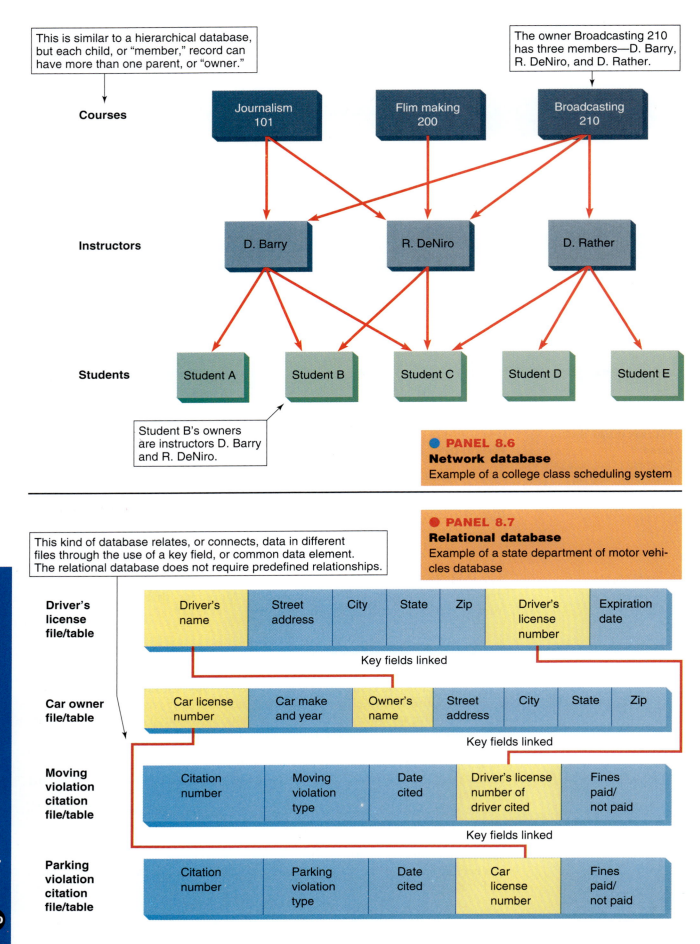

This is similar to a hierarchical database, but each child, or "member," record can have more than one parent, or "owner."

The owner Broadcasting 210 has three members—D. Barry, R. DeNiro, and D. Rather.

Courses

Journalism 101

Flim making 200

Broadcasting 210

Instructors

D. Barry

R. DeNiro

D. Rather

Students

Student A

Student B

Student C

Student D

Student E

Student B's owners are instructors D. Barry and R. DeNiro.

● PANEL 8.6
Network database
Example of a college class scheduling system

● PANEL 8.7
Relational database
Example of a state department of motor vehicles database

This kind of database relates, or connects, data in different files through the use of a key field, or common data element. The relational database does not require predefined relationships.

Driver's license file/table

| Driver's name | Street address | City | State | Zip | Driver's license number | Expiration date |

Key fields linked

Car owner file/table

| Car license number | Car make and year | Owner's name | Street address | City | State | Zip |

Key fields linked

Moving violation citation file/table

| Citation number | Moving violation type | Date cited | Driver's license number of driver cited | Fines paid/ not paid |

Key fields linked

Parking violation citation file/table

| Citation number | Parking violation type | Date cited | Car license number | Fines paid/ not paid |

Object-Oriented Database

An **_object-oriented database_ uses "objects," software written in small, reusable chunks, as elements within database files.** An *object* consists of (1) data in any form, including graphics, audio, and video, and (2) instructions on the action to be taken on the data.

A hierarchical or network database might contain only numeric and text data about a student—identification number, name, address, and so on. By contrast, an object-oriented database might also contain the student's photograph, a "sound bite" of his or her voice, and even a short piece of video. Moreover, the object would store operations, called *methods*, the programs that objects use to process themselves. For example, these programs might indicate how to calculate the student's grade-point average or how to display or print the student's record.

CONCEPT CHECK

Describe the four types of database organization.

8.5 Features of a Database Management System

KEY QUESTION

What are the features of a database management system?

A database management system may have a number of components, including the following. *(See ● Panel 8.8.)*

Data Dictionary

Some databases have a **_data dictionary_, a procedures document or disk file that stores the data definitions or a description of the structure of data used in the database.** The data dictionary may monitor the data being entered to make sure it conforms to the rules defined during data definition, such as field name, field size, type of data (text, numeric, date, and so on). The data dictionary may also help protect the security of the database by indicating who has the right to gain access to it.

Utilities

The **_DBMS utilities_ are programs that allow you to maintain the database** by creating, editing, and deleting data, records, and files. The utilities allow people to monitor the types of data being input and to adjust display screens for data input, for example.

● PANEL 8.8
Some important features of a database management system

Component	Description
Data dictionary	Describes files and fields of data
Utilities	Help maintain the database by creating, editing, and monitoring data input
Query language	Enables users to make queries to a database and retrieve selected records
Report generator	Enables nonexperts to create readable, attractive onscreen or hardcopy reports of records
Access security	Specifies user access privileges
Data recovery	Enables contents of database to be recovered after system failure

Query Language

Also known as a *data manipulation language,* a **_query language_ is an easy-to-use computer language for making queries to a database and for retrieving selected records,** based on the particular criteria and format indicated. Typically, the query is in the form of a sentence or near-English command, using such basic words as SELECT, DELETE, or MODIFY. There are several different query languages, each with its own vocabulary and procedures. One of the most popular is Structured Query Language, or SQL. An example of an SQL query is as follows:

SELECT PRODUCT-NUMBER, PRODUCT-NAME
FROM PRODUCT
WHERE PRICE < 100.00

This query selects all records in the product file for products that cost less than $100.00 and displays the selected records according to product number and name—for example:

A-34   Mirror
C-50   Chair
D-168   Table

One feature of most query languages is *query by example.* Often a user will seek information in a database by describing a procedure for finding it. However, in **_query by example (QBE),_ the user asks for information in a database by using a sample record to define the qualifications he or she wants for selected records.**

For example, a university's database of student-loan records of its students all over the United States and the amounts they owe might have the column headings (field names) NAME, ADDRESS, CITY, STATE, ZIP, AMOUNT OWED. When you use the QBE method, the database would display an empty record with these column headings. You would then type in the search conditions that you want in the appropriate columns.

Thus, if you wanted to find all Beverly Hills, California, students with a loan balance due of $3000 or more, you would type BEVERLY HILLS under the CITY column, CA under the STATE column, and >=3000 ("greater than or equal to $3000") in the AMOUNT OWED column.

Some DBMSs, such as Symantec's Q&A, use natural language interfaces, which allow users to make queries in any spoken language, such as English. With this software, you could ask your questions—either typing or speaking (if the system has voice recognition)—in a natural way, such as "How many sales reps sold more than one million dollars worth of books in the Western Region in January?"

Report Generator

A **_report generator_ is a program for producing an onscreen or printed document from all or part of a database.** You can specify the format of the report in advance—row headings, column headings, page headers, and so on. With a report generator, even nonexperts can create attractive, readable reports on short notice.

Access Security

At one point in the Michael Douglas/Demi Moore movie *Disclosure,* the Douglas character, the beleaguered division head suddenly at odds with his company, types SHOW PRIVILEGES into his desktop computer, which is tied to the corporate network. To his consternation, the system responds by showing him downgraded from PRIOR USER LEVEL: 5 to CURRENT USER LEVEL: 0, shutting him out of files to which he formerly had access.

This is an example of the use of access security, a feature allowing database administrators to specify different access privileges for different users of a DBMS. For instance, one kind of user might be allowed only to retrieve (view) data, whereas another might have the right to update data and delete records. The purpose of this security feature, of course, is to protect the database from unauthorized access and sabotage.

Physical security is also important. Simply isolating a database system is an effective means of protecting it from threats. For example, backup copies of databases on removable magnetic disks could be stored in a guarded vault, with authorized employees admitted only by producing a badge with their encoded personal voice prints.[25]

System Recovery

Database management systems should have *system recovery features* that enable the database administrator to recover contents of the database in the event of a hardware or software failure. For instance, the feature may recover transactions that appear to have been lost since the last time the system was backed up.

Performing a recovery may be difficult because it is often impossible to just fix the problem and resume processing where it was interrupted. Even if no data was lost during the failure—an unrealistic assumption, usually, because some computer memories will be volatile—the timing and scheduling of computer processing are too complex to be accurately re-created.

Four approaches are possible for system recovery—*mirroring, reprocessing, rollforward,* and *rollback:*[26]

- **Mirroring—two copies in different locations:** Database *mirroring* relies on frequent simultaneous copying of the database to maintain two or more complete copies of the database online but in different locations. Although expensive, mirroring is essential when recovery is needed quickly—say, in seconds or minutes—as with an airline reservation system's database.

- **Reprocessing—redoing the processing from a known past point:** In *reprocessing,* the database administrator goes back to a known point of database activity before the failure and reprocesses the workload from there.

    For this purpose, periodic database copies (called database saves) must be made. Also, records must be kept of all the transactions made since each save. When there is a failure, the database can be restored from the save. Then all the transactions made since that save are re-entered and reprocessed.

    This type of recovery can be time-consuming, and the processing of new transactions must be delayed until the database recovery is completed.

- **Rollforward—a variant on reprocessing:** *Rollforward,* also called *forward recovery,* is somewhat similar to reprocessing. Here, too, the current database is re-created, using a previous database state. However, in this case, transactions made since the last save are not re-entered and then reprocessed all over again. Rather, the lost data is recovered using a more sophisticated version of a transaction log that contains what are called after-image records and also some processing information.

- **Rollback—undoing unwanted changes:** *Rollback,* or *backward recovery,* is used to undo unwanted changes to the database—for example, when some failure interrupts a half-completed transaction.

What is a data dictionary?

What do DBMS utilities do?

Explain query language, query by example, report generator.

How does access security work?

Explain the four types of system recovery.

8.6 The Ethics of Using Databases: Concerns about Accuracy & Privacy

KEY QUESTION

What are some ethical concerns about the uses of databases?

The enormous capacities of today's storage devices has given photographers, graphics professionals, and others a new tool—the ability to manipulate images at the pixel level. For example, photographers can easily do *morphing*—transforming one image into another. In **morphing**, **a film or video image is displayed on a computer screen and altered pixel by pixel, or dot by dot. As a result, the image metamorphoses into something else**—a pair of lips into the front of a Toyota, for example.

The ability to manipulate digitized output—images and sounds—has brought a wonderful new tool to art. However, it has created some big new problems in the area of credibility, especially for journalism. How can we know that what we're seeing or hearing is the truth? Consider the following.

Morphing

Manipulation of Sound

Frank Sinatra's 1994 album *Duets* paired him through technological tricks with singers like Barbra Streisand, Liza Minnelli, and Bono of U2. Sinatra recorded solos in a recording studio. His singing partners, while listening to his taped performance on earphones, dubbed in their own voices. These second voices were recorded not only at different times but often, through distortion-free phone lines, from different places. The illusion in the final recording is that the two singers are standing shoulder to shoulder.

Newspaper columnist William Safire called *Duets* "a series of artistic frauds." Said Safire, "The question raised is this: When a performer's voice and image can not only be edited, echoed, refined, spliced, corrected, and enhanced—but can be transported and combined with others not physically present—what is performance? . . . Enough of additives, plasticity, virtual venality; give me organic entertainment."[27] Some listeners feel that the technology changes the character of a performance for the better. Others, however, think the practice of assembling bits and pieces in a studio drains the music of its essential flow and unity.

Whatever the problems of misrepresentation in art, however, they pale beside those in journalism. What if, for example, a radio station were to edit a stream of digitized sound so as to misrepresent what actually happened?

Manipulation of Photos

When O. J. Simpson was arrested in 1994 on suspicion of murder, the two principal American newsmagazines both ran pictures of him on their covers.[28] *Newsweek* ran the mug shot unmodified, as taken by the Los Angeles Police Department. At *Time*, an artist working with a computer modified the shot with special effects as a "photo-illustration." Simpson's image was dark-

ened so that it still looked like a photo but, some critics said, with a more sinister cast to it.

Should a magazine that reports the news be taking such artistic license? Should *National Geographic* in 1982 have photographically moved two Egyptian pyramids closer together so that they would fit on a vertical cover? Was it even right for *TV Guide* in 1989 to run a cover showing Oprah Winfrey's head placed on Ann-Margret's body? In another case, to show what can be done, a photographer digitally manipulated the famous 1945 photo showing the meeting of the leaders of the wartime Allied powers at Yalta. Joining Stalin, Churchill, and Roosevelt are some startling newcomers: Sylvester Stallone and Groucho Marx. The additions are so seamless that it is impossible to tell the photo has been altered. *(See ● Panel 8.9.)*

The potential for abuse is clear. "For 150 years, the photographic image has been viewed as more persuasive than written accounts as a form of 'evidence,'" says one writer. "Now this authenticity is breaking down under the assault of technology."[29] Asks a former photo editor of the *New York Times Magazine*, "What would happen if the photograph appeared to be a straightforward recording of physical reality, but could no longer be relied upon to depict actual people and events?"[30]

Many editors try to distinguish between photos used for commercialism (advertising) versus for journalism, or for feature stories versus for news stories. However, this distinction implies that the integrity of photos is only important for some narrow category of news. In the end, it can be argued, altered photographs pollute the credibility of all of journalism.

Manipulation of Video & Television

The technique of morphing, used in still photos, takes a massive jump when used in movies, videos, and television commercials. Digital image manipulation has had a tremendous impact on filmmaking. Director and digital pioneer Robert Zemeckis *(Death Becomes Her)* compares the new technology to the advent of sound in Hollywood.[31] It can be used to erase jet contrails from the sky in a western and to make digital planes do impossible stunts. It can even be used to add and erase actors.

Films and videotapes are widely thought to accurately represent real scenes (as evidenced by the reaction to the amateur videotape of the Rodney King beating by police in Los Angeles). Thus, the possibility of digital alterations raises some real problems. Videotapes supposed to represent actual events could easily be doctored. Another concern is for film archives: Because digital videotapes suffer no loss in resolution when copied, there are no "generations." Thus, it will be impossible for historians and archivists to tell whether the videotape they're viewing is the real thing or not.[32]

Virtual advertising
The oil company 76 logo doesn't really appear on this wall but looks as though it does.

Indeed, it is possible to create virtual images during live television events. These images—such as a Coca-Cola logo in the center of a soccer field—don't exist in reality but millions of viewers see them on their TV screens.[33]

Accuracy & Completeness

Databases—including public data banks such as Nexis/Lexis—can provide you with *more* facts and *faster* facts but not always *better* facts. Penny Williams, professor of broadcast journalism at Buffalo State College in New York and formerly a television anchor and reporter, suggests five limitations to bear in mind when using databases for research:[34]

- **You can't get the whole story:** For some purposes, databases are only a foot in the door. There may be many facts or facets of the topic that are not in a database. Reporters, for instance, find a database is a starting point. It may take intensive investigation to get the rest of the story.
- **It's not the gospel:** Just because you see something on a computer screen doesn't mean it's accurate. Numbers, names, and facts must be verified in other ways.
- **Know the boundaries:** One database service doesn't have it all. For example, you can find full text articles from the *New York Times* on Lexis/Nexis, from *The Wall Street Journal* on Dow Jones News Retrieval, and from the *San Jose Mercury News* on America Online, but no service carries all three.
- **Find the right words:** You have to know which keywords (search words) to use when searching a database for a topic. As Lynn Davis, a professional researcher with ABC News, points out, if you're searching for stories on guns, the keyword "can be guns, it can be firearms, it can be handguns, it can be pistols, it can be assault weapons. If you don't cover your bases, you might miss something."[35]
- **History is limited:** Most public databases, Davis says, have information going back to 1980, and a few into the 1970s, but this poses problems if you're trying to research something that happened or was written about earlier.

Matters of Privacy

Privacy is the right of people to not reveal information about themselves. Who you vote for in a voting booth and what you say in a letter sent through the U.S. mail are private matters. However, the ease of pulling together and disseminating information via databases and communications lines has put privacy under extreme pressure.

As you've no doubt discovered, it's no trick at all to get your name on all kinds of mailing lists. Theo Theoklitas, for instance, has received applications for credit cards, invitations to join video clubs, and notification of his finalist status in Ed McMahon's $10 million sweepstakes. Theo is a black cat who's been getting mail ever since his owner sent in an application for a rebate on cat food. A whole industry has grown up of professional information gatherers and sellers, who collect personal data and sell it to fundraisers, direct marketers, and others.

In the 1970s, the Department of Health, Education, and Welfare developed a set of five Fair Information Practices regarding the use and misuse of information. These rules have since been adopted by a number of public and private organizations and have led to the enactment of a number of laws to protect individuals from invasion of privacy. (See ● *Panel 8.10*.)

Another concern is privacy in communications. Although the government is constrained by several laws on acquiring and disseminating information and listening in on private conversations, privacy advocates still worry. In

Fair Information Practices

1. There must be no personal data record-keeping systems whose existence is a secret from the general public.

2. People have the right to access, inspect, review, and amend data about them that is kept in an information system.

3. There must be no use of personal information for purposes other than those for which it was gathered without prior consent.

4. Managers of systems are responsible and should be held accountable and liable for the reliability and security of the systems under their control, as well as for any damage done by those systems.

5. Governments have the right to intervene in the information relationships among private parties to protect the privacy of individuals.

Important Federal Privacy Laws

Freedom of Information Act (1970): Gives you the right to look at data concerning you that is stored by the federal government. A drawback is that sometimes a lawsuit is necessary to pry it loose.

Fair Credit Reporting Act (1970): Bars credit agencies from sharing credit information with anyone but authorized customers. Gives you the right to review and correct your records and to be notified of credit investigations for insurance or employment. A drawback is that credit agencies may share information with anyone they reasonably believe has a "legitimate business need." Legitimate is not defined.

Privacy Act (1974): Prohibits federal information collected about you for one purpose from being used for a different purpose. Allows you the right to inspect and correct records. A drawback is that exceptions written into the law allow federal agencies to share information anyway.

Family Educational Rights and Privacy Act (1974): Gives students and their parents the right to review, and to challenge and correct, students' school and college records; limits sharing of information in these records.

Right to Financial Privacy Act (1978): Sets strict procedures that federal agencies must follow when seeking to examine customer records in banks; regulates financial industry's use of personal financial records. A drawback is that the law does not cover state and local governments.

Privacy Protection Act (1980): Prohibits agents of federal government from making unannounced searches of press offices if no one there is suspected of a crime.

Cable Communications Policy Act (1984): Restricts cable companies in the collection and sharing of information about their customers.

Computer Fraud and Abuse Act (1986): Allows prosecution for unauthorized access to computers and databases. A drawback is that people with legitimate access can still get into computer systems and create mischief without penalty.

Electronic Communications Privacy Act (1986): Makes eavesdropping on private conversations illegal without a court order.

Computer Security Act (1987): Makes actions that affect the security of computer files and telecommunications illegal.

Computer Matching and Privacy Protection Act (1988): Regulates computer matching of federal data; allows individuals a chance to respond before government takes adverse actions against them. A drawback is that many possible computer matches are not affected, such as those done for law-enforcement or tax reasons.

Video Privacy Protection Act (1988): Prevents retailers from disclosing video-rental records without the customer's consent or a court order.

Is that me on that card?

Nurse Law Wai Fong, of Hong Kong's Queen Elizabeth Hospital, holds up a wallet-size card. The card can contain the equivalent of an entire filing cabinet of a patient's medical history—about 2000 pages. The ability to concentrate information typified by smart cards and databases has many people concerned about whether such records can be kept private.

recent times, the government has tried to impose new technologies that would enable law-enforcement agents to gather a wealth of personal information. Proponents have urged that Americans must be willing to give up some personal privacy in exchange for safety and security. This topic will be discussed further in Chapter 9.

Monopolizing Information

"We want to capture the entire human experience throughout history," says Corbis Corporation chief executive officer Doug Rowan.[36] Corbis was formed by software billionaire Bill Gates to acquire digital rights to fine art and photographic images that can be viewed electronically—in everything from electronic books to computerized wall hangings.[37] In 1995 Corbis acquired the Bettmann Archive of 17 million photographs, for scanning into its digital database.[38] The founder of the archive, Dr. Otto Bettmann, called his famous collection a "visual story of the world," and indeed many of the images are unique. They include tintypes of black Civil War soldiers, the 1937 crash of the *Hindenburg* dirigible, and John F. Kennedy Jr. saluting the casket of his assassinated father. Corbis also owns digital imaging rights to the National Gallery in London and the State Hermitage Museum in St. Petersburg, Russia.

However, when Rowan says Corbis wants to capture all of human experience, he doesn't just mean photos, paintings, and sculpture. "Film, video, audio," he says. "We are interested in those fields too."

Are there any ethical problems with one company having the exclusive digital rights to our visual and audio history? Like many museums and libraries (such as the Library of Congress), Corbis joins a trend toward democratizing art and scholarship by converting the images and texts of the past into digital form and making them available to people who could never travel to, say, London or St. Petersburg.

However, when Gates acquired the Bettmann images, for example, the move put their future use "into the hands of an aggressive businessman who, unlike Dr. Bettmann, is planning his own publishing ventures," points out one reporter. "While Mr. Gates's initial plans will make Bettmann images more widely accessible, this savvy competitor now ultimately controls who can use them—and who can't."[39]

CONCEPT CHECK

Discuss how computers can manipulate sounds and images.

What are the limitations of using databases for research?

Discuss concerns about privacy and the monopolization of information.

Experience Box
Preventing Your Identity from Getting Stolen

One day, Kathryn Rambo, 28, of Los Gatos, California, learned that she had a new $35,000 sports utility vehicle listed in her name, along with five credit cards, a $3000 loan, and even an apartment—none of which she'd asked for. "I cannot imagine what would be weirder, or would make you angrier, than having someone pretend to be you, steal all this money, and then leave you to clean up all their mess later," said Rambo, a special-events planner.[40] Added to this was the eerie matter of constantly having to prove that she was, in fact, herself: "I was going around saying, 'I am who I am!'"[41]

Identity Theft: Stealing Your Good Name—and More

Theft of identity (TOI) is a crime in which thieves hijack your very name and identity and use your good credit rating to get cash or to buy things. To begin, all they need is your full name or Social Security number. Using these, they tap into Internet databases and come up with other information—your address, phone number, employer, driver's license number, mother's maiden name, and so on. Then they're off to the races, applying for credit everywhere.

In Rambo's case, someone had used information lifted from her employee-benefits form. The spending spree went on for months, unbeknownst to her. The reason it took so long was that Rambo never saw any bills. They went to the address listed by the impersonator, a woman, who made a few payments to keep creditors at bay while she ran up even more bills. For Rambo, straightening out the mess required months of frustrating phone calls, time off from work, court appearances, and legal expenses.

How Does Identity Theft Start?

Identity theft typically starts in one of several ways:[42]

- **Wallet or purse theft:** There was a time when a thief would steal a wallet or purse, take the cash, and toss everything else. No more. Everything from keys to credit cards can be parlayed into further thefts.

- **Mail theft:** Thieves also consider mailboxes fair game. The mail will yield them bank statements, credit-card statements, new checks, tax forms, and other personal information.

- **Mining the trash:** You might think nothing of throwing away credit-card offers, portions of utility bills, or old cancelled checks. But "dumpster diving" can produce gold for thieves. Credit-card offers, for instance, may have limits of $5000 or so.

- **Telephone solicitation:** Prospective thieves may call you up and pretend to represent a bank, credit-card company, government agency, or the like in an attempt to pry loose essential data about you.

- **Insider access to databases:** You never know who has, or could have, access to databases containing your personnel records, credit records, car-loan applications, bank documents, and so on. This is one of the harder TOI methods to guard against.

What to Do Once Theft Happens

If you're the victim of a physical theft (or even loss), as when your wallet is snatched, you should immediately contact—first by phone, and then in writing—all your credit card companies, other financial institutions, the Department of Motor Vehicles, and any other organization whose cards you use that are now compromised. Be sure to call utility companies—telephone, electricity, and gas; identity thieves can run up enormous phone bills. Also call the local police and your insurance company to report the loss.

It's important to notify financial institutions *within two days of learning of your loss* because then you are legally responsible for only the first $50 of any theft. If you become aware of fraudulent transactions, immediately contact the fraud units of the three major credit bureaus: Equifax, Experian, and TransUnion. (See ● *Panel 8.11*.)

If your Social Security number has been fraudulently used, alert the Social Security Administration (800-772-1213). It's possible, as a last resort, to have your Social Security Number changed.

If you have a check guarantee card that was stolen, if your checks have been lost, or if a new checking account has been opened in your name, there are two organizations to notify so that payment on any fraudulent checks will be denied. They are Telecheck (800-366-2424) and National Processing Company (800-526-5380).

If your mail has been used for fraudulent purposes or if an identity thief filed a change of address form, look in the phone directory under U.S. Government Postal Service for the local Postal Inspector's office.

● PANEL 8.11
The three major credit bureaus

Equifax	Experian	TransUnion
To check your credit report: 800-685-1111 To report fraud: 800-525-6285	To check your credit report: 800-682-7654 To report fraud: 800-301-7195	To check your credit report: 800-916-8800 To report fraud: 800-680-7289

How to Prevent Identity Theft

One of the best ways to keep your finger on the pulse of your financial life is, on a regular basis—once a year, say—to get a copy of your credit report from one or all three of the main credit bureaus. This will show you whether there is any unauthorized activity. Reports cost $8.

In addition, there are some specific measures you can take to guard against personal information getting into the public realm.

- **Check your credit-card billing statements:** If you see some fraudulent charges, report them immediately. If you don't receive your statement, call the creditor first. Then call the post office to see if a change of address has been filed under your name.

- **Treat credit cards and other important papers with respect:** Make a list of your credit cards and other important documents, and the list of numbers to call if you need to report them lost. (You can photocopy the cards front and back, but make sure the numbers are legible.)

  Carry only one or two credit cards at a time. Carry your Social Security card, passport, or birth certificate only when needed.

  Don't dispose of credit card receipts in a public place.

Don't give out your credit-card numbers or Social Security number over the phone, unless you have some sort of trusted relationship with the party on the other end.

Tear up credit-card offers before you throw them away.

Keep tax records and other financial documents in a safe place.

- **Treat passwords with respect:** Memorize passwords and PINs. Don't use your birth date, mother's maiden name, or similar common identifiers, which thieves may be able to guess.

- **Treat checks with respect:** Pick up new checks at the bank. Shred cancelled checks before throwing them away. Don't let merchants write your credit-card number on the check.

- **Watch out for "shoulder surfers" when using phones and ATMs:** When using PINs and passwords at public telephones and automated teller machines, shield your hand so that anyone watching through binoculars or using a video camera—"shoulder surfers"—can't read them.

Visual Summary

business-to-business (B2B) system (p. 319, KQ 8.1) Direct sales between businesses, using the Internet or private network to cut transaction costs and increase efficiencies. Why it's important: *Business-to-business activity is expected to balloon to a $1 trillion industry by 2002 and a $2.8 trillion industry by 2004.*

character (p. 321, KQ 8.2) A single letter, number, or special character. Why it's important: *Characters—such as A, B, C, 1, 2, 3, #, $, %—are part of the data storage hierarchy.*

database (p. 322, KQ 8.2) Organized collection of related (integrated) files. Why it's important: *Businesses and organizations build databases to help them keep track of and manage their affairs. In addition, online database services put enormous research resources at the user's disposal.*

database administrator (DBA) (p. 328, KQ 8.3) Person who coordinates all related activities and needs for an organization's database. Why it's important: *The DBA determines user access privileges; sets standards, guidelines, and control procedures; assists in establishing priorities for requests; prioritizes conflicting user needs; and develops user documentation and input procedures. He or she is also concerned with security—setting up and monitoring a system for preventing unauthorized access and making sure that the system is regularly backed up and that data can be recovered should a failure or disaster occur.*

database management system (DBMS) (p. 325, KQ 8.3) Also called a *database manager;* software that controls the structure of a database and access to the data. Allows users to manipulate more than one file at a time. Why it's important: *This software enables: sharing of data (same information is available to different users); economy of files (several departments can use one file instead of each individually maintaining its own files, thus reducing data redundancy, which in turn reduces the expense of storage media and hardware); data integrity (changes made in the files in one department are automatically made in the files in other departments); security (access to specific information can be limited to selected users).*

data dictionary (p. 331, KQ 8.5) File that stores data definitions and descriptions of database structure. It may also monitor new entries to the database as well as user access to the database. Why it's important: *The data dictionary monitors the data being entered to make sure it conforms to the rules defined during data definition. The data dictionary may also help protect the security of the database by indicating who has the right to gain access to it.*

data files (p. 322, KQ 8.2) Files that contain data—words, numbers, pictures, sounds, and so on. Why it's important: *Unlike program files, data files don't instruct the computer to do anything. Rather, data files are there to be acted on by program files. Examples of common extensions in data files are .txt (text) and .xls (spreadsheets). Certain proprietary software programs have their own extensions, such as .ppt for PowerPoint and .mdb for Access.*

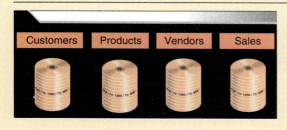

data mining (DM) (p. 317, KQ 8.1) Computer-assisted process of sifting through and analyzing vast amounts of data in order to extract meaning and discover new knowledge. Why it's important: *The purpose of DM is to describe past trends and predict future trends. Thus, data-mining tools might sift through a company's immense collections of customer, marketing, production, and financial data and identify what's worth noting and what's not.*

Type of data	Contains	E
Database	Several files	Yc ad pa
File	Several records	Fr Ar St
Record	Several fields	Ar

data storage hierarchy (p. 321, KQ 8.2) The ranked levels of data stored in a computer: bits, bytes (characters), fields, records, files, and databases. Why it's important: *Understanding the data storage hierarchy is necessary to understand how to use a database.*

data warehouse (p. 318, KQ 8.1) A database containing cleaned-up data and meta-data (information about the data). Stored using high-capacity disk storage. Why it's important: *Data warehouses combined vast amounts of data from many sources in a database form that can be searched, for example, for patterns not recognizable with smaller amounts of data.*

DBMS utilities (p. 331, KQ 8.5) Programs that allow the maintenance of databases by creating, editing, and deleting data, records, and files. Why it's important: *DBMS utilities allow people to establish what is acceptable input data, to monitor the types of data being input, and to adjust display screens for data input.*

direct access storage (p. 324, KQ 8.2) Storage system that allows the computer to go directly to the desired information. The data is retrieved (accessed) according to a unique data identifier called a key field. It also uses a file allocation table (FAT), a hidden on-disk table that records exactly where the parts of a given file are stored. Why it's important: *This method of file organization, used with hard disks and other types of disks, is ideal for applications where there is no fixed pattern to the requests for data—for example, in airline reservation systems or computer-based directory-assistance operations. Direct access storage is much faster than sequential access storage.*

distributed database (p. 326, KQ 8.3) Database that is stored on different computers in different locations connected by a client/server network. Why it's important: *Data need not be centralized in one location.*

e-commerce (p. 315, KQ 8.1) Electronic commerce; the buying and selling of products and services through computer networks. Why it's important: *By 2003, total U.S. e-commerce sales to consumers are expected to reach $108 billion, or 6% of consumer retail spending; online shopping is growing even faster than the increase in computer use, which has been fueled by the falling price of personal computers.*

field (p. 321, KQ 8.2) Unit of data consisting of one or more characters (bytes). An example of a field is your name, your address, or your Social Security number. Why it's important: *A collection of fields makes up a record. Also see* key field.

file (p. 322, KQ 8.2) Collection of related records. An example of a file is data collected on everyone employed in the same department of a company, including all names, addresses, and Social Security numbers. Why it's important: *A file is the collection of data or information that is treated as a unit by the computer; a collection of related files makes up a database.*

filename (p. 322, KQ 8.2) The name given to a file. Why it's important: *Files are given names so that they can be differentiated. Filenames also have extension names. These extensions of up to three letters are added after a period following the filename—for example, the .doc in Psychreport.doc is recognized by Microsoft Word as the extension for "document." Extensions are usually inserted automatically by the application software.*

hierarchical database (p. 328, KQ 8.4) Database in which fields or records are arranged in related groups resembling a family tree, with child (lower-level) records subordinate to parent (higher-level) records. The parent record at the top of the database is called the *root record.* Why it's important: *The hierarchical database is one of the common database structures.*

individual database (p. 326, KQ 8.3) Collection of integrated files used by one person. Why it's important: *Microcomputer users can set up their own individual databases using popular database management software; the information is stored on the hard drives of their personal computers. Today the principal database programs are Microsoft Access, Corel Paradox, and Lotus Approach. In addition, types of individual databases known as personal information managers (PIMs) can help users keep track of and manage information used on a daily basis, such as addresses, telephone numbers, appointments, to-do lists, and miscellaneous notes. Popular PIMs are Microsoft Outlook, Lotus Organizer, and Act.*

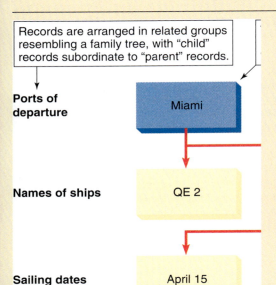

Records are arranged in related groups resembling a family tree, with "child" records subordinate to "parent" records.

Ports of departure — Miami

Names of ships — QE 2

Sailing dates — April 15

key field (p. 322, KQ 8.2) Field that is chosen to uniquely identify a record so that it can be easily retrieved and processed. The key field is often an identification number, Social Security number, customer account number, or the like. Why it's important: *The primary characteristic of the key field is that it is unique and thus can be used to identify one specific record.*

master file (p. 323, KQ 8.2) Data file containing records that are generally updated periodically. Why it's important: *Master files contain relatively permanent information used for reference purposes. They are updated through the use of transaction files.*

morphing (p. 334, KQ 8.6) Altering a film or video image displayed on a computer screen pixel by pixel, or dot by dot. Why it's important: *Morphing and other techniques of digital manipulation can produce images that misrepresent reality.*

network database (p. 329, KQ 8.4) Database similar in structure to a hierarchical database; however, each child record can have more than one parent record. Thus, a child record, which in network database terminology is called a *member*, may be reached through more than one parent, which is called an *owner*. Why it's important: *The network database is one of the common database structures.*

object-oriented database (p. 331, KQ 8.4) Database that uses "objects," software written in small, reusable chunks, as elements within database files. An object consists of (1) data in any form, including graphics, audio, and video, and (2) instructions on the action to be taken on the data. Why it's important: *A hierarchical or network database might contain only numeric and text data. By contrast, an object-oriented database might also contain photographs, sound bites, and video clips. Moreover, the object would store operations, called* methods, *the programs that objects use to process themselves.*

offline storage (p. 324, KQ 8.2) System in which stored data is not directly accessible for processing until the tape or disk it's on has been loaded onto an input device. Why it's important: *The storage medium and data are not under the direct, immediate control of the central processing unit.*

online storage (p. 324, KQ 8.2) System in which stored data is randomly (directly) accessible for processing. Why it's important: *The storage medium and data is under the direct, immediate control of the central processing unit. There's no need to wait for a tape or disk to be loaded onto an input device.*

privacy (p. 336, KQ 8.6) Right of people to not reveal information about themselves. Why it's important: *The ease of pulling together and disseminating information via databases and communications lines has put the basic democratic right of privacy under extreme pressure.*

program files (p. 322, KQ 8.2) Files containing software instructions. Why it's important: *Contrast data files.*

public databank (p. 327, KQ 8.3) Compilation of data available to the public. Why it's important: *The public databank is one of the basic types of database.*

query by example (QBE) (p. 332, KQ 8.5) Feature of query-language programs whereby the user asks for information in a database by using a sample record to define the qualifications he or she wants for selected records. Why it's important: *QBE further simplifies database use.*

query language (p. 332, KQ 8.5) Easy-to-use computer language for making queries to a database and retrieving selected records. To retrieve information from a database, users make queries—that is, they use a query language. These languages have commands such as SELECT, DELETE, and MODIFY. Why it's important: *Query languages make it easier for users to deal with databases.*

record (p. 322, KQ 8.2) Collection of related fields. An example of a record would be your name and address and Social Security number. Why it's important: *Related records make up a file.*

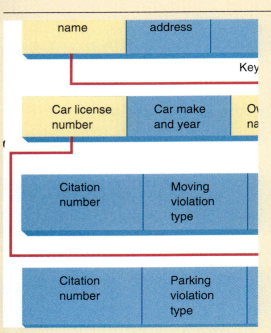

name	address	

Key

Car license number	Car make and year	Ov na

Citation number	Moving violation type

Citation number	Parking violation type

relational database (p. 329, KQ 8.4) Common database structure that relates, or connects, data in different files through the use of a key field, or common data element. In this arrangement there are no access paths down through a hierarchy. Instead, data elements are stored in different tables made up of rows and columns. In database terminology, the tables are called *relations* (files), the rows are called *tuples* (records), and the columns are called *attributes* (fields). All related tables must have a key field that uniquely identifies each row; that is, the key field must be in all tables. Why it's important: *The relational database is one of the common database structures; it is more flexible than hierarchical and network database models.*

report generator (p. 332, KQ 8.5) In a database management system, a program users can employ to produce on-screen or printed-out documents from all or part of a database. Why it's important: *Report generators allow users to produce finished-looking reports without much fuss.*

sequential storage (p. 324, KQ 8.2) Storage system whereby data is stored and retrieved in sequence, such as alphabetically. Why it's important: *An inexpensive form of storage, sequential storage is the only type of storage provided by tape, which is used mostly for archiving and backup. The disadvantage of sequential file organization is that searching for data is slow.* Compare direct access storage.

shared database (p. 326, KQ 8.3) Also called a *company database;* a database shared by users in one company or organization in one location. The organization owns the database, which may be stored on a server such as a mainframe. Users are linked to the database via a local area or wide area network; the users access the network through terminals or microcomputers. Why it's important: *Shared databases, such as those you find when surfing the Web, are the foundation for a great deal of electronic commerce, particularly B2B commerce.*

transaction file (p. 323, KQ 8.2) Temporary holding file that holds all changes to be made to the master file: additions, deletions, revisions. Why it's important: *The transaction file is used to periodically update the master file.*

Chapter Review

Self-Test Questions

1. According to the data storage hierarchy, databases are composed of _____, _____, _____, _____, and _____.

2. An individual piece of data within a record is called a _____.

3. A(n) _____ coordinates all activities related to an organization's database.

4. _____ is the right of people not to reveal information about themselves.

5. The four types of databases are _____, _____, _____, and _____.

6. The buying and selling of products and services through computer networks is called _____.

7. _____ files contain software instructions; _____ files contain data.

8. _____ storage means that the computer can go directly to the information you want.

Multiple-Choice Questions

1. Which of the following is not an advantage of a DBMS?
   a. file sharing
   b. reduced data redundancy
   c. increased data redundancy
   d. improved data integrity
   e. increased security

2. Which of the following database models stores data in any form, including graphics, audio, and video?
   a. hierarchical
   b. network
   c. object-oriented
   d. relational
   e. offline

True/False Questions

T F 1. The use of key fields makes it easier to locate a record in a database.

T F 2. A transaction file contains permanent records that are periodically updated.

T F 3. A database is an organized collection of integrated files.

T F 4. A shared database is stored on different computers in different locations connected by a client/server network.

T F 5. A directory (folder) is a storage place for files in one of your drives.

Short-Answer Questions

1. Name three responsibilities of a database administrator.
2. What is the difference between master files and transaction files?
3. What is data mining?
4. What is the difference between offline and online storage?
5. Briefly explain what a data warehouse is.
6. What is an ASCII file?

Concept Mapping

On a separate sheet of paper, draw a concept map, or visual diagram, linking concepts. Show how the following terms are related.

B2B system	hierarchical database
character	key field
database	master file
DBA	network database
DBMS	object-oriented database
data files	privacy
data storage hierarchy	program files
direct access storage	query language
e-commerce	record
field	relational database
file	sequential storage
filename	transaction file

Knowledge in Action

Interview someone who works with or manages an organization's database. What types of records make up the database? Which departments use it? What database structure is used? What are the types and sizes of storage devices? Are servers used? Was the database software custom written?

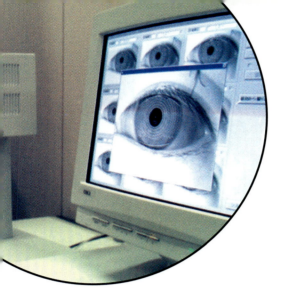

The Challenges of the Digital Age

Society & Information Technology Today

Key Questions
You should be able to answer the following questions.

9.1 Security Issues: Threats to Computers & Communications Systems What are some characteristics of the key security issues for information technology?

9.2 Security: Safeguarding Computers & Communications What are the characteristics of the four components of security?

9.3 Quality-of-Life Issues: The Environment, Mental Health, & the Workplace How does information technology create environmental, mental-health, and workplace problems?

9.4 Economic Issues: Employment & the Haves/Have-Nots How may technology affect the unemployment rate and the gap between rich and poor?

9.5 The Digital Environment: Is There a Grand Design? What are the NII, the new Internet, the Telecommunications Act, the 1997 White House plan, and ICANN?

A Visual Overview of This Chapter

1 **Security Issues: Threats to Computers & Communications Systems.**
Among the threats to the security of computers are the following. (1) Errors and accidents, such as human errors, procedural errors, software errors, electromechanical problems, and "dirty data" problems. (2) Natural hazards, such as earthquakes, hurricanes, and fires, and civil strife and terrorism. (3) **Computer crimes,** which can be either illegal acts perpetrated against computers or the use of computers to accomplish illegal acts. Crimes against computers include theft of hardware, software, time and services, or information, or crimes of malice and destruction. (4) Crimes using computers include credit-card theft and investment fraud. (5) **Worms,** programs that copy themselves repeatedly into a computer's memory or disk drive, and **viruses,** deviant programs stored on hard drives that can destroy data. Worms and viruses are passed by exchange of infected floppy disks or infected data sent over a network; **antivirus software** can detect viruses. (6) Computer criminals may be an organization's employees, outside users, hackers and crackers, and professional criminals. **Hackers** gain unauthorized access to computers often just for the challenge, **crackers** for malicious purposes.

2 **Security: Safeguarding Computers & Communications. Security,** the system of safeguards for protecting computers against disasters, failure, and unauthorized access, has four components. (1) Computer systems try to determine authorized users by three criteria: by what they have (keys, badges, signatures); by what they know (as with **PINs,** or personal identification numbers, and **passwords,** or special words or codes); and by who they are (as by physical traits, as determined perhaps through **biometrics,** the science of measuring individual body characteristics). (2) **Encryption,** altering data so it is not usable unless the changes are undone, tries to make computer messages more secure. (3) Software and data are protected by controlling access to files, by audit controls that track the programs used, and by people controls that screen job applicants and other users. (4) **Disaster-recovery plans** are methods for restoring computer operations after destruction or accident.

3 **Quality-of-Life Issues: The Environment, Mental Health, & the Workplace.** Some quality-of-life issues related to information technology are as follows.
(1) Computers may create environmental problems, such as putting a lot of telecommunications equipment in natural settings. (2) Computer-related mental-health problems include isolation, online gambling, and stress. (3) Problems affecting workplace productivity include misuse of technology, as when employees waste company time going online for personal purposes; fussing with computers because of hardware/software problems; and information overload.

4 **Economic Issues: Employment & the Haves/Have-Nots.** Two charges by economic critics of information technology are as follows. (1) Technology replaces humans in countless tasks, for millions of workers into temporary or part-time employment or unemployment. (2) Technology widens the gap between rich and poor, between information "haves" and "have-nots."

5 **The Digital Environment: Is There a Grand Design?** Some factors affecting the shape of the digital environment are as follows. (1) The *National Information Infrastructure* is a grand design relying on private companies and the Internet. (2) The old Internet is being replaced by new Internet networks—VBNS, Internet2, and the NGI. The **VBNS** is the main government component to upgrade the backbone, or primary hubs of data transmission. **Internet2** is a cooperative university-business program to enable high-end users to quickly move data, using VBNS. The **Next Generation Internet (NGI)** is designed to help tie campus high-performance backbones to the federal infrastructure. (3) The *1996 Telecommunications Act* was designed to increase competition between telecommunications businesses, so different companies are no longer restricted from offering different services. (4) The 1997 White House plan stresses that government should stay out of the way of Internet commerce. (5) **ICANN** is the Internet Corporation for Assigned Names and Numbers, which is a nonprofit corporation established to regulate Internet domain names.

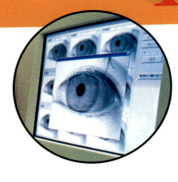

"fom. the Internet is on its way to becoming the dominant mode of information exchange, then it is no longer a luxury but, like the telephone, a necessity."

And it follows, in this analyst's opinion, that "anyone without it is in danger of being shut out."[1]

This "digital divide" between those with and without access to information technology is growing steadily wider as the Information Age continues to expand productivity and wealth. Addressing that divide is one of the most important challenges of our time. Internationally, the digital divide is between rich countries and poor countries. In the United States, 10% of the economy is devoted to the purchase of hardware and software, as compared with, say, Bangladesh, where it is one-tenth of 1%.[2] Of the world's approximately 1 billion Web pages, more than 80% are in English.[3] With only 5% of the world's population, the U.S. has 50% of the world's Internet-linked home computers.[4] Even in the United States, however, great numbers of people are shut out of the new system of information exchange, on account of their low income and/or limited knowledge of English.[5]

The digital divide is only one of the many challenges confronting us as infotech sweeps the world. Elsewhere in the book we have considered ergonomics (Chapter 6) and privacy (Chapters 7 and 8). In this chapter, we consider some other major issues:

- Security issues—accidents, hazards, crime, viruses—and security safeguards
- Quality-of-life issues—environment, mental health, the workplace
- Economic issues—employment and the haves/have-nots
- Is there a grand design for the digital environment?

9.1 Security Issues: Threats to Computers & Communications Systems

KEY QUESTION

What are some characteristics of the key security issues for information technology?

Security issues go right to the heart of the workability of computer and communications systems. Here we discuss the following threats to computers and communications systems:

- Errors and accidents
- Natural and other hazards
- Crime against computers and communications
- Crime using computers and communications
- Worms and viruses
- Computer criminals

Errors & Accidents

In general, errors and accidents in computer systems may be classified as human errors, procedural errors, software errors, electromechanical problems, and "dirty data" problems.

- **Human errors:** Quite often, when experts speak of the "unintended effects of technology," what they are referring to are the unexpected things people do with it. Among the ways in which people can complicate the workings of a system are the following:[6]

  (1) Humans often are not good at assessing their own information needs. For example, many users will acquire a computer and commu-

nications system that either is not sophisticated enough or is far more complex than they need.

(2) *Human emotions affect performance.* For example, one frustrating experience with a computer is enough to make some people abandon the whole system. But throwing your computer out the window isn't going to get you any closer to learning how to use it better.

(3) *Humans act on their perceptions,* which in modern information environments are often too slow to keep up with the equipment. Decisions influenced by information overload, for example, may be just as faulty as those based on too little information.

- **Procedural errors:** Some spectacular computer failures have occurred because someone didn't follow procedures. In 1999, the $125 million Mars Climate Orbiter was fed data expressed in pounds, the English unit of force, instead of newtons, the metric unit (about 22% of a pound). As a result, the spacecraft flew too close to the surface of Mars and broke apart.[7] A few years earlier, Nasdaq, the nation's second largest stock market, was shut down for 2½ hours by an effort, ironically, to make the computer system more user friendly. Technicians were phasing in new software, adding technical improvements a day at a time. A few days into this process, technicians trying to add more features to the software flooded the data-storage capability of the computer system. The result was to delay the opening of the stock market and shorten the trading day.[8]

- **Software errors:** We are forever hearing about "software glitches" or "software bugs." A software bug is an error in a program that causes it not to work properly. In a column lamenting the absence of a bug-free Web browser, *Business Week* technology journalist Stephen Wildstrom wrote: "I have used every browser produced by Netscape, . . . and every one of them has been buggy." But its major rival is no better. "Where Netscape is crash-prone," Wildstrom said, Microsoft Internet Explorer 5.0 "has gotten downright scary," requiring numerous software "patches" to close the security holes in the software that allow strangers to access your computer.[9]

- **Electromechanical problems:** Mechanical systems, such as printers, and electrical systems, such as circuit boards, don't always work. They may be faultily constructed, get dirty or overheated, wear out, or become damaged in some other way. Power failures (brownouts and blackouts) can shut a system down. Power surges can also burn out equipment.

  Modern systems are made up of thousands of parts, all of which interrelate in ways that are impossible to anticipate. Because of that complexity, argues Yale University sociologist Charles Perrow, what he calls "normal accidents" are inevitable. That is, it is almost certain that some combinations of minor failures will eventually amount to something catastrophic. Indeed, it is just such collections of small failures that led to catastrophes such as the blowing up of the *Challenger* space shuttle in 1986 and the near-meltdown of the Three Mile Island nuclear-power plant in 1979.[10] In the Digital Age, "normal accidents" will not be anomalies but are to be expected.

- **"Dirty data" problems:** When keyboarding a research paper, you undoubtedly make a few typing errors (which, hopefully, you clean up). So do all the data-entry people around the world who feed a continual stream of raw data into computer systems. A lot of problems are caused by this kind of "dirty data." *Dirty data* is incomplete, outdated, or otherwise inaccurate data.

  A good reason for having a look at your records—credit, medical, school—is so that you can make any corrections to them before they cause you complications. As the president of a firm specializing in

business intelligence writes, "Electronic databases, while a time-saving resource for the information seeker, can also act as catalysts, speeding up and magnifying bad data."[11]

Natural & Other Hazards

Some disasters do not merely lead to temporary system downtime; they can wreck the entire system. Examples are natural hazards, and civil strife and terrorism.

- **Natural hazards:** Whatever is harmful to property (and people) is harmful to computers and communications systems. This certainly includes natural disasters: fires, floods, earthquakes, tornadoes, hurricanes, blizzards, and the like. If they inflict damage over a wide area, as have ice storms in eastern Canada or hurricanes in Florida, natural hazards can disable all the electronic systems we take for granted. Without power and communications connections, automated teller machines, credit-card verifiers, and bank computers are useless.

- **Civil strife and terrorism:** We may take comfort in the fact that wars and insurrections seem to take place in other parts of the world. Yet we are not immune to civil unrest, such as the riot in Los Angeles in 2000 after the hometown Lakers won the National Basketball Association championship. Nor are we immune, apparently, to acts of terrorism, such as the 1993 bombing of New York's World Trade Center. In that case, companies found themselves frantically moving equipment to new offices and reestablishing their computer networks. The Pentagon (which has 650,000 terminals and workstations, 100 WANs, and 10,000 LANs) has been taking steps to reduce its own systems' vulnerability to intruders, although hackers have managed to penetrate top-secret operations.

Crimes against Computers & Communications

Ethics

A _computer crime_ can be of two types. (1) It can be an illegal act perpetrated against computers or telecommunications. Or (2) it can be the use of computers or telecommunications to accomplish an illegal act.

Crimes against information technology include theft—of hardware, of software, of computer time, of cable or telephone services, or of information. Other illegal acts are crimes of malice and destruction. Some examples are as follows:

Laptop left on a car seat

Computers left in cars are temptations for smash-and-grab thieves.

- **Theft of hardware:** Hardware theft can range from shoplifting an accessory in a computer store to removing a laptop or cellular phone from someone's car. Professional criminals may steal shipments of microprocessor chips off a loading dock or even pry cash machines out of shopping-center walls.

  Eric Avila, 26, a history student at the University of California at Berkeley, had his doctoral dissertation—involving six years of painstaking research—stored on the hard drive of his Macintosh laptop when a thief stole it out of his apartment. Although he had copied an earlier version of his dissertation (70 pages entitled "Paradise Lost: Politics and Culture in Post-War Los Angeles") onto a diskette, the thief stole that, too. "I'm devastated," Avila said. "Now it's gone, and there is no way I can recover it other than what I have in my head." To make matters worse, he had no choice but to pay off the $2000 loan for a computer he did not have anymore.[12] The moral, as we've emphasized in this book: _Always make backup copies of your important data, and store them in a safe place—away from your computer._

- **Theft of software:** Generally, software theft involves illegally copying programs, rather than physically taking someone's floppy disks. Software makers secretly prowl electronic bulletin boards in search of purloined products, then try to get a court order to shut down the bulletin boards. They also look for organizations that "softlift"—companies, colleges, or other institutions that buy one copy of a program and make copies for many computers.

  Many such so-called pirates are reported by co-workers or fellow students to the "software police," the Software Publishers Association. The SPA has a toll-free number (800-388-7478) for reporting illegal copying of software. In the 1990s, two New England college students were indicted for allegedly using the Internet to encourage the exchange of copyrighted software.[13]

  Another type of software theft is copying or counterfeiting of well-known software programs. These pirates often operate in China, Taiwan, Mexico, Russia, and various parts of Asia and Latin America. In some countries, most of the U.S. microcomputer software in use is thought to be illegally copied.

- **Theft of time and services:** The theft of computer time is more common than you might think. Probably the biggest instance is people using their employer's computer time to play games, do online shopping or stock trading, or dip into Web pornography. Some people even operate sideline businesses.

  For years "phone phreaks" have bedeviled the telephone companies. For example, they have found ways to get into company voice-mail systems, then use an extension to make long-distance calls at the company's expense. They have also found ways to tap into cellular phone networks and dial for free.

- **Theft of information:** "Information thieves" have infiltrated the files of the Social Security Administration, stolen confidential personal records, and sold the information. On college campuses, thieves have snooped on or stolen private information such as grades. Thieves have also broken into computers of the major credit bureaus and stolen credit information. They have then used the information to charge purchases or have resold it to other people. One thief broke into the systems of Internet retailer CD Universe and stole 300,000 customers' credit-card numbers. When the company's executives refused to pay a ransom demand, he sold them off piecemeal on the Internet (for others to illegally charge on) until he was stopped.

- **Crimes of malice and destruction:** Sometimes criminals are more interested in abusing or vandalizing computers and telecommunications systems than in profiting from them. For example, a student at a Wisconsin campus deliberately and repeatedly shut down a university computer system, destroying final projects for dozens of students. A judge sentenced him to a year's probation, and he left the campus.

Ethics

Crimes Using Computers & Communications

Just as a car can be used to perpetrate or assist in a crime, so can information technology. For example, Craig Pribila, 18, a student at the University of Nevada, Reno, faced a possible sentence of a year in jail after being convicted of charges of using his computer to counterfeit $20 bills and fake driver's licenses.[14]

In addition, investment fraud has come to cyberspace. Many people now use online services to manage their stock portfolios through brokerages hooked into the services. Scam artists have followed, offering nonexistent investment deals and phony solicitations, and manipulating stock prices.

Worms & Viruses

Worms and viruses are forms of high-tech maliciousness. A **_worm_ is a program that copies itself repeatedly into a computer's memory or onto a disk drive.** Sometimes it will copy itself so often it will cause a computer to crash. An example was Melissa, a 1999 worm program that infected perhaps a million PCs. Melissa launched a file of porno Web sites via e-mail and sent them on to people on the recipient's e-mail address list.

A **_virus_ is a "deviant" program, stored on a computer hard drive, that can cause unexpected and often undesirable effects, such as destroying or corrupting data.** (See ● *Panel 9.1.*) The famous e-mail Love Bug (its subject line was I LOVE YOU), which originated in the Philippines in May 2000 and did perhaps as much as $10 billion in damage worldwide, was both a worm and a virus. A variation on the Melissa worm, it spread faster and caused more damage than any other bug before it.[15] The Love Bug was followed almost immediately by a variant virus. This new Love Bug didn't reveal itself with an ILOVEYOU line but changed to a random word or words each time a new computer was infected.[16]

Worms and viruses are passed in two ways:

- **By diskette:** The first way is via an infected diskette, perhaps obtained from a friend or a repair person.
- **By network:** The second way is via a network, as from e-mail or an electronic bulletin board. This is why, when taking advantage of all the freebie games and other software available online, you should use virus-scanning software to check downloaded files.

The virus usually attaches itself to your hard disk. It might then display annoying messages ("Your PC is stoned—legalize marijuana") or cause Ping-Pong balls to bounce around your screen and knock away text. More seri-

- **Boot-sector virus:** The boot sector is that part of the system software containing most of the instructions for booting, or powering up, the system. The boot sector virus replaces these boot instructions with some of its own. Once the system is turned on, the virus is loaded into main memory before the operating system. From there it is in a position to infect other files. Any diskette that is used in the drive of the computer then becomes infected. When that diskette is moved to another computer, the contagion continues. Examples of boot-sector viruses: AntCMOS, AntiEXE, Form.A, NYB (New York Boot), Ripper, Stoned.Empire.Monkey.

- **File virus:** File viruses attach themselves to executable files—those that actually begin a program. (In DOS these files have the extensions .com and .exe.) When the program is run, the virus starts working, trying to get into main memory and infecting other files.

- **Multipartite virus:** A hybrid of the file and boot-sector types, the multipartite virus infects both files and boot sectors, which makes it better at spreading and more difficult to detect. Examples of multipartite viruses are Junkie and Parity Boot.

    A type of multipartite virus is the *polymorphic virus*, which can mutate and change form just as human viruses can. Such viruses are especially troublesome because they can change their profile, making existing antiviral technology ineffective.

    A particularly sneaky multipartite virus is the *stealth virus*, which can temporarily remove itself from memory to elude capture. An example of a multipartite, polymorphic stealth virus is One Half.

- **Macro virus:** Macro viruses take advantage of a procedure in which miniature programs, known as macros, are embedded inside common data files, such as those created by e-mail or spreadsheets, which are sent over computer networks. Until recently, such documents have typically been ignored by antivirus software. Examples of macro viruses are Concept, which attaches to Word documents and e-mail attachments, and Laroux, which attaches to Excel spreadsheet files. Fortunately, the latest versions of Word and Excel come with built-in macro virus protection.

- **Logic bomb:** Logic bombs, or simply bombs, differ from other viruses in that they are set to go off at a certain date and time. A disgruntled programmer for a defense contractor created a bomb in a program that was supposed to go off two months after he left. Designed to erase an inventory tracking system, the bomb was discovered only by chance.

- **Trojan horse:** The Trojan horse covertly places illegal, destructive instructions in the middle of a legitimate program, such as a computer game. Once you run the program, the Trojan horse goes to work, doing its damage while you are blissfully unaware. An example of a Trojan horse is FormatC.

ously, it might add garbage to your files, then erase or destroy your system software. It may evade your detection and spread its havoc elsewhere, since an infected hard disk will infect every floppy disk used by the system.

If you look in the utility section of any software store, you'll see a variety of virus-fighting programs. **<u>Antivirus software</u> scans a computer's hard disk, floppy disks, and main memory to detect viruses and, sometimes, to destroy them.** Such virus watchdogs operate in two ways. First, they scan disk drives for "signatures," characteristic strings of 1s and 0s in the virus that uniquely identify it. Second, they look for suspicious virus-like behavior, such as attempts to erase or change areas on your disks. Examples of antivirus programs are Norton Anti-Virus 2000, McAfee Virus Scan for Windows, and Virex for Macs.

Computer Criminals

Ethics

What kind of people are perpetrating most of the information-technology crime? Over 80% may be employees; the rest are outside users, hackers and crackers, and professional criminals.

- **Employees:** Says Michigan State University criminal justice professor David Carter, who surveyed companies about computer crime, "Seventy-five to 80% of everything happens from inside."[17] Most common frauds, Carter found, involved credit cards, telecommunications, employees' personal use of computers, unauthorized access to confidential files, and unlawful copying of copyrighted or licensed software.

Workers may use information technology for personal profit, or steal hardware or information to sell. They may also use it to seek revenge for real or imagined wrongs, such as being passed over for promotion. Sometimes they may use the technology simply to demonstrate to themselves they have power over people.

- **Outside users:** Suppliers and clients may also gain access to a company's information technology and use it to commit crimes. This becomes more likely as electronic connections such as intranets and extranets become more commonplace.

- **Hackers and crackers:** *Hacker* is so overused it has come to be applied to anyone who breaks into a computer system. Some people think it means almost any computer lover. In reality, there is a difference between hackers and crackers, although the term *cracker* has never caught on with the general public.

    <u>Hackers</u> **are people who gain unauthorized access to computer or telecommunications systems, often just for the challenge of it.** Some hackers even believe they are performing a service by exposing security flaws. Whatever the motivation, network system administrators view any kind of unauthorized access as a threat, and they usually try to pursue offenders vigorously. The most flagrant cases of hacking are met with federal prosecution. Former computer science student Ikenna Iffih, who studied at Northeastern University in Boston, could have been sentenced to 20 years in prison for hacking against private and government targets, but under a plea bargain was allowed to serve no more than six months. As a Nigerian national, he also faced possible deportation.[18]

    <u>Crackers</u> **are people who illegally break into computers for malicious purposes—to obtain information for financial gain, shut down hardware, pirate software, or alter or destroy data.** Sometimes you hear of "white hat" hackers (who aren't malicious) versus "black-hat" hackers (who are). The Federal Bureau of Investigation estimates that cybercrime costs Americans more than $20 billion a year—and that more than 60% of such computer crime goes unreported. Says one article:

    > On the low end, Web vandals, mostly teenagers derided by older [crackers] as "script kiddies," have attacked and shut down hundreds of busy Web sites, including those of the NASDAQ stock exchange, ABC, the White House, the Senate, and the FBI itself. On the more sophisticated end, a single program triggered by a New Jersey hacker—[the] Melissa virus—caused an estimated $80 million in damage to computer users.[19]

- **Professional criminals:** Members of organized crime rings don't just steal information technology. They also use it the way that legal businesses do—as a business tool, but for illegal purposes. For instance, databases can be used to keep track of illegal gambling debts and stolen goods. Not surprisingly, the old-fashioned illegal booking operation has gone high-tech, with bookies using computers and fax machines in place of betting slips and paper tally sheets.

As information-technology crime has become more sophisticated, so have the people charged with preventing it and disciplining its outlaws. Campus administrators are no longer being quite as easy on offenders and are turning them over to police. Industry organizations such as the Software Publishers Association are going after software pirates large and small. (Commercial software piracy is now a felony, punishable by up to five years in prison and fines of up to $250,000 for anyone convicted of stealing at least 10 copies of a program, or more than $2500

Police work
A Texas police officer uses a laptop computer for field reports.

worth of software.) Police departments as far apart as Medford, Massachusetts, and San Jose, California, now have officers patrolling a "cyber beat." They regularly cruise online bulletin boards and chat rooms looking for pirated software, stolen trade secrets, child molesters, and child pornography.

In 1988, after the last widespread Internet break-in, the U.S. Defense Department created the Computer Emergency Response Team (CERT). Although it has no power to arrest or prosecute, CERT provides round-the-clock international information and security-related support services to users of the Internet. Whenever it gets a report of an electronic snooper, whether on the Internet or on a corporate e-mail system, CERT stands ready to lend assistance. It counsels the party under attack, helps thwart the intruder, and evaluates the system afterward to protect against future break-ins.

Fake-buck buster
This electronic device can quickly (0.7 second) detect whether a U.S. bill is counterfeit or not.

CONCEPT CHECK

Explain some of the errors, accidents, and hazards that can affect computers.

What are the principal crimes against computers?

What are some crimes using computers?

Describe some types of computer criminals.

9.2 Security: Safeguarding Computers & Communications

KEY QUESTION
What are the characteristics of the four components of security?

The ongoing dilemma of the Digital Age is balancing convenience against security. **Security** is a system of safeguards for protecting information technology against disasters, systems failure, and unauthorized access that can result in damage or loss. We consider four components of security:

- Identification and access
- Encryption
- Protection of software and data
- Disaster-recovery plans

Identification & Access

Are you who you say you are? The computer wants to know.

There are three ways a computer system can verify that you have legitimate right of access. Some security systems use a mix of these techniques. The systems try to authenticate your identity by determining (1) what you have, (2) what you know, or (3) who you are.

- **What you have—cards, keys, signatures, badges:** Credit cards, debit cards, and cash-machine cards all have magnetic strips or built-in computer chips that identify you to the machine. Many require you to display your signature, which may be compared with any future signature you write. Computer rooms are always kept locked, requiring a key. Many people also keep a lock on their personal computers. A computer room may also be guarded by security officers, who may need to see an authorized signature or a badge with your photograph before letting you in.

  Of course, credit cards, keys, and badges can be lost or stolen. Signatures can be forged. Badges can be counterfeited.

The Challenges of the Digital Age

357

- **What you know—PINs, passwords, and digital signatures:** To gain access to your bank account through an automated teller machine (ATM), you key in your PIN. A **_PIN (personal identification number)_ is the security number known only to you that is required to access the system.** Telephone credit cards also use a PIN. If you carry either an ATM or a phone card, never carry the PIN written down elsewhere in your wallet (even disguised).

  A **_password_ is a special word, code, or symbol required to access a computer system.** Passwords are one of the weakest security links, says AT&T security expert Steven Bellovin. Passwords (and PINs, too) can be guessed, forgotten, or stolen. To foil a stranger's guesses, Bellovin recommends never choosing a real word or variations of your name, birthdate, or those of your friends or family. Instead you should mix letters, numbers, and punctuation marks in an oddball sequence of no fewer than eight characters.[20] Some good passwords: 2b/orNOT2b%. Alfred!E!Newman7. You can also choose an obvious and memorable password but shift the position of your hands on the keyboard, creating a meaningless string of characters—the best kind of password. (Thus, ELVIS becomes R;BOD when you move your fingers one position right on the keyboard.)

Fingerprint check
This system can be mounted on a door or on a safe. Entry is granted only after the individual's fingerprint is verified.

- **Who you are—physical traits:** Some forms of identification can't be easily faked—such as your physical traits. Biometrics tries to use these in security devices. **_Biometrics_ is the science of measuring individual body characteristics.**

  For example, before University of Georgia students can use the all-you-can-eat plan at the campus cafeteria, they must have their hands read. As one writer describes the system, "a camera automatically compares the shape of a student's hand with an image of the same hand pulled from the magnetic strip of an ID card. If the patterns match, the cafeteria turnstile automatically clicks open. If not, the would-be moocher eats elsewhere."[21]

Besides handprints, other biological characteristics read by biometric devices are fingerprints (computerized "finger imaging"), voices, the blood vessels in the back of the eyeball (retinal scan), the lips, and even the entire face.

Some computer security systems have a "call-back" provision. In a _call-back system,_ the user calls the computer system, punches in the password, and hangs up. The computer then calls back a certain preauthorized number. This measure will block anyone who has somehow got hold of a password but is calling from an unauthorized telephone.

Encryption

Ethics

PGP is a computer program written for encrypting computer messages—putting them into secret code. **_Encryption_ is the altering of data so it is not usable unless the changes are undone.** _PGP_ (for _Pretty Good Privacy_) is so good that it is practically unbreakable; even government experts can't crack it. Another encryption system is _DES_ (for _Data Encryption Standard_), adopted as a federal standard in 1976.

Iris scan

This system verifies the identity of computer users by iris scans.

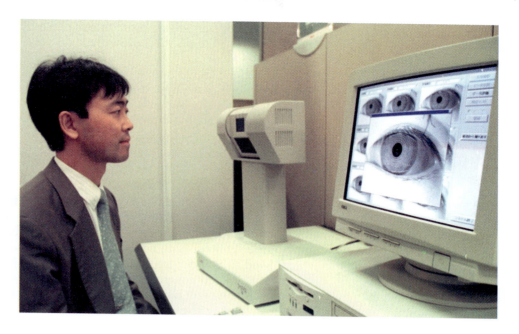

Encryption is clearly useful for some organizations, especially those concerned with trade secrets, military matters, and other sensitive data. Some maintain that encryption will determine the future of e-commerce, because transactions cannot flourish over the Internet unless they are secure.[22] However, from the standpoint of our society, encryption is a two-edged sword. For instance, police in Sacramento, California, found that PGP blocked them from reading the computer diary of a convicted child molester and finding links to a suspected child pornography ring. Should the government be allowed to read the coded e-mail of its citizens? What about government surveillance of overseas terrorists, drug dealers, and other enemies?

The government maintains that it needs access to scrambled data for national security and law enforcement. At present, there are limitations on encryption technology governing the export of hardware and software containing data-scrambling features. U.S. technology firms complain they are losing the export market because

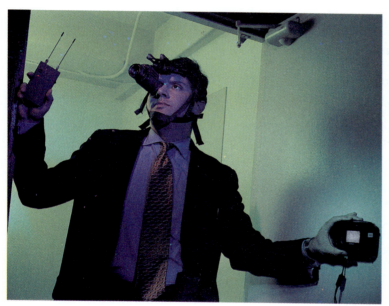

Spies R Us

Robert Wendt is fully equipped. His walkie-talkie scrambles communication, making eavesdropping impossible. A wireless receiver in his ear allows him to listen to instructions from afar. His eyepiece provides him with perfect night vision. The tie conceals a video camera capable of transmitting images to either the monitor in his left hand or a remote receiver. The pen in his breast pocket is a sensor that detects hidden electronic devices. The watch is actually a digital camera.

of these controls. In addition, however, the FBI and the National Security Agency have been pushing to require all encryption products sold within the United States to include a "back door" allowing the government to peep at any message. Encryption is opposed not only by civil libertarians who fear "back door" access could lead to government snooping but also by most information technology companies and the majority of the regional telephone companies. In mid-2000, the Clinton administration announced it would loosen controls on the export of encryption software.[23]

Protection of Software & Data

Organizations go to tremendous lengths to protect their programs and data. As might be expected, this includes educating employees about making backup disks, protecting against viruses, and so on. Other security procedures include the following:

- **Control of access:** Access to online files is restricted to those who have a legitimate right to access—because they need them to do their jobs. Many organizations have a system of transaction logs for recording all accesses or attempted accesses to data.

- **Audit controls:** Many networks have *audit controls* for tracking which programs and servers were used, which files opened, and so on. This creates an *audit trail*, a record of how a transaction was handled from input through processing and output.

- **People controls:** Because people are the greatest threat to a computer system, security precautions begin with the screening of job applicants. Résumés are checked to see if people did what they said they did. Another control is to separate employee functions, so people are not allowed to wander freely into areas not essential to their jobs. Manual and automated controls—input controls, processing controls, and output controls—are used to check if data is handled accurately and completely during the processing cycle. Printouts, printer ribbons, and other waste that may contain passwords and trade secrets to outsiders is disposed of through shredders or locked trash barrels.

Disaster-Recovery Plans

A *disaster-recovery plan* **is a method of restoring information processing operations that have been halted by destruction or accident.** "Among the countless lessons that computer users have absorbed in the hours, days, and weeks after the [1993 New York] World Trade Center bombing," wrote one reporter, "the most enduring may be the need to have a disaster-recovery plan. The second most enduring lesson may be this: Even a well-practiced plan will quickly reveal its flaws."[24]

Mainframe computer systems are operated in separate departments by professionals, who tend to have disaster plans. Whereas mainframes are usually backed up, many personal computers, and even entire local area networks, are not, with potentially disastrous consequences. It has been reported that, on average, a company loses as much as 3% of its gross sales within eight days of a sustained computer outage. In addition, the average company struck by a computer outage lasting more than 10 days never fully recovers.[25]

A disaster-recovery plan is more than a big fire drill. It includes a list of all business functions and the hardware, software, data, and people to support those functions, as well as arrangements for alternate locations. The disaster-recovery plan includes ways for backing up and storing programs and data in another location, ways of alerting necessary personnel, and training for those personnel.

CONCEPT CHECK

How do computer systems try to authenticate your identity?

Describe encryption.

What security procedures are used to protect software and data?

What is a disaster-recovery plan?

The worrisome effects of technology on intellectual property rights and truth in art and journalism, on censorship, on health matters and ergonomics, and on privacy were explained earlier in this book. Here are some other quality-of-life issues related to information technology.

Environmental Problems

"This county will do peachy fine without computers," says Micki Haverland, who has lived in rural Hancock County, Tennessee, for 20 years.[26] Telecommunications could bring jobs to an area that badly needs them, but several people moved there precisely because they like things the way they are—pristine rivers, unspoiled forests, and mountain views.

But it isn't just people in rural areas who are concerned. Suburbanites worry that lofty metal poles topped by cellular-transmitting equipment will destroy views and property values. City dwellers everywhere are concerned that deregulation of the telecommunications industry is leading to a rat's nest of roof antennas, satellite dishes, and above-ground transmission stations. As a result, telecommunications companies are now experimenting with hiding transmitters in the "foliage" of fake trees made of metal.

Political scientist James Snider, of Northwestern University, believes that the problems of the cities could expand well beyond the cities, if telecommuting triggers a massive movement of people to rural areas. "If all Americans succeed in getting their dream homes with several acres of land," he writes, "the forests and open lands across the entire continental United States will be destroyed" as they become carved up with subdivisions and roads.[27] But a 2000 study by the Center for Energy and Climate Solutions (which consults for companies trying to become more "green") found that energy consumption remained the same during 1997–98 and energy prices stayed low even as the economy expanded. One suggested explanation is that the Internet is making commerce more efficient, reducing the use of natural resources.[28] Regardless, the disposal of old PCs and other high-tech gear is becoming a problem, as toxic materials—lead, cadmium, mercury, chromium, plastics—pile up in landfills.[29]

Mental-Health Problems: Isolation, Gambling, Net Addiction, & Stress

Some of the mental health problems linked to information technology are the following:

- **Isolation:** Automation allows us to go for days without actually speaking with or touching another person, from buying gas to playing games. Even the friendships we make online in cyberspace, some believe, "are likely to be trivial, short lived, and disposable—junk friends." Says one writer, "We may be overwhelmed by a continuous static of information and casual acquaintance, so that finding true soul mates will be even harder than it is today."[30]

  A Stanford University survey found that, as people spend more time online, they can spare less time for real-life relationships with family and friends. "As Internet use becomes more widespread," says a study coauthor, "it will have an increasingly isolating effect on society."[31] On the other hand, many find the Net empowering. If e-mail is the most popular online activity—as it is, for example, for many women and teenage girls—that may mean that many people are keeping online company with friends and relatives.[32] Moreover, the Internet has led to a great many Web communities, ranging from cancer

E-mailed prayers

A messenger stuffs a prayer into a crack in Jerusalem's Wailing Wall, which is sacred to Jews, who believe prayers delivered there reach God more quickly. The message is one of 200 e-mailed prayers that arrives each day at Virtual Jerusalem, an organization that not only helps far-flung Jews talk to God but also offers live video of the Wall, as well as chat rooms.

survivors to Arab immigrants. In the opinion of one sociologist, "The Internet gives us a different kind of social life—not better or worse than before, but just different."[33]

- **Gambling:** Gambling is already widespread in North America, but information technology makes it almost unavoidable. Instead of driving 40 minutes to a Connecticut casino, for example, Rhode Island resident Beverly Richard discovered that slots, roulette, and blackjack were just a mouse click away, and she quickly found herself $13,000 in debt. "It was too convenient," she said. "I don't have to leave home."[34] Although gambling by wire is illegal in the U.S., host computers for Internet casinos and sports books have been established in Caribbean tax havens. Satellites, decoders, and remote-control devices allow TV viewers to do racetrack wagering from home. In these circumstances, law enforcement is extremely difficult.

- **Stress:** In one survey of 2802 American PC users, three-quarters of the respondents (ranging in age from children to retirees) said personal computers had increased their job satisfaction and were a key to success and learning. However, many found PCs stressful: In particular, 59% admitted getting angry at their PCs within the previous year, and 41% said they thought computers had reduced job opportunities rather than increased them.[35] Another survey found that 83% of corporate network administrators reported "abusive and violent behavior" by employees toward computers—including smashing of monitors, throwing of mice, and kicking of system units.[36]

Workplace Problems: Impediments to Productivity

First the mainframe computer, then the desktop stand-alone PC, and now the networked computer were all brought into the workplace for one reason only: to improve productivity. How is it working out? Let's consider three aspects: misuse of technology, fussing with computers, and information overload.

- **Misuse of technology:** "For all their power," says an economics writer, "computers may be costing U.S. companies tens of billions of dollars a year in downtime, maintenance and training costs, useless game playing, and information overload."[37]

  Employees may look busy, as they stare into their computer screens with brows crinkled. But sometimes they're just hard at work playing Quake. Or browsing online malls (forcing corporate mail rooms to cope with a deluge of privately ordered parcels). Or looking at their investments or pornography sites.[38] Indeed, one study found that recreational Web surfing accounts for nearly *one-third* of office workers' time online.[39]

- **Fussing with computers:** Another reason for so much wasted time is all the fussing that employees do with hardware, software, and online connections. One study in the early 1990s estimated microcomputer users wasted 5 billion hours a year waiting for programs to run, checking computer output for accuracy, helping coworkers use their applications, organizing cluttered disk storage, and calling for technical support.[40] And that was *before* most people had to get involved with making online connections work.

How People Shirk at Work: Recreational Web Surfing	
29.1%	General news
22.5%	Investment
9.7%	Pornography
8.2%	Travel
6.6%	Entertainment
6.1%	Sports
3.1%	Shopping
14.7%	Other

Here's a description of the attempt by one person (a top-level Microsoft techie, no less!) to install a program and run the Internet on a new IBM laptop:

> The connection kept dropping out. Then came a weird error message he attributed to the IBM machines. And then there were things that were not errors, but simply the sorts of annoying user-hostile phenomena that are all too familiar: endless dialogue boxes, loud unwanted music, annoying rebooting, cluttered menus, even tough-to-open shrink-wrapped software boxes.[41]

Comments technology writer Dan Gillmor, "We would never buy a TV that forced us to reboot the set once a month, let alone once a week or every other day."[42] But this is the current evolutionary stage of the computer age.

- **Information overload:** "It used to be considered a status symbol to carry a laptop computer on a plane," says futurist Paul Saffo. "Now anyone who has one is clearly a working dweeb who can't get the time to relax. Carrying one means you're on someone's electronic leash."[43]

The new technology is definitely a two-edged sword. Cellular phones, pagers, fax machines, and modems may untether employees from the office, but these employees tend to work longer hours under more severe deadline pressure than do their tethered counterparts who stay at the office, according to one study.[44] Moreover, the gadgets that once promised to do away with irksome business travel by ushering in a new era of communications have done the opposite. They have created the office-in-a-bag that allows business travelers to continue to work from airplane seats, hotel desks, and their own kitchen tables.

To avoid information overload, some people install so-called *Bozo filters*, software for screening out trivial e-mail messages and cellular calls and assigning priorities to the remaining files. Still others are beginning to employ programs called *intelligent agents* to help them make decisions, as we discuss in Chapter 10. But the real change may come as people realize that they need not always be tied to the technological world, that solitude is a scarce resource, and that seeking serenity means streamlining the clutter and reaching for simpler things.

Government statistics show a substantial pickup in U.S. productivity growth since 1990, particularly after 1996. This has been attributed to the gradual diffusion of information technologies across the economy, particularly into older industries such as manufacturing.[45] But, says Stephen Roach, chief economist of Morgan Stanley Dean Witter, "The dirty little secret of the information age is that an increasingly large slice of work goes on outside the official work hours the government recognizes. . . . The '24/7' culture of nearly round-the-clock work is endemic to the wired economy. . . . But improving productivity is not about working longer; it's about adding more value per unit of work time."[46]

Chapter 9

PRACTICAL ACTION BOX

When the Internet Isn't Productive: Online Addiction & Other Time Wasters

There are a handful of activities that can drain hours of time, putting studying—and therefore college—in serious jeopardy. They include excessive television watching, partying, and working too many hours while going to school. They also include misuse of the computer.

The Great Campus Goof-Off Machine?

"I have friends who have spent whole weekends doing nothing but playing Quake or Warcraft or other interactive computer games," reports Swarthmore College sophomore Nate Stulman, in an article headed "The Great Campus Goof-Off Machine." He goes on: "And many others I know have amassed overwhelming collections of music on their computers. It's the searching and finding they seem to enjoy: some of them have more music files on their computers than they could play in months." [a]

In Stulman's opinion, having a computer in the dorm is more of a distraction than a learning tool for students. "Other than computer science or mathematics majors, few students need more than a word processing program and access to e-mail in their rooms."

Most educators wouldn't banish computers completely from student living quarters. Nevertheless, it's important to be aware that your PC can become a gigantic time sink, if you let it. Reports Rutgers communication professor Robert Kubey: "About 5% to 10% of students, typically males and more frequently first- and second-year students, report staying up late at night using chat lines and e-mail and then feeling tired the next day in class or missing class altogether." [b]

Internet Addiction/Dependency

"A student e-mails friends, browses the World Wide Web, blows off homework, botches exams, flunks out of school." [c] This is a description of the downward spiral of the "Net addict," often a college student—because schools give students no-cost/low-cost linkage to the Internet—but it can be anyone. Some become addicted (although until recently some professionals felt "addiction" was too strong a word) to chat groups, some to online pornography, some simply to the escape from real life. [d]

Stella Yu, 21, a college student from Carson, California, was rising at 5 A.M. to get a few hours online before school, logging on to the Internet between classes and during her part-time job, and then going home to Web surf until 1 A.M. Her grades dropped and her father was irate over her phone bills. "I always make promises I'm going to quit; that I'll just do it for research," she said. "But I don't. I use it for research for 10 minutes, then I spend two hours chatting." [e]

College students are unusually vulnerable to Internet addiction, which is defined as "a psychological dependence on the Internet, regardless of type of activity once 'logged on,'" according to psychologist Jonathan Kandell. [f] The American Psychological Association, which officially recognized "Pathological Internet Use" as a disorder in 1997, defines *Internet addict* as anyone who spends an average of 38 hours a week online. [g] (The average Interneter spends 5½ hours on the activity. [h]) More recently, psychologist Keith J. Anderson of Rensselaer Polytechnic Institute found that Internet-dependent students, who make up at least 10% of college students, spent an average of 229 minutes a day online for nonacademic reasons, compared with 73 minutes a day for other students. As many as 6% spend an average of more than 400 minutes a day—almost seven hours—using the Internet. [i]

What are the consequences of Internet addiction disorder? A study of the freshman dropout rate at Alfred University in New York found that nearly half the students who quit the preceding semester had been engaging in marathon, late-night sessions on the Internet. [j] The University of California, Berkeley, found some students linked to excessive computer use neglected their course work. [k] A survey by Viktor Brenner of State University of New York at Buffalo found that some Internet addicts had "gotten into hot water" with their school for Internet-related activities. [l] "Grades decline, mostly because attendance declines," says psychologist Anderson about Internet-dependent students. "Sleep patterns go down. And they become socially isolated."

The accompanying box lists questions that may yield insights as to whether you or someone you know is Internet-dependent. [m]

Online Gambling

A particularly risky kind of Internet dependence is online gambling. David, a senior at the University of Florida, Gainesville, owes $1500 on his credit cards as a result of his online gambling habit, made possible by easy access to offshore casinos in cyberspace. He is not alone. A survey of 400 students at Southern Methodist University found that 5% said they gambled frequently via the Internet. [n]

Other students do the kind of de facto gambling known as day-trading—buying and selling stocks on the Internet. This can be risky, too. "It's a lot like going to a casino," says one finance professor. "You can make or lose money within a few seconds, and there are going to be some people who are addicted to it." [o]

9.4 Economic Issues: Employment & the Haves/Have–Nots

KEY QUESTION

How may technology affect the unemployment rate and the gap between rich and poor?

In recent times, a number of critics have provided a counterpoint to the hype and overselling of information technology to which we have long been exposed. Some critics find that the benefits of information technology are balanced by a real downside. Other critics make the alarming case that technological progress is actually no progress at all—indeed, it is a curse. The two biggest charges (which are related) are, first, that information technology is killing jobs and, second, that it is widening the gap between the rich and the poor.

Technology, the Job Killer?

Certainly, ATMs do replace bank tellers, E-Z pass electronic systems do replace turnpike-toll takers, and Internet travel agents do lure customers away from small travel agencies. The contribution of technological advances to social progress is not purely positive.

But is it true, as technology critic Jeremy Rifkin says, that intelligent machines are replacing humans in countless tasks, "forcing millions of blue-collar and white-collar workers into temporary, contingent, and part-time employment and, worse, unemployment"?[47]

This is too large a question to be fully considered in this book. We can say for sure that the U.S. economy is undergoing powerful structural changes, brought on not only by the widespread diffusion of technology but also by greater competition, increased global trade, the shift from manufacturing to service employment, the weakening of labor unions, more flexible labor markets, more rapid immigration, partial deregulation, and other factors.[48]

A counterargument is that jobs don't disappear, they just change. According to some observers, the jobs that do disappear represent drudgery. "If your job has been replaced by a computer," says Stewart Brand, "that may have been a job that was not worthy of a human."[49]

High-tech farmers
Greg Hermesmeyer employs ultrasound to gauge this cow's meat and fat content.

Gap between Rich & Poor

"In the long run," says M.I.T. economist Paul Krugman, "improvements in technology are good for almost everyone. . . . Unfortunately, what is true in the long run need not be true over shorter periods."[50] We are now, he believes, living through one of those difficult periods in which technology doesn't produce widely shared economic gains but instead widens the gap between those who have the right skills and those who don't.

A U.S. Department of Commerce survey of "information have-nots" reveals that about 20% of the poorest households in the U.S. do not have telephones (although phone subsidies are available). Moreover, only a fraction of those poor homes that do have phones will be able to afford the information technology that most economists agree is the key to a comfortable future.[51] The richer the family, the more likely it is to have and use a computer.

Education—especially college—makes a great difference. Every year of formal schooling after high school adds 5–15% to annual earnings later in life.[52] Being well educated is only part of it, however; it's essential to be technologically literate. Employees with technology skills "earn roughly 10–15% higher pay," according to the chief economist for the U.S. Labor Department.[53]

CONCEPT CHECK

What are some potential environmental consequences of information technology?

Discuss four types of mental-health problems linked to information technology.

Describe some workplace problems associated with computers.

What are some key economic issues related to information technology?

9.5 The Digital Environment: Is There a Grand Design?

KEY QUESTIONS
What are the NII, the new Internet, the Telecommunications Act, the 1997 White House plan, and ICANN?

In Chapter 1, we mentioned the *information superhighway,* a term that since has lost its luster in favor of other coinages such as the *digital environment.* The presumed goal of this worldwide system of computers and telecommunications is to give us lightning-fast (high-bandwidth) voice and data exchange, multimedia, interactivity, and near-universal, low-cost access—and to do so reliably and securely. Whether you're a Russian astronaut aloft in a

Wearable PC mini-monitor with earphones
Designed by IBM, this device is intended mainly for airplane mechanics and engineers. Users can receive data from a central database via a computer attached to the belt. Here it is being tried out by German Chancellor Gerhard Schroeder during a computer fair in Hanover.

spacecraft, a Bedouin tribesman in the desert with a PDA/cell phone, or a Canadian work-at-home mother with her office in a spare bedroom, you'll be able, it is hoped, to connect with nearly anything or anybody anywhere. You'll be able to participate in telephony, teleconferencing, telecommuting, teleshopping, telemedicine, tele-education, televoting, and even telepsychotherapy (already available), to name a few possibilities.

What shape will this digital environment take? Some government officials hope it will follow a somewhat orderly model, such as that envisioned in the National Information Infrastructure (of which new versions of the Internet are a part). Others hope it will evolve out of competition intended by the passage of the 1996 Telecommunications Act. Still others hope that a White House document, *A Framework for Global Electronic Commerce*, offers a realistic policy. Finally, a nonprofit organization, ICANN, is concerned with Internet addresses. What can these attempts to create an all-encompassing design do?

The National Information Infrastructure

As portrayed by government officials, the *National Information Infrastructure (NII)* is a kind of grand vision for existing networks and technologies, as well as technologies yet to be deployed. Services would be delivered by telecommunications companies, cable-television companies, and the Internet for a range of applications—education, health care, information access, electronic commerce, and entertainment.

Who would put the pieces of the NII together? The current national policy is to let private industry do it, with the government trying to ensure fair competition among the carriers—phone, cable, and satellite companies—and compatibility among various technological systems. In addition, NII envisions open access to people of all income levels.

The New Internet: VBNS, Internet2, & NGI

Lately less is being said about NII and more about new Internet networks: *VBNS, Internet2,* and the *Next Generation Internet.*

Does this mean three new networks will be built? Actually, all three names refer to the same network. This high-speed Internet is designed to relieve the congested electronic highway of the older Internet. Here's what the three efforts represent:

- **VBNS: Linking supercomputers and other banks of computers across the nation, _VBNS (Very-High-Speed Backbone Network Service)_ is the main U.S. government component to upgrade the "backbone," or primary hubs of data transmission.** Speeds are 1000 times conventional Internet speeds.

  Financed by the National Science Foundation and managed by the telecommunications giant MCI, VBNS is somewhat exclusive. It is being used by just 101 top universities and other research institutions for network-intensive applications, such as transmitting high-quality video for distance education.[54] (Internet2, by contrast, will eventually touch a great many more members.) VBNS has been underway since 1996, and most of the present members are also members of Internet2.

- **Internet2: _Internet2_ is a cooperative university/business program to enable high-end users to quickly and reliably move huge amounts of data, using VBNS as the official backbone.** Whereas VBNS provides data transfer at 1000 times commercial Web speeds, Internet2 operates at only 100 times those speeds.

Call your office
A traveling salesman contacts his office via cell phone and laptop computer.

In effect, Internet2 adds "toll lanes" to the older Internet to speed things up. The purpose is to advance videoconferencing, research, and academic collaboration—to enable a kind of "virtual university." Presently more than 150 universities in partnership with companies such as Cisco Systems and IBM are participants.

- **Next Generation Internet:** The ***Next Generation Internet (NGI)*** **is the U.S. government's program to parallel the university/business–sponsored effort of Internet2. It is designed to provide money to six government agencies to help tie the campus high-performance backbones into the broader federal infrastructure.** The intent is that NGI will connect at least 100 sites, including universities, federal national laboratories, and other research organizations. Speeds are 100 times those of the older Internet, and 10 sites are connected at speeds that are 1000 times as fast.

All of these networks are modeled after the original Internet, except that they will use high-speed fiber-optic circuits and more sophisticated software. NGI and Internet 2 should be available to the public by 2003.

The 1996 Telecommunications Act

After years of legislative attempts to overhaul the 1934 Communications Act, in February 1996 President Bill Clinton signed into law the *Telecommunications Act of 1996,* undoing 60 years of federal and state communications regulations. The act is designed to let phone, cable, and TV businesses compete and combine more freely. The purpose of the law was to cultivate greater competition between local and long-distance telephone companies, as well as between the telephone and cable industries. Under this legislation, different carriers can offer the same services—for example, cable companies can offer telephone services, phone companies can offer cable.

Is the law successful? "The only point on which all parties agree," says Laurence Tribe, Harvard professor of constitutional law, "is that the law isn't working as intended, and that American consumers are still waiting for free and healthy competition in communications services."[55]

No batteries needed
Boeing's new 777 airplanes feature a digital phone service, allowing laptop users to send and receive faxes and log onto the Internet. Power outlets on the back of the seats eliminate users' need to change batteries when their laptops run low on power during the flight.

The 1997 White House Plan for Internet Commerce

In 1997, the Clinton administration unveiled a document, authored by a White House group, that is significant because it endorses a governmental hands-off approach to the Internet—or, as the report more grandly calls it, "the Global Information Infrastructure."[56] Behind the title *A Framework for Global Electronic Commerce* was a plan whose gist is this: Government should stay out of the way of Internet commerce.

"Where government is needed," it states, "its aim should be to support and enforce a predictable, minimalist, consistent, and legal environment for commerce."[57] Otherwise, the plan states that private companies, not government, should take the lead in promoting the Internet as an electronic marketplace, in adopting self-regulation, and in devising ratings systems to help parents guide their children away from objectionable online content.

The policy "literally follows the first rule of the Hippocratic oath: Do no harm," says one observer. "It lets businesses and consumers determine the Internet's growth."[58]

Eli Noam, professor of finance and economics at the Columbia Business School, remains a skeptic, believing that cyberspace will be regulated no matter what the White House says:

> For all the rhetoric of an Internet "free trade zone," will the United States readily accept an Internet that includes Thai child pornographers, Albanian tele-doctors, Cayman Island tax dodges, Monaco gambling, Nigerian blue sky stock schemes, Cuban mail-order catalogues? Or, for that matter, American violators of privacy, purveyors of junk e-mail or "self-regulating" price-fixers? Unlikely. And other countries will feel the same on matters they care about.[59]

We consider some of these concerns in the following sections.

ICANN: The Internet Corporation for Assigned Names & Numbers

Acting on the belief that the Internet moves too quickly to be regulated by government, the White House decided to let Internet users govern themselves. In June 1998, it proposed the creation of a series of nonprofit corporations to manage such complex issues as fraud prevention, privacy, and intellectual property protection. The first such group, **_ICANN (Internet Corporation for Assigned Names & Numbers)_ was established to regulate Internet domain names,** those addresses ending with .com, .org, .net, and so on, that identify a Web site. ICANN got off to a rocky start, but if it succeeds, according to one report, "it could evolve into the preeminent regulatory body on the Internet, with power extending far beyond the arena of domain names."[12]

CONCEPT CHECK

What is the National Information Infrastructure?

Distinguish among VBNS, Internet2, and the Next Generation Internet.

What is the purpose of the 1996 Telecommunications Act?

Describe the 1997 White House plan for Internet commerce and its offshoot, ICANN.

In a World of Breakneck Change, Can You Still Thrive?

Clearly, information technology is driving the new world of jobs, services, and leisure, and nothing is going to stop it. People pursuing careers find the rules are changing very rapidly. Up-to-date skills are becoming ever more crucial. Job descriptions of all kinds are metamorphosing, and even familiar jobs are becoming more demanding.

Where will you be in all this? Today, experts advise, you must be willing to continually upgrade your skills, to specialize, and to market yourself. In a world of breakneck change, you _can_ thrive.

Experience Box

Succeeding at Distance Learning

Distance learning, one writer points out, is "a contemporary version in some ways of the correspondence [education-by-mail] schools of the 1950s."[61] Students take courses given at a distance, using television or a computer network. The teacher may be close by in another room or on the other side of the world. (For testing, however, students may have to go to classroom for an examination in the presence of the instructor or proctor. Or a student may be required to have a $20 video camera atop his or her computer that sends the professor a stream of images of the student taking a test.[62])

Studying online has its advantages: You can log on from anywhere, at any time of day or night, and participate in discussions to whatever extent you choose. In return for giving up the live classroom setting, students gain the convenience of being able to take courses not offered locally, often at times that suit their own schedules, not the logistical needs of an educational institution. This is a particular boon to the growing numbers of part-time and nontraditional (older) students, particularly mothers with children and people working full time. Part-timers presently make up about 45% of all college enrollments. This, says one writer, is "a group for whom 'anytime, anywhere' education holds special appeal."[63] In addition, says tele-education entrepreneur Glenn Jones, distance learning can give students computer access to virtual libraries of print, photos, recordings, and movies.[64]

Despite the benefits, distance learning has certain attributes that all cyber-students should be aware of.

- **Distance learning may work better for some subjects than others:** "Some courses just don't translate well at a distance," says Pam Dixon, author of *Virtual College*. Sometimes it's better to get your hands dirty and experience the course in a physical setting. Would you have wanted your dentist to have learned his or her craft entirely by videoconferencing?"[65] Science instructors also wonder about whether they can teach the hands-on aspects of their disciplines, such as using a microscope, from afar.[66] But, Dixon points out, courses in business, writing, computers, mathematics, and library science, to name just a few, are well suited to teaching from afar.

- **Classes may cost more than regular classes:** In distance-learning programs run by degree-granting institutions, fees may be higher than they are for regular chalk-and-talk classes. Rates are higher to offset the cost of creating online courses, which require a great deal of preparation and teaching time by faculty.

- **Online classes require responsibility and accountability:** "Online school is a venue for self-starters," says one writer, "which may be one reason that, though 10 people enrolled in the class I took, only half were ever heard from after the first few weeks."[67] Says another observer, "I think the type of student who takes an online class is one that is trying really hard, but sometimes gets in over his head."[68] Some administrators say that course-completion rates are often 10 to 20 percentage points higher in traditional courses than in distance learning.[69] "The online model of higher education requires having students who are willing to assume responsibility for their education," says one graduate student who wrote his doctoral research project on distance education.[70]

- **Class size can be important:** According to one study, class size should be limited to 40 people or fewer.[71] If an instructor is to teach well online, it takes a lot of time; therefore classes should be reasonably small.

- **Contact with instructors and fellow students is different:** With online learning, "students typically can't buttonhole a teacher's aide to answer a question immediately, can't turn professors into mentors, and can't gather spontaneously to bounce ideas off classmates," points out one report.[72] The downside of distance learning, reports one student, is that "it removes the human element. Without face-to-face contact with my professors and fellow students, I found the course work to be much less stimulating [than in a regular classroom], and I think I learned less."[73] On the other hand, students may get more attention from teachers online than they would from teachers in a large classroom with hundreds of students. Indeed, online proponents say students actually interact *more* with their instructors and fellow students. "There are no wallflowers in online classes," suggests one article.[74] Most interaction takes place on electronic bulletin boards that everyone in the class can see and respond to.

- **Distance learning takes different communication skills:** Joseph Walter, professor of communication studies at Northwestern University, found "that the computer changes the dynamics of communication. It takes four or five times as long as speaking when you have to type out what you want to say."[75] Richard Clark, professor of educational psychology at the University of Southern California, says online classes require more work for instructors because they have to be extremely clear and direct. "Engaging a student with a computer is much different than engaging them live," he says.[76]

If you can't get the educational offerings you want locally, where do you start your search for the distance-learning equivalent? You might begin by looking at Distance Learning on the Net (*http://homepage.interaccess.com/~ghoyle*) and the Comprehensive Distance Education List of Resources (*http://talon.extramural.uiuc.edu/ramage/disted.html*).

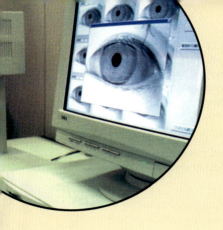

Visual Summary

antivirus software (p. 355, KQ 9.1) Program that scans a computer's hard disk, floppy disks, and main memory to detect viruses and, sometimes, to destroy them. *Why it's important: Computer users must find out what kind of antivirus software to install in their systems—and how to keep it up to date—for protection against damage or shutdown.*

biometrics (p. 358, KQ 9.2) Science of measuring individual body characteristics. *Why it's important: Biometrics is used in some computer security systems, to restrict user access.*

computer crime (p. 353, KQ 9.1) Crime of two types: (1) an illegal act perpetrated against computers or telecommunications; (2) the use of computers or telecommunications to accomplish an illegal act. *Why it's important: Crimes against information technology include theft—of hardware, of software, of computer time, of cable or telephone services, or of information. Other illegal acts are crimes of malice and destruction.*

crackers (p. 356, KQ 9.1) People who illegally break into computers for malicious purposes. *Why it's important: Crackers attempt to break into computers to obtain information for financial gain, shut down hardware, pirate software, or alter or destroy data.*

disaster-recovery plan (p. 360, KQ 9.2) Method of restoring information processing operations that have been halted by destruction or accident. *Why it's important: Such a plan is important if an organization desires to resume computer operations quickly.*

encryption (p. 358, KQ 9.2) Altering data so it is not usable unless the changes are undone. *Why it's important: Encryption is clearly useful for some organizations, especially those concerned with trade secrets, military matters, and other sensitive data. Some maintain that encryption will determine the future of e-commerce, because transactions cannot flourish over the Internet unless they are secure.*

hackers (p. 356, KQ 9.1) People who gain unauthorized access to computer or telecommunications systems, often just for the challenge of it. *Why it's important: Hackers create problems not only for the institutions that are victims of break-ins but also for ordinary users of the systems.*

ICANN (Internet Corporation for Assigned Names & Numbers) (p. 369, KQ 9.5) Organization established to regulate Internet domain names, those addresses ending with .com, .org, .net, and so on, that identify a Web site. *Why it's important: ICANN could evolve into the preeminent regulatory body on the Internet, with power extending far beyond the arena of domain names.*

Internet2 (p. 367, KQ 9.5) Cooperative university/business program established to enable high-end users to quickly and reliably move huge amounts of data, using VBNS as the official backbone. *Why it's important: Whereas VBNS provides data transfer at 1000 times commercial Web speeds, Internet2 operates at only 100 times those speeds. In effect, Internet2 adds "toll lanes" to the older Internet to speed things up. The purpose is to advance videoconferencing, research, and academic collaboration—to enable a kind of "virtual university."*

The Challenges of the Digital Age

Next Generation Internet (NGI) (p. 368, KQ 9.5) U.S. government's program to parallel the university/business–sponsored effort of Internet2. Why it's important: *NGI is designed to provide money to six government agencies to help tie the campus high-performance backbones into the broader federal infrastructure. The intent is that NGI will connect at least 100 sites, including universities, federal national laboratories, and other research organizations. Speeds are 100 times those of the older Internet, and 10 sites are connected at speeds that are 1000 times as fast.*

password (p. 358, KQ 9.2) Special word, code, or symbol required to access a computer system. Why it's important: *Passwords are one of the weakest security links; they can be guessed, forgotten, or stolen.*

PIN (personal identification number) (p. 358, KQ 9.2) Security number known only to the user; it is required to access a system. Why it's important: *PINS are required to access many computer systems, as well as automated teller machines.*

security (p. 357, KQ 9.2) System of safeguards for protecting information technology against disasters, systems failure, and unauthorized access that can result in damage or loss. Four components of security are identification and access, encryption, protection of software and data, and disaster-recovery plans. Why it's important: *With proper security, organizations and individuals can minimize information technology losses from disasters, system failures, and unauthorized access.*

VBNS (Very-High-Speed Backbone Network Service) (p. 367, KQ 9.5) Main U.S. government component to upgrade the backbone, or primary hubs of data transmission, of the Internet. Why it's important: *Linking supercomputers and other banks of computers across the United States, the VBNS is 1000 times as fast as the conventional Internet. Financed by the National Science Foundation and managed by the telecommunications giant MCI, VBNS is somewhat exclusive. It is being used by just 101 top universities and other research institutions for network-intensive applications, such as transmitting high-quality video for distance education.*

virus (p. 354, KQ 9.1) Deviant program that can cause unexpected and often undesirable effects, such as destroying or corrupting data. Why it's important: *Viruses can cause users to lose data and/or files or can shut down entire computer systems.*

worm (p. 354, KQ 9.1) Program that copies itself repeatedly into a computer's memory or onto a disk drive until no space is left. Why it's important: *Worms can shut down computers.*

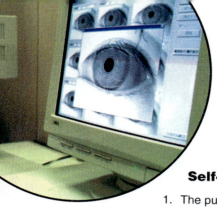

Chapter Review

Self-Test Questions

1. The purpose of _____ is to scan a computer's disk devices and memory to detect viruses and, sometimes, to destroy them.

2. So that information processing operations can be restored after destruction or accident, organizations should adopt a _____.

3. _____ is the altering of data so that it is not usable unless the changes are undone.

4. _____ is incomplete, outdated, or otherwise inaccurate data.

5. An error in a program that causes it not to work properly is called a _____.

Multiple-Choice Questions

1. Which of the following are crimes against computers and communications?

   a. natural hazards

   b. software theft

   c. information theft

   d. software bugs

   e. procedural errors

2. Which of the following are methods or means of safeguarding computer systems?

   a. signatures

   b. keys

   c. physical traits of users

   d. worms

   e. VBNS

True/False Questions

T F   1. Viruses cannot be passed from computer to computer through a network.

T F   2. One of the weakest links in a security system is biometrics.

T F   3. The VBNS is faster than the conventional Internet.

Short-Answer Questions

1. What is the difference between a hacker and a cracker?

2. What does a worm do?

3. Briefly describe and differentiate VBNS, Internet2, and Next Generation Internet.

4. What was the 1996 U.S. Telecommunications Act intended to do?

5. What does ICANN stand for, and what does it do?

6. Name five threats to computers and communications systems.

7. The definition of computer crime distinguishes between two types. What are they?

Concept Mapping

On a separate sheet of paper, draw a concept map, or visual diagram, linking concepts. Show how the following terms are related.

antivirus software	Internet2
biometrics	NGI
computer crime	password
crackers	PIN
disaster-recovery plan	security
encryption	VBNS
hacker	virus
ICANN	worm

Knowledge in Action

1. What, in your opinion, are the most significant disadvantages of using computers? What do you think can be done about these problems?

2. What's your opinion about the issue of free speech on an electronic network? Research some recent legal decisions in various countries, as well as some articles on the topic. Should the contents of messages be censored? If so, under what conditions?

3. Research the problems of stress and isolation experienced by computer users in the United States, Japan, and one other country. Write a brief report on your findings.

10

The Promises of the Digital Age

Society & Information Technology Tomorrow

Key Questions

You should be able to answer the following questions.

10.1 **Emerging Global Telecommunications** What are the two models of telecommunications?

10.2 **Artificial Intelligence** What are the main areas of artificial intelligence?

10.3 **Information & Education** How is information technology being used in education?

10.4 **Health, Medicine, & Science** How is information technology being used in health, medicine, and science?

10.5 **Commerce & Money** How is information technology being used in sales and marketing, retailing, banking, and stock trading?

10.6 **Entertainment & the Arts** How is information technology being used in music and movies?

10.7 **Government & Electronic Democracy** What benefits does information technology offer government and democracy?

10.8 **Jobs & Careers** How can you use the Internet to find a job, and what are some hot jobs in information technology?

A Visual Overview of This Chapter

Graphical Interface

1 Emerging Global Communications. Two models of telecommunications are prevalent. In the **tree-and-branch telecommunications model,** a centralized information provider sends out messages through many channels to many consumers, as in many mass media. In the **switched-network telecommunications model,** people on the system are not only consumers of information but also possible providers of it; this model, embodied in the Internet, is much more participatory.

2 Artificial Intelligence. Artificial intelligence consists of technologies used for developing machines to emulate human qualities, including the following: (1) **Virtual reality**, devices that project a person into a sensation of three-dimensional space, is used in arcade-type games and also in **simulators,** devices that represent the behavior of physical or abstract systems and are used in training, as of airplane pilots. (2) **Robotics**, the development and study of machines that can perform work normally done by people, has produced **robots,** automatic devices that perform functions usually performed by people. (3) **Natural language processing** is the study of ways for computers to recognize and understand human language. (4) **Fuzzy logic** is a method of dealing with imprecise data and uncertainty, with problems that have many answers rather than one. (5) **Expert systems** are interactive computer programs used to solve problems normally requiring the assistance of human experts. An expert system has three components: a **knowledge base,** a database of knowledge about a particular subject; an **inference engine,** the software that controls the knowledge base and produces conclusions; and a *user interface.* (6) **Neural networks** use physical electronic devices or software to mimic the neurological structure of the human brain. (7) **Genetic algorithms** are programs that use Darwinian principles of random mutation to improve themselves.

Artificial life is the study of "creatures"—computer instructions, or pure information—that, like live organisms, are created, replicate, evolve, and die. A-life raises the question of how we can know when computers can be said to possess "intelligence" or "self-awareness." One answer is suggested by the **Turing test,** in which a human judge converses by means of a computer terminal with two entities—one a person, one a computer—hidden in another location to try to determine which seems to have the most human qualities.

3 Information & Education. The challenge of making sense of vast stores of information is being addressed with **intelligent agents,** programs that roam networks and compile data and perform work tasks on your behalf. In education, students at all levels are finding computers helpful. Their use in distance learning over the Internet is increasing.

4 **Health, Medicine, & Science.** *Telemedicine*—medical care delivered via telecommunications—is one way computers and communications are changing health and medicine. The digitizing of medical information is affecting everything from psychotherapy to implants. Patients' use of health-care databases is changing their relationship with doctors. A new idea in science is the "collaboratory," an Internet-based collaborative laboratory of researchers around the world, such as that among space physicists. In archaeology, computer technology may be used to avoid invasive excavations.

5 **Commerce & Money.** Information technology erases boundaries in business between company departments, suppliers, and customers. Consequently, the idea of what constitutes an organization is changing. There are new developments in sales and marketing and retailing, as with online sales, and in banking and cybercash, as with electronic payment systems.

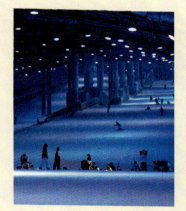

6 **Entertainment & the Arts.** Information technology is producing changes in music and movies. In music, new digitized instruments offer a wide range of sounds, while the Internet is reshaping the marketing of songs. In movies, computers are used for all kinds of animation and other special effects; digital equipment permits better film editing, and enables amateurs to make movies more cheaply.

7 **Government & Electronic Democracy.** The Internet has potential for civic betterment because it is free of government intrusion, is fast and cheap, and facilitates communication among citizens. Examples are found in cities in California, Colorado, Texas, and Nevada. Online voting has been tried and may be expanded. The government itself is making increasing use of computers, as in electronic tax filing.

8 **Jobs & Careers.** Job seekers can now use employer databases to get leads on jobs, and they can post résumés with electronic job registries so employers can find them. The five information-technology job categories projected to have the largest percentage increase in the near future are computer engineers, computer support specialists, systems analysts, database administrators, and desktop publishing specialists.

s the computer just a convenient new gadget that makes things easier? Or is it a truly revolutionary tool whose long-range result will be to change all the rules?

In 1869, steam engines delivered 1.2 million horsepower to U.S. manufacturers. Seventy years later, the electric motor, which replaced the steam engine, delivered 45 million horsepower—a 40-fold increase in mechanical power.

The electric motor, suggests economist Stephen Cohen, was a *tool* as opposed to a *gadget*. The difference? Gadgets are inventions that make life easier but do not change a country's economic dynamics. An automatic transmission may make driving easier but, Cohen argues, "a car with an automatic transmission still does what a car does."[1] Startling innovations such as television in the 1950s and air transport in the 1960s did not change society's basic economic forces. The electric motor, by contrast, is a tool. It took 40 years for industry to learn how to take full advantage of it, but among the revolutionary results was mass production. "The long-run consequences—industrial, organizational, social—were enormous," says Cohen.

In the late 1950s, the world's 2000 computers had an average processing power of 10,000 instructions per second, points out Cohen's University of California, Berkeley colleague Brad DeLong. Today there are 200 million computers with an average processing power of 100 million instructions per second. "This is a million-fold increase in 40 years," writes DeLong.[2]

So are computers and communications technology gadgets or tools? The tools of the industrial revolution amplified muscle power, so that we don't have to rely on a human or a horse anymore. Today's information technology amplifies brain power. "We no longer need to rely on human memory or on human eyes for scanning, or human brains for organizing and computing information and data," DeLong and Cohen suggest. Because the forces of this technological tool thread through everything, they are "important in every single economic activity." Thus, the two economists think, a century from now, people will look back and conclude that today's innovations in information technology constituted life-changing tools.

What are the promises of this future Digital Age? We consider the following areas:

- Emerging global telecommunications
- Artificial intelligence
- Information and education
- Health, medicine, and science
- Commerce and money
- Entertainment and the arts
- Government and electronic democracy
- Jobs and careers

10.1 Emerging Global Telecommunications

KEY QUESTION

What are the two models of telecommunications?

The hallmark of great civilizations has been their systems of communications.

In the beginning, communications was based on transportation: The Roman Empire had its network of roads, the European powers had their far-flung navies, and the United States and Canada built their transcontinental railroads. Then transportation yielded to the electronic exchange of information. Beginning in 1844, the telegraph ended the short existence of the

Pony Express. In 1876, the telephone came along as a competitor to the telegraph.

In 1889, electromagnetic radio waves were transmitted the width of the English Channel. The amplifying vacuum tube, invented in 1906, led to commercial radio, beginning with Pittsburgh's KDKA in 1920. Television came into being in England in 1925; one of the first scheduled TV shows, broadcast from New York on July 21, 1931, introduced Kate Smith singing "When the Moon Comes Over the Mountain."

During the 1950s and 1960s, as TV sets appeared seemingly everywhere, Canadian communications philosopher Marshall McLuhan posed the notion of a "global village." The global village refers to the "shrinking of the world society because of the ability to communicate." This shrinking has resulted from the global reach of mass media such as television. People throughout the world can now view similar shows and so have common points of reference.

Recently, technology has become portable, giving individuals even more power over communications and making the global village even more closely knit. Two decades ago, cell phones, pagers, and portable computers with communications links barely existed; now they are commonplace. Moreover, as computer and communications technology go digital, the computer, communications, consumer electronics, entertainment, and mass media industries are undergoing a technological convergence.

Two Models of Telecommunications: Tree-&-Branch versus Switched-Network

Telecommunications can be organized through two kinds of arrangements: the "tree-and-branch" model or the "switched-network" model.

- **Tree-and-branch model:** In the _**tree-and-branch telecommunications model**_, **a centralized information provider sends out messages through many channels to thousands of consumers.** This is the model of most mass media, such as radio and TV broadcasting. It is also the model envisioned by some cable and entertainment companies looking to provide movies and other services from a centralized library or a core of "content providers."

- **Switched-network model:** In the _**switched-network telecommunications model**_, **people on the system are not only consumers of information ("content") but also possible providers of it.** This is the model of the telephone system, of course, and also of most computer networks, including, most certainly, the Internet.

Videoconferencing
Armanda Holmes, shown here with her daughter at Fort Bragg, N.C., talks with her husband, Frank, an Army lieutenant stationed 5000 miles away in Sarajevo, Bosnia, using ProShare Videoconferencing technology.

The Potential Effect of the Switched-Network Model: Napster & the Rise of File Sharing

Whereas the tree-and-branch allows only few-to-many communications, the switched-network allows many-to-many communications. Thus, the more the switched-network becomes prominent, the more participatory our systems of communications become.

For example, the file-sharing software Napster has tapped into the participatory potential of the switched-network model. Napster, like Gnutella and similar programs enables users to go on the Web and download songs that reside in MP3 format on the hard drives of other music lovers. (MP3 is a simple way of condensing a sound file.) These music files, which are free, can be shared regardless of their location on the Internet. There is no way to police this file swapping, and many such files were originally taken—pirated—from copyrighted music released by music companies on CDs. (Warning: When you share music using Napster, this means that nearly anyone can access your hard disk through the phone line, making you vulnerable to worms and viruses.)[3] The few-to-many system has given way to a many-to-many system.

A majority of college students, according to one survey, used the controversial Napster music Web service at least once a month.[4] Because most Napster users got their downloaded music free from other fans' hard disks, not from record company servers, they have changed the economics of music distribution. Most of the 7000 new CD titles made available each year by record companies are by not-yet-famous bands. These bands are subsidized by megasellers such as Metallica and Dr. Dre. Record companies may charge $16 for a CD that costs only 60 cents to manufacture. But their net profit is only about 59 cents per CD after they subtract out the costs of supporting all the artists they carry.[5] The danger of free content, then, is that without financial incentives, ultimately the source of content will dry up.

Napster, Gnutella, Scour Exchange, CuteMX, and similar arrangements may have been vastly modified or even shut down by the time you read this. The record industry is not inclined to leave their copyrights unprotected for long. (By comparison, anyone who creates an Internet site to share copyrighted software is likely to get a visit from the FBI and be liable for some serious jail time.[6]) Nevertheless, the switched-network file-swapping model is probably here to stay, and it could well change the basic economics of the music business and, perhaps, of the online book-publishing business. In the wake of Napster, the five major record companies, for instance, have rushed to make songs available for download. These, of course, you have to pay for.[7]

10.2 Artificial Intelligence

KEY QUESTION

What are the main areas of artificial intelligence?

You're having trouble with your new software program. You call the customer "help desk" at the software maker. Do you get a busy signal or get put on hold to listen to music (or, worse, advertising) for several minutes? Technical support lines are often swamped, and waiting is commonplace. Or, to deal with your software difficulty, do you find yourself dealing with . . . other software?

The odds are good that you will. Programs that can walk you through a problem and help solve it are called *expert systems*. As the name suggests, these are systems imbued with knowledge by a human expert. Expert systems are one of the most useful applications of artificial intelligence.

Artificial intelligence (AI) **is a group of related technologies used for developing machines to emulate human qualities, such as learning, reasoning, communicating, seeing, and hearing.** Today the main areas of AI are:

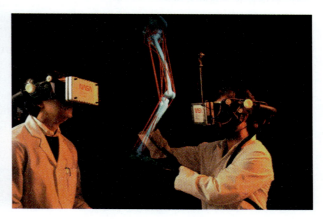

Top left: In a navigation simulator, a captain at the helm tries to steer a container ship away from a collision course with an oil tanker. *Top right:* A man uses a treadmill to walk through a virtual reality experiment in Chapel Hill, North Carolina. *Lower left:* Using the Dismounted Infrared Combat Simulation Treadmill, Sgt. March Turchin practices using the latest combat equipment in a virtual city. *Lower right:* Two medical students study the leg bones of a "virtual cadaver."

- Virtual reality
- Robotics
- Natural language processing
- Fuzzy logic
- Expert systems
- Neural networks
- Genetic algorithms

We will consider these and also an area known as *artificial life.*

Virtual-Reality & Simulation Devices

Virtual reality (VR), **a computer-generated artificial reality, projects a person into a sensation of three-dimensional space.** *(See ● Panel 10.1.)* To put yourself into virtual reality, you need software and special headgear; then you can add gloves, and later perhaps a special suit. The headgear—which is called a *head-mounted display*—has two small video display screens, for each eye, to create the sense of three-dimensionality. Headphones pipe in stereophonic sound or even 3-D sound; so that you think you are hearing sounds not only near each ear but also in various places all around you. The glove has sensors for collecting data about your hand movements. Once you are wearing this equipment, software gives you interactive sensory feelings similar to real-world experiences.

WABOT-2
This robot, created in 1984, can read sheet music well enough to play simple tunes on an electric organ.

You may have seen virtual reality used in arcade-type games, such as Atlantis, a computer simulation of The Lost Continent. However, there are far more important uses—for example, in simulators for training. **_Simulators are devices that represent the behavior of physical or abstract systems._** Virtual-reality simulation technologies are applied a great deal in training. For instance, to train bus drivers, they create lifelike bus control panels and various scenarios such as icy road conditions. They are used to train pilots on various aircraft and to prepare air-traffic controllers for equipment failures. Surgeons-in-training can develop their skills through simulation on "digital patients." Virtual-reality therapy has been used for autistic children and in the treatment of phobias, such as extreme fear of public speaking or of being in public places or high places.

Robotics

More than 40 years ago, in the film _Forbidden Planet_, Robby the Robot could sew, distill bourbon, and speak 187 languages. We haven't caught up with science-fiction movies, but maybe we'll get there yet.

Robotics is the development and study of machines that can perform work normally done by people. The machines themselves are called robots. _(See_ ● _Panel 10.2.)_ Basically, a **_robot_ is an automatic device that performs functions ordinarily executed by human beings or that operates with what appears to be almost human intelligence.** ScrubMate—a robot equipped with computerized controls, ultrasonic "eyes," sensors, batteries, three different cleaning and scrubbing tools, and a self-squeezing mop—can clean bathrooms. Rosie the HelpMate delivers special-order meals from the kitchen to nursing stations in hospitals. Robodoc is used in surgery to bore the thighbone so that

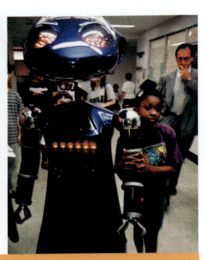

● **PANEL 10.2**
Robots
Top left: The Human Extender Robot allows people to pick up very heavy packages. _Top middle:_ NASA remote-controlled research robot. _Top right_ Jamie Quinones, 11, strolls down the hall with the robot SICO in Governor Hospital in New York City. Robots like SICO help children with emotional problems come out of their shells. _Bottom left:_ Security robot on patrol at the Los Angeles County Museum of Art. _Bottom right:_ A police robot handles a live bomb via remote control.

a hip implant can be attached. A driverless harvester, guided by satellite signals and artificial vision system, is used to harvest alfalfa and other crops.

Robots are also used for more exotic purposes such as fighting oil-well fires, doing nuclear inspections and cleanups, and checking for mines and booby traps. An eight-legged, satellite-linked robot called Dante II was used to explore the inside of Mount Spurr, an active Alaskan volcano, sometimes without human guidance. A six-wheeled robot vehicle named Sojourner was used in NASA's 1997 Pathfinder exploration of Mars to sample the planet's atmosphere and soil and to radio data and photos back to Earth. A similar robot has been designed for use on the next Mars expedition.

Natural Language Processing

Natural languages are ordinary human languages, such as English. (A second definition, discussed in the Appendix, is that they are fifth-generation programming languages.) **_Natural language processing_ is the study of ways for computers to recognize and understand human language,** whether in spoken or written form.

Think how challenging it is to make a computer translate English into another language. In one instance, the English sentence "The spirit is willing, but the flesh is weak" came out in Russian as "The wine is agreeable, but the meat is spoiled." The problem with human language is that it is often ambiguous; different listeners may arrive at different interpretations.

Most existing language systems run on large computers, although scaled-down versions are now available for microcomputers. A product called Intellect uses a limited English vocabulary to help users orally query databases on both mainframes and microcomputers. LUNAR, developed to help analyze moon rocks, answers questions about geology on the basis of an extensive database. Verbex, used by the U.S. Postal Service, lets mail sorters read aloud an incomplete address and will reply with the correct ZIP code.

Fuzzy Logic

A relatively new concept being used in the development of natural languages is fuzzy logic. The traditional logic behind computers is based on either-or, yes-no, true-false reasoning. Such computers make "crisp" distinctions, leading to precise decision making. **_Fuzzy logic_ is a method of dealing with imprecise data and uncertainty, with problems that have many answers rather than one.** Unlike classical logic, fuzzy logic is more like human reasoning: it deals with probability and credibility. That is, instead of being simply true or false, a proposition is *mostly* true or *mostly* false, or *more* true or *more* false.

For example, one place fuzzy logic has been applied is in running elevators. How long will most people wait for an elevator before getting antsy? About a minute and a half, say researchers at the Otis Elevator Company. The Otis artificial intelligence division has thus done considerable research into how elevators may be programmed to reduce waiting time.[8] Ordinarily, when someone on a floor in the middle of the building pushes the call button, the system will send whichever elevator is closest. However, that car might be filled with passengers, who will be delayed by the new stop (perhaps making them antsy), whereas another car that is farther away might be empty. In a fuzzy-logic system, the computer assesses not only which car is nearest but also how full the cars are before deciding which one to send.

Fuzzy logic circuitry also enables autofocus cameras to focus properly.

Expert Systems

An **_expert system_ is an interactive computer program used in solving problems that would otherwise require the assistance of a human expert.** Fundamental to an expert system is a knowledge base constructed by experts that can "learn" by adding new knowledge. People fearful about machines taking over need to understand that expert systems are designed to be users' *assistants,* not replacements. Also, the success of these systems depends on the quality of the data and rules obtained from the human experts.

The expert system MYCIN helps diagnose infectious diseases. PROSPECTOR assesses geological data to locate mineral deposits. DENDRAL identifies chemical compounds. Home-Safe-Home evaluates the residential environment of an elderly person. Business Insight helps businesses find the best strategies for marketing a product. REBES (Residential Burglary Expert System) helps detectives investigate crime scenes. CARES (Computer Assisted Risk Evaluation System) helps social workers assess families for risks of child abuse. CLUES (Countrywide Loan Underwriting Expert System) evaluates home-mortgage-loan applications. Crush takes a body of expert advice and combines it with worksheets reflecting a user's business situation to come up with a customized strategy to "crush competitors."

All these programs simulate the reasoning process of experts in certain well-defined areas. That is, professionals called knowledge engineers interview the expert or experts and determine the rules and knowledge that must go into the system. For example, to develop Muckraker, an expert system to assist with investigative reporting, the knowledge engineers interviewed journalists.

Programs incorporate not only the experts' surface knowledge ("textbook knowledge") but also deep knowledge ("tricks of the trade"). What, exactly, is deep knowledge? "An expert in some activity has by definition reduced the world's complexity by his or her specialization," say some authorities. One result is that "much of the knowledge lies outside direct conscious awareness."[9]

An expert system consists of three components. *(See ● Panel 10.3.)*

- ● **Knowledge base:** A ***knowledge base*** **is an expert system's database of knowledge about a particular subject,** including relevant facts, information, beliefs, assumptions, and procedures for solving problems. The basic unit of knowledge is expressed as an IF-THEN-ELSE rule. ("IF this happens, THEN do this, ELSE do that.") Programs can have as many as 10,000 rules. A system called ExperTAX, for example, which helps accountants figure out a client's tax options, consists of over 2000 rules.

- ● **Inference engine:** The ***inference engine*** **is the software that controls the search of the expert system's knowledge base and produces conclusions.** It takes the problem posed by the user and fits it into the rules in the knowledge base. It then derives a conclusion from the facts and rules contained in the knowledge base.

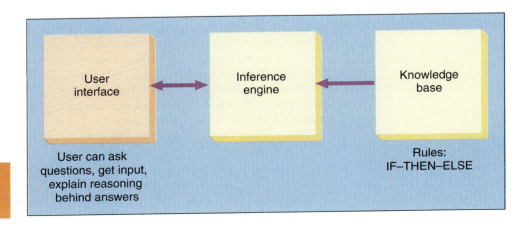

User interface

User can ask questions, get input, explain reasoning behind answers

Inference engine

Knowledge base

Rules: IF–THEN–ELSE

- **User interface:** The *user interface* is the display screen that the user deals with. It gives the user the ability to ask questions and get answers. It also explains the reasoning behind the answer.

Expert systems include stand-alone microcomputers, as well as workstations and terminals connected to servers and larger computer systems in local area networks or wide area networks.

Neural Networks

Artificial intelligence and fuzzy logic principles are being applied to the development of neural networks. **_Neural networks_ use physical electronic devices or software to mimic the neurological structure of the human brain.** Because they are structured to mimic the rudimentary circuitry of the cells in the human brain, they learn from example and don't require detailed instructions.

To understand how neural networks operate, let's compare them to the operation of the human brain.

- **The human neural network:** The human body has a neural network consisting of neurons, or nerve cells. The neurons are connected by a three-dimensional lattice called *axons*. Electrical connections between neurons are activated by *synapses*.

  The human brain is made up of about 100 billion neurons. However, these cells do not act as "computer memory" sites. No cell holds a picture of your dog or the idea of happiness. You could eliminate any cell—or even a few million—in your brain and not alter your "mind." Where do memory and learning lie? In the electrical connections between cells, the synapses. Using electrical pulses, the neurons send on/off messages along the synapses.

- **The computer neural network:** In a hardware neural network, the nerve cell is replaced by a transistor, which acts as a switch. Wires connect the cells (transistors) with each other. The synapse is replaced by an electronic component called a resistor, which determines whether a cell should activate the electricity to other cells. A software neural network emulates a hardware neural network, although it doesn't work as fast.

  Computer-based neural networks use special AI software and complicated fuzzy-logic processor chips to take inputs and convert them to outputs with a kind of logic similar to human logic.

Neural networks are already being used in a variety of situations. One helped a mutual-fund manager to outperform the stock market by 2.3–5.6 percentage points over three years.[10] At a San Diego hospital emergency room in which patients complained of chest pains, the same information was given to a neural-network program and to doctors. The network correctly diagnosed patients with heart attacks 97% of the time, compared to 78% for the human physicians.[11] In Chicago, a neural-net used to look for signs of breast cancer on patient X-rays outperformed most doctors in distinguishing malignant tumors from benign ones.[12] Banks that use neural-network software to spot irregularities in purchasing patterns within individual accounts often notice when a credit card is stolen before its owner does.

Genetic Algorithms

A **_genetic algorithm_ is a program that uses Darwinian principles of random mutation to improve itself.** The algorithms are lines of computer code that act like living organisms. Different sections of code haphazardly come together, producing programs. As in Darwin's rules of evolution, many chunks of code compete to see which can best fulfill the goal of the program. Some chunks will even become extinct. Those that survive will combine with other survivors to produce offspring programs.

Expert systems can capture and preserve the knowledge of expert specialists, but they may be slow to adapt to change. Neural networks can sift through mountains of data and discover obscure causal relationships, but if there is too much data, or too little, they may be ineffective. Genetic algorithms, by contrast, use endless trial and error to learn from experience—to discard unworkable approaches and grind away at promising approaches with the kind of tireless energy of which humans are incapable.

In 2000, scientists at Brandeis University reached a major milestone when they created a computerized robot that designs and builds other robots, automatically evolving without any significant human intervention.[13] The "robotic life forms" are only a few inches long and are composed of a few plastic parts with rudimentary nervous systems made of wire. They do only one thing: inch themselves, worm-like, along a horizontal surface, using miniature motors. With no idea what a successful design might look like, the Brandeis computer was given the goal of moving on a horizontal surface; a list of possible parts to work with; a group of 200 randomly constructed, nonworking designs; and the physical laws of gravity and friction. Mimicking evolution through classic survival-of-the-fittest selection, the computer changed pieces in the designs, mutated the programming instructions for controlling movements, and ran simulations to test the designs. After 300–600 generations of evolution, the computer sent the design to a machine to build the robot. The robots currently have the brainpower of bacteria, say researchers, who say they hope to get up to insect level in a couple of years.

Artificial Life, the Turing Test, & AI Ethics

Ethics

Genetic algorithms would seem to lead us away from mechanistic ideas of artificial intelligence and into more fundamental questions: What is life, and how can we replicate it out of silicon chips, networks, and software? We are dealing now not with artificial intelligence but with artificial life. **Artificial life, or A-life, is a field of study concerned with "creatures"—computer instructions, or pure information—that are created, replicate, evolve, and die as if they were living organisms.** Thus, A-life software (such as LIFE) tries to simulate the responses of a human being.

Of course, "silicon life" does not have two principal attributes associated with true living things—it is not water- and carbon-based. Yet in other respects such creatures mimic life: If they cannot learn or adapt, then they perish.

How can we know when we have reached the point where computers have achieved human intelligence? How will you know, say, whether you're talking to a human being on the phone or to a computer? Clearly, with the strides made in the fields of artificial intelligence and artificial life, this question is no longer just academic.

Interestingly, Alan Turing, an English mathematician and computer pioneer, addressed this very question in 1950. Turing predicted that by the end of the century computers would be able to mimic human thinking and to conduct conversations indistinguishable from a person's. Out of these observations came the Turing test, which is intended to determine whether a computer possesses "intelligence" or "self-awareness."

In the **Turing test, a human judge converses by means of a computer terminal with two entities hidden in another location—one a person typing on a keyboard, the other a software program. Following the conversation, the judge must decide which entity is human. In this test, intelligence—the ability to think—is demonstrated by the computer's success in fooling the judge.** (See ● Panel 10.4.)

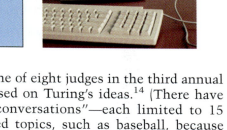
Judith Anne Gunther participated as one of eight judges in the third annual Loebner Prize Competition, which is based on Turing's ideas.[14] (There have been other competitions since.) The "conversations"—each limited to 15 minutes—are restricted to predetermined topics, such as baseball, because even today's best programs have neither the databases nor the syntactical ability to handle an unlimited number of subjects.

Gunther found that she wasn't fooled by any of the computer programs. The winning program, for example, relied as much on deflection and wit as it did on responding logically and conversationally. (For example, to a judge trying to discuss a federally funded program, the computer said: "You want logic? I'll give you logic: shut up, shut up, shut up, shut up, shut up, now go away! How's that for logic?") However, Gunther *was* fooled by one of the five humans, a real person discussing abortion. "He was so uncommunicative," wrote Gunther, "that I pegged him for a computer."

Ethics

Behind everything to do with artificial intelligence and artificial life—just as it underlies everything we do—is the whole matter of *ethics*. In his book *Ethics in Modeling*, William A. Wallace, professor of decision sciences at Rensselaer Polytechnic Institute, points out that computer software, including expert systems, is often subtly shaped by the ethical judgments and assumptions of the people who create it.[15] In one instance, he notes, a bank had to modify its loan-evaluation software on discovering that the software rejected certain applications because it unduly emphasized old age as a negative factor. Another expert system, used by health maintenance organizations (HMOs), tells doctors when they should opt for expensive medical procedures, such as magnetic resonance imaging tests. HMOs like such systems because they help control expenses, but critics are concerned that doctors will have to base decisions not on the best medicine but simply on "satisfactory" medicine combined with cost constraints.[16]

Clearly, there is no such thing as completely "value-free" technology. Human beings build it, use it, and have to live with the results.

Is this fish for real?
This sea bream is about 1½ feet long, weighs 5½ pounds, and can swim up to 38 minutes—before recharging. The robot fish, created by Mitsubishi, looks and swims exactly like the real thing.

CONCEPT CHECK

Distinguish between the tree-and-branch model and the switched-network model.

Distinguish among the main areas of AI.

What is artificial life?

10.3 Information & Education

KEY QUESTION
How is information technology being used in education?

Getting the right kind of information is a major challenge. Can everything in the Library of Congress be made available online to citizens and companies? What about government records, patents, contracts, and other legal documents? Or geographical maps photographed from satellites?

An even greater challenge is getting information to make sense. What the online world really needs, therefore, is a terrific librarian. "What bothers me most," says Christine Borgman, chair of the UCLA Department of Library and Information Science, "is that computer people seem to think that if you have access to the Web, you don't need libraries."[17] But what's in a library is standardized and well organized, whereas what's on the Web is often overwhelming, unstandardized, and chaotic.

As one solution, computer scientists have been developing so-called intelligent agents to find information on computer networks and filter it. An ***intelligent agent* program performs work tasks—such as roaming networks and compiling data— on your behalf.** A software agent is a kind of electronic assistant that will filter messages, scan news services, and perform similar secretarial chores. An agent will also travel over communications lines to computer databases, collecting files to add to a personalized database. As you might suspect, one popular use for such agents is as "shopping bots" that consumers can use to find online bargains.

With their potential for finding and processing information, computers are pervasive on campus. More than a third of college courses require the use of e-mail, and many have their own dedicated Web pages.[18] College students spend an average of 5.6 hours a week on the Internet.[19] And as computer prices drop, more and more students have their own PCs. (For the rest, colleges often make public microcomputers available.)

Many students have been exposed to computers since the lower grades. In 1997, for example, 75% of elementary schools (and 89% of secondary schools) were wired to the Internet, according to the U.S. Department of Education.[20] The results have not always been productive, especially in elementary

Science class
Schoolchildren in Wales track the wind's speed, direction, and temperature. Their findings will be e-mailed to other students in Wales.

Learning center

Children who come from around the world seeking medical treatment at St. Jude's Children's Hospital in London have access to the hospital's learning center to help keep up with their classmates back home.

schools. However, when used selectively by trained teachers in middle schools, according to research, computers can significantly enhance performance in math learning.[21] In art classes, computers can be used to encourage creativity without mess, as children learn to make art digitally.[22]

Thousands of schoolchildren have tapped into a Web site by a Wake Forest University biology researcher that tracks albatrosses across the Pacific Ocean. The Web site offers the students a chance to test their own hypotheses about the tracking date.[23] Students exposed to the Internet in high school say they think the Web has helped them improve the quality of their academic research and of their written work.[24] There is even software such as Intelligent Essay Assessor and E-Rater that grades essays, although its effectiveness is still debated.[25]

One revolution in education—before, during, and after the college years—is the advent of distance learning, or "cyberclasses," along with the explosion of Internet resources. The home-schooling movement, for example, has come of age, thanks to Internet resources.[26] Many colleges—both individually and in associations such as that of the Western Governors University (backed by 17 states and Guam) and the Community College Distance Learning Network—are offering a variety of Internet and/or video-based online courses.[27] Corporations are also offering training classes via the Internet or corporate intranet.[28]

10.4 Health, Medicine, & Science

KEY QUESTION

How is information technology being used in health, medicine, and science?

Treatment for acrophobia

Barbara Teixera suffers from acrophobia, the great fear of heights. Using 3-D computer simulation, she peeks over the edge of a virtual Golden Gate Bridge in San Francisco at the water 250 feet below. After practicing in this simulated environment, Teixera was able to step out onto the real thing.

Viktor Yazykov, competing in the perilous Around Alone solo sailing competition, found himself in the stormy South Atlantic with a seriously infected arm that needed emergency surgery. So, with the help of step-by-step instructions sent by e-mail from Boston-based Dr. Daniel Carlin to his solar-powered laptop computer, Yazykov operated on his own arm.[29]

Here we have a dramatic example of *telemedicine*—medical care delivered via telecommunications. For some time, physicians in rural areas lacking local access to radiologists have used "teleradiology" to exchange digital images such as X-rays via telephone-linked networks with expert physicians in metropolitan areas. Now telemedicine is moving to an exciting new level, as the use of digital cameras and sound, in effect, moves patients to doctors rather than the reverse. Already telemedicine is being embraced by administrators in the American prison system, where by law inmates are guaranteed medical treatment—and where the 8% increase in prisoners every year has led to the need to control health costs.[30]

Computer technology is also radically changing the tools of medicine. "All medical information—X-ray or lab test or blood pressure or pulse monitor—

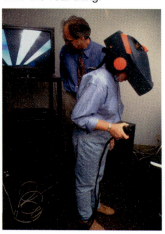

OR output

A technician monitors the heart rate and blood pressure of a patient undergoing surgery in a hospital operating room.

can now be digitized," says surgeon Rick Satava. "It can all be brought to the physician or surgeon in digital format."[31] Although only a small percentage of such information is digitized at present, some breakthroughs are producing results that would have been science fiction only 10 years ago. Software is now able to compute a woman's breast cancer risk.[32] Mental-health researchers are using computers to screen troubled teenagers in need of psychotherapy.[33] And psychotherapy that is done online has been found to provide short-term relief.[34] Epidemiologists are using the Web to improve public health in developing nations.[35] MIT has developed a "micropill," a pharmacy on a chip that can be implanted in the body to deliver tiny amounts of medicine on a controlled-release basis.[36] Hydraulics and computers are being used to help artificial limbs get "smarter."[37] And a patient paralyzed by a stroke has received an implant that allows communication between his brain and a computer; as a result, he can move a cursor across a screen by brainpower and convey simple messages—as in "Star Trek."[38]

Besides physicians, patients and health consumers are also going online. Although online health information can be misleading and even dangerous, growing numbers of consumers are seeking medical advice on the Internet—by tapping into health-care databases, e-mailing health professionals, or communicating with people who have similar conditions in online chat rooms. For instance, hours after 10-year-old Robert Lord of San Diego fractured his spine in a fall from a tree, his father found an experimental drug on the Internet, saving the boy from lifetime paralysis.[39] Often patients are already steeped in information about their conditions when they arrive in the offices of health-care professionals. "It's a fundamental shift of knowledge, and therefore power, from physicians to patients,"says one consultant.[40] In addition, health-care consumers are able to share experiences and information with each other. Inquisitive mothers-to-be, for example, can find a gathering spot at the Pregnancy Today Web site.[41]

Scientists have long been users of information technology. Because science is so vast, we can only sketch a couple of instances here—looking to the future in space and to the past beneath the earth. We also show how even an amateur can use information technology to advance environmental science.

A new adventure among scientists is the idea of a "collaboratory," an Internet-based collaborative laboratory, in which researchers all over the world can work easily together even at a distance. An example is the Space Physics & Aeronomy Research Collaboratory, based at the University of Michigan at Ann Arbor. This arrangement allows space physicists to band together to measure the earth's ionosphere from instruments on four satellites and in Massachusetts, Norway, Puerto Rico, and Peru; a supercomputer processes the information, and the results can be transmitted at once to scientists in many places.[42] The Puerto Rico observatory is also participating in the Serendip program at the University of California at Berkeley, which is analyzing radio signals from the cosmos for evidence of intelligent life. Part of a project known as "SETI@home" (SETI stands for Search for Extraterrestrial Intelligence), Serendip is enlisting the help of thousands of home computer users to help crunch the numbers. "The plan is to give everybody a different chunk of the sky," said the director of the Berkeley program.[43]

On an entirely different scientific front, archaeologists are using ultrasensitive underwater sensors and satellite tracking to find the fabled Queen Cleopatra's palace in the waters off the Egyptian city of Alexandria; the palace sank beneath the waves after an earthquake 1600 years ago.[44] Archaeologists are also using ground-penetrating radar and other remote-sensing tools coupled to software and high-powered computers to find underground

Swallow this

MIT student Bradley Geilfuss prepares to swallow a radio transmitter that will send data about his metabolism via wireless modem to fellow students during his participation in a San Francisco marathon.

objects and the remains of old buildings, thereby avoiding invasive excavations of prehistoric burial sites.[45] And they are using digital cameras and infrared cameras to take pictures of American Indian rock art in Wyoming and 32,000-year-old cave rock paintings in France. The pictures will be shared in databases to create a permanent record of what may vanish because of erosion and vandalism.[46]

Whatever their feelings about the far-distant past or the unforeseeable future, even nonscientists should be concerned about environmental issues, since we all live in the world of the present. A great deal of information about the environment—from global warming to guidelines on "socially responsible travel"—is available on the World Wide Web.[47] For example, the Environmental Defense Fund's Web site called Scorecard combines 150 government and university databases, by means of which users can locate polluters in their community, compile pollution rankings, or research the dangers of common household products.[48]

10.5 Commerce & Money

KEY QUESTION

How is information technology being used in sales and marketing, retailing, banking, and stock trading?

Businesses clearly see the Internet as a way to enhance productivity and competitiveness. However, as we have already observed, the changes go well beyond this.

The thrust of the original Industrial Revolution was separation—to break work up into its component parts so as to permit mass production. The effect of computer networks in the Digital Revolution, however, is unification—to erase boundaries between company departments, suppliers, and customers.[49] Indeed, the parts of a company can now as easily be scattered around the globe as down the hall from one another. Thus, designs for a new product can be tested and exchanged with factories in remote locations. With information flowing faster, goods can be sent to market faster and inventories reduced. Says an officer of the Internet Society, "Increasingly you have people in a wide variety of professions collaborating in diverse ways in other places. The whole notion of 'the organization' becomes a blurry boundary around a set of people and information systems and enterprises."[50]

Some areas of business that are undergoing rapid change are sales and marketing, retailing, banking, and stock trading.

Sales & Marketing

In the old days, "you could earn a good living" as a salesperson if you could talk and read a price list, says Ted Urpens, the head of sales training for Caradon Everest, a British manufacturer of replacement windows. Now sales representatives not only need to be better educated and more knowledgeable about their customers' businesses but also must be comfortable with computer technology.

Virtual shopping

In Shanghai, China, the New World Department Store saves on dressing-room space by letting shoppers try on virtual clothes right on the sales floor. After their video image is captured on screen, customers can see themselves in any outfit in any color at the click of a mouse.

Caradon Everest, for example, equips its sales staff with laptops containing software that "configures" customized windows and calculates the prices on the spot, a process that was once handled by the company's technical people and took a week. "The company can also load on product images for multimedia presentations and training programs," says one account. "Using a digital camera, pictures of the customer's house can be loaded into the computer, which can superimpose the company's windows and print out a color preview."[51]

PRACTICAL ACTION BOX

Tips for Dealing with Online Auctions[a]

There are generally two types of auction sites: (1) *Person-to-person auctions,* such as eBay, connect buyers and sellers for a listing fee and a commission on sold items. (2) *Vendor-based auctions,* such as Bid.com, Surplus Auction, and OnSale, buy merchandise and sell it at discount. Some sites are shown in the accompanying table.[b]

To become a bidder, you register with a site, providing your name, address, and sometimes a credit-card number. Sellers provide a description or scanned photo of what they want to sell and name a starting bid. Buyers submit bids over periods of time ranging from hours to weeks, and the highest bidder wins. Buyer and seller then arrange for payment and delivery. The auction site, which catalogues the merchandise and may provide a search engine to find items, then collects a percentage. For instance, eBay has listing fees up to $2.00 and seller's commissions ranging from 1.25% to 5%.

What's to prevent ripoffs? Actually, complaints about online auction fraud have exploded (from 107 in 1997 to 10,700 in 1999, according to the Federal Trade Commission). More than 80% of the complaints related to the failure of the seller to deliver the item to the buyer.[c] Here are some precautions to take:[d]

- **Know the seller:** Many auction sites offer a profile of sellers based on comments from other buyers. Sellers who receive too many complaints are barred from the site. If the seller is a company, you can contact the Better Business Bureau to check it out.

- **Pay by credit card, if you can:** Credit-card users by law are allowed to dispute charges for goods or services that they never received or that were misrepresented. You have to dispute the charge within 60 days of it appearing on your credit-card statement.

- **Get an insurance policy:** Most items on eBay, for instance, are covered by insurance for up to $200, excluding the $25 deductible.

- **Use an online escrow service:** For items worth more than $100, after you've made a deal, consider using an online escrow service to act as an impartial third party that holds funds until the transaction is complete. Escrow services generally charge about 5% of the transaction (and the fee is not refundable). Thus, if you're a buyer, you would send your money to the escrow service (or let them charge your credit card). The seller would ship you merchandise by a reputable method, such as UPS or FedEx. If you're satisfied with the goods, you then authorize the escrow service to release the funds.

- **Document everything about the transaction:** Save the auction listing, e-mail correspondence, credit-card receipts or cancelled checks, and phone bills and faxes.

Some online auction sites

Site and Web Address	Comments
Auction Universe www.auctionuniverse.com	Strong in collectibles, especially sports items
Buy Collectibles www.buycollectibles.com	Has a searchable database
Collector Online www.collectoronline.com	Matches buyers and sellers
eBay www.ebay.com	The largest Web auction site—12 million registered users
Egghead.com www.egghead.com	Office supplies and equipment
eHammer www.ehammer.com	Antiques by established antique dealers
Interactive Collector www.icollector.com	International dealers
uAuction.com www.uauction.com	Wide variety
Yahoo!Auctions Auctions.yahoo.com	Strong in toys and collectibles; teamed with auctioneer Butterfield & Butterfield

Other companies are also taking advantage of information technology in marketing. An Atlanta company called The Mattress Firm, for instance, uses geographical information systems (GIS) software called MapLinx that plots drop-off points for a driver's most efficient delivery of mattresses to customers. Moreover, MapLinx depicts sales by neighborhood, showing owner Darin Lewin where his mattresses are selling well and thus where he should spend money to market them. "We can pinpoint our customer," says Lewin.[52]

The Internet and World Wide Web have also, of course, become popular marketing tools. According to Forester Research, online sales, along with the cost of tools necessary to generate and support them, will constitute a $2 trillion annual market in 2003.[53] For example, Lou Ann Hammond's company, Car-List Inc., of San Francisco, uses telecommunications links to provide customers with access to a database describing features, prices, and availability of new and used cars at selected dealerships and among individual sellers. A similar service, Auto-By-Tel, in Irvine, California, which is free to consumers, slashes the marketing cost to $25 per car—as compared to the average of $425 and 3–5 salespeople that is required to market a car the old-fashioned way. "At traditional dealerships you talk to lots of people," says president Peter Ellis. "Through Auto-By-Tel, you're already predisposed to buying so all the dealer does is write up the contract."[54]

Retailing Becomes E-tailing

Are there any limitations to what can be sold on the Internet? Computers, books, CDs, and airline tickets seem obvious. But what about groceries?

"You can't exactly sniff cantaloupes over your modem," points out one writer.[55] Even so, it's predicted that, by 2007, online grocery shopping will be a $60 billion to $85 billion business, reaching some 15% of U.S. households.[56] Two pioneer online supermarkets, Peapod and NetGrocer, have been joined by HomeGrocer.com (partly owned by Amazon.com), Webvan, and WholeFoods.com.

The challenges for all these Internet food ventures are daunting: Even designing a Web site shopping menu can be difficult, because the profusion of groceries makes them hard to categorize. Figuring out how to deliver items, particularly perishable ones (drop off at customer's workplace? put in refrigerated box in the garage?), is one of the biggest headaches. And many people don't want to pay surcharges to support grocery delivery. Even filling orders can be an expensive task, since most online grocers employ clerks to wheel carts through a warehouse and fill individual customer orders. Webvan has tried to solve this problem by building a mechanized warehouse in which machines zip groceries around so that a single worker can fill several orders.[57]

Handheld ATM

In the remote Australian outback, more than 200 miles away from the nearest bank, 8-month-old Shanley Malbunka watches as his mother Celina withdraws money from her bank account. Handheld automated teller machines enable the aboriginal residents in the town of Yuendumu to conduct their banking electronically.

Banking & E-money

"This is the beginning of the end of cash," says San Francisco secretary Valerie Baptiste, as she hands over her smart card—plastic embedded with a microchip containing stored-up funds like an electronic purse—to pay for her morning coffee and bagel. The cashier inserts the smart card into a machine that, in less than five seconds, deducts $2.15 from the stored value. Later, when Baptiste finds her card running low, she can stick it into a kind of modem and dial up her bank account to "refuel" the card. [58]

The world of cybercash has come to banking—not only smart cards, but Internet banking, electronic deposits, electronic bill paying, online stock and bond trading, and online insurance shopping.

Some banks are backing an electronic-payment system that will allow Internet users to buy tiny online goods and services with "micropayments" of as little as 25 cents, from participating merchants. For instance, publishers could charge buyers a quarter to buy an article or listen to a song online, a small transaction that until now has not been practical. "It allows you to buy things by the sip rather than the gulp," says futurist Paul Saffo. [59]

More than 1000 banks now have World Wide Web sites, offering services that include account access, funds transfer, bill payment, loan and credit card applications, and investments. In one interactive application, you can apply for a Visa card called NextCard online and get approved (or turned down) within about two minutes. The system then shows you the account balances on your other credit cards so that, if you wish, you can transfer your balances to NextCard and benefit from a lower interest rate. [60]

Only about 46% of U.S. workers have their paychecks electronically deposited into their bank accounts (as opposed to 95% or more in Japan, Norway, and Germany, for example), but this is sure to change as Americans discover that direct deposit is actually safer and faster. [61] Getting them used to electronic bill paying will probably take longer, but that day too is arriving. For more than two decades, it has been possible to pay your bills online with special software and direct modem hookups to your bank, but only about 3% of U.S. households have tried it. The main delay has been that scarcely any bills are delivered online, but now banks have convinced corporations to electronically deliver bills to them, which will then be sent to customers via the Web. [62]

Stock Trading

Computer technology is unquestionably changing the nature of stock trading. Only a few years ago, hardly anyone had the occupation of "day trader," an investor who relies on quick market fluctuations to turn a profit. Now, points out technology observer Denise Caruso, "anyone with a computer, a connection to the global network, and the requisite ironclad stomach for risk has the information, tools, and access to transaction systems required to play the stock market, a game that was once the purview of an elite few."[63] (Caruso is right that it takes a strong stomach to be a day trader; many have lost their shirts.)

Online trading has also changed the nature of trading institutions themselves. The nation's two largest stock exchanges—the New York Stock Exchange (NYSE) and the Nasdaq—have felt the pressure from ECNs, or electronic trading networks, with names such as Island and Instinet.[64] ECNs "offer investors a way to trade for a fraction of what it costs to trade on the NYSE or Nasdaq," explains one report. "Their secret is that computers post, match, and execute all trades. No specialists or market makers serve as middlemen, as they do on the floor of the NYSE or in Nasdaq trading rooms."[65] In response to this challenge, both exchanges have taken steps such as extending their trading day into the evening hours to accommodate investors on the West Coast.[66]

Competition from such electronic discount brokerage firms as E*trade, Ameritrade, Schwab Online, and DLJ Direct has forced full-service brokerage firms such as the 85-year-old Merrill Lynch & Co. to offer online trading. As a *USA Today* editorial observed, "With the Internet, investors can quickly amass boatloads of company information, pick stocks, and execute trades on their own and on the cheap. Who needs brokers, particularly when their advice can easily cost 100 times as much?"[67] But Allan Sloan, *Newsweek*'s Wall Street editor, begs to differ: "The more information that's available online, the more valuable competent advisers and middlemen become. . . . Someone has to filter the information—much of which may not be true— and do something useful with it."[68] Indeed, major scandals involving stock fraud show that anyone considering buying securities online should make sure the seller is trustworthy.

CONCEPT CHECK

Describe the benefits of an intelligent agent.

What are some uses of computers in education, health, medicine, and science?

Discuss some uses of information technology in commerce and money.

10.6 Entertainment & the Arts

KEY QUESTION

How is information technology being used in music and movies?

Information technology is being used for all kinds of entertainment, ranging from video games to tele-gambling. It is also being used in the arts, from painting to photography. Let's consider just two examples, music and film.

Music

Yamaha's Disklavier GranTouch looks like an acoustic grand piano, and the keyboard action is the same as in concert grands. However, this new instrument has no strings and hence no heavy iron frame to support them.[69] The sounds themselves are drawn from a huge 30-megabyte database of digitized sounds. Because the sounds are synthesized, the Disklavier GranTouch never needs tuning. In addition, it can summon the voices of more than 700 instruments, from harp to soprano sax.

More profoundly, the World Wide Web is standing the system of music recording and distribution on its head—and, in the process, is changing the financial underpinnings of the music industry. As we mentioned earlier, Internet retailers compete with brick-and-mortar record stores by providing easy online shopping for music CDs. Spinner Networks, owned by America

Indoor year-round skiing

This indoor computer-driven "winter" sports facility—the largest in the world—enables skiers to ski all year. Located outside Tokyo, the system uses microprocessors to keep lifts running, snow falling, and temperature at 26 degrees Fahrenheit.

Ethics

Online, "broadcasts" about 2 million popular songs each day through the Net.[70]

Not only do Web sites offer guitar chords and sheet music, but promotional sites feature signed and unsigned artists, both garage bands and established professional musicians.[71] There are legal MP3 sites on the Internet but also many illegal ones, which operate in violation of copyright laws protecting ownership of music. Many of these will probably cease to exist as the recording industry develops new Web technologies that frustrate the pirating of music and instead offer fee-based services.[72]

Some musicians, dissatisfied with the current structure of the industry, see the Internet as a force that can democratize the market, bypassing record companies and radio stations and offering music directly to listeners.[73] Many people who avoid buying CDs by using Napster or other file-sharing software to download music as MP3 files think they are striking a blow against greedy record companies.[74] But if these companies are destroyed by changing economics, will buyers have lost the crucial function of *winnowing*? "With the demise of business as usual," writes one critic, "every garage band in existence will have equal standing in the undifferentiated mass of millions of titles thrown up on the Web. Yes, downloads will be inexpensive, but how will one find the good stuff?"[75]

Movies

"In computers, we know there are going to continue to be bazillions of jobs," says Jan Millsapps, chair of the cinema department at San Francisco State University, where cinema arts programs lead to degrees in multimedia and animation. "But artists are expanding the possibilities. . . . More and more the art of these students will be the movie stars of the future."[76]

Now that blockbuster movies routinely meld live action and animation, computer artists are in big demand. *Star Wars: Episode I*, for instance, had fully 1965 digital shots out of about 2200 shots. Even when film was used, it was scanned into computers to be tweaked with animated effects, lighting, and the like. Entire beings were created on computers by artists working on designs developed by producer George Lucas and his chief artist.[77]

What is driving the demand for computer artists? One factor is that animation, though not cheap, looks more and more like a bargain, because hiring movie actors costs so much—some make $20 million a film. Moreover, special effects are readily understood by audiences in other countries, and major studios increasingly count on revenues from foreign markets to make a film profitable. Digital manipulation also allows a crowd of extras to be multiplied into an army of thousands. Computer techniques have even been used to develop digitally created actors—called "synthespians." The late John

Wayne was recruited for a Coors beer ad.[78] With the soaring demand for computer-generated imagery—not only for movies but also for TV ads and video games—colleges and trade schools are expanding their digital-animation training programs.

But animation is not the only area in which computers are revolutionizing movies. Digital editing has radically transformed the way films are assembled. Says one report, "Traditional film editing was always a funky, hands-on proposition: reeling and unreeling spools of film, cutting and gluing pieces of celluloid together, working amid a sea of film that sometimes got trampled underfoot." Today, by contrast, "150 miles of film can be stored on hard drives and an editor with the press of a key or the click of a mouse can instantly access any visual or audio moment in the film."[79] Thus, hundreds of variations of a scene can be stored and called up for review and comparison.

Even nonprofessionals can get into movie making as new computer-related products come to market. Now that digital video capture and editing systems are available for under $1000, amateurs can turn home videotapes into digital data and edit them. Also, digital camcorders, which offer outstanding picture and sound quality, are dropping in price.

10.7 Government & Electronic Democracy

KEY QUESTION

What benefits does information technology offer government and democracy?

"In general, the Internet is being defined not by its civic or political content," says one writer, "but by merchandising and entertainment, much like television."[80] Even so, a Rutgers University study suggests that the Internet has great potential for civic betterment because it is free of government intrusion, is fast and cheap for users (once connected), and facilitates communication among citizens better than mass media such as radio and TV.[81]

The potential has been demonstrated since 1989 by an experiment in online democracy called the Public Electronic Network (PEN) operated by the city of Santa Monica, California. In its first year, when the major topic on PEN was homelessness, Donald Paschal, himself homeless, logged on from a public terminal and said he would like to get a job but was dirty, had nowhere to wash, and nowhere to store his clothing. Out of the subsequent online discussions came a program called Swashlock, for "showers, washers, and lockers." Today 50% of Santa Monica's households have Internet access, and one in 10 residents is a registered user of PEN.[82]

Similar strides have been made elsewhere. Some cities have adopted Neighborhood Link, a free, easy-to-use system of neighborhood Web sites in which residents can communicate among themselves and with local governments. In Denver, for instance, the system serves 1186 neighborhoods. In Austin, Texas, an entrepreneur formed E-The People, which describes itself as "America's Interactive Town Hall" and is designed to connect citizens with their government officials, local and national. In Nevada, citizens visit the state legislature's hearings and floor voting sessions by accessing the legislators' Web site, where they can either listen to Internet broadcasts or read the text of legislation.

If we can shop online, could we not vote online for political candidates? Ten states have approved testing of online voting. In 1999, voters in the small city of Piedmont, California, were the first in the state to vote by "touchscreen" technology at their polling sites; the results were tallied in 29 minutes instead of the usual 3 hours. In 2000, 86,000 citizens in Arizona also voted in that state's Democratic presidential primary via the Internet.[83]

As you might expect, politicians and political candidates have found information technology to be of considerable use. Computer databases, for instance, enable the two major political parties and their allies to lock on targets with unprecedented precision. In some cases, online surveys have taken the place of phone polls, although there is some doubt as to their accuracy.

Politicians also do online campaigning, even announcing their candidacies on the Internet. Every presidential candidate "has to have a Web site," says one research professor. "If you don't, it's like not having a telephone number."[84] In summer 2000, President Clinton announced that the government was consolidating its 20,000 Web sites into a single Internet location: *www.first-gov.gov.*[85] (If you're a student interested in information about financial aid, however, you can still go directly to *www.students.gov.*)

Government itself has been deeply involved with computers from the beginning. Some noteworthy examples: You can have a tax preparer file your federal tax return online. You can renew your passport online. Eighteen-year-old males can register with the Selective Service online. The Savings Bond Marketing Office sells savings bonds online, which you can charge to your credit card.

The market for e-government is huge. Local governments collect $450 billion annually in fines alone, mostly in person or by mail.[86] Linking citizens to city hall electronically could provide 24-hour-a-day, seven-day-a-week solutions to such matters as paying traffic citations and property taxes, getting building permits, renewing driver's licenses, and digging up deeds. Two Internet companies, govWorks.com and ezgov.com, have established links to various local governments to help them make information available to citizens.[87]

CONCEPT CHECK

What are some uses of information technology in health, medicine, and science?

How are computers being used in music and movies?

Describe some ways computers are being used in the civic realm.

10.8 Jobs & Careers

KEY QUESTION

How can you use the Internet to find a job, and what are some hot jobs in information technology?

It's 3 A.M. Even at this hour you can, if you wish, find job-search advice, tips on interviewing and résumé writing, and postings of employment opportunities around the world. For example, you can direct your Web browser to a portal such as Yahoo! (*www.yahoo.com*) to obtain a list of popular Web sites. In the menu, you can click on Business, then Employment, then Jobs. This will bring up a list of sites that offer career advice, résumé postings, job listings, research about specific companies, and other services. (Caution: As might be expected, there is also a fair amount of junk out there: get-rich-quick offers, résumé-preparation firms, and other attempts to separate you from your money.)

Advice about careers, occupational trends, employment laws, and job hunting is also available through on-line chat groups, and other portals—America Online, Alta Vista, Microsoft Network, and Prodigy.

Ways for You to Find Employers

As you might expect, the first to use cyberspace as a job bazaar were companies seeking people with technical backgrounds and technical people seeking employment. However, as the public's interest in commercial services and the Internet has exploded, the focus of online job exchanges has broadened. Now, says one writer, "interspersed among all the ads for programmers on the Internet are openings for English teachers in China, forest rangers in New York, physical therapists in Atlanta, and models in Florida."[88] Most Web sites are free to job seekers, although many require you to fill out an online registration form. *(See Panel ● 10.5.)*

America's Job Bank: *www.ajb.dni.us.*

Career Builder: *www.careerbuilder.com*

Career Mosaic: *www.careermosaic.com*

CareerPath.com: *www.careerpath.com*

College Grad Job Hunter: *www.collegegrad.com*

Cruel World: *www.cruelworld.com*

Dice.com: *www.dice.com*

Erecruiting: *www.erecruiting.com*

4Work.com: *www.4work.com*

FedWorld (U.S. Government jobs): *www.fedworld.gov*

Headhunter: *www.headhunter.net*

Hot Jobs: *www.hotjobs.com*

Jobs.com: *www.jobs.com*

Job-Hunt.Org: *www.job-hunt.org*

Jobs on Line: *www.jobsonline.com*

JobTrak.com: *www.jobtrak.com*

JobWeb: *www.jobweb.org*

Monster.com: *www.monster.com*

NationJob Network: *www.nationjob.com*

Ways for Employers to Find You

Posting your résumé on line for prospective employers to view is attractive because of its low (or zero) cost and wide reach. But does it have any disadvantages? Certainly it might if the employer who sees your posting happens to be the one you're already working for. In addition, you have to be aware that you lose control over anything broadcast into cyberspace. You're putting your credentials out there for the whole world to see, and you need to be somewhat concerned about who might gain access to them.

If you have a technical background, it's definitely worth posting your résumé with an electronic jobs registry, since technology companies in particular find this an efficient way of screening and hiring. However, it may also benefit people with less technical backgrounds. Online recruitment "is popular with companies because it pre-screens applicants for at least basic computer skills," says one writer. "Anyone who can master the Internet is likely to know something about word processing, spreadsheets, or database searches, knowledge required in most good jobs these days."[89]

The latest wrinkle in job seeking is to prepare a résumé with hypertext links and/or clever graphics and multimedia effects and then put it on a Web site to entice employers to chase after you. If you don't know how to do this, there are many companies that—for a fee—can convert your résumé to HTML and publish it on their own Web sites. Some of these services can't dress it up with fancy graphics or multimedia, but since complex pages take longer for employers to download anyway, the extra pizzazz is probably not worth the effort. A number of Web sites allow you to post your résumé for free. (See ● *Panel 10.6, next page.*)

Companies are also beginning to replace their recruiters' campus visits with online interviewing. For example, the firm VIEWnet Inc. of Madison, Wisconsin, offers first-round screenings or interviews for summer internships through its teleconferencing "InterVIEW" technology, which allows video signals to be transmitted via telephone lines.

Further information about online career strategies is given in the Experience Box at the end of this chapter.

What Do You Want to Do?

Even low-tech people can find high-tech jobs—positions that don't require a technological background in fields such as human resources, product marketing, and contract management. But high tech leads the employment parade, offering the most opportunities—"jobs that are fun, well-paying, and with benefits galore," as career writer Belle Wise puts it.[90] The five jobs projected to have the largest percentage increase from 1998 to 2008, according to the Bureau of Labor Statistics, are as follows.

● PANEL 10.6
Sites on which you can post your résumé

General

America's Job Bank: *www.ajb.dni.us*

Career Magazine: *www.careermag.com*

Career Mosaic: *www.careermosaic.com*

CareerPath.com: *www.careerpath.com*

College Grad Job Hunter: *www.collegegrad.com*

The Employment Guide's CareerWeb: *www.cweb.com*

4Work: *www.4work.com*

Job-Hunt.Org: *www.job-hunt.org*

JobOptions: *www.joboptions.com*

JobTrak.com: *www.jobtrak.com/jobguide*

Monster.com: *www.monster.com*

The Occupational Outlook Handbook: *www.jobweb.org/occhandb.htm*

Specialized

eAttorney (jobs in field of law): *www.attorneysatwork.com*

EngineeringJobs.com: *www.engineeringjobs.com*

FINANCIALjobs.com (accounting and finance jobs): *www.financialjobs.com*

Health Search USA (health and medicine): *www.healthsearchusa.com*

Industrial Light & Magic (George Lucas): *www.ilm-jobs.com*

- Computer engineers—108% increase
- Computer support specialists—102% increase
- Systems analysts—94% increase
- Database administrators—77% increase
- Desktop publishing specialists—73% increase

According to chief information officers at companies with more than 100 employees, the information technology skills most in demand are: Internet/intranet development (23%), networking (21%), help desk/end-user support (14%), and applications development (9%).[91]

All these hot jobs, those with the fastest employment growth, require a good education. But other careers that you might want to pursue in the Knowledge Age probably demand a solid—and perpetual—education as well.

"Knowledge is the hot ticket now," says Karyne Conley, a vice president of SBC Communications. "Although it's hard to stay on top of all the new information, it seems I at least have to try. Really, there aren't many excuses left not to."[92]

CONCEPT CHECK

In what ways can you find employers?

In what ways can employers find you?

What are the fastest-growing jobs in information technology?

Experience Box
Career Strategies for the Digital Age

"The average person will go job hunting *eight* times in his or her life," points out Richard Bolles, author of the best-selling job-hunting book *What Color Is Your Parachute?*[93] Thus, you will need to train for the task of *finding and getting* a job as much as for the ability to do the job itself.

For one thing, you need to know résumé databases work. In a high-tech résumé-scanning system such as Resumix, an optical scanner inputs hundreds of pages of résumés a day, compiling a computerized database. The system can search for up to 60 key factors, such as job titles, technical expertise, education, geographic location, and employment history. Resumix can also track race, religion, gender, and other factors to help companies diversify their workforce. These descriptors are matched against available openings.

Résumé scanners allow organizations with vacancies to more efficiently search their existing pool of applicants before turning to advertising or executive-search ("head-hunter") firms. For applicants, however, résumé banks and other electronic systems have turned job hunting into a whole new ball game.

Writing a Computer-Friendly & Recruiter-Friendly Résumé

Some of the old rules for presenting yourself in a résumé might no longer benefit you at all. The latest advice is as follows.

Use the Right Paper & Print In the past, job seekers have used tricks such as colored paper and fancy typefaces in their résumés to try to catch a bored personnel officer's eye. However, optical scanners have trouble reading type on colored or gray paper and are confused by unusual typefaces. They even have difficulty reading underlining and poor-quality dot-matrix printing.[94] Thus, you need to be aware of new format rules for résumé writing. *(See ● Panel 10.7.)*

Use Keywords for Skills or Attributes Just as important as the format of a résumé today are the *words* used in it. In the past, résumé writers tried to clearly present their skills. Now it's necessary to use as many of the buzzwords or keywords of your profession or industry as you can.

Action words ("managed," "created," "developed") should still be used, but they are less important than nouns—job titles, capabilities, languages spoken, type of degree, and the like ("sales representative," "Spanish," "Unix," "B.A."). The reason, of course, is that a computer will scan for keywords applicable to the job that is to be filled.

Because résumé-screening programs sort and rank the number of keywords found, those with the most rise to the top of the electronic pile. Consequently, you should pack your résumé with every keyword that applies to you—especially those of the sort that appear in help-wanted ads.[95]

If you are looking for a job in desktop publishing, for instance, a number of specific keywords will make you stand out, such as *Adobe Illustrator, Pagemaker, PhotoShop, Quark.*

Sending Résumés as E-Mail Don't send your résumé as an e-mail attachment, which recipients—particularly if they get 100 résumés a day—will find tedious to download. Rather, make the résumé part of the main body of the e-mail message, with the first page as a cover letter (see below, "Writing a Good Cover Letter"). Note that many formatting features—indentation, boldface, italics—might not show up in an e-mail message.

Make the Résumé Impress People, Too Your résumé shouldn't just be pages of keywords. It has to impress a human recruiter, too, who may still have some fairly traditional ideas about résumés. Moreover, many recipients find Internet résumés less personal.

Here are some tips for organizing résumés, from reporter Kathleen Pender, who interviewed numerous professional résumé writers.[96]

- **The beginning:** Start with your name, address, and phone number. (These days you should also add your fax number, e-mail address, and your Web site address, if you have one.)

Follow with a clear objective stating what you want to do. (Example: "Sales representative in computer industry.")

Under the heading "Summary" give three compelling reasons why you are the ideal person for the job. (Example of one line: "Experienced sales representative to corporations and small businesses.")

After the beginning, your résumé can follow either a *chronological* format or a *functional* format.

- **The chronological résumé:** The chronological résumé works best for people who have stayed in the same line of work and moved steadily upward, with no gaps in work history. Start with your most recent job and work backwards, and say more about your recent jobs than earlier ones.

  The format is to list the years you worked at each place down one side of the page. Opposite indicate your job title, employer name, and a few of your accomplishments. Omit accomplishments that have nothing to do with the job you're applying for.

- **The functional résumé:** The functional résumé works best for people who are changing careers, or re-entering the job market. It may also suit people who need to emphasize skills from earlier in their careers or who want to emphasize their volunteer experience. It's recommended, too, for people who have had responsibility but never an important job title.

  The format is to highlight the skills, then follow with a brief chronological work history featuring dates, job titles, and employer names.

- **The conclusion:** Both types of résumés should have a concluding section showing college, degree, and graduation date; professional credentials or licenses; and professional affiliations and awards if relevant.

- **The biggest mistakes on résumés:** The biggest mistake is to lie. Sooner or later a lie will catch up with you and may get you fired—or even sued.

  The second mistake is sloppy spelling. Spelling mistakes communicate to prospective employers a basic carelessness.

Writing a Good Cover Letter Write a targeted cover letter to accompany your résumé. This is especially important if you're responding to an ad.

Most people, say San Francisco employment experts Howard Bennett and Chuck McFadden, "tend to talk about what *they* are looking for in a job. This is a major turn-off for employers."[97] Employers don't care very much about your dreams and aspirations, only about finding the best candidate for the job.

Bennett and McFadden suggest the following strategy for a cover letter:

- **Emphasize how you will meet the employer's needs:** Employers advertise because they have needs to be met. "You will get much more attention," say Bennett and McFadden, "if you demonstrate your ability to fill those needs."

  How do you find out what those needs are? *You read the ad.* By reading the ad closely you can find out how the company talks about itself. You can also discover what attributes it is looking for in employees and what the needs are for the particular position. (Incidentally, in any cover letter, *be sure to specify the job for which you are applying.* Don't say "I saw your job posting on the Internet." The company may be listing more than one job.)

- **Use the language of the ad:** In your cover letter, use as much of the ad's language as you can. "Use the same words as much as possible," advise Bennett and McFadden. "Feed the company's language back to them." The effect of this will be to produce "an almost subliminal realization in the company that you are the person they've been looking for."

- **Take care with the format of the letter:** Keep the letter to one page and use bullets or dashes to emphasize the areas where you meet the needs described in the ad. Make sure the sentences read well and—very important—that no word or name is misspelled.

The intent of both cover letter and résumé is to get you an interview, which means you are in the top 10–15% of candidates. A different set of skills is needed for the interview; you should research these on your own. Richard Bolles suggests that, aside from looking clean and well-groomed, you need to tell the employer what distinguishes you from the 20 other candidates. "If you say you are a very thorough person, don't just say it," suggests Bolles. "Demonstrate it by telling them what you know about their company, which you learned beforehand by doing your homework."[98]

Visual Summary

artificial intelligence (AI) (p. 380, KQ 10.2) Group of related technologies used for developing machines to emulate human-like qualities, such as learning, reasoning, communicating, seeing, and hearing. Why it's important: *Today the main areas of AI are virtual reality, robotics, natural language processing, fuzzy logic, expert systems, neural networks, and genetic algorithms, plus artificial life.*

artificial life (p. 386, KQ 10.2) Also called *A-life;* field of study concerned with "creatures"—computer instructions, or pure information—that are created, replicate, evolve, and die as if they were living organisms. Why it's important: *A-life software (such as LIFE) tries to simulate the responses of a human being.*

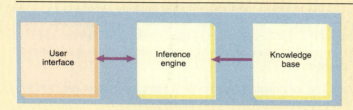

expert system (p. 383, KQ 10.2) Interactive computer program that helps solve problems that would otherwise require the assistance of a human expert. Fundamental to an expert system is a knowledge base constructed by experts that can "learn" by adding new knowledge. Why it's important: *Expert systems are designed to be users' assistants, not replacements—to help them more easily perform their jobs.*

fuzzy logic (p. 383, KQ 10.2) Method of dealing with imprecise data and uncertainty, with problems that have many answers rather than one. Why it's important: *Unlike traditional "crisp," yes/no digital logic, fuzzy logic deals with probability and credibility.*

genetic algorithm (p. 385, KQ 10.2) Program that uses Darwinian principles of random mutation to improve itself. Why it's important: *Genetic algorithms use trial and error to learn from experience, thus constantly improving themselves.*

inference engine (p. 384, KQ 10.2) Software that controls the search of the expert system's knowledge base and produces conclusions. Why it's important: *An inference engine fits the user's problem into the knowledge base and derives a conclusion from the rules and facts it contains.*

intelligent agent (p. 388, KQ 10.3) Program that performs work tasks—such as roaming networks and compiling data—on the user's behalf. Why it's important: *A software agent is a kind of electronic assistant that will filter messages, scan news services, and perform similar secretarial chores. An agent will also travel over communications lines to computer databases, collecting files to add to a personalized database.*

knowledge base (p. 384, KQ 10.2) Expert system's database of knowledge about a particular subject. Why it's important: *A knowledge base includes relevant facts, information, beliefs, assumptions, and procedures for solving problems.*

natural language processing (p. 383, KQ 10.2) Study of ways for computers to recognize and understand human language, whether in spoken or written form. Why it's important: *Natural languages make it easier to work with computers.*

neural networks (p. 385, KQ 10.2) Artificial intelligence networks that use physical electronic devices or software to mimic the neurological structure of the human brain, with, for instance, transistors for nerve cells and resistors for synapses. Why it's important: *Neural networks are able to mimic human learning behavior and pattern recognition.*

robot (p. 382, KQ 10.2) Automatic device that performs functions ordinarily executed by human beings or that operates with what appears to be almost human intelligence. Why it's important: *Robots are performing more and more functions in business and the professions.*

robotics (p. 382, KQ 10.2) Development and study of machines that can perform work normally done by people. Why it's important: *See robot.*

simulator (p. 382, KQ 10.2) Device that represents the behavior of physical or abstract systems. Why it's important: *Virtual-reality simulation technologies are widely applied for training purposes.*

switched-network telecommunications model (p. 379, KQ 10.1) Telecommunications system in which people on the system are not only consumers of information ("content") but also possible providers of it. This is the model of the telephone system and of most computer networks, including the Internet. Why it's important: *Whereas the tree-and-branch allows only few-to-many communications, the switched-network allows many-to-many communications. Thus, the more the switched-network becomes prominent, the more participatory our systems of communications become.*

tree-and-branch telecommunications model (p. 379, KQ 10.1) Telecommunications system in which a centralized information provider sends out messages through many channels to thousands of consumers. This is the model of most mass media, such as radio and TV broadcasting. Why it's important: *Unlike the switched-network model, this model allows fewer participants in the system. This is the model envisioned by some cable and entertainment companies looking to provide movies and other services over the Internet from a centralized library or a core of "content providers."*

Turing test (p. 386, KQ 10.2) A test for determining whether a computer possesses "intelligence" or "self-awareness." In the Turing test, a human judge converses by means of a computer terminal with two entities hidden in another location. Why it's important: *Some experts believe that once a computer has passed the Turing test, it will be judged to have achieved a level of human intelligence.*

virtual reality (VR) (p. 381, KQ 10.2) Computer-generated artificial reality that projects a person into a sensation of three-dimensional space. Users need software and special headgear, gloves, and perhaps a special suit. Why it's important: *With virtual reality, users can experience almost anything they want without ever leaving their chairs. Virtual reality is employed in simulators for training programs of many types.*

Chapter Review

Self-Test Questions

1. A(n) _____ is an automatic device that performs functions ordinarily performed by human beings.

2. The goal of _____ is to enable the computer to communicate with the user in the user's native language.

3. In the _____ model, used by most mass media, a centralized information provider sends out messages through many channels to thousands of consumers; the _____ model, in contrast, is used by the telephone system and computer networks.

4. _____ is a group of related technologies used for developing machines to emulate human qualities such as learning, reasoning, communicating, seeing, and hearing.

5. _____ use physical electronic devices or software to mimic the neurological structure of the human brain.

6. A(n) _____ is a program that uses Darwinian principles of random mutation to improve itself.

7. Devices that represent the behavior of physical or abstract systems are called _____.

Multiple-Choice Questions

1. Which of the following are main areas of AI?
   a. genetic algorithms
   b. fuzzy logic
   c. robotics
   d. natural language processing
   e. all of the above

2. Which of the following are components of the Turing test?
   a. human judge
   b. A-life
   c. computer terminal
   d. software program
   e. human typing on keyboard

True/False Questions

T F 1. Expert systems are not interactive.

T F 2. You need only software to create virtual reality.

T F 3. A knowledge base is part of a natural language system.

T F 4. The user interface is the display screen that the user deals with.

Short-Answer Questions

1. What are intelligent agents used for?

2. What equipment do you need to experience virtual reality?

3. What is the Turing test?

4. What are the four main areas of artificial intelligence?

5. What is e-tailing? e-money?

Concept Mapping

On a separate sheet of paper, draw a concept map, or visual diagram, linking concepts. Show how the following terms are related.

AI	neural networks
artificial life	robot
expert system	robotics
fuzzy logic	simulator
genetic algorithm	switched-network model
inference engine	tree-and-branch model
intelligent agent	Turing test
knowledge base	virtual reality
natural language processing	

Knowledge in Action

1. If you could design a robot, what kind would you create? What would it do?

2. If you could build an expert system, what would it do? What kinds of questions would you ask experts in order to elicit the appropriate information?

Information Systems

Information Management & Systems Development

Key Questions

You should be able to answer the following questions.

11.1 Organizations, Managers, & Information What are the departments, tasks, and levels of managers in an organization, and what types of decisions do they make?

11.2 Computer-Based Information Systems What are the six computer-based information systems—and what are their purposes?

11.3 Systems Development: The Six Phases of Systems Analysis & Design What are the six phases of the systems development life cycle?

chapter

11

A Visual Overview of This Chapter

Graphical Interface

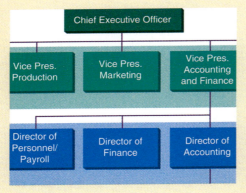

1 Organizations, Managers, & Information. To understand how information flows in an organization, we need to understand how organizations work. Information flows horizontally between the five departments of an organization: *research and development, production, marketing, accounting and finance,* and *human resources.* It also flows vertically between the layers of managements.

There are three levels of management corresponding to three kinds of decisions, as reflected in the **organization chart,** a schematic drawing showing the hierarchy of formal relationships among an organizations' employees. (1) **Top managers** are concerned with long-range, or strategic, planning and decisions. (2) **Middle-level managers** make tactical decisions to implement the strategic goals of the organization. (3) **Supervisory managers** make operational decisions, predictable decisions that can be made by following a well-defined set of routine procedures.

Information has three distinct properties: level of summarization, degree of accuracy, and timeliness. To make the appropriate decisions—strategic, tactical, operation—the different levels of managers need the right kind of information: structured, semistructured, and unstructured. **Structured information** is detailed, current, concerned with past events, records a narrow range of facts, and covers an organization's internal activities. **Unstructured information** is summarized, less current, concerned with future events, records a broad range of facts, and covers activities outside as well as inside an organization. **Semistructured information** includes some structured information and some unstructured information.

2 Computer-Based Information Systems. Six types of computer-based information systems provide managers with appropriate information for making decisions: (1) A **transaction processing system (TPS)** is used by supervisory managers to keep track of **transactions**—recorded events having to do with routine business activities—needed to conduct business. A TPS produces *detail reports,* which contain specific information about routine activities. (2) A **management information system (MIS)** is used by middle managers. An MIS uses data from a TPS to produce routine reports—*summary reports* to show totals and trends, *exception reports* to show out-of-the-ordinary data, *periodic reports* produced on a regular schedule, and *demand reports* to produce information in response to an unscheduled demand. (3) A **decision support system (DSS)** is also used by middle managers. A DSS provides **models**—mathematical representations of real systems—that gives managers a tool for analysis and helps them focus on the future. (4) An **executive support system (ESS)** is used by top managers to support strategic decision making. (5) An **office automation system (OAS)** is used by all levels of managers as well as nonmanagers. An OAS combines various technologies, such as word processing, scheduling software, e-mail, and the like, on a network to reduce the manual labor required in operating an efficient office. (6) An **expert system** helps users solve problems that would otherwise require the assistance of a human expert.

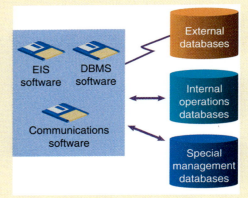

3 **Systems Development: The Six Phases of Systems Analysis & Design.** A powerful tool for helping organizations keep up with new information needs is systems analysis and design. In general, a **system** is a collection of related components that interact to perform a task in order to accomplish a goal. Participants in an information-system project should be users, managers, and technical staff, including **systems analysts,** information specialists who perform systems analysis, design, and implementation.

Systems analysis and design is a six-phase problem-solving procedure for examining an information system and improving it. The six phases make up the **systems development life cycle (SDLC),** the step-by-step process that organizations follow during systems analysis and design. The six steps are preliminary investigation followed by systems analysis, design, development, implementation, and maintenance.

(1) The objective of **preliminary investigation** is to conduct a preliminary analysis, propose alternative solutions, describe costs and benefits, and submit a preliminary plan with recommendations.

(2) The objective of **systems analysis** is to gather data, analyze the data, and write a report. Several tools are used to analyze the data. **Modeling tools** enable an analyst to present graphic representations of a system. **Data flow diagrams,** for example, graphically show the flow of data through a system.

(3) The objective of **systems design** is to do a **preliminary design,** which describes the general functional capabilities of a proposed information system; then do a **detail design,** which describes how the system will deliver the capabilities described in the preliminary design; and then to write a report. Tools used in the preliminary design are CASE tools and project management software. **CASE** (computer-aided software engineering) **tools** are programs that automate the various activities of the SDLC in several phases. **Prototyping** refers to using workstations, CASE tools, and other software applications to build working models of system components that can be quickly tested. A **prototype** is just such a limited working system developed to test out design concepts. *Project management software* consists of programs used to plan, schedule, and control the people, costs, and resources required to complete a project.

(4) The objective of **systems development** is to develop or acquire the software, acquire the hardware, and then test the system. In considering what software to acquire, the systems analyst must make a **make-or-buy decision**—decide whether to create a program or buy existing software.

(5) **Systems implementation** consists of converting the hardware, software, and files to the new system and training the users. Conversion to the next system may be by *direct implementation* (stop the old, start the new), *parallel implementation* (operate both old and new concurrently for a while), *phased implementation* (phase in new system in stages), or *pilot implementation* (try out new system by some users).

(6) **Systems maintenance** adjusts and improves the system by having system audits and periodic evaluations and by making changes based on new conditions.

That was what the old so-called 50-Foot Rule stated. But modern communications technology has now given us global collaboration—through what are known as virtual teams.

Virtual teams, according to experts Jessica Lipnack and Jeffrey Stamps, are groups of people who work closely together even though they are separated by space, time, and organizational barriers.[1]

Collaborative networks—combinations of local-area and wide-area communications networks—linking hundreds or thousands of people can allow businesses to form and dissolve clusters of workers on a moment's notice.

Forming a small virtual team—a handful people on the East and West coasts working on a print ad, for example—might seem relatively easy. But the NCR Corporation created a virtual team of more than a thousand people spread out over 17 locations to develop a new computer system. "Using a high-speed full-bandwidth audio/video/data link," says one account, "the virtual team completed the project on budget and ahead of schedule."[2]

With communications technology, space and time are no longer the biggest hurdles for virtual teams; the greatest ones are organizational—that is, the pyramid hierarchy and Main Street business norms like caution, continuity, and conservatism.

Part of the redefinition of the workplace comes with handing employees laptop computers, portable phones, and beepers and telling them to work from their homes, cars, or customers' offices. A grab bag of electronic information organizers, personal communicators, personal digital assistants, and similar gadgets helps untether people from a fixed office.

"Flex-time" shift hours and voluntary part-time telecommuting programs have been around for a few years. Unlike them, however, the new high-tech tools are forcing some profound changes in the way people work. Many people, of course, like the flexibility of a mobile office. However, others resent having to work at home or chafe at their ever-expanding work hours. One computer-company vice president worries about getting her staff to stop sending faxes to each other in the middle of the night. Some employees may work 90 hours a week and still feel they are falling short. In great part, this is because their managers' skills have not kept pace with the trend.[3] At some point, a lifestyle of constant work becomes counterproductive.

Are there ways to prepare ourselves for the restructuring of work? In this chapter, we describe the traditional shape and functions of an organization. We also discuss the layers of managers and their information needs. Next we show how computer-based information systems—TPSs, MISs, DSSs, and the like—can help managers make decisions. We then describe a strategy for rethinking business and information systems—specifically, the six-phase problem-solving procedure called systems analysis and design. The Appendix describes programming and programming languages.

At the heart of an organization is information and how it is used. To understand how to bring about change in an organization, we need to understand how organizations and their managers work—how they need, organize, and use information.

The Flow of Information within an Organization

Consider any sizable organization you are familiar with. Its purpose is to perform a service or deliver a product. If it's nonprofit, for example, it may deliver the service of educating students or the product of food for famine victims. If it's profit-oriented, it may, for example, sell the service of fixing computers or the product of computers themselves.

Information—whether computer-based or not—has to flow within an organization in a way that will help managers, and the organization, achieve their goals. To this end, organizations are often structured horizontally and vertically—horizontally to reflect functions and vertically to reflect management levels.

BOOKMARK IT!

PRACTICAL ACTION BOX
Too Much, Too Fast—Staying Focused to Avoid Information Overload

The Internet would seem perfectly suited to today's information consumer. But with the Web offering *1 billion* or more home pages, the technology presents all the possibilities for being simply a stupifying waste of time. Says well-known computer journalist John Dvorak, "the Web and the Net revolution have removed the natural barriers between us and the carloads of information we would normally never see."[a]

How, then, to keep yourself from being overwhelmed by information from the Internet—and everything else clamoring for your attention? Some suggestions:

- **Avoid "multitasking":** As we mentioned in Chapter 7, nearly four-fifths of Internet users spend five hours or more a week online "multitasking"—engaging in another activity while online: eating and drinking, listening to the radio, watching TV, talking on the phone, or chatting with others in the room.[b] But, experts say, just because we *can* do everything doesn't mean we have to—or should.[c] Studies show that trying to do two or more mentally demanding tasks (such as homework and something else) is actually self-defeating, burning up more total brainpower than if you did them one at a time. And studies also show that people are becoming less happy and more error-prone in the process.[d]

- **Eliminate the inessential:** It may seem cool to have a lot of electronic gadgets—PC, PDA, TV, VCR, CD player, pager, cell phone, fax machine, and so on. But what do we need? "Which machines really serve you?" asks productivity specialist Odette Pollar. "Adjust your attitude to all this abundance by making sure these machines conform to your pace."[e]

- **Stay focused on what you're trying to find:** If you're doing research on the Web, the biggest problem, especially for members of the channel-changing MTV generation, is *staying focused.* Dvorak gives this example of how following random links can be a time killer: "You begin your session looking for a coffee distributor, . . . and at that site, you see a link to a site on rare chocolates. You jump to that site, and before you know it, you've learned more than you ever wanted to know about the history of the chocolates. Enough already." Thus, remember why you're online in the first place.

- **Limit your time online:** Plan to go on the Web just for 45 minutes or an hour, and *stick with* the plan. If you don't, you'll find yourself up in front of the computer until 4 A.M.

- **Use your printer frequently:** When you see interesting material online, don't try to read it all then and there. Rather, use your computer's printer to print it out immediately for reading later. "That way," says Dvorak, "you can gather as much information as possible without spending all night online or chasing useless links."

- **Be ruthless about deciding what to read:** If you're deluged with e-mail, unsubscribe from mailing lists. If you're letting publications or printed-out articles pile up, decide what is essential for you to read. "Just because an item is interesting is not good enough," says Pollar. "Will you use it again in the next 3 months? If not, pass it by."[f]

Departments: R&D, Production, Marketing, Accounting, Human Resources

Depending on the services or products they provide, most organizations have departments that perform five functions: *research and development (R&D)*, *production, marketing, accounting and finance*, and *human resources (personnel)*.

- **Research and development:** The research and development (R&D) department does two things: (1) It conducts basic research, relating discoveries to the organization's current or new products. (2) It does product development and tests and modifies new products or services created by researchers. Special software programs are available to aid in these functions.

- **Production:** The production department makes the product or provides the service. In a manufacturing company, it takes the raw materials and has people or machinery turn them into finished goods. In many cases, this department uses CAD/CAM software and workstations, as well as robotics. In another type of company, this department might manage the purchasing, handle the inventories, and control the flow of goods and services.

- **Marketing:** The marketing department oversees advertising, promotion, and sales. The people in this department plan, price, advertise, promote, package, and distribute the services or goods to customers or clients. The sales reps may use laptop computers, cell phones, wireless e-mail, and faxes in their work while on the road.

- **Accounting and finance:** The accounting and finance department handles all financial matters. It handles cash management, pays bills and taxes, issues paychecks, records payments, makes investments, and compiles financial statements and reports. It also produces financial budgets and forecasts financial performance after receiving information from other departments.

- **Human resources:** The human resources, or personnel, department finds and hires people and administers sick leave and retirement matters. It is also concerned with compensation levels, professional development, employee relations, and government regulations.

Management Levels: Three Levels, Three Kinds of Decisions

Within each of the five departments we have described, there are three traditional levels of management—top, middle, and lower. These levels are reflected in the organization chart. An **_organization chart_ is a schematic drawing showing the hierarchy of formal relationships among an organization's employees.** Managers on each of the three levels have different levels of responsibility and are therefore required to make different kinds of decisions. (See ● *Panel 11.1, pages 414–415.*)

- **Top managers—strategic decisions:** The chief executive officer (CEO) or president is the very top manager. However, for our purposes, "top management" refers to the vice presidents, one of whom heads each department. Typical titles found at the top management level are treasurer, director, controller (chief accounting officer), and senior partner.

    Top managers are concerned with long-range, or strategic, planning and decisions. Strategic decisions are complex decisions rarely based on predetermined routine procedures; they involve the subjective judgment of the decision maker. For instance, strategic decisions might relate to how growth should be financed and what new markets should be tackled first. Determining the company's 5-year goals, evaluating future financial resources, and formulating a response to competitors' actions are also strategic decisions.

An AT&T vice president of marketing might have to make strategic decisions about promotional campaigns to sell a new cable-modem service. The top manager who runs an electronics store might have to make strategic decisions about stocking a new line of personal digital assistants (PDAs).

- **Middle managers—tactical decisions:** Examples of middle managers are plant manager, division manager, sales manager, branch manager, and director of personnel. **_Middle-level managers_ make tactical decisions to implement the strategic goals of the organization.** A tactical decision is made without a base of clearly defined informational procedures; it may require detailed analysis and computations. An example might be deciding how many units of a specific product (PDAs, say) should be kept in inventory. Another is whether or not to purchase a larger computer system.

  The director of sales, who reports to the vice president of marketing for AT&T, sets sales goals for district sales managers throughout the country. They in turn feed him or her weekly and monthly sales reports.

- **Supervisory managers—operational decisions:** An example of a supervisory manager is a warehouse manager in charge of inventory restocking. **_Supervisory managers_ make operational decisions—predictable decisions that can be made by following well-defined sets of routine procedures.** These managers focus principally on supervising nonmanagement employees, monitoring day-to-day events, and taking corrective action where necessary.

  Determining not to restock inventory is an operation decision. (The guideline on when to restock may be determined at the level above.) A district sales manager for AT&T would monitor the promised sales and orders for cable modems coming in from sales representatatives. When sales begin to drop off, the supervisor would need to take immediate action.

Types of Information: Unstructured, Semistructured, & Structured

To make the appropriate decisions—strategic, tactical, operational—the different levels of managers need the right kind of information: structured, semistructured, and unstructured.

In general, *all* information to support intelligent decision making at all three levels must be correct—that is, accurate. It must also be complete, including *all* relevant data, yet concise, including *only* relevant data. It must be cost effective, meaning efficiently obtained, yet understandable. It must be current, meaning timely, yet also time sensitive, based on historical, current, or future information needs. This shows that information has three distinct properties:

1. Level of summarization
2. Degree of accuracy
3. Timeliness

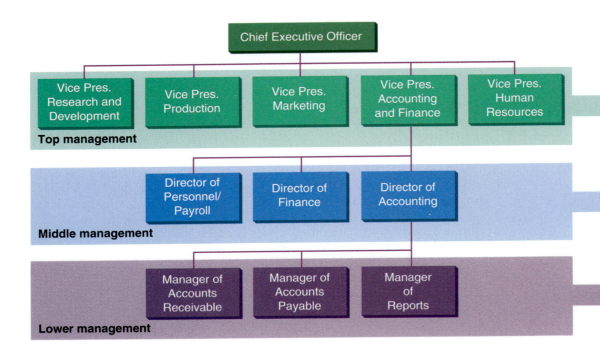

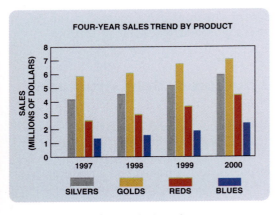

Strategic-Level
Inquiry/Report

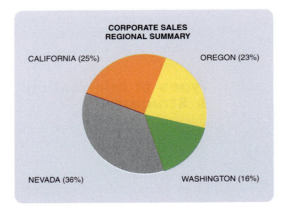

Tactical-Level
Inquiry/Report

● PANEL 11.1
An organization chart, and management levels and responsibilities

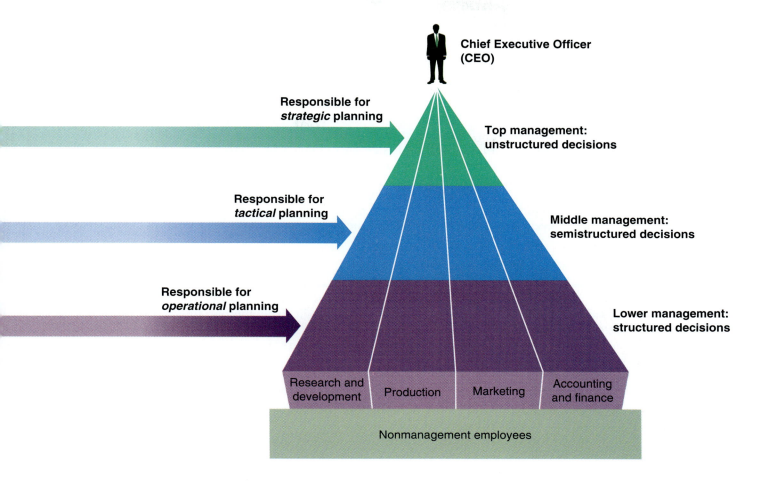

Chief Executive Officer (CEO)

Responsible for *strategic* planning

Top management: unstructured decisions

Responsible for *tactical* planning

Middle management: semistructured decisions

Responsible for *operational* planning

Lower management: structured decisions

Research and development | Production | Marketing | Accounting and finance

Nonmanagement employees

SALES DEPARTMENT — NEVADA
SALES SUMMARY($1000) — 1ST QUARTER

SALESPERSON	SILVERS	GOLDS	REDS	BLUES	TOTAL
VINE	70	10	14	65	159
WU	90	85	99	110	384
HERNANDEZ	95	126	111	115	447
WASHINGTON	120	98	103	28	349
LEE	60	225	219	180	684
OGG	93	33	35	68	229
WILLIAMS	32	24	28	10	94
TOTALS	560	601	609	576	2,346

SALES DEPARTMENT— NEVADA
SALES SUMMARY($1000) —1ST QUARTER
SALESPERSONS WITH SALES < $15,000 FOR ANY PRODUCT

SALESPERSON	SILVERS	GOLDS	REDS	BLUES	TOTAL
VINE	70	10	14	65	159
WILLIAMS	32	24	28	10	94

Operational-Level Inquiry/Report

These properties will be different for structured and unstructured information. Whether structured or unstructured information is more appropriate depends on the level of management and the type of decision making required. **_Structured information_ is detailed, current, not subjective, concerned with past events, records a narrow range of facts, and covers an organization's internal activities.** Unstructured information is the opposite. **_Unstructured information_ is summarized, less current, highly subjective, concerned with future events, records a broad range of facts, and covers activities outside as well as inside an organization. _Semistructured information_ includes some structured information and some unstructured information.**

Now that we've covered some basic concepts about how organizations are structured and what kinds of information are needed at different levels of management, we need to examine what types of management information systems provide the information.

11.2 Computer-Based Information Systems

KEY QUESTIONS
What are the six computer-based information systems—and what are their purposes?

The purpose of a computer-based information system is to provide managers (and various categories of employees) with the appropriate kind of information to help them make decisions. There are six types of computer-based information systems, which serve different levels of management. *(See ● Panel 11.2.)*

- **For lower managers:** Transaction processing systems (TPSs)
- **For middle managers:** Management information systems (MISs) and decision support systems (DSSs)
- **For top managers:** Executive support system (ESSs)
- **For all levels, including nonmanagement:** Office automation systems (OASs) and expert systems (ESs)

Let us consider these.

Transaction Processing Systems: For Supervisory Managers

In most organizations, particularly business organizations, most of what goes on consists largely of transactions. A **_transaction_ is a recorded event having to do with routine business activities.** This includes everything concerning the product or service in which the organization is engaged: production, distribution, sales, orders. It also includes materials purchased, employees hired, taxes paid, and so on. Today in most organizations, the bulk of such transactions are recorded in a computer-based information system. These systems tend to have clearly defined inputs and outputs, and there is an emphasis on efficiency and accuracy. Transaction processing systems record data but do little in the way of converting data into information.

A **_transaction processing system (TPS)_ is a computer-based information system that keeps track of the transactions needed to conduct business.** Some features of a TPS are as follows:

- **Input and output:** The inputs to the system are transaction data: bills, orders, inventory levels, and the like. The output consists of processed transactions: bills, paychecks, and so on.
- **For lower managers:** Because the TPS deals with day-to-day matters, it is principally of use to supervisory managers. That is, the TPS helps in making tactical decisions. Such systems are not usually helpful to middle or top managers.

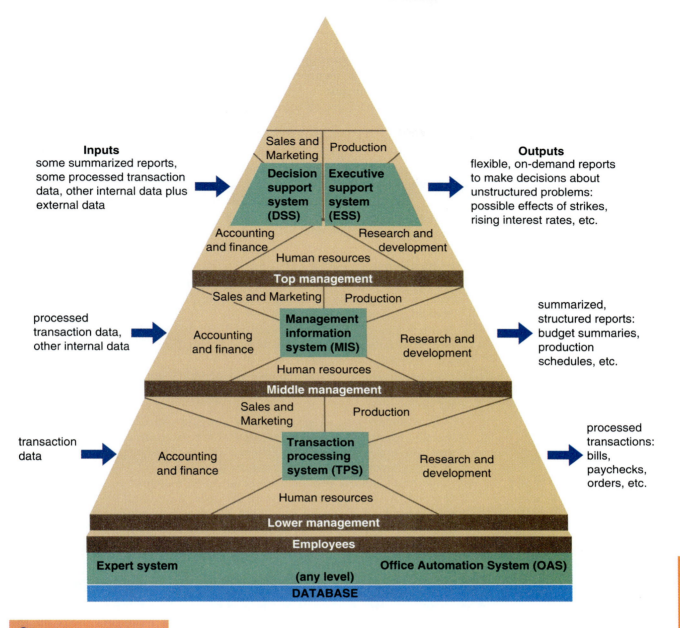

Inputs
some summarized reports, some processed transaction data, other internal data plus external data

Outputs
flexible, on-demand reports to make decisions about unstructured problems: possible effects of strikes, rising interest rates, etc.

Sales and Marketing | Production
Decision support system (DSS) | **Executive support system (ESS)**
Accounting and finance | Research and development
Human resources
Top management

processed transaction data, other internal data

summarized, structured reports: budget summaries, production schedules, etc.

Sales and Marketing | Production
Accounting and finance | **Management information system (MIS)** | Research and development
Human resources
Middle management

transaction data

processed transactions: bills, paychecks, orders, etc.

Sales and Marketing | Production
Accounting and finance | **Transaction processing system (TPS)** | Research and development
Human resources
Lower management

Employees

Expert system | **Office Automation System (OAS)**
(any level)
DATABASE

● **PANEL 11.2**
Six information systems for three levels of management
Executive support systems are for top managers. Management information systems and decision support systems are for middle managers. Transaction processing systems are for supervisory managers. Office automation systems and expert systems are for all levels, including nonmanagement.

- **Produces detail reports:** A manager at this level typically will receive information in the form of detail reports. A *detail report* contains specific information about routine activities. An example might be the information needed to decide whether to restock inventory.

- **One TPS for each department:** Each department or functional area of an organization—research and development, production, marketing, accounting and finance, and human resources—usually has its own TPS. For example, the accounting and finance TPS handles order processing, accounts receivable, inventory and purchasing, accounts payable, order processing, and payroll.

- **Basis for MIS and DSS:** The database of transactions stored in a TPS provides the basis for management information systems and decision support systems, as we describe next.

Management Information Systems: For Middle Managers

A *management information system (MIS)* **is a computer-based information system that uses data recorded by TPS as input into programs that produce routine reports as output.**

Features of an MIS are as follows:

- **Input and output:** Inputs consist of processed transaction data, such as bills, orders, and paychecks, plus other internal data. Outputs consist of summarized, structured reports: budget summaries, production schedules, and the like.

- **For middle managers:** An MIS is intended principally to assist middle managers—specifically to help them with tactical decisions. It enables them to spot trends and get an overview of current business activities.

- **Draws from all departments:** The MIS draws from all five departments or functional areas, not just one.

- **Produces several kinds of reports:** Managers at this level usually receive information in the form of several kinds of reports: *summary, exception, periodic, demand.*

  Summary reports show totals and trends. An example would be a report showing total sales by office, by product, and by salesperson, as well as total overall sales.

  Exception reports show out-of-the-ordinary data. An example would be an inventory report listing only those items of which fewer than 10 are in stock.

  Periodic reports are produced on a regular schedule. Such daily, weekly, monthly, quarterly, or annual reports may contain sales figures, income statements, or balance sheets. They are usually produced on paper, such as computer printouts.

  Demand reports produce information in response to an unscheduled demand. A director of finance might order a demand credit-background report on an unknown customer who wants to place a large order. Demand reports are often produced on a terminal or microcomputer screen, rather than on paper.

Decision Support Systems: Also for Middle Managers

A *decision support system (DSS)* **is a computer-based information system that provides a flexible tool for analysis and helps managers focus on the future.** Whereas a TPS records data and an MIS summarizes data, a DSS *analyzes* data. To reach the DSS level of sophistication in information technology, an organization must have established TPS and MIS systems first.

Some features of a DSS are as follows:

- **Inputs and outputs:** Inputs include internal data—such as summarized reports and processed transaction data—and also data that is external to the organization. External data may be produced by trade associations, marketing research firms, the U.S. Bureau of the Census, and other government agencies.

  The outputs are demand reports on which a top manager can make decisions about unstructured problems.

- **Mainly for middle managers:** A DSS is intended principally to assist middle managers in making tactical decisions. Questions addressed by the DSS might be, for example, whether interest rates will rise or whether there will be a strike in an important materials-supplying industry.

- **Produces analytic models:** The key attribute of a DSS is that it uses models. A **_model_ is a mathematical representation of a real system.** The models use a DSS database, which draws on the TPS and MIS files, as well as external data such as stock reports, government reports, and national and international news. The system is accessed through DSS software.

The model allows the manager to do a simulation—play a "what-if" game—to reach decisions. Thus, the manager can simulate an aspect of the organization's environment in order to decide how to react to a change in conditions affecting it. By changing the hypothetical inputs to the model, the manager can see how the model's outputs are affected.

Many DSSs are developed to support the types of decisions faced by managers in specific industries, such as airlines or real estate. Curious how airlines decide how many seats to sell on a flight when so many passengers are no-shows? American Airlines developed a DSS, the yield management system, that helps managers decide how much to overbook and how to set prices for each seat so that a plane is filled and profits are maximized. Wonder how owners of those big apartment complexes set rents and lease terms? Investors in commercial real estate use a DSS called RealPlan to forecast property values up to 40 years into the future, based on income, expense, and cash-flow projections. Ever speculate about how insurance carriers set different rates or how Arby's and McDonald's decide where to locate a store? Many companies use DSSs called geographic information systems (GISs), such as MapInfo and Atlas GIS, which integrate geographic databases with other business data and display maps. (See ● *Panel 11.3.*)

Executive Support Systems: For Top Managers

An **_executive support system (ESS)_ is an easy-to-use DSS made especially for top managers; it specifically supports strategic decision making.** An ESS is also called an *executive information system (EIS)*. It draws on data not only

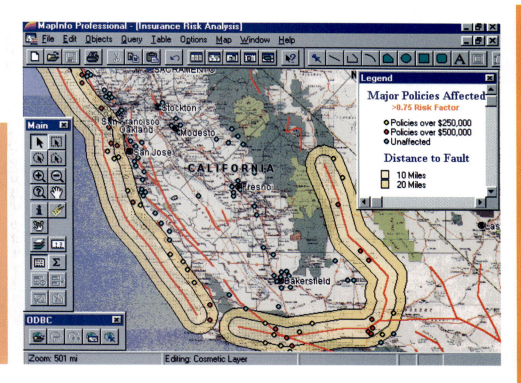

● **PANEL 11.3**

Geographic DSS for earthquake insurance

Using geographic information systems, such as MapInfo, insurance underwriters can set rates and examine potential liability in the event of a natural disaster. This one presents a visual analysis of policy-holders living near earthquake fault lines in California.

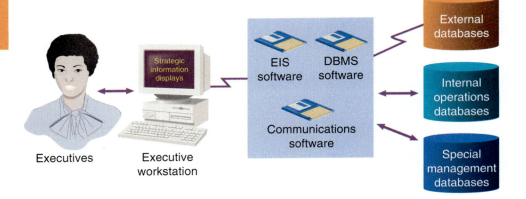

Executives | Executive workstation | EIS software | DBMS software | Communications software | External databases | Internal operations databases | Special management databases

from systems internal to the organization but also from those outside, such as news services or market-research databases. *(See ● Panel 11.4.)*

An ESS might allow senior executives to call up predefined reports from their personal computers, whether desktops or laptops. They might, for instance, call up sales figures in many forms—by region, by week, by anticipated year, by projected increases. The ESS includes capabilities for analyzing data and doing "what-if" scenarios. ESSs also have the capability to browse through summarized information on all aspects of the organization and then zero in on ("drill down" to) detailed areas the manager believes require attention.

Office Automation & Expert Systems: Information for Everyone

TCPs, MISs, DSSs, ESSs—the alphabet soup of information systems discussed so far—are designed for managers of various levels. But there are two types of information systems that are intended for workers of all levels, including those who aren't managers: *office automation systems* and *expert systems.*

- **Office automation systems: _Office automation systems (OASs)_ combine various technologies to reduce the manual labor required in operating an efficient office environment.** Used throughout all levels of an organization, OAS technologies include fax, voice mail, e-mail, scheduling software, word processing, and desktop publishing, among others. *(See ● Panel 11.5.)*

  The backbone of an OAS is a network—LAN, intranet, extranet—that connects everything. All office functions—dictation, typing, filing, copying, fax, microfilm and records management, telephone calls and switchboard operations—are candidates for integration into the network.

- **Expert systems: An _expert system_, or knowledge-based system, is a set of interactive computer programs that helps users solve problems that would otherwise require the assistance of a human expert.** Expert systems are created on the basis of knowledge collected on specific topics from human specialists, and they imitate the reasoning process of a human being. As we described in Chapter 10, expert systems have emerged from the field of artificial intelligence, the branch of computer science that is devoted to the creation of computer systems that simulate human reasoning and sensation.

  Expert systems are used by both management and nonmanagement personnel to solve specific problems, such as how to reduce production costs, improve workers' productivity, or reduce environmental impact. Because of their giant appetite for memory, expert systems are

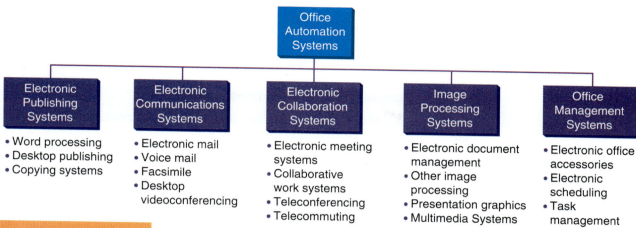

Office Automation Systems				
Electronic Publishing Systems	**Electronic Communications Systems**	**Electronic Collaboration Systems**	**Image Processing Systems**	**Office Management Systems**
• Word processing • Desktop publishing • Copying systems	• Electronic mail • Voice mail • Facsimile • Desktop videoconferencing	• Electronic meeting systems • Collaborative work systems • Teleconferencing • Telecommuting	• Electronic document management • Other image processing • Presentation graphics • Multimedia Systems	• Electronic office accessories • Electronic scheduling • Task management

PANEL 11.5

Office automation systems

The backbone is a network linking these technologies.

usually run on large computers, although some microcomputer expert systems also exist. For example, Negotiator Pro for IBM and Macintosh computers helps executives plan effective negotiations by examining the personality types of the other parties and recommending negotiating strategies.

Now that we have seen how managers work within an organization and what their information needs are, let's look at how changes can be made to keep up with the new demands. A very powerful tool for this purpose is systems analysis and design.

11.3 Systems Development: The Six Phases of Systems Analysis & Design

KEY QUESTION

What are the six phases of the systems development life cycle?

Organizations can make mistakes, of course, and big organizations can make *really big* mistakes.

California's state Department of Motor Vehicles' databases needed to be modernized, and in 1988 Tandem Computers said it could do it. "The fact that the DMV's database system, designed around an old IBM-based platform, and Tandem's new system were as different as night and day seemed insignificant at the time to the experts involved," said one writer who investigated the project.[4] The massive driver's license database, containing the driving records of more than 30 million people, first had to be "scrubbed" of all information that couldn't be translated into the language used by Tandem computers. One such scrub yielded 600,000 errors. Then the DMV had to translate all its IBM programs into the Tandem language. "Worse, DMV really didn't know how its current IBM applications worked anymore," said the writer, "because they'd been custom-made decades before by long-departed programmers and rewritten many times since." Eventually the project became a staggering $44 million loss to California's taxpayers.

Needless to say, not all mistakes are so huge. Computer foul-ups can range from minor to catastrophic. But this example shows how important planning is, especially when an organization is trying to launch a new kind of system. The best way to avoid such mistakes is to employ systems analysis and design.

But, you may say, you're not going to have to wrestle with problems on the scale of motor-vehicle departments. That's a job for computer professionals. You're mainly interested in using computers and communications to increase your own productivity. Why, then, do you need to know anything about systems analysis and design?

In many types of jobs, you may find your department or your job the focus of a study by a systems analyst. Knowing how the procedure works will help you better explain how your job works or what goals your department is supposed to achieve. In progressive companies, management is always interested in suggestions for improving productivity. Systems analysis provides a method for developing such ideas.

The Purpose of a System

A _system_ **is defined as a collection of related components that interact to perform a task in order to accomplish a goal.** A system may not work very well, but it is nevertheless a system. The point of systems analysis and design is to ascertain how a system works and then take steps to make it better.

An organization's computer-based information system consists of hardware, software, people, procedures, and data, as well as communications setups. These work together to provide people with information for running the organization.

Getting the Project Going: How It Starts, Who's Involved

A single individual who believes that something badly needs changing is all it takes to get the project rolling. An employee may influence a supervisor. A customer or supplier may get the attention of someone in higher management. Top management may decide independently to take a look at a system that seems inefficient. A steering committee may be formed to decide which of many possible projects should be worked on.

Participants in the project are of three types:

- **Users:** The system under discussion should _always_ be developed in consultation with users, whether floor sweepers, research scientists, or customers. Indeed, if user involvement in analysis and design is inadequate, the system may fail for lack of acceptance.
- **Management:** Managers within the organization should also be consulted about the system.
- **Technical staff:** Members of the company's information systems (IS) department, consisting of systems analysts and programmers, need to be involved. For one thing, they may have to execute the project. Even if they don't, they will have to work with outside IS people contracted to do the job.

Complex projects will require one or several systems analysts. A _systems analyst_ **is an information specialist who performs systems analysis, design, and implementation.** The analyst's job is to study the information and communications needs of an organization and determine what changes are required to deliver better information to people who need it. "Better" information means information that is summarized in the acronym "CART"—complete, accurate, relevant, and timely. The systems analyst achieves this goal through the problem-solving method of systems analysis and design.

The Six Phases of Systems Analysis & Design

Systems analysis and design **is a six-phase problem-solving procedure for examining an information system and improving it.** The six phases make up what is called the systems development life cycle. The _systems development life cycle (SDLC)_ **is the step-by-step process that many organizations follow during systems analysis and design.**

Whether applied to a Fortune 500 company or a three-person engineering business, the six phases in systems analysis and design are as shown in the illustration. _(See ● Panel 11.6.)_ Phases often overlap, and a new one may

● **PANEL 11.6**

The systems development life cycle
An SDLC typically includes six phases.

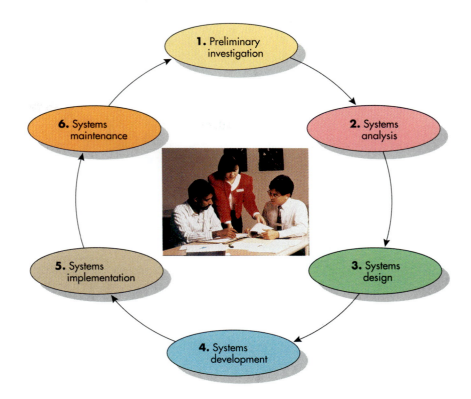

1. Preliminary investigation

2. Systems analysis

3. Systems design

4. Systems development

5. Systems implementation

6. Systems maintenance

start before the old one is finished. After the first four phases, management must decide whether to proceed to the next phase. *User input and review is a critical part of each phase.*

The First Phase: Conduct a Preliminary Investigation

The objective of Phase 1, *preliminary investigation*, is to conduct a preliminary analysis, propose alternative solutions, describe costs and benefits, and submit a preliminary plan with recommendations. *(See* ● *Panel 11.7.)*

- **Conduct the preliminary analysis:** In this step, you need to find out what the organization's objectives are and the nature and scope of the problem under study. Even if a problem pertains only to a small segment of the organization, you cannot study it in isolation. You need to find out what the objectives of the organization itself are. Then you need to see how the problem being studied fits in with them.

- **Propose alternative solutions:** In delving into the organization's objectives and the specific problem, you may have already discovered some solutions. Other possible solutions can come from interviewing people inside the organization, clients or customers affected by it, suppliers, and consultants. You can also study what competitors are doing. With this data, you then have three choices. You can leave the system as is, improve it, or develop a new system.

- **Describe the costs and benefits:** Whichever of the three alternatives is chosen, it will have costs and benefits. In this step, you need to indicate what these are. Costs may depend on benefits, which may offer savings. A broad spectrum of benefits may be derived. A process may be speeded up, streamlined through elimination of unnecessary steps, or combined with other processes. Input errors or redundant output may be reduced. Systems and subsystems may be better integrated. Users may be happier with the system. Customers' or suppliers' interactions with the system may be more satisfactory. Security may be improved. Costs may be cut.

1. Conduct preliminary analysis. This includes stating the objectives, defining nature and scope of the problem.
2. Propose alternative solutions: leave system alone, make it more efficient, or build a new system.
3. Describe costs and benefits of each solution.
4. Submit a preliminary plan with recommendations.

- **Submit a preliminary plan:** Now you need to wrap up all your findings in a written report. The readers of this report will be the executives who are in a position to decide in which direction to proceed—make no changes, change a little, or change a lot—and how much money to allow the project. You should describe the potential solutions, costs, and benefits and indicate your recommendations.

The Second Phase: Do an Analysis of the System

The objective of Phase 2, _systems analysis_, is to gather data, analyze the data, and write a report. (See ● _Panel 11.8_) In this second phase of the SDLC, you will follow the course that management has indicated after having read your Phase 1 feasibility report. We are assuming that they have ordered you to perform Phase 2—to do a careful analysis or study of the existing system in order to understand how the new system you proposed would differ. This analysis will also consider how people's positions and tasks will have to change if the new system is put into effect.

- **Gather data:** In gathering data, you will review written documents, interview employees and managers, develop questionnaires, and observe people and processes at work.
- **Analyze the data:** Once the data has been gathered, you need to come to grips with it and analyze it. Many analytical tools, or modeling tools, are available. **_Modeling tools_ enable a systems analyst to present graphic, or pictorial, representations of a system.** An example of a modeling tool is a **_data flow diagram_, which graphically shows the flow of data through a system**—that is, the essential processes of a system, along with inputs, outputs and files. (See ● _Panel 11.9_.)
- **Write a report:** Once you have completed the analysis, you need to document this phase. This report to management should have three parts. First, it should explain how the existing system works. Second, it should explain the problems with the existing system. Finally, it should describe the requirements for the new system and make recommendations on what to do next.

  At this point, not a lot of money will have been spent on the systems analysis and design project. If the costs of going forward seem prohibitive, this is a good time for the managers reading the report to call a halt. Otherwise, you will be asked to move to Phase 3.

1. Gather data, using tools of written documents, interviews, questionnaires, and observations.
2. Analyze the data, using modeling tools: grid charts, decision tables, data flow diagrams, systems flowcharts, connectivity diagrams.
3. Write a report.

The Third Phase: Design the System

The objective of Phase 3, _systems design_, is to do a preliminary design and then a detail design, and write a report. *(See* ● *Panel 11.10, next page.)* In this third phase of the SDLC, you will essentially create a "rough draft" and then a "detail draft" of the proposed information system.

- **Do a preliminary design:** A _**preliminary design**_ **describes the general functional capabilities of a proposed information system.** It reviews the system requirements and then considers major components of the system. Usually several alternative systems (called *candidates*) are considered, and the costs and the benefits of each are evaluated.

  Some tools that may be used in the design are *CASE tools* and *project management software.*

  **CASE (computer-aided software engineering) tools** **are programs that automate various activities of the SDLC in several phases.** This technology is intended to speed up the process of developing systems and to improve the quality of the resulting systems. These tools, which are also known as *automated design tools,* may be used at

● **PANEL 11.9**
Data flow diagram
Example of data flow diagram and explanation of symbols.

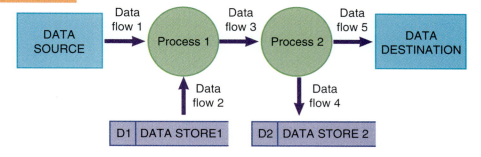

Explanation of standard data flow diagram symbols

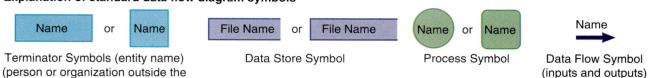

Terminator Symbols (entity name) (person or organization outside the system boundaries) Data Store Symbol Process Symbol Data Flow Symbol (inputs and outputs)

1. Do a preliminary design, using CASE tools, prototyping tools, and project management software, among others.
2. Do a detail design, defining requirements for output, input, storage, and processing and system controls and backup.
3. Write a report.

other stages of the SDLC as well. Examples of such programs are Excelerator, Iconix, System Architect, and Powerbuilder.

**Prototyping** **refers to using workstations, CASE tools, and other software applications to build working models of system components, so that they can be quickly tested and evaluated.** Thus, a _**prototype**_ **is a limited working system developed to test out design concepts.** A prototype, which may be constructed in just a few days, allows users to find out immediately how a change in the system might benefit them. For example, a systems analyst might develop a menu as a possible screen display, which users could try out. The menu can then be redesigned or fine-tuned, if necessary.

Project management software consists of programs used to plan, schedule, and control the people, costs, and resources required to complete a project on time.

- **Do a detail design:** A _**detail design**_ **describes how a proposed information system will deliver the general capabilities described in the preliminary design.** The detail design usually considers the following parts of the system in this order: output requirements, input requirements, storage requirements, processing requirements, and system controls and backup.

- **Write a report:** All the work of the preliminary and detail designs will end up in a large, detailed report. When you hand over this report to senior management, you will probably also make some sort of presentation or speech.

The Fourth Phase: Develop the System

In Phase 4, _systems development_**, the systems analyst or others in the organization develop or acquire the software, acquire the hardware, and then test the system.** _(See_ ● _Panel 11.11.)_ Depending on the size of the project, this phase will probably involve the organization in spending substantial sums of money. It could also involve spending a lot of time. However, at the end you should have a workable system.

- **Develop or acquire the software:** During the design stage, the systems analyst may have had to address what is called the "make-or-buy" decision, but that decision certainly cannot be avoided now. In the _**make-or-buy decision**_**, you decide whether you have to create a program—have it custom-written—or buy it, meaning simply purchase an existing software package.** Sometimes programmers decide they can buy an existing program and modify it rather than write it from scratch.

1. Acquire software.
2. Acquire hardware.
3. Test the system.

If you decide to create a new program, then the question is whether to use the organization's own staff programmers or to hire outside contract programmers (outsource it). Whichever way you go, the task could take many months.

Programming is an entire subject unto itself, which we discuss in the Appendix, along with programming languages.

- **Acquire hardware:** Once the software has been chosen, the hardware to run it must be acquired or upgraded. It's possible your new system will not require any new hardware. It's also possible that the new hardware will cost millions of dollars and involve many items: microcomputers, mainframes, monitors, modems, and many other devices. The organization may find it's better to lease rather than to buy some equipment, especially since, as we mentioned (Moore's law), chip capability has traditionally doubled every 18 months.

- **Test the system:** With the software and hardware acquired, you can now start testing the system. Testing is usually done in two stages: unit testing, then system testing.

  In *unit testing*, the performance of individual parts is examined, using test (made-up, or sample) data. If the program is written as a collaborative effort by multiple programmers, each part of the program is tested separately.

  In *system testing*, the parts are linked together, and test data is used to see if the parts work together. At this point, actual organization data may be used to test the system. The system is also tested with erroneous and massive amounts of data to see if the system can be made to fail ("crash").

  At the end of this long process, the organization will have a workable information system, one ready for the implementation phase.

The Fifth Phase: Implement the System

Whether the new information system involves a few handheld computers, an elaborate telecommunications network, or expensive mainframes, the fifth phase will involve some close coordination in order to make the system not just workable but successful. **Phase 5, _systems implementation_, consists of converting the hardware, software, and files to the new system and training the users.** *(See ● Panel 11.12, next page.)*

- **Convert to the new system:** *Conversion,* the process of transition from an old information system to a new one, involves converting hardware, software, and files. There are four strategies for handling conversion: *direct, parallel, phased,* and *pilot.*

  Direct implementation means that the user simply stops using the old system and starts using the new one. The risk of this method should be evident: What if the new system doesn't work? If the old system has truly been discontinued, there is nothing to fall back on.

● **PANEL 11.12**
Fifth phase: systems implementation

1. Convert hardware, software, and files through one of four types of conversions: direct, parallel, phased, or pilot.
2. Compile final documentation.
3. Train the users.

Parallel implementation means that the old and new systems are operated side by side until the new system has shown it is reliable, at which time the old system is discontinued. Obviously there are benefits in taking this cautious approach. If the new system fails, the organization can switch back to the old one. The difficulty with this method is the expense of paying for the equipment and people to keep two systems going at the same time.

Phased implementation means that parts of the new system are phased in separately—either at different times (parallel) or all at once in groups (direct).

Pilot implementation means that the entire system is tried out but only by some users. Once the reliability has been proved, the system is implemented with the rest of the intended users. The pilot approach still has its risks, since all of the users of a particular group are taken off the old system. However, the risks are confined to a small part of the organization.

- **Train the users:** Various tools are available to familiarize users with a new system—from documentation (instruction manuals) to videotapes to live classes to one-on-one, side-by-side teacher-student training. Sometimes training is done by the organization's own staffers; at other times it is contracted out.

The Sixth Phase: Maintain the System

Phase 6, _systems maintenance_, adjusts and improves the system by having system audits and periodic evaluations and by making changes based on new conditions. (See ● *Panel 11.13.*) Even with the conversion accomplished and the users trained, the system won't just run itself. There is a sixth—and never-ending—phase in which the information system must be monitored to ensure that it is successful. Maintenance includes not only keeping the machinery running but also updating and upgrading the system to keep pace with new products, services, customers, government regulations, and other requirements.

The sixth phase is to keep the system running through system audits and periodic evaluations.

● **PANEL 11.13**
Sixth phase: systems maintenance

Experience Box
Critical Thinking Tools

Clear thinkers aren't born that way. They work at it.

The systems development life cycle is basically an exercise in clear thinking—critical thinking. Critical thinking is fundamental to systems analysis and design—particularly in the first phase, preliminary analysis. Reaching for the truth may not come easily; it is a stance toward the world, developed with practice. To achieve this, we have to wrestle with obstacles that are mostly of our own making: mindsets. By the time we are grown, our minds have become "set" in various patterns of thinking that affect the way we respond to new situations and new ideas. Such mindsets determine what ideas we think are important and, conversely, what ideas we ignore.

To break past mindsets, we need to learn to think critically. *Critical thinking* means sorting out conflicting claims, weighing the evidence for them, letting go of personal biases, and arriving at reasoned conclusions. Critical thinking means actively seeking to understand, analyze, and evaluate information in order to solve specific problems.

Learning to identify fallacious (incorrect) arguments will help you avoid patterns of faulty thinking in your own writing and thinking and to identify it in others'.

Jumping to Conclusions In the fallacy called *jumping to conclusions,* also known as *hasty generalization,* a decision maker reaches a conclusion before all the facts are available. *Example:* A company instituted the strategy of total quality management (TQM) 12 months ago. As a new manager of the company, you see that TQM has not improved profitability over the past year, and you order TQM junked, in favor of more traditional business strategies. However, what you don't know is that the traditional business strategies employed prior to TQM had an even *worse* effect on profitability.

Irrelevant Reason or False Cause In the faulty reasoning known as *non sequitur* (Latin for "It does not follow"), which might better be called *false cause* or *irrelevant reason,* the conclusion does not follow logically from the supposed reasons stated earlier. There is no *causal* relationship. *Example:* You receive an A on a test. However, because you felt you hadn't been well prepared, you attribute your success to your friendliness with the professor. Or to your horoscope. Or to wearing your "lucky shirt." None of these supposed reasons have anything to do with the result.

Irrelevant Attack on a Person or Opponent Known as an *ad hominem* argument (Latin for "to the person"), the *irrelevant attack on an opponent* attacks a person's reputation or beliefs rather than his or her argument. *Example:* Your boss insists you may not hire a certain person as a programmer because he or she has been married and divorced nine times, although the person's marital history plainly has no bearing on his or her skills as a programmer.

Slippery Slope The *slippery slope* is a failure to see that the first step in a possible series of steps does not lead inevitably to the rest. *Example:* The "Domino theory," under which the United States waged wars against Communism for half a century, was a slippery-slope argument. It assumed that if Communism triumphed in one country (for example, Nicaragua), then it would inevitably triumph in other regions (the rest of Central America), finally threatening the borders of the United States itself.

Appeal to Authority The *appeal to authority* argument (known in Latin as *argumentum ad verecundiam*) uses authorities in one area in an effort to validate claims in another area where the person is not an expert. *Example:* You see the appeal-to-authority argument used all the time in advertising. But how qualified is a professional golfer to speak about headache remedies?

Circular Reasoning In *circular reasoning,* a statement to be proven true is rephrased, and then the new formulation is offered as supposed proof that the original statement is in fact true. *Example:* You declare that you can drive safely at high speeds with only inches separating you from the car ahead because you have driven this way for years without an accident.

Straw Man Argument In the *straw man* argument, you misrepresent your opponent's position to make it easier to attack, or you attack a weaker position while ignoring a stronger one. In other words, you sidetrack the argument from the main discussion. *Example:* Politicians use this argument all the time. If you attack a legislator for being "fiscally irresponsible" in supporting funds for a gun-control bill, when what you really object to is the fact of gun control, you're using a straw man argument.

Appeal to Pity In the *appeal to pity* argument, the advocate appeals to mercy rather than making an argument on the merits of the case itself. *Example:* Begging the dean not to expel you for cheating because your parents are poor and made sacrifices to put you through college would be a blatant appeal to pity.

Questionable Statistics Statistics can be misused in many ways as supporting evidence. The statistics may be unknowable, drawn from an unrepresentative sample, or otherwise suspect. *Examples:* Stating how much money is lost to taxes because of illegal drug transactions is speculation because such transactions are hidden or unrecorded.

Visual Summary

CASE (computer-aided software engineering) tools (p. 425, KQ 11.3) Software that provides computer-automated means of designing and changing systems. Why it's important: *CASE tools may be used in almost any phase of the SDLC, not just design. So-called* front-end CASE tools *are used during the first three phases—preliminary analysis, systems analysis, systems design—to help with the early analysis and design. So-called* back-end CASE tools *are used during two later phases—systems development and implementation—to help in coding and testing, for instance.*

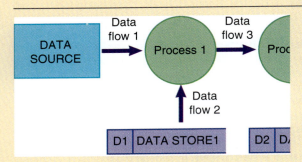

data flow diagram (DFD) (p. 424, KQ 11.3) Modeling tool that graphically shows the flow of data through a system. Why it's important: *A DFD diagrams the processes that change data into information. DFDs have only four symbols—for source or destination of data, data flow, data processing, and data storage—which makes them easy to use.*

decision support system (DSS) (p. 418, KQ 11.2) Computer-based information system that helps managers with nonroutine decision-making tasks. Inputs consist of some summarized reports, some processed transaction data, and other internal data. They also include data from sources outside the organization—for example, data may be produced by trade associations, marketing research firms, and government agencies. The outputs are flexible, on-demand reports from which a top manager can make decisions about unstructured problems. Why it's important: *A DSS is installed to help top managers and middle managers make strategic decisions—decisions about unstructured problems, those involving events and trends outside the organization (for example, rising interest rates). The key attribute of a DSS is that it uses* models. *The DSS database, which draws on the TPS and MIS files, as well as outside data, is accessed through DSS software.*

detail design (p. 426, KQ 11.3) Second stage of Phase 3 of the SDLC; describes how a proposed information system will deliver the general capabilities described in the preliminary design phase. The detail design usually considers the following system requirements: *output, input, storage, processing,* and *system controls and backup.* Why it's important: *A new system must be designed in detail before any hardware and software can be developed/purchased.*

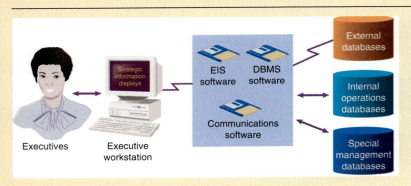

executive support system (ESS) (p. 419, KQ 11.2) Also called an *executive information system (ESS);* DSS made especially for top managers that specifically supports strategic decision making. It draws on data both from inside and outside the organization (for example, news services, market-research databases). Why it's important: *The EIS includes capabilities for analyzing data and doing "what if" scenarios.*

expert system (p. 420, KQ 11.2) Set of computer programs that perform a task at the level of a human expert. Why it's important: *Expert systems are used by management and non-management personnel to solve sophisticated problems.*

make-or-buy decision (p. 426, KQ 11.3) Decision made in Phase 4 (programming) of the SDLC concerning whether the organization has to make a program—have it custom-written—or buy it, meaning simply purchase an existing software package. Why it's important: *The decision taken affects the costs and time required to develop the system.*

management information system (MIS) (p. 418, KQ 11.2) Computer-based information system that derives data from all an organization's departments and produces *summary, exception, periodic,* and *on-demand* reports of the organization's performance. Why it's important: *A MIS principally assists middle managers, helping them make tactical decisions—spotting trends and getting an overview of current business activities.*

middle-level managers (p. 413, KQ 11.1) One of the three types of managers; they implement the goals of the organization. Their job is to oversee the supervisors and to make tactical decisions. Why it's important: *Middle managers require information that is both structured and unstructured.*

model (p. 419, KQ 11.2) Mathematical representation of a real system; models are often used in a DSS. Why it's important: *A model allows the manager to do a simulation—play a "what if" game—to reach decisions. By changing the hypothetical inputs to the model, one can see how its outputs are affected.*

modeling tools (p. 424, KQ 11.3) Analytical tools like charts, tables, and diagrams used by systems analysts. Examples are data flow diagrams, decision tables, systems flowcharts, and object-oriented analysis. Why it's important: *Modeling tools enable a systems analyst to present graphic, or pictorial, representations of a system.*

Electronic Publishing Systems	Electronic Communications Systems	Electronic Collaboration Systems	Image Processing Systems	Office Management Systems
• Word processing • Desktop publishing • Copying systems	• Electronic mail • Voice mail • Facsimile • Desktop videoconferencing	• Electronic meeting systems • Collaborative work systems • Teleconferencing • Telecommuting	• Electronic document management • Other image processing • Presentation graphics • Multimedia Systems	• Electronic office accessories • Electronic scheduling • Task management

office automation systems (OAS) (p. 420, KQ 11.2) Computer information system that combines various technologies to reduce the manual labor needed to operate an office efficiently; used at all levels of an organization. Why it's important: *An OAS uses a network to integrate such technologies as fax, voice mail, e-mail, scheduling software, word processing, and desktop publishing and make them available throughout the organization.*

organization chart (p. 412, KQ 11.1) Schematic drawing showing the hierarchy of relationships among an organization's employees. Why it's important: *Organization charts show levels of management and formal lines of authority.*

1. Conduct preliminary analysis. This includes stating the objectives, defining nature and scope of the problem.
2. Propose alternative solutions: leave system alone, make it more efficient, or build a new system.
3. Describe costs and benefits of each solution.
4. Submit a preliminary plan with recommendations.

preliminary design (p. 425, KQ 11.3) First stage of Phase 3 of the SDLC; describes the general functional capabilities of a proposed information system. Three tools that may be used are *prototyping tools, CASE tools,* and *project management software.* Why it's important: *During the preliminary design phase, staff reviews the system requirements and then considers major components of the system. Usually several alternative systems (called candidates) are considered, and the costs and the benefits of each are evaluated.*

preliminary investigation (p. 423, KQ 11.3) Phase 1 of the SDLC; the purpose is to conduct a preliminary analysis (determine the organization's objectives, determine the nature and scope of the problem), propose alternative solutions (leave the system as is, improve the efficiency of the system, or develop a new system), describe costs and benefits, and submit a preliminary plan with recommendations. Why it's important: *The preliminary investigation lays the groundwork for the other phases of the SDLC.*

prototype (p. 426, KQ 11.3) A limited working system, or part of one. It is developed to test out design concepts. Why it's important: *A prototype, which may be constructed in just a few days, allows users to find out immediately how a change in the system might benefit them.*

prototyping (p. 426, KQ 11.3) Involves building a model or experimental version of all or part of a system so that it can be quickly tested and evaluated; uses workstations, CASE tools, and other applications software. Why it's important: *Prototyping is part of the preliminary design stage of Phase 3 of the SDLC.*

semistructured information (p. 416, KQ 11.1) Information that does not necessarily result from clearly defined, routine procedures. Middle managers need semistructured information that is detailed and more summarized than information for operating managers. Why it's important: *Semistructured information involves review, summarization, and analysis of data to help plan and control operations and implement policy formulated by upper managers.*

structured information (p. 416, KQ 11.1) Detailed, current information concerned with past events; it records a narrow range of facts and covers an organization's internal activities. Why it's important: *Lower-level managers need easily defined information that relates to the current status and activities within the basic business functions.*

supervisory managers (p. 413, KQ 11.1) Also called *low-level managers;* the lowest level in the hierarchy of the three types of managers. Their job is to make operational decisions, monitoring day-to-day events, and, if necessary, taking corrective action. Why it's important: *Lower managers need information that is structured—that is, detailed, current, and past-oriented, covering a narrow range of facts and events inside the organization.*

system (p. 422, KQ 11.3) Collection of related components that interact to perform a task in order to accomplish a goal. Why it's important: *Understanding a set of activities as a system allows one to look for better ways to reach the goal.*

1. Gather data, using tools of written documents, interviews, questionnaires, and observations.
2. Analyze the data, using modeling tools: grid charts, decision tables, data flow diagrams, systems flow-charts, connectivity diagrams.
3. Write a report.

systems analysis (p. 424, KQ 11.3) Phase 2 of the SDLC; the purpose is to gather data (using written documents, interviews, questionnaires, observation, and sampling), analyze the data, and write a report. Why it's important: *The results of systems analysis will determine whether the system should be redesigned.*

systems analysis and design (p. 422, KQ 11.3) Problem-solving procedure for examining an information system and improving it; consists of the six-phase *systems development life cycle.* Why it's important: *The point of systems analysis and design is to ascertain how a system works and then take steps to make it better.*

systems analyst (p. 422, KQ 11.3) Information specialist who performs systems analysis, design, and implementation. Why it's important: *The systems analyst studies the information and communications needs of an organization to determine how to deliver information that is more accurate, timely, and useful. The systems analyst achieves this goal through the problem-solving method of systems analysis and design.*

1. Do a preliminary design, using CASE tools, prototyping tools, and project management software, among others.
2. Do a detail design, defining requirements for output, input, storage, and processing and system controls and backup.
3. Write a report.

systems design (p. 425, KQ 11.3) Phase 3 of the SDLC; the purpose is to do a preliminary design and then a detail design, and write a report. Why it's important: *Systems design is one of the most crucial phases of the SDLC.*

systems development (p. 426, KQ 11.3) Phase 4 of the SDLC; hardware and software for the new system are acquired and tested. The fourth phase begins once management has accepted the report containing the design and has approved the way to development. Why it's important: *This phase may involve the organization in investing substantial time and money.*

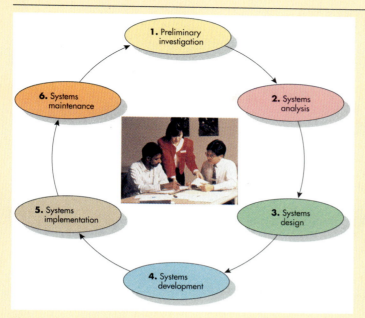

systems development life cycle (SDLC) (p. 422, KQ 11.3) Six-phase process that many organizations follow during systems analysis and design: (1) *preliminary investigation;* (2) *systems analysis;* (3) *systems design;* (4) *systems development;* (5) *systems implementation;* (6) *systems maintenance.* Phases often overlap, and a new one may start before the old one is finished. After the first four phases, management must decide whether to proceed to the next phase. User input and review is a critical part of each phase. Why it's important: *The SDLC is a comprehensive tool for solving organizational problems, particularly those relating to the flow of computer-based information.*

systems implementation (p. 427, KQ 11.3) Phase 5 of the SDLC; consists of converting the hardware, software, and files to the new system and training the users. Why it's important: *This phase is important because it involves putting design ideas into operation.*

systems maintenance (p. 428, KQ 11.3) Phase 6 of the SDLC; consists of keeping the system working by having system audits and periodic evaluations. Why it's important: *This phase is important for keeping a new system operational and useful.*

top managers (p. 412, KQ 11.1) One of the three types of managers; also called *strategic managers,* they are concerned with long-range planning and strategic decisions. Why it's important: *Top managers need information that is unstructured—that is, summarized, less current, future-oriented, covering a broad range of facts, and concerned with events outside as well as inside the organization.*

transaction (p. 416, KQ 11.2) Recorded event having to do with routine business activities (for example, materials purchased, employees hired, or taxes paid). Why it's important: *Today in most organizations the bulk of transactions are recorded in a computer-based information system.*

transaction processing system (TPS) (p. 416, KQ 11.2) Computer-based information system that keeps track of the transactions needed to conduct business. Inputs are transaction data (for example, bills, orders, inventory levels, production output). Outputs are processed transactions (for example, bills, paychecks). Each functional area of an organization—Research and Development, Production, Marketing, and Accounting and Finance—usually has its own TPS. Why it's important: *The TPS helps supervisory managers in making operational decisions. The database of transactions stored in a TPS are used to support a management information system and a decision support system.*

unstructured information (p. 416, KQ 11.1) Summarized, less current information concerned with future events; it records a broad range of facts and covers activities outside as well as inside an organization. Why it's important: *Top managers need information in the form of highly unstructured reports. The information should cover large time periods and survey activities outside as well as inside the organization.*

Chapter Review

Self-Test Questions

1. _____ managers are concerned with long-range, or strategic, planning and decisions.

2. _____ managers make tactical decisions to implement the strategic goals of an organization.

3. _____ managers make operational decisions—predictable decisions that can be made by following well-defined sets of routine decisions.

3. A _____ is a recorded event having to do with routine business activities.

4. A _____ is a mathematical representation of a real system.

5. _____ combine various technologies to reduce the manual labor required in operating an efficient office environment.

6. A _____ is a collection of related components that interact to perform a task in order to accomplish a goal.

Multiple-Choice Questions

1. One of the following activites is not an objective of phase 1 of the SDLC, preliminary investigation. Which one?

   a. conduct preliminary analysis

   b. describe costs and benefits

   c. acquire new software and hardware

   d. submit a preliminary plan

   e. propose alternative solutions

2. One of the following activities is not an objective of phase 4 of the SDLC, systems development. Which one?

   a. convert files to the new system

   b. acquire software

   c. acquire hardware

   d. test the system

   e. address the make-or-buy decision

True/False Questions

T F 1. Summary reports enable systems analysts to present graphic, or pictorial, representations of a new system.

T F 2. CASE tools—programs that automate various activities of the SDLC—are used only in phase 3.

T F 3. Four methods of systems implementation are direct, parallel, phased, and pilot.

T F 4. User training takes place during phase 1 of the SDLC.

Short-Answer Questions

1. What does an organization chart show?

2. What is the difference between structured information and unstructured information?

3. Briefly define the following: detail report, summary report, exception report, periodic report, demand report.

4. What is an expert system, and what can it be used for?

5. What are the six phases of the SDLC?

6. What does a systems analyst do?

7. What is a prototype, and what does it do?

Concept Mapping

On a separate sheet of paper, draw a concept map, or visual diagram, linking concepts. Show how the following terms are related.

CASE	structured information
data flow diagram	supervisory managers
DSS	system
detail design	systems analysis
ESS	systems design
make-or-buy decision	systems development
MIS	systems implementation
middle-level managers	systems maintenance
model	top managers
modeling tools	transaction
OAS	TPS
organization chart	unstructured information
preliminary design	
preliminary investigation	
prototype	
SDLC	

Knowledge in Action

1. Using an Internet search tool, identify a company that develops CASE tools. In a few paragraphs, describe what this company's CASE tools are used for.

2. Design a system that would handle the input, processing, and output of a simple form of your choice. Use a data flow diagram to illustrate the system.

3. Have you participated in a project that failed? Why did it fail? Based on what you know now, what might you have done to help the project succeed?

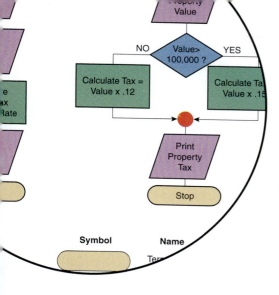

Software Development

Programming & Languages

Key Questions

You should be able to answer the following questions.

A.1   **Programming: A Five-Step Procedure** What is programming, and what are the five steps in accomplishing it?

A.2   **The First Step: Clarify the Programming Needs** How are programming needs clarified?

A.3   **The Second Step: Design the Program** How is a program designed?

A.4   **The Third Step: Code the Program** What is involved in coding a program?

A.5   **The Fourth Step: Test the Program** How is a program tested?

A.6   **The Fifth Step: Document & Maintain the Program** What is involved in documenting and maintaining a program?

A.7   **Five Generations of Programming Languages** What are the five generations of programming languages?

A.8   **Object-Oriented & Visual Programming** How do OOP and visual programming work?

A.9   **Internet Programming: HTML, XML, VRML, Java, ActiveX, & Scripting Languages** What are the features of HTML, XML, VRML, Java, ActiveX, and Scripting Languages?

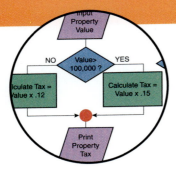

A.1 Programming: A Five-Step Procedure

To see how programming works, we must understand what a program is. A **_program_** **is a list of instructions that the computer must follow in order to process data into information.** The instructions consist of *statements* used in a programming language, such as BASIC. Examples are programs that do word processing, desktop publishing, or payroll processing.

The decision whether to buy or create a program forms part of Phase 4 in the systems development life cycle. *(See ● Panel A.1.)* Once the decision is made to develop a new system, the programmer goes to work.

● **PANEL A.1**

Where programming fits in the systems development life cycle

The fourth phase of the six-phase systems development life cycle includes a five-step procedure of its own. These five steps constitute the problem-solving process called *programming.*

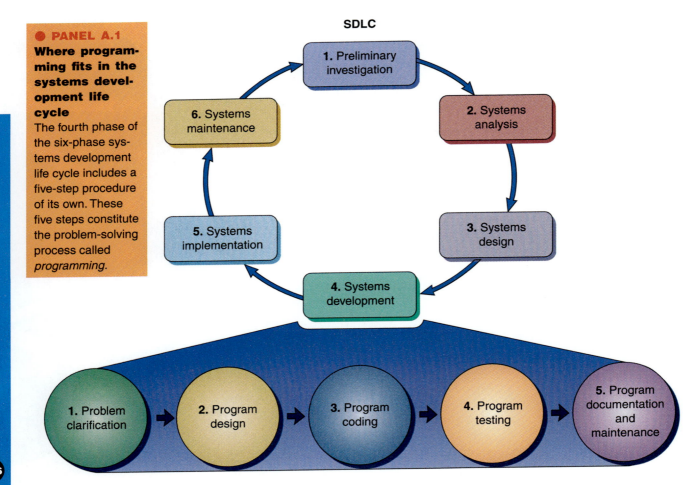

A program, we said, is a list of instructions for the computer. **_Programming_**, also called software engineering, is a multistep process for creating that list of instructions.

The five steps are as follows.

1. Clarify the problem—include needed output, input, processing requirements.

2. Design a solution—use modeling tools to chart the program.

3. Code the program—use a programming language's syntax, or rules, to write the program.

4. Test the program—get rid of any logic errors, or "bugs," in the program ("debug" it).

5. Document and maintain the program—include written instructions for users, explanation of the program, and operating instructions.

Coding—sitting at the keyboard and typing words into a computer—is what many people imagine programming to be. As we see, however, it is only one of the five steps. Coding consists of translating the logic requirements into a programming language—the letters, numbers, and symbols that make up the program.

A.2 The First Step: Clarify the Programming Needs

KEY QUESTION

How are programming needs clarified?

The **_problem clarification_** step consists of six mini-steps—clarifying program objectives and users, outputs, inputs, and processing tasks; studying the feasibility of the program; and documenting the analysis. *(See ● Panel A.2.)* Let us consider these six mini-steps.

1. Clarify Objectives & Users

You solve problems all the time. A problem might be deciding whether to take a required science course this term or next, or selecting classes that allow you also to fit a job into your schedule. In such cases, you are specifying your *objectives.* Programming works the same way. You need to write a statement of the objectives you are trying to accomplish—the problem you are trying to solve. If the problem is that your company's systems analysts have designed a new computer-based payroll processing program and brought it to you as the programmer, you need to clarify the programming needs.

You also need to make sure you know who the users of the program will be. Will they be people inside the company, outside, or both? What kind of skills will they bring?

2. Clarify Desired Outputs

Make sure you understand the outputs—what the system designers want to get out of the system—before you specify the inputs. For example, what kind

● **PANEL A.2**

First step: clarify programming needs

1. Specify program objectives and program users.
2. Specify output requirements.
3. Specify input requirements.
4. Specify processing requirements.
5. Study feasibility of implementing program.
6. Document the analysis.

of hardcopy is wanted? What information should the outputs include? This step may require several meetings with systems designers and users to make sure you're creating what they want.

3. Clarify Desired Inputs

Once you know the kind of outputs required, you can then think about input. What kind of input data is needed? What form should it appear in? What is its source?

4. Clarify the Desired Processing

Here you make sure you understand the processing tasks that must occur in order for input data to be processed into output data.

5. Double-Check the Feasibility of Implementing the Program

Is the kind of program you're supposed to create feasible within the present budget? Will it require hiring a lot more staff? Will it take too long to accomplish?

Sometimes programmers decide they can buy an existing program and modify it rather than write it from scratch.

6. Document the Analysis

Throughout program clarification, programmers must document everything they do. This includes writing objective specifications of the entire process being described.

A.3 The Second Step: Design the Program

KEY QUESTION
How is a program designed?

Assuming the decision is to make, or custom-write, the program, you then move on to design the solution specified by the systems analysts. In the **_program design_** step, the software is designed in three mini-steps. First, the program logic is determined through a top-down approach and modularization, using a hierarchy chart. Then it is designed in detail, either in narrative form, using pseudocode, or graphically, using flowcharts. *(See ● Panel A.3.)*

It used to be that programmers took a kind of a seat-of-the-pants approach to programming. Programming was considered an art, not a science. Today, however, most programmers use a design approach called structured programming. **_Structured programming_ takes a top-down approach that breaks programs into modular forms. It also uses standard logic tools called control structures (sequential, selection, case, and iteration).**

● **PANEL A.3**
Second step: program design

1. Determine program logic through top-down approach and modularization, using a hierarchy chart.
2. Design details using pseudocode and/or flowcharts, preferably involving control structures.

The point of structured programming is to make programs more efficient (with fewer lines of code) and better organized (more readable), and to have better notations so that they have clear and correct descriptions.

The three mini-steps of program design are as follows.

1. Determine the Program Logic, Using a Top-Down Approach

Determining the program logic is like outlining a long term paper before you proceed to write it. **_Top-down program design_ proceeds by identifying the top element, or module, of a program and then breaking it down in hierarchical fashion to the lowest level of detail. The top-down program design is used to identify the program's processing steps, or modules.** After the program is designed, the actual coding proceeds from the bottom up, using the modular approach.

The concept of modularization is important. _Modularization dramatically simplifies program development, because each part can be developed and tested separately._

A **_module_ is a processing step of a program. Each module is made up of logically related program statements.** (Sometimes a module is called a _subprogram_ or _subroutine_.) An example of a module might be a programming instruction that simply says "Open a file, find a record, and show it on the display screen." It is best if each module has only a single function, just as an English paragraph should have a single, complete thought. This rule limits the module's size and complexity.

Top-down program design can be represented graphically in a hierarchy chart. A **_hierarchy chart_, or structure chart, illustrates the overall purpose of the program, by identifying all the modules needed to achieve that purpose and the relationships among them.** (See ● _Panel A.4, next page._) The program must move in sequence from one module to the next until all have been processed. There must be three principal modules corresponding to the three principal computing operations—input, processing, and output. (In Panel A.4 they are "Read input," "Calculate pay," and "Generate output.")

2. Design Details, Using Pseudocode and/or Flowcharts

Once the essential logic of the program has been determined, through the use of top-down programming and hierarchy charts, you can go to work on the details.

There are two ways to show details—write them or draw them; that is, use _pseudocode_ or use _flowcharts._ Most projects use both methods.

- Pseudocode: **_Pseudocode_ is a method of designing a program using normal human-language statements to describe the logic and processing flow.** (See ● _Panel A.5, next page._) Pseudocode is like an outline or summary form of the program you will write.

    Sometimes pseudocode is used simply to express the purpose of a particular programming module in somewhat general terms. With the use of such terms as IF, THEN, or ELSE, however, the pseudocode follows the rules of _control structures,_ an important aspect of structured programming, as we shall explain.

- Program flowcharts: We described system flowcharts in the previous chapter. Here we consider program flowcharts. A **_program flowchart_ is a chart that graphically presents the detailed series of steps (algorithm, or logical flow) needed to solve a programming problem.** The flowchart uses standard symbols—called _ANSI symbols,_ after the American National Standards Institute, which developed them. (See ● _Panel A.6, p. 441._)

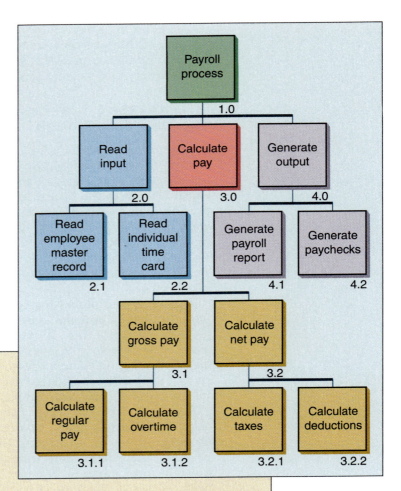

A hierarchy chart
This represents a top-down design for a payroll program. Here the modules, or processing steps, are represented from the highest level of the program down to details. The three principal processing operations—input, processing, and output—are represented by the modules in the second layer: "Read input," "Calculate pay," and "Generate output." Before tasks at the top of the chart can be performed, all the ones below must be performed. Each module represents a logical processing step.

1. Each module must be of manageable size.

2. Each module should be independent and have a single function.

3. The functions of input and output are clearly defined in separate modules.

4. Each module has a single entry point (execution of the program module always starts at the same place) and a single exit point (control always leaves the module at the same place).

5. If one module refers to or transfers control to another module, the latter module returns control to the point from which it was "called" by the first module.

```
START
DO WHILE (so long as) there are records
        Read a customer billing account record
        IF today's date is greater than 30 days from
        date of last customer payment
            Calculate total amount due
            Calculate 5% interest on amount due
            Add interest to total amount due to calculate
            grand total
            Print on invoice overdue amount
        ELSE
            Calculate total amount due
        ENDIF
        Print out invoice
END DO
END
```

Pseudocode

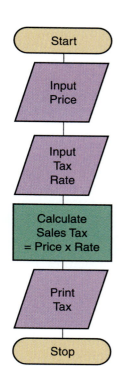

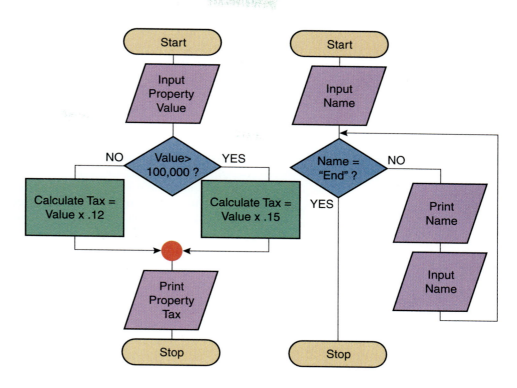

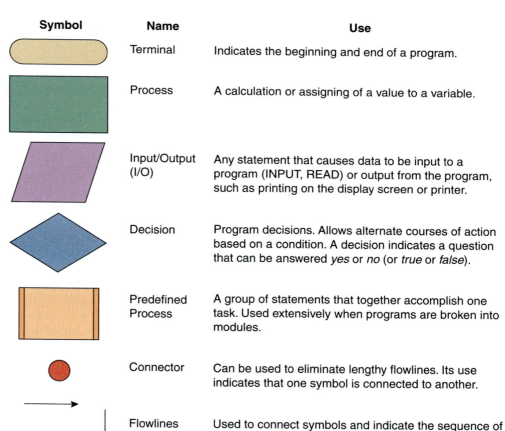

Symbol	Name	Use
Terminal	Terminal	Indicates the beginning and end of a program.
Process	Process	A calculation or assigning of a value to a variable.
Input/Output	Input/Output (I/O)	Any statement that causes data to be input to a program (INPUT, READ) or output from the program, such as printing on the display screen or printer.
Decision	Decision	Program decisions. Allows alternate courses of action based on a condition. A decision indicates a question that can be answered *yes* or *no* (or *true* or *false*).
Predefined Process	Predefined Process	A group of statements that together accomplish one task. Used extensively when programs are broken into modules.
Connector	Connector	Can be used to eliminate lengthy flowlines. Its use indicates that one symbol is connected to another.
Flowlines and Arrowheads	Flowlines and Arrowheads	Used to connect symbols and indicate the sequence of operations. The flow is assumed to go from top to bottom and from left to right. Arrowheads are only required when the flow violates the standard direction.

● PANEL A.6

Example of a program flowchart and explanation of flowchart symbols
This example represents a flowchart for a payroll program.

The symbols at the left of the drawing might seem clear enough. But how do you figure out the *logic* of a program? How do you reason the program out so it will really work? The answer is to use control structures, as explained next.

- **Control structures:** When you're trying to determine the logic behind something, you use words like "if" and "then" and "else." (For example, without using these exact words, you might reason along these lines: "*If* she comes over, *then* we'll go out to a movie, *else* I'll just stay in and watch TV.") Control structures make use of the same words. A **_control structure_, or logic structure, is a structure that controls the logical sequence in which computer program instructions are executed. In structured program design, three control structures are used to form the logic of a program: sequence, selection, and iteration (or loop).** *(See ● Panel A.7.)* These are the tools with which you can write structured programs and take a lot of the guesswork out of programming. (Additional variations of these three basic structures are also used.)

  One thing that all three control structures have in common is *one entry* and *one exit*. The control structure is entered at a single point and exited at another single point. This helps simplify the logic so that it is easier for others following in a programmer's footsteps to make sense of the program. (In the days before this requirement was instituted, programmers could have all kinds of variations, leading to the kind of incomprehensible program known as *spaghetti code*.)

  Let us consider the three control structures:

  (1) In the **_sequence control structure_, one program statement follows another in logical order.** For example, in the example shown in Panel A.7, there are two boxes ("statement" and "statement"). One box could say "Open file," the other "Read a record." There are no decisions to make, no choices between "yes" or "no." The boxes logically follow one another in sequential order.

  (2) The **_selection control structure_—also known as an _IF-THEN-ELSE structure_—represents a choice. It offers two paths to follow when a decision must be made by a program.** An example of a selection structure is as follows:

  IF a worker's hours in a week exceed 40
  THEN overtime hours equal the number of hours exceeding 40
  ELSE the worker has no overtime hours.

  A variation on the usual selection control structure is the *case control structure*. This offers more than a single yes-or-no decision. The case structure allows several alternatives, or "cases," to be presented. "IF Case 1 occurs, THEN do thus-and-so. IF Case 2 occurs, THEN follow an alternative course . . ." And so on.) The case control structure saves the programmer the trouble of having to indicate a lot of separate IF-THEN-ELSE conditions.

  (3) In the **_iteration_, or _loop, control structure_, a process may be repeated as long as a certain condition remains true.** There are two types of iteration structures—*DO UNTIL* and *DO WHILE*. Of these, DO UNTIL is more often encountered.

  An example of a DO UNTIL structure is as follows:

  DO read in employee records UNTIL there are no more employee records.

  An example of a DO WHILE structure is as follows:

  DO read in employee records WHILE—that is, as long as—there continue to be employee records.

Sequence control structure
(one program statement follows another
in logical order)

Iteration control structures:
DO UNTIL and DO WHILE

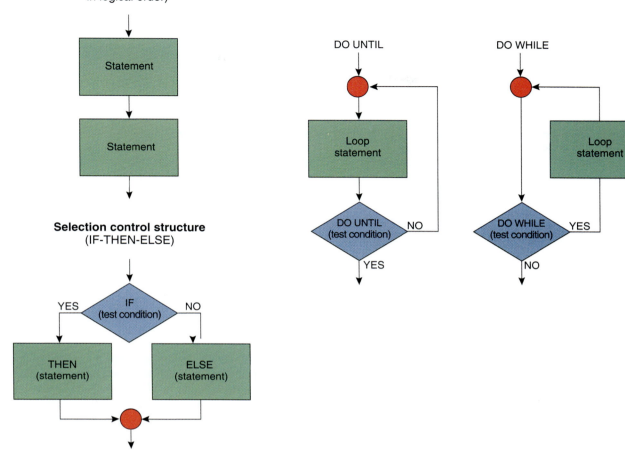

Selection control structure
(IF-THEN-ELSE)

Variation on selection: the case control structure
(more than a single yes-or-no decision)

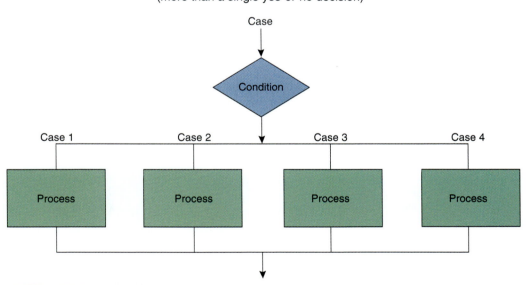

● **PANEL A.7**

The three control structures
The three structures used in structured program design to
form the logic of a program are *sequence, selection,* and
iteration. Case is an important version of selection.

What is the difference between the two iteration structures? It is simply this: If several statements need to be repeated, you must decide when to *stop* repeating them. You can decide to stop them at the *beginning* of the loop, using the DO WHILE structure. Or you can decide to stop them at the *end* of the loop, using the DO UNTIL structure. The DO UNTIL iteration means that the loop statements will be executed at least once, because in this case the iteration statements are executed *before* the program checks whether to stop.

A.4   The Third Step: Code the Program

KEY QUESTION

What is involved in coding a program?

Once the design has been developed, the actual writing of the program begins. **Writing the program is called _coding_.** *(See ● Panel A.8.)* Coding is what many people think of when they think of programming, although it is only one of the five steps. Coding consists of translating the logic requirements from pseudocode or flowcharts into a programming language—the letters, numbers, and symbols that make up the program.

1. Select the Appropriate Programming Language

A **_programming language_ is a set of rules that tells the computer what operations to do.** Examples of well-known programming languages are BASIC, COBOL, and C. These are called "high-level languages," as we explain in a few pages.

Not all languages are appropriate for all uses. Some, for example, have strengths in mathematical and statistical processing. Others are more appropriate for database management. Thus, in choosing the language, you need to consider what purpose the program is designed to serve and what languages are already being used in your organization or in your field. We consider these matters in the second half of this Appendix.

2. Follow the Syntax

In order for a program to work, you have to follow the **_syntax_, the rules of the programming language.** Programming languages have their own grammar just as human languages do. But computers are probably a lot less forgiving if you use these rules incorrectly.

A.5   The Fourth Step: Test the Program

KEY QUESTION

How is a program tested?

Program testing involves running various tests and then running real-world data to make sure the program works. *(See ● Panel A.9.)* Two principal

1. Select the appropriate high-level programming language.
2. Code the program in that language, following the syntax carefully.

● **PANEL A.8**

Third step: program coding

The third step in programming is to translate the logic of the program worked out from pseudocode or flowcharts into a high-level programming language, following its grammatical rules.

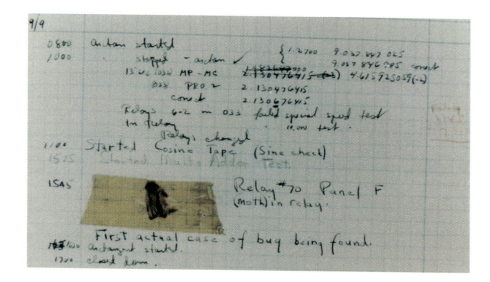

activities are *desk-checking* and *debugging*. These steps are called *alpha testing*.

1. Perform Desk-Checking

Desk-checking **is simply reading through, or checking, the program to make sure that it's free of errors and that the logic works.** In other words, desk-checking is like proofreading. This step should be taken before the program is actually run on a computer.

2. Debug the Program

Once the program has been desk-checked, further errors, or "bugs," will doubtless surface. To **debug** **means to detect, locate, and remove all errors in a computer program.** Mistakes may be syntax errors or logical errors. **Syntax errors** **are caused by typographical errors and incorrect use of the programming language.** **Logic errors** **are caused by incorrect use of control structures.** Programs called *diagnostics* exist to check program syntax and display syntax-error messages. Diagnostic programs thus help identify and solve problems.

3. Run Real-World Data

After desk-checking and debugging, the program may run fine—in the laboratory. However, it needs to be tested with real data; this is called *beta testing*. Indeed, it is even advisable to test the program with *bad data*—data that is faulty, incompete, or in overwhelming quantities—to see if you can make the system crash. Many users, after all, may be far more heavy-handed, ignorant, and careless than programmers have anticipated.

Several trials using different test data may be required before the programming team is satisfied that the program can be released. Even then, some bugs may persist, because there comes a point where the pursuit of errors is uneconomical. This is one reason why many users are nervous about using the first version (version 1.0) of a commercial software package.

A.6 The Fifth Step: Document & Maintain the Program

Writing the program documentation is the fifth step in programming. The resulting **documentation** **consists of written descriptions of what a program is and how to use it.** Documentation is not just an end-stage process of programming. It has been (or should have been) going on throughout all

1. Write user documentation.
2. Write operator documentation.
3. Write programmer documentation.
4. Maintain the program.

programming steps. Documentation is needed for people who will be using or be involved with the program in the future. *(See* ● *Panel A.10.)*

Documentation should be prepared for several different kinds of readers—users, operators, and programmers.

1. Prepare User Documentation

When you buy a commercial software package, such as a spreadsheet, you normally get a manual with it. This is *user documentation.*

2. Prepare Operator Documentation

The people who run large computers are called *computer operators.* Because they are not always programmers, they need to be told what to do when the program malfunctions. The *operator documentation* gives them this information.

3. Write Programmer Documentation

Long after the original programming team has disbanded, the program may still be in use. If, as is often the case, a fifth of the programming staff leaves every year, after 5 years there could be a whole new bunch of programmers who know nothing about the software. *Program documentation* helps train these newcomers and enables them to maintain the existing system.

4. Maintain the Program

A word about maintenance: *Maintenance* includes any activity designed to keep programs in working condition, error-free, and up to date—adjustments, replacements, repairs, measurements, tests, and so on. The rapid changes in modern organizations—in products, marketing strategies, accounting systems, and so on—are bound to be reflected in their computer systems. Thus, maintenance is an important matter, and documentation must be available to help programmers make adjustments in existing systems.

The five steps of the programming process are summarized in the accompanying table. *(See* ● *Panel A.11.)*

A.7 Five Generations of Programming Languages

As we've said, a *programming language* is a set of rules that tells the computer what operations to do. Programmers, in fact, use these languages to create other kinds of software. Many programming languages have been written, some with colorful names (SNOBOL, HEARSAY, DOCTOR, ACTORS, JOVIAL). Each is suited to solving particular kinds of problems. What do all these languages have in common? Simply this: Ultimately they must be reduced to digital form—a 1 or 0, electricity on or off—because that is all the computer can work with.

PANEL A.11
Summary of the five programming steps

Step	Activities
Step 1: Problem definition	1. Specify program objectives and program users. 2. Specify output requirements. 3. Specify input requirements. 4. Specify processing requirements. 5. Study feasibility of implementing program. 6. Document the analysis.
Step 2: Program design	1. Determine program logic through top-down approach and modularization, using a hierarchy chart. 2. Design details using pseudocode and/or using flowcharts, preferably on the basis of control structures. 3. Test design with structured walkthrough.
Step 3: Program coding	1. Select the appropriate high-level programming language. 2. Code the program in that language, following the syntax carefully.
Step 4: Program testing	1. Desk-check the program to discover errors. 2. Run the program and debug it (alpha testing). 3. Run real-world data (beta testing).
Step 5: Program documentation and maintenance	1. Prepare user documentation. 2. Write operator documentation. 3. Write programmer documentation. 4. Maintain the program.

To see how this works, it's important to understand that there are five *levels* or *generations* of programming languages, ranging from low-level to high-level. **The five _generations of programming languages_ start at the lowest level with (1) machine language. They then range up through (2) assembly language, (3) high-level languages (procedural languages), and (4) very high-level languages (problem-oriented languages). At the highest level are (5) natural languages.** Programming languages are said to be *lower level* when they are closer to the language that the computer itself uses—the 1s and 0s. They are called *higher level* when they are closer to the language people use—more like English, for example.

Beginning in 1945, the five levels or generations have evolved over the years, as programmers gradually adopted the later generations. The births of the generations are as follows. *(See ● Panel A.12.)*

- First generation, 1945—*Machine language*
- Second generation, mid-1950s—*Assembly language*
- Third generation, early 1960s—*High-level languages (procedural languages):* Examples are COBOL, BASIC, and C
- Fourth generation, early 1970s—*Very high-level languages (problem-oriented languages):* SQL, Intellect, NOMAD, FOCUS
- Fifth generation, early 1980s—*Natural languages*

Let's consider these five generations.

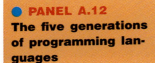

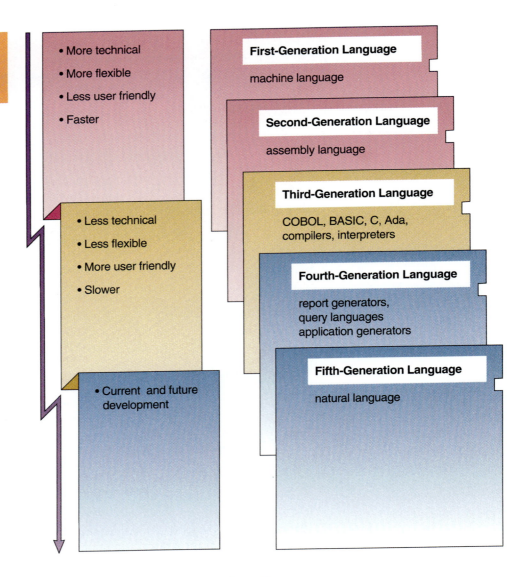

PANEL A.12
The five generations of programming languages

- More technical
- More flexible
- Less user friendly
- Faster

- Less technical
- Less flexible
- More user friendly
- Slower

- Current and future development

First-Generation Language

machine language

Second-Generation Language

assembly language

Third-Generation Language

COBOL, BASIC, C, Ada, compilers, interpreters

Fourth-Generation Language

report generators, query languages application generators

Fifth-Generation Language

natural language

First Generation: Machine Language

Machine language **is the basic language of the computer, representing data as 1s and 0s.** *(See* ● *Panel A.13.)* Machine language programs varied from computer to computer; that is, they were *machine-dependent.*

These binary digits, which correspond to the on and off electrical states of the computer, are clearly not convenient for people to read and use. Believe it or not, though, programmers *did* work with these mind-numbing digits. There must have been great sighs of relief when the next generation of programming languages—assembly language—came along.

Second Generation: Assembly Language

Assembly language **is a low-level programming language that allows a computer user to write a program using abbreviations or more easily remembered words instead of numbers.** *(Refer to* ● *Panel A.13 again.)* For example, the letters MP could be used to represent the instruction MULTIPLY and STO to represent STORE.

As you might expect, a programmer can write instructions in assembly language more quickly than in machine language. Nevertheless, it is still not an easy language to learn, and it is so tedious to use that mistakes are frequent. Moreover, assembly language has the same drawback as machine language in that it varies from computer to computer—it is machine-dependent.

First generation
Machine language

```
11110010 01110011 1101 001000010000 0111 000000101011
11110010 01110011 1101 001000011000 0111 000000101111
11111100 01010010 1101 001000010010 1101 001000011101
11110000 01000101 1101 001000010011 0000 000000111110
11110011 01000011 0111 000001010000 1101 001000010100
10010110 11110000 0111 000001010100
```

Second generation
Assembly language

```
PACK 210(8,13),02B(4,7)
PACK 218(8,13),02F(4,7)
MP   212(6,13),21D(3,13)
SRP  213(5,13),03E(0),5
UNPK 050(5,7),214(4,13)
OI   054(7),X FO
```

Third generation
COBOL

```
MULTIPLY HOURS-WORKED BY PAY-RATE GIVING GROSS-PAY ROUNDED
```

We now need to introduce the concept of *language translator.* Because a computer can execute programs only in machine language, a translator or converter is needed if the program is written in any other language. A ***language translator*** **is a type of systems software that translates a program written in a second-, third-, or higher-generation language into machine language.**

Language translators are of three types:

- Assemblers
- Compilers
- Interpreters

An ***assembler***, **or assembler program, is a program that translates the assembly-language program into machine language.** We describe compilers and interpreters in the next section.

Third Generation: High-Level or Procedural Languages

A ***high-level*** **or** ***procedural language*** **resembles some human language such as English;** an example is COBOL, which is used for business applications. *(Refer again to* ● *Panel A.13.)* A procedural language allows users to write in a familiar notation, rather than numbers or abbreviations. Also, unlike machine and assembly languages, most are not machine-dependent—that is, they can be used on more than one kind of computer. Familiar languages of this sort include FORTRAN, COBOL, BASIC, Pascal, and C.

For a procedural language to work on a computer, it needs a *language translator* to translate it into machine language. Depending on the procedural language, either of two types of translators may be used—a *compiler* or an *interpreter*.

- **Compiler—execute later:** A ***compiler*** **is a language translator that converts the entire program of a high-level language into machine language BEFORE the computer executes the program.** The programming instructions of a procedural language are called the *source code.* The compiler translates it into machine language, which in this case is called the *object code.* The important point here is that the object code *can be saved* and thus can be executed later (as many times as desired), rather than run right away. *(See ● Panel A.14.)*

  Examples of procedural languages using compilers are COBOL, FORTRAN, Pascal, and C.

- **Interpreter—execute immediately:** An ***interpreter*** **is a language translator that converts each procedural language statement into machine language and executes it IMMEDIATELY, statement by statement.** In contrast to the compiler, no object code is saved. Therefore, interpreted code generally runs more slowly than compiled code. However, code can be tested line by line.

  BASIC is a procedural language using an interpreter.

Who cares, you might say, whether you can run a program now or later? (After all, "later" could be only a matter of seconds or minutes.) Here's the significance: When a compiler is used, it requires *two* steps (the source code and the object code) before the program can be executed. The interpreter, on the other hand, requires only *one* step. The advantage of a compiler language is that, once you have obtained the object code, *the program executes faster.* The advantage of an interpreter language, on the other hand, is that *programs are easier to develop.*

Some of the most popular procedural languages are *COBOL, BASIC* (and *Visual BASIC*), and *C* (and *C++*). *(See ● Panel A.15, on the next two pages.)*

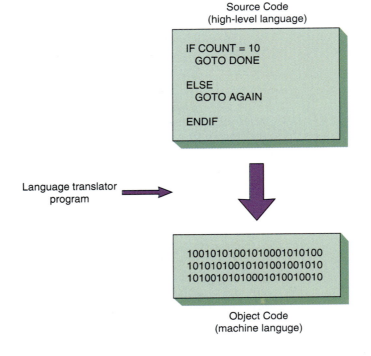

● PANEL A.14
Compiler
This language translator converts the procedural language (source code) into machine language (object code) before the computer can execute the program.

Source Code
(high-level language)

IF COUNT = 10
  GOTO DONE

ELSE
  GOTO AGAIN

ENDIF

Language translator
program

100101010010100010101000
101010100101010010010100
101001010100010100100100

Object Code
(machine languge)

COBOL: The Language of Business

Formally adopted in 1960, *COBOL* (for COmmon Business Oriented Language) was originally the most frequently used business programming language for large computers. Its most significant attribute is that it is extremely readable. For example, a COBOL line might read: MULTIPLY HOURLY-RATE BY HOURS-WORKED GIVING GROSS-PAY

Writing a COBOL program resembles writing an outline for a research paper. The program is divided into four divisions—Identification, Environment, Data, and Procedure. The divisions in turn are divided into sections, which are divided into paragraphs, which are divided into sentences.

COBOL has both advantages and disadvantages.

- **Advantages:** (1) It is machine-independent. (2) Its English-like statements are easy to understand, even for a nonprogrammer. (3) It can handle many files, records, and fields. (4) It easily handles input-output operations.
- **Disadvantages:** (1) Because it is so readable, it is wordy. Thus, even simple programs are lengthy, and programmer productivity is slowed. (2) It cannot handle mathematical processing as well as FORTRAN.

BASIC: The Easy Language

BASIC was developed by John Kemeny and Thomas Kurtch in 1965 for use in training their students at Dartmouth College. By the late 1960s, it was widely used in academic settings on all kinds of computers, from mainframes to PCs. Now its use has extended to business.

BASIC (for Beginner's All-purpose Symbolic Instruction Code) has been the most popular microcomputer language and is considered the easiest programming language to learn. Although it is available in compiler form, the interpreter form is more popular with first-time and casual users. This is because it is interactive, meaning that user and computer can communicate with each other during the writing and running of the program. Today there are competing versions of BASIC. One of the current evolutions is Visual BASIC.

The advantage and disadvantages of BASIC are as follows:

- **Advantage:** BASIC is easy to use.
- **Disadvantages:** (1) Its processing speed is relatively slow, although compiler versions are faster than interpreter versions. (2) There is no single dominant version of BASIC, although in 1987 ANSI adopted a new standard that eliminated portability problems.

C: For Portability & Scientific Use

"C" is the language's entire name; it does not "stand" for anything. Developed at Bell Laboratories, *C* is a general-purpose, compiled language that works well for microcomputers and is portable among many computers. It is widely used in commercial software development, including games, robotics, and graphics.

Here are the advantages and disadvantages of C:

- **Advantages:** (1) C works well with microcomputers. (2) It has a high degree of portability—it can be run without change on a variety of computers. (3) It is fast and efficient. (4) It enables the programmer to manipulate individual bits in main memory.
- **Disadvantages:** (1) C is considered difficult to learn. (2) Because of its conciseness, the code can be difficult to follow. (3) It is not suited to applications that require a lot of report formatting.

This shows how three languages handle the same statement. The statement specifies that a customer gets a discount of 7% of the invoice amount if the invoice is greater than $500; if the invoice is lower, there is no discount.

COBOL

```
OPEN-INVOICE-FILE.
    OPEN I-O INVOICE FILE.

READ-INVOICE-PROCESS.
    PERFORM READ-NEXT-REC THROUGH READ-NEXT-REC-EXIT UNTIL END-OF-FILE.
    STOP RUN.

READ-NEXT-REC.
    READ INVOICE-REC
        INVALID KEY
            DISPLAY 'ERROR READING INVOICE FILE'
            MOVE 'Y' TO EOF-FLAG
            GOTO READ-NEXT-REC-EXIT.
    IF INVOICE-AMT > 500
        COMPUTE INVOICE-AMT = INVOICE-AMT – (INVOICE-AMT * .07)
        REWRITE INVOICE-REC.

READ-NEXT-REC-EXIT.
    EXIT.
```

BASIC

```
10  REM        This Program Calculates a Discount Based on the Invoice Amount
20  REM             If Invoice Amount is Greater Than 500, Discount is 7%
30  REM             Otherwise Discount is 0
40  REM
50  INPUT "What is the Invoice Amount"; INV.AMT
60  IF INV.AMT A> 500 THEN LET DISCOUNT = .07 ELSE LET DISCOUNT = 0
70  REM             Display results
80  PRINT "Original Amt", "Discount", "Amt after Discount"
90  PRINT INV.AMT, INV.AMT * DISCOUNT, INV.AMT – INV.AMT * DISCOUNT
100 END
```

C

```
if (invoice_amount > 500.00)

    DISCOUNT = 0.07 * invoice_amount;

else

    discount = 0.00;

invoice_amount = invoice_amount – discount;
```

Fourth Generation: Very High Level or Problem-Oriented Languages

Very high level or problem-oriented languages, also called fourth-generation languages (4GLs), are much more user-oriented and allow users to develop programs with fewer commands compared with procedural languages, although they require more computing power. These languages are called _problem-oriented_ because they are designed to solve specific problems, whereas procedural languages are more general-purpose languages.

Two types of problem-oriented languages are query languages and application generators.

- **Query languages:** A _query language_ **is an easy-to-use language for retrieving data from a database management system.** The query may be expressed in the form of a sentence or near-English command. Or the query may be obtained from choices on a menu.

  Examples of query languages are SQL (for Structured Query Language) and Intellect. For example, with Intellect, which is used with IBM mainframes, you can ask an English-language question such as "Tell me the number of employees in the sales department."

- **Application generators:** An _application generator_ **is a programmer's tool consisting of modules that have been preprogrammed to accomplish various tasks.** The benefit is that the programmer can generate applications programs from descriptions of the problem rather than by traditional programming, in which he or she has to specify how the data should be processed.

Programmers use application generators to help them create parts of other programs. For example, the software is used to construct onscreen menus or types of input and output screen formats. NOMAD and FOCUS, two database management systems, include application generators.

Fifth Generation: Natural Languages

Natural languages **are of two types. The first are ordinary human languages: English, Spanish, and so on. The second are programming languages that use human language to give people a more natural connection with computers.**

With a problem-oriented language, you can type in some rather routine inquiries, such as (in the language known as FOCUS) the following: SUM SHIPMENTS BY STATE BY DATE. Natural languages, by contrast, allow questions or commands to be framed in a more conversational way or in alternative forms—for example,

> I WANT THE SHIPMENTS OF PERSONAL DIGITAL ASSISTANTS FOR ALABAMA AND MISSISSIPPI BROKEN DOWN BY CITY FOR JANUARY AND FEBRUARY. ALSO, MAY I HAVE JANUARY AND FEBRUARY SHIPMENTS LISTED BY CITIES FOR PERSONAL COMMUNICATORS SHIPPED TO WISCONSIN AND MINNESOTA.

Natural languages are part of the field of study known as _artificial intelligence_ (discussed in detail in Chapter 10). Artificial intelligence (AI) is a group of related technologies that attempt to develop machines capable of emulating human-like qualities, such as learning, reasoning, communicating, seeing, and hearing.

Consider how it was for the computer pioneers, programming in machine language or assembly language. Novices putting together programs in BASIC or C can breathe a collective sigh of relief that they weren't around at the dawn of the Computer Age. Even some of the simpler third-generation languages represent a challenge, because they are procedure-oriented, forcing the programmer to follow a predetermined path.

Fortunately, two new developments have made things easier—*object-oriented programming* and *visual programming.*

Object-Oriented Programming: Block by Block

Imagine you're programming in a traditional third-generation language, such as BASIC, creating your coded instructions one line at a time. As you work on some segment of the program (such as how to compute overtime pay), you may think, "I'll bet some other programmer has already written something like this. Wish I had it. It would save a lot of time." Fortunately, a kind of recycling technique now exists. This is object-oriented programming.

1. **What OOP is:** In ***object-oriented programming*** (**OOP, pronounced "oop"**), **data and the instructions for processing that data are combined into a self-sufficient "object" that can be used in other programs.** The important thing here is the object.

2. **What an "object" is:** An *object* is a self-contained module consisting of preassembled programming code. The module contains, or encapsulates, both (1) a chunk of data and (2) the processing instructions that may be performed on that data.

3. **When an object's data is to be processed—sending the "message":** Once the object becomes part of a program, the processing instructions may or may not be activated. A particular set of instructions is activated only when the corresponding "message" is sent. A *message* is an alert sent to the object when an operation involving that object needs to be performed.

4. **How the object's data is processed— the "methods":** The message need only identify the operation. How it is actually to be performed is embedded within the processing instructions that are part of the object. These processing instructions within the object are called the *methods.*

Once you've written a block of program code (that computes overtime pay, for example), it can be reused in any number of programs. Thus, with OOP, unlike traditional programming, you don't have to start from scratch—that is, reinvent the wheel—each time.

Conventional Programs

Object-Oriented Programs

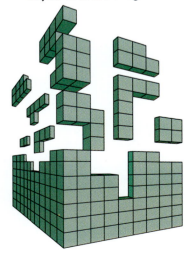

Object-oriented programming takes longer to learn than traditional programming because it means training oneself to a new way of thinking. However, the beauty of OOP is that an object can be used repeatedly in different applications and by different programmers, speeding up development time and lowering costs.

Three Important Concepts of OOP

Object-oriented programming has three important concepts, which go under the jaw-breaking names of *encapsulation, inheritance,* and *polymorphism.* Actually, these are not as fearsome as they look:

- **Encapsulation:** *Encapsulation* means an object contains (encapsulates) both (1) data and (2) the relevant processing instructions, as we have seen. Once an object has been created, it can be reused in other programs. An object's uses can also be extended through concepts of *class* and *inheritance.*

- **Inheritance:** Once you have created an object, you can use it as the foundation for similar objects that have the same behavior and characteristics. All objects that are derived from or related to one another are said to form a *class.* Each class contains specific instructions (methods) that are unique to that group.

   Classes can be arranged in hierarchies—classes and subclasses. *Inheritance* is the method of passing down traits of an object from classes to subclasses in the hierarchy. Thus, new objects can be created by *inheriting* traits from existing classes.

   Writer Alan Freedman gives this example: "The object MACINTOSH could be one instance of the class PERSONAL COMPUTER, which could inherit properties from the class COMPUTER SYSTEMS."[1] If you were to add a new computer, such as COMPAQ, you would need to enter only what makes it *different* from other computers. The *general* characteristics of personal computers could be inherited.

- **Polymorphism:** Polymorphism means the presence of "many shapes." In object-oriented programming, *polymorphism* means that a message (generalized request) produces different results based on the object that it is sent to.

   Polymorphism has important uses. It allows a programmer to create procedures about objects whose exact type is not known in advance but will be at the time the program is actually run on the computer. Freedman gives this example: "A screen cursor may change its shape from an arrow to a line depending on the program mode." The processing instructions "to move the cursor on screen in response to mouse movement would be written for 'cursor,' and polymorphism would allow that cursor to be whatever shape is required at runtime." It would also allow a new cursor shape to be easily integrated into the program.

   An example of an OOP language is **<u>C++</u>—the plus signs stand for "more than C"—which combines the traditional C programming language with object-oriented capability.** C++ was created by Bjarne Stroustrup. With C++, programmers can write standard code in C without the object-oriented features, use object-oriented features, or do a mixture of both.

Visual Programming: The Example of Visual BASIC

Essentially, visual programming takes OOP to the next level. The goal of visual programming is to make programming easier for programmers and more accessible to nonprogrammers, by borrowing the object orientation of

OOP languages but exercising it in a graphical or visual way. Visual programming enables users to think more about the problem solving than about handling the programming language. There is no learning of syntax or actual writing of code.

Visual programming is a method of creating programs in which the programmer makes connections between objects by drawing, pointing, and clicking on diagrams and icons and by interacting with flowcharts. Thus, the programmer can create programs by clicking on icons that represent common programming routines.

An example of visual programming is **_Visual BASIC_, a Windows-based, object-oriented programming language from Microsoft that lets users develop Windows and Office applications by (1) creating command buttons, text boxes, windows, and toolbars, which (2) then may be linked to small BASIC programs that perform certain actions.** Visual BASIC is "event-driven," which means that the program waits for the user to do something (an "event"), such as click on an icon, and then the program responds. At the beginning, for example, the user can use drag-and-drop tools to develop a graphical user interface, which is created automatically by the program. Because of its ease of use, Visual BASIC allows even novice programmers to create impressive Windows-based applications.

A.9 Internet Programming: HTML, XML, VRML, Java, ActiveX, & Scripting Languages

KEY QUESTION

What are the features of HTML, XML, VRML, Java, ActiveX, and Scripting Languages?

Many of the thousands of Internet data and information sites around the world are text-based only; that is, the user sees no graphics, animation, or video and hears no sound. The World Wide Web, however, permits all of this.

One way to build such multimedia sites on the Web is to use some fairly recently developed programming languages and standards: HTML, XML, VRML, Java, ActiveX, and Scripting Languages.

HTML: For Creating 2-D Web Documents & Links

HTML (Hypertext Markup Language, discussed in Chapter 2) is a markup language that lets people create onscreen documents for the Internet that can easily be linked by words and pictures to other documents. HTML is a type of code that embeds simple commands within standard ASCII text documents to provide an integrated, two-dimensional display of text and graphics. In other words, a document created in any word processor and stored in ASCII format can become a Web page with the addition of a few HTML commands.

One of the main features of HTML is the ability to insert hypertext links into a document. Hypertext links enable you to display another Web document simply by clicking on a link area—usually underlined or highlighted—on your current screen. One document may contain links to many other related documents. The related documents may be on the same server as the first document, or they may be on a computer halfway around the world. A link may be a word, a group of words, or a picture.

XML: For Making the Web Work Better

The chief characteristics of HTML are its simplicity and its ease in combining plain text and pictures. But, in the words of journalist Michael Krantz, "HTML simply lacks the software muscle to handle the business world's endless and complex transactions."[2]

Enter XML. Whereas HTML makes it easy for humans to read Web sites, **_XML (extensible markup language)_ makes it easy for machines to read Web sites by enabling Web developers to add more "tags" to a Web page.** At present, when you use your browser to click on a Web site, search engines can turn up too much, so that it's difficult to find the specific site you want—say, one with a recipe for a low-calorie chicken dish for 12. According to Krantz, "XML makes Web sites smart enough to tell other machines whether they're looking at a recipe, an airline ticket, or a pair of easy-fit blue jeans with a 34-inch waist." XML lets Web site developers put "tags" on their Web pages that describe information in, for example, a food recipe as "ingredients," "calories," "cooking time," and "number of portions." Thus, your browser no longer has to search the entire Web for a low-calorie poultry recipe for 12.

VRML: For Creating 3-D Web Pages

VRML rhymes with "thermal." **_VRML (Virtual Reality Modeling Language)_ is a type of programming language used to create three-dimensional Web pages.** Even though VRML's designers wanted to let nonprogrammers create their own virtual spaces quickly and painlessly, it's not as simple to describe a three-dimensional scene as it is to describe a page in HTML. However, many existing modeling and CAD tools now offer VRML support, and new VRML-centered software tools are arriving.

Java: For Creating Interactive Web Pages

Available from Sun Microsystems and derived from C++, Java is a major departure from the HTML coding that makes up most Web pages. Sitting atop markup languages such as HTML and XML, **_Java_ is an object-oriented programming language allowing programmers to build applications that can run on any operating system.** With Java, big applications programs can be broken into mini-applications, or "applets," that can be downloaded off the Internet and run on any computer.

ActiveX: Also for Creating Interactive Web Pages

ActiveX was developed by Microsoft as an alternative to Java for creating interactivity on Web pages. Indeed, Java and ActiveX are the two major contenders in the Web-applet war for transforming the Web into a complete interactive environment.

 ActiveX is a set of controls, or reusable components, that enables programs or content of almost any type to be embedded within a Web page. Whereas Java requires you to download an applet each time you visit a Web site, with ActiveX the component is downloaded only once, then stored on your hard drive for later and repeated use.

 Thus, the chief characteristic of ActiveX is that it features _reusable_ components—small modules of software code that perform specific tasks (such as a spelling checker), which may be plugged seamlessly into other applications. With ActiveX, you can obtain from your hard disk any file that is suitable for the Web—such as a Java applet, animation, or pop-up menu—and insert it directly into an HTML document.

 Programmers can create ActiveX controls or components in a variety of programming languages, including C, C++, Visual Basic, and Java. Thousands of ready-made ActiveX components are now commercially available from numerous software development companies.

Scripting Languages

HTML is an example of what is called a scripting language. A **_scripting language_, also called a _macro language_, is a simple programming language that allows users to create programs—called scripts or macros—to automate limited or repetitive tasks.** On the Web, scripts are commonly used to customize or add interactivity to Web pages. However, scripting languages are also included with word processing and spreadsheet software, and they allow non-programmers who want to customize or automate tasks performed by their software. For example, you might use a scripting language in a spreadsheet to automatically update an end-of-month worksheet that draws on other worksheets for current data.

Besides HTML, examples of scripting languages are Microsoft's VBScript, Netscape's JavaScript, and PERL (based on C).

Factors Affecting Communications Among Devices

Key Question

You should be able to answer the following question.

B.1   Factors Affecting How Data Is Transmitted What factors affect data transmission?

B.1   Factors Affecting How Data Is Transmitted

KEY QUESTION

What factors affect data transmission?

Things are changing, and changing fast. It's not enough to know about the types of communications channels and network configurations available. As the technology moves forward, you'll also want to know what's happening behind the scenes. This appendix provides a brief survey.

Several factors affect how data is transmitted. They include the following:

- Transmission rate—frequency and bandwidth
- Line configurations—point-to-point versus multipoint
- Serial versus parallel transmission
- Direction of transmission—simplex, half-duplex, and full-duplex
- Transmission mode—asynchronous versus synchronous
- Packet switching
- Multiplexing
- Protocols

Transmission Rate: Higher Frequency, Wider Bandwidth, More Data

Transmission rate is a function of two variables: frequency and bandwidth.

The amount of data that can be transmitted on a channel depends on the wave *frequency*—the cycles of waves per second (expressed in hertz). The more cycles per second, the more data that can be sent through that channel.

The greater a channel's *bandwidth*—the difference (range) between the highest and lowest frequencies—the more frequencies it has available and hence the more data that can be sent through that channel (expressed in bits per second, or bps).

A twisted-pair telephone wire of 4000 hertz might send only 1 kilobyte per second of data. A coaxial cable of 100 megahertz might send 10 megabytes per second. And a fiber-optic cable of 200 trillion hertz might send 1 gigabyte per second.

Line Configurations: Point-to-Point & Multipoint

There are two principal line configurations, or ways of connecting communications lines: point-to-point and multipoint.

- Point-to-point: A ***point-to-point line*** **directly connects the sending and receiving devices,** such as a terminal with a central computer. This

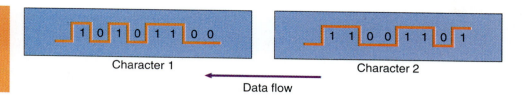

● PANEL B.1
Serial data transmission
Data resembles cars moving down a one-lane road.

Character 1   Character 2

← Data flow

arrangement is appropriate for a private line whose sole purpose is to keep data secure by transmitting it from one device to another.

- **Multipoint:** A **_multipoint line_ is a single line that interconnects several communications devices to one computer.** Often on a multipoint line only one communications device, such as a terminal, can transmit at any given time.

Serial & Parallel Transmission

Data is transmitted in two ways: serially and in parallel.

- **Serial data transmission:** In **_serial data transmission_, bits are transmitted sequentially, one after the other.** This arrangement resembles cars proceeding down a one-lane road. *(See ● Panel B.1.)*

    Serial transmission is the way most data flows over a twisted-pair telephone line. Serial transmission is found in communications lines and modems. When you send a command through your mouse, it will probably be conveyed by serial transmission. The plug-in board for a microcomputer modem usually has a serial port.

- **Parallel data transmission:** In **_parallel data transmission_, bits are transmitted through separate lines simultaneously.** The arrangement resembles cars moving in separate lanes at the same speed on a multilane freeway. *(See ● Panel B.2.)*

    Parallel lines move information faster than serial lines do, but they are efficient for up to only 15 feet. Thus, parallel lines are used, for example, to transmit data from a computer's CPU to a printer.

    Parallel transmission may also be used within a company's facility for transmitting data between terminals and the main computer.

Direction of Transmission: Simplex, Half-Duplex, & Full-Duplex

When two computers are in communication, data can flow in three ways: simplex, half-duplex, or full-duplex. These are fancy terms for easily understood processes. *(See ● Panel B.3, next page.)*

● PANEL B.2
Parallel data transmission
Data resembles cars moving in separate lanes at the same speed on a multilane freeway.

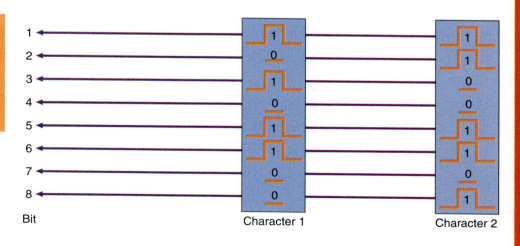

Bit   Character 1   Character 2

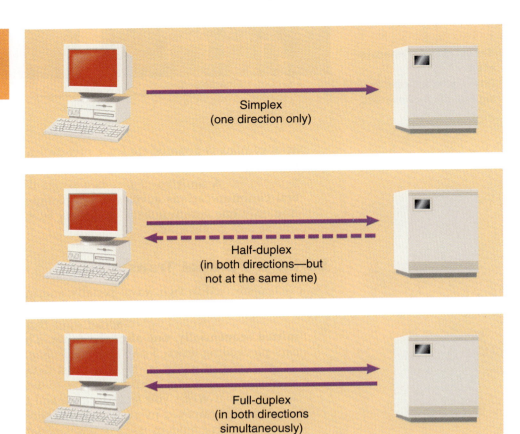
Simplex
(one direction only)

Half-duplex
(in both directions—but
not at the same time)

Full-duplex
(in both directions
simultaneously)

- **Simplex transmission:** In _**simplex transmission**_, **data can travel in only one direction.**

  An example is a traditional television broadcast, in which the signal is sent from the transmitter to your TV antenna. There is no return signal. Some computerized data collection devices also work this way, such as seismograph sensors that measure earthquakes.

- **Half-duplex transmission:** In _**half-duplex transmission**_, **data travels in both directions but only in one direction at a time.** This arrangement resembles traffic on a one-lane bridge; the separate streams of cars heading in both directions must take turns.

  Half-duplex transmission is seen with CB or marine radios, in which both parties must take turns talking. This is the most common mode of data transmission used today.

- **Full-duplex transmission:** In _**full-duplex transmission**_, **data is transmitted back and forth at the same time.** This arrangement resembles automobile traffic on a two-way street.

  An example is two people on the telephone talking and listening simultaneously. Full-duplex is sometimes used in large computer systems. It is also available in newer microcomputer modems to support truly interactive workgroup computing.

Transmission Mode: Asynchronous versus Synchronous

- Suppose your computer sends the word CONGRATULATIONS! to someone as bits and bytes over a communications line. How does the receiving equipment know where one byte (or character) ends and another begins? This matter is resolved through either asynchronous transmission or synchronous transmission. _(See ● Panel B.4.)_

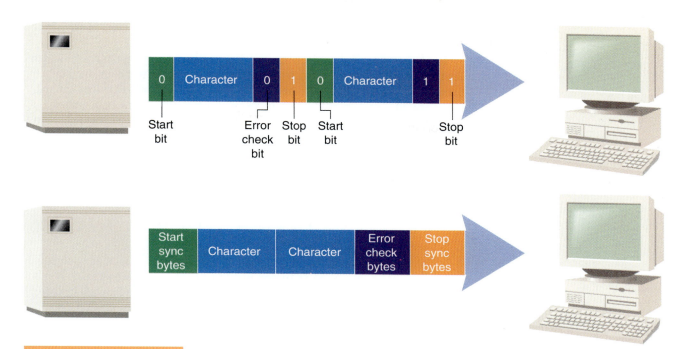

● **PANEL B.4**

Asynchronous and synchronous transmission

Top: Each character is preceded by a "start" bit and followed by a "stop" bit. *Bottom:* Messages are sent in blocks, with start and stop patterns of bits, called synch bytes, before and after the blocks. The synch bytes synchronize the timing of the internal clocks between sending and receiving devices.

- **Asynchronous transmission:** This method, used with most microcomputers, is also called *start-stop transmission*. In **_asynchronous transmission_, data is sent one byte (or character) at a time. Each string of bits making up the byte is bracketed, or marked off, with special control bits.** That is, a "start" bit represents the beginning of a character, and a "stop" bit represents its end.

  Because only one byte is transmitted at a time, this method is relatively slow. As a result, asynchronous transmission is not used when great amounts of data must be sent rapidly. Its advantage is that the data can be transmitted whenever it is convenient for the sender.

- **Synchronous transmission:** Instead of using start and stop bits, **_synchronous transmission_ sends data in blocks. Start and stop bit patterns, called synch bytes, are transmitted at the beginning and end of the blocks.** These start and end bit patterns synchronize internal clocks in the sending and receiving devices so that they are in time with each other.

  This method is rarely used with microcomputers because it is more complicated and more expensive than asynchronous transmission. It also requires careful timing between sending and receiving equipment. It is appropriate for computer systems that need to transmit great quantities of data quickly.

Circuit Switching, Packet Switching, & Asynchronous Transfer Mode: For Voice, Data, & Both

What is the most efficient way to send messages over a telephone line? That depends on whether the messages are voice, data, or both.

- **Circuit switching—best for voice:** Circuit switching is used by the telephone company for its voice networks to guarantee steady, consistent service for telephone conversations. In **_circuit switching_, the transmitter has full use of the circuit until all the data has been transmitted and the circuit is terminated.**

- **Packet switching—best for data:** A **_packet_ is a fixed-length block of data for transmission.** The packet also contains instructions about the destination of the packet. In **_packet switching_, electronic messages**

are divided into packets for transmission over a wide area network to their destination, through the most expedient route.

Here's how packet switching works: A sending computer breaks an electronic message apart into packets. The various packets are sent through a communications network—often by different routes, at different speeds, and sandwiched in between packets from other messages. Once the packets arrive at their destination, the receiving computer reassembles them into proper sequence to complete the message.

The benefit of packet switching is that it can handle high-volume traffic in a network. It also allows more users to share a network, thereby offering cost savings. The method is particularly appropriate for sending data long distances, such as across the country. Accordingly, it is used in large data networks such as Telenet, Tymnet, and AT&T's Accunet.

- **Asynchronous transfer mode (ATM)—best for both:** A newer technology, called ***asynchronous transfer mode (ATM)*,** **combines the efficiency of packet switching with some aspects of circuit switching,** thus enabling it to handle both data and real-time voice and video. ATM is designed to run on high-bandwidth fiber-optic cables.

Multiplexing: Enhancing Communications Efficiencies

Communications lines nearly always have far greater capacity than a single microcomputer or terminal can use. Because operating such lines is expensive, it's more efficient if several communications devices can share a line at the same time. This is the rationale for multiplexing. ***Multiplexing*** **is the transmission of multiple signals over a single communications channel.**

Three types of devices are used to achieve multiplexing—*multiplexers, concentrators,* and *front-end processors:*

- **Multiplexers:** A ***multiplexer*** **is a device that merges several low-speed transmissions into one high-speed transmission.** *(See* ● *Panel B.5)*

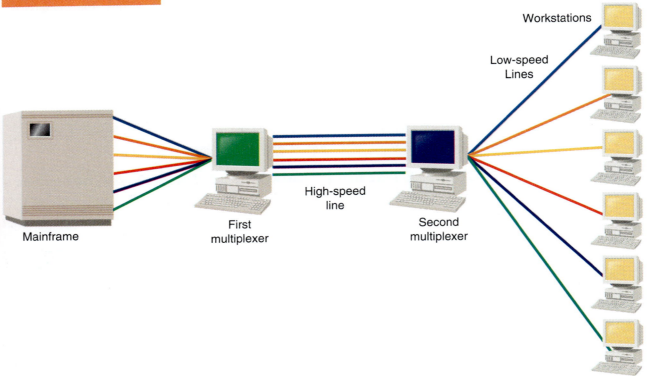

Workstations

Low-speed Lines

Mainframe   First multiplexer   High-speed line   Second multiplexer

Depending on the multiplexer, 32 or more devices may share a single communications line. Messages sent by a multiplexer must be received by a multiplexer of the same type. The receiving multiplexer sorts out the individual messages and directs them to the proper recipient.

High-speed multiplexers called *T1 multiplexers*, which use high-speed digital lines, can carry as many messages, both voice and data, as 24 analog telephone lines.

- **Concentrators:** Like a multiplexer, a concentrator is a piece of hardware that enables several devices to share a single communications line. However, unlike a multiplexer, **a _concentrator_ collects data in a temporary storage area.** Whereas a multiplexer spreads the signals back out again on the receiving end, the concentrator has a receiving computer perform that function.

- **Front-end processors:** The most sophisticated of these communications-management devices is the front-end processor, a computer that handles communications for mainframes. A **_front-end processor_ is a smaller computer that is connected to a larger computer and assists with communications functions.** It transmits and receives messages over the communications channels, corrects errors, and relieves the larger computer of routine computational tasks.

Protocols: The Rules of Data Transmission

Does the foregoing information in this section seem unduly technical for an ordinary computer user? Although you should understand these details, fortunately you won't have to think about them much. Experts will already have taken care of them for you in sets of rules called protocols.

The word *protocol* is used in the military and in diplomacy to express rules of precedence, rank, manners, and other matters of correctness. (An example would be the protocol for who will precede whom into a formal reception.) Here, however, a **_protocol_, or communications protocol, is a set of conventions governing the exchange of data between hardware and/or software components in a communications network.**

Protocols are built into the hardware or software you are using. The protocol in your communications software, for example, will specify how receiver devices will acknowledge sending devices, a matter called handshaking. Handshaking establishes the fact that the circuit is available and operational. It also establishes the level of device compatibility and the speed of transmission. Protocols will also specify the type of electrical connections used, the timing of message exchanges, error-detection techniques, and so on.

In the past, not all hardware and software developers subscribed to the same protocols. As a result, many kinds of equipment and programs have not been able to work with one another. In recent years, more developers have agreed to subscribe to a standard of protocols called **_OSI (Open Systems Interconnection)_. Backed by the International Standards Organization, OSI is an international standard that defines seven layers of protocols for worldwide computer communications.** *(See ● Panel B.6, next page.)*

CONCEPT CHECK

Distinguish frequency and bandwidth.

What are two line configurations?

What is serial versus parallel transmission?

Describe simplex, half-duplex, and full-duplex.

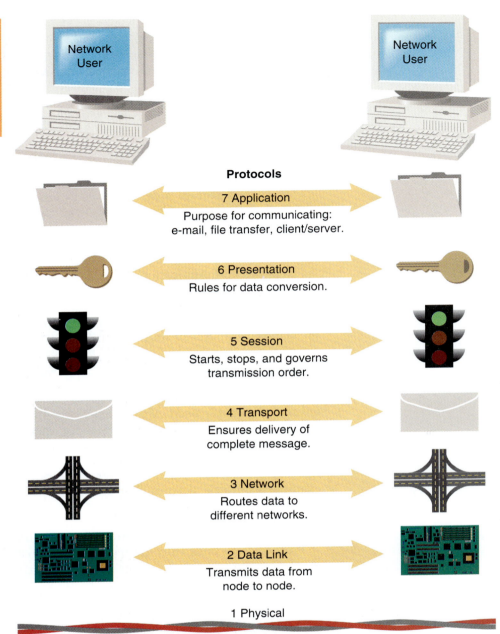

● PANEL B.6
OSI
The seven layers of the ISO standard for world-wide communications that defines a framework for implementing protocols.

Network User

Network User

Protocols

7 Application
Purpose for communicating: e-mail, file transfer, client/server.

6 Presentation
Rules for data conversion.

5 Session
Starts, stops, and governs transmission order.

4 Transport
Ensures delivery of complete message.

3 Network
Routes data to different networks.

2 Data Link
Transmits data from node to node.

1 Physical

Passes bits onto connecting medium.

Distinguish asynchronous from synchronous transmission mode.

What is packet switching?

What are three types of multiplexers?

Discuss protocols and OSI.

Notes

Chapter 1

1. Frederick Allen, "Technology at the End of the Century," *Invention & Technology*, Winter 2000, pp. 11–16.
2. Bruce Sterling, "Learning to Love Obsolescence," *Newsweek*, January 1, 2000, p. 68.
3. Fred Abatemarco, "Hand Weapons of the Modern Age," *Popular Science*, September 1999, pp. 9–10.
4. Vivian S. Toy, "Teen-Agers and Cell Phones: It's All Talk, and All the Time," *New York Times*, August 2, 1999, pp. A1, A17.
5. Michael Specter, "Your Mail Has Vanished," *New Yorker*, December 6, 1999, pp. 96–103.
6. Data from International Data Corp., in "Like It or Not, You've Got Mail," *Business Week*, October 4, 1999, pp. 178–184.
7. "Like It or Not, You've Got Mail," 1999.
8. Robert Rossney, "E-Mail's Best Asset–Time to Think," *San Francisco Chronicle*, October 5, 1995, p. E7.
9. Adam Gopnik, "The Return of the Word," *New Yorker*, December 6, 1999, pp. 49–50.
10. Peter H. Lewis, "The Good, the Bad, and the Truly Ugly Faces of Electronic Mail," *New York Times*, September 6, 1994, p. B7.
11. Gopnik, 1999.
12. David A. Whittle, *Cyberspace: The Human Dimension* (New York: W. H. Freeman, 1997).
13. David A. Whittle, quoted in "Living Online," *The Futurist*, July–August 1997, p. 54.
14. Edward Iwata, "The Net at 30," *USA Today*, December 14, 1999, pp. 1A, 2A.
15. Kevin Maney, "The Net Effect: Evolution or Revolution?" *USA Today*, August 9, 1999, pp. 1B, 2B.
16. Center for Communication Policy (www.ccp.ucla.edu), reported in David Plotnikoff, "Study Asks How Wired World Has Changed the Way We Live," *San Jose Mercury News*, June 13, 1999, pp. 1F, 7F.
17. December 10–13, 1999, survey by The Strategis Group, reported in Dru Sefton, "The Big Online Picture: Daily Web Surfing Now the Norm," *USA Today*, March 22, 2000, p. 3D.
18. Teenage Research Unlimited, reported in Eryn Brown, "The Future of Net Shopping? Your Teens," *Fortune*, April 12, 1999, p. 152.
19. Brown, 1999.
20. Kevin Murphy, Gartner Group, quoted in Timothy J. Mullaney, "Death to E-Words Everywhere," *Business Week*, November 15, 1999, p. 10.
21. Adapted from Dan Gillmor, "Electronic Appliances We Have—We Need Embedded Values, Too," *San Jose Mercury News*, October 3, 1999, pp. 1E, 3E.
22. Andy Reinhardt, Steven V. Brull, Peter Burrows, and Catherine Yang, "The Soul of a New Refrigerator," *Business Week*, January 17, 2000, p. 42; Cox News Service, "Internet Can Be Found in Tiny Places," *San Francisco Chronicle*, October 11, 1999; and Steven Butler, "Smart Toilets and Wired Refrigerators," *Newsweek*, June 7, 1999, p. 48.
23. David Einstein, "Custom Computers," *San Francisco Chronicle*, April 15, 1999, pp. B1, B3.
24. Einstein, 1999.
25. Tammy Joyner, "Turning Old Jobs into Hot Ones," *San Francisco Chronicle*, January 17, 2000, p. B3.
26. Laurence Hooper, "No Compromises," *Wall Street Journal*, November 16, 1992, p. R8.
27. Tom Forester and Perry Morrison, *Computer Ethics: Cautionary Tales and Ethical Dilemmas in Computing* (Cambridge, MA: The MIT Press, 1990), pp. 1–2.
28. Norman Solomon, "The Media's Role in the Commercialization of Cyberspace," *San Francisco Chronicle*, January 27, 2000, p. A25.
29. Kevin Maney, "What's on a Snack Web Site? A Whole Lot of Nothing," *USA Today*, February 2, 2000, p. 3B.
30. Joseph N. Pelton, "The Fast-Growing Global Brain," *The Futurist*, August–September, 1999, pp. 24–27.
31. E. Z. Zechmeister and S. E. Nyberg, *Human Memory: An Introduction to Research and Theory* (Pacific Grove, CA: Brooks/Cole, 1982).
32. F. P. Robinson, *Effective Study*, 4th ed. (New York: Harper & Row, 1970).
33. B. K. Broumage and R. E. Mayer, "Quantitative and Qualitative Effects of Repetition on Learning from Technical Text," *Journal of Educational Psychology*, 1982, 78, 271–278.
34. R. J. Palkovitz and R. K. Lore, "Note Taking and Note Review: Why Students Fail Questions Based on Lecture Material," *Teaching of Psychology*, 1980, 7:159–161.
35. J. Langan and J. Nadell, *Doing Well in College: A Concise Guide to Reading, Writing, and Study Skills* (New York: McGraw-Hill, 1980), pp. 93–100.

Bookmark It! Box

a. Study by Impulse Research for *Iconoclast* newsletter, reported in D. Plotnikoff, "E-Mail: A Critical Medium Has Reached Critical Mass," *San Jose Mercury News*, April 4, 1999, pp. 1F, 2F.
b. K. Clark, "At Least the Coffee and Pens Are Still Free," *U.S. News & World Report*, June 7, 1999, p. 70.
c. S. Shostak, "You Call This Progress?" *Newsweek*, January 18, 1999, p. 16.
d. A. Markels, "Don't Manage by E-Mail," *San Francisco Examiner*, August 11, 1996, p. B-5, reprinted from *Wall Street Journal;* "Don't Overuse Your E-Mail," *CPA Client Bulletin*, March 1999, p. 3; B. Fryer, "E-Mail: Backbone of the Info Age or Smoking Gun?" *Your Company*, July/August 1999, pp. 73–76; S. Armour, "Boss: It's in the E-Mail," *USA Today*, August 10, 1999, p. 3B; L. Guernsey, "Attachments #@%&#@ Are Full #+@&`¢# of Surprises," *New York Times*, July 22, 1999, p. D11.

Chapter 2

1. Greg Miller, "Ethernet Is Changing Dorm Life," *Los Angeles Times*, January 14, 2000, pp. A1, A10.
2. Miller, 2000.
3. Virginia Brooks, quoted in Andy Reinhardt, Peter Elstrom, and Paul Judged, "Zooming Down the I-Way," *Business Week*, April 7, 1997, pp. 76–87.
4. Some of this discussion was adapted from Kate Murphy, "Cruising the Net—in Hyperdrive," *Business Week*, January 24, 2000, pp. 170–172.
5. Joel Dreyfuss, "Set Yourself Free: You Have Nothing to Lose but Your Wires," *Fortune*, January 24, 2000, pp. 139–140.
6. William J. Holstein and Fred Vogelstein, "You've Got a Deal!" *U.S. News & World Report*, January 24, 2000, pp. 34–40.
7. Paul Davidson, "Companies Rush to Bring You Net for Free," *USA Today*, December 16, 1999, p. 1B; and Larry Armstrong, "Something for Nothing," *Business Week*, April 26, 1999, p. 94.
8. Paul Davidson, "Start-Up Pitches Free High-Speed Net Access," *USA Today*, January 4, 2000, p. 1B.
9. Stephen H. Wildstrom, "What Keeps AOL on Top," *Business Week*, January 24, 2000, p. 25.
10. Lawrence Magid, "AOL Not Only Way America Can Get Online," *San Francisco Examiner*, January 23, 2000, pp. B-%, B-6.
11. Study by Impulse Research for Iconoclast newsletter, reported in D. Plotnikoff, "E-Mail: A Critical Medium Has Reached Critical Mass," *San Jose Mercury News*, April 4, 1999, pp. 1F, 2F.
12. Kelley, 1999.
13. Elizabeth Weise, "Instant Battles Keep Raging Over Messaging," *USA Today*, August 4, 1999, p. 5D.
14. Deborah Kong, "Cross E-Mail with a Phone, That's Instant Messaging," *San Jose Mercury News*, August 1, 1999, pp. 1E, 3E.
15. Tom Stein, "Instant Message Nirvana," *San Francisco Chronicle*, August 2, 1999, pp. B1, B6.
16. Michelle Slatalla, "The Office Meeting that Never Ends," *New York Times*, September 23, 1999, pp. D1, D8.
17. Jeanne Hinds, quoted in Slatalla , 1999.
18. Pitney Bowes study, cited in Slatalla, 1999.
19. Editors of *PC Computing*, "End E-Mail Insanity Forever," *PC Computing*, November 1999, pp. 170–198; Jennifer Powell, "E-Mail: Common Sense Plays a Big Role," *Smart Computing*, March 2000, p. 47; and Kenneth M. Morris, *User's Guide to the Information Age* (New York: Lightbulb Press, 1999), pp. 111, 153.
20. David Lazarus, "Fan Spam Is Hard to Shake," *San Francisco Chronicle*, February 7, 2000, pp. C1, C2.
21. Julian Haight, quoted in Lazarus, 2000, p. C1.
22. Calculations by NEC Research and Inktomi, reported in Elizabeth Weise, "Web Changes Direction to People Skills," *USA Today*, January 24, 2000, pp. 1D, 2D.
23. Jennifer Powell, "Lessons from the Browser War (1995–2000), *PC Computing*, March 2000, p. 64.
24. Weise, "Web Changes Direction to People Skills," 2000; and Saul Hansell, "Obsessively Independent, Yahoo Is the Web's Switzerland," *New York Times*, August 23, 1999, pp. C1, C10.
25. Janet Kornblum, "Portals Suffer as Internet Surfers Get More Savvy," *USA Today*, July 2, 1999, p. 1B; and Dan Gillmor, "Small Portals Prove that Size Matters," *San Jose Mercury News*, December 6, 1998, pp. 1E, 3E..
26. Study by Steve Lawrence and C. Lee Giles in journal *Nature*, reported in Peter Svensson, "Search Engines Can't Keep Up," *San Francisco Chronicle*, July 8, 1999, p. B3; and in Mike

Snider, "Web Growth Outpacing Search Power," *USA Today*, July 8, 1999, p. 1A;

27. Weise, "Web Changes Direction to People Skills," 2000.

28. Elizabeth Weise, "Successful Net Search Starts with a Need," *USA Today*, January 24, 2000, p. 3D; Timothy Hanrahan, "The Best Way to . . . Search Online," *Wall Street Journal*, December 6, 1999, p. R25; and Weise, "Web Changes Direction to People Skills," 2000.

29. Timothy Hanrahan, "The Best Way to Search Online," *Wall Street Journal*, December 6, 1999, p. R25; Re;va Basch, "Cutting Through the Clutter," *Smart Computing*, July 1999, pp. 88–91; Matt Lake and Dylan Tweney, "Find It on the Web," *PC World*, June 1999, pp. 168–182; and Matt Lake, "Desperately Seeking Susan OR Suzie NOT Sushi," *New York Times*, September 3, 1998, p. D1

30. Lake, 1998.

31. Lake and Tweney, 1999.

32. Jon Van, "Internet Firms Take Aim at Phone Giants," *San Francisco Examiner*, February 6, 2000, p. B-3; reprinted from *Chicago Tribune*.

33. Laurie Bryan, "Wired for Shopping," *San Francisco Chronicle*, February 9, 2000, zone 7, pp. 1, 4.

34. Morris, 1999, p. 100.

35. Megan Doscher, "The Best Way to Find Love," *Wall Street Journal*, December 6, 1999, p. R34.

36. Sharon Cleary, "The Best Way to Find an Old Friend," *Wall Street Journal*, December 6, 1999, pp. R26, R45.

37. Elizabeth Weise, "Cultivating a Sense of Community On Line," *USA Today*, June 30, 1999, p. 6D.

38. Wendy M. Grossman, "Language Is a Virus," *PC Computing*, March 2000, p. 62.

39. William M. Bulkeley, "The Best Way to Go to School," *Wall Street Journal*, December 6, 1999, pp. R18, R22.

40. Bulkeley, 1999; and Faith Bremner, "On-line College Classes Get High Marks Among Students," *USA Today*, November 16, 1998, p. 16E.

41. Jane Manners, "Web Health Checkup," *Brill's Content*, February 2000, pp. 114–116; and Lisa Bransten, "The Best Way to Stay Healthy," *Wall Street Journal*, December 6, 1999, pp. R36, R45.

42. Study by Cyber Dialogue, reported in Evan Ramstad, "The Best Way to Amuse Yourself," *Wall Street Journal*, December 6, 1999, p. R43.

43. Lorrie Grant, "Internet Has Become Integral Part of Everyday Life," *USA Today*, April 21, 1999, p. 6B.

44. Haya El Nasser, "Main Street Enters Mainstream," *USA Today*, November 16, 1999, p. 3A.

45. Denise Caruso, "On-Line Day Traders Are Starting to Have an Impact on a Few Big-Cap Internet Stocks in What Some Call a 'Feeding Frenzy,'" *New York Times*, December 14, 1998, C3.

46. Gary McWilliams, "The Best Way to Find a Job," *Wall Street Journal*, December 6, 1999 pp. R16, R22.

47. Del Jones, "E-Purchasing Saves Businesses Billions," *USA Today*, February 7, 2000, pp. 1B, 2B.

48. John Rubino, "Measures," *Individual Investor*, March 2000, pp. 37–42.

49. Rubino, 2000.

Bookmark It! Box

a. Karen Jacobs, "The Best Way to Pick a Provider," *Wall Street Journal*, December 6, 1999, pp. R8, R10; Tracy Baker, "Free Access to the Internet," *Smart Computing*, June 1999, pp. 44–46; Tina Kelley, "Choosing an ISP: Convenience, Cost, and Service," *San Jose Mercury News*, May 30, 1999, pp. 1F, 4F; reprinted from the *New York Times*; and "Going Online: Best Ways to Get Started," *Consumer Reports*, July 1998, p. 64.

Chapter 3

1. Alan Blinder, quoted in Associated Press, "Internet Isn't the Only Factor in Productivity," *San Francisco Chronicle*, July 14, 2000, p. B3.

2. Robert M. Solow, quoted in Louis Uchitelle, "Productivity Finally Shows the Impact of Computers," *New York Times*, March 12, 2000, sec. 3, p. 4.

3. Stephen D. Oliner and Daniel E. Sichel, Federal Reserve, reported in Uchitelle, 2000.

4. Dan Gillmor, "Online Reliability Will Carry a Price," *San Jose Mercury News*, July 18, 1999, p. 1E, 7E.

5. Margaret Trejo, quoted in Richard Atcheson, "A Woman for Lear's," *Lear's*, November 1993, p. 87.

6. John Ennis, quoted in Peter Plagens and Ray Sawhill, "Throw Out the Brushes," *Newsweek*, September 1, 1997, pp. 76 – 77.

7. Glenn Rifkin, "Designing Tools for Designers," *New York Times*, June 18, 1992, p. C6.

Chapter 4

1. Alan Robbins, "Why There's Egg on Your Interface," *New York Times*, December 1, 1996, sec. 3, p. 12.

2. Rick Jurgens, "Apple Pushes Back to January Operating-System Release Date," *Wall Street Journal*, May 6, 2000, p. B8; Anthony B. Perkins, "Gates vs. Jobs: The Rematch," *Wall Street Journal*, March 13, 2000, p. A46; Stephen H. Wildstrom, "Mac Hits Another Home Run," *Business Week*, February 28, 2000, p. 24; Joshua Quittner, "Aqua: The Movie," *Time*, January 31, 2000, p. 82; and Lee Gomes, "Apple Revamps Web Strategy, Operating System," *Wall Street Journal*, January 6, 2000, p. B8.

3. Quittner, 2000.

4. Brent Schlender, "Steve Jobs' Apple Gets Way Cooler," *Fortune*, January 24, 2000, pp. 66–71.

5. Guy Richardson, "Pondering World Domination via Microsoft," *Reno Gazette-Journal*, April 3, 2000, p. 3E.

6. Suzanne Kantra Kirschner, "Windows Gets Playful," *Popular Science*, August 2000, pp. 68–69; Lawrence Magid, "Microsoft Cleans Up Windows," *San Francisco Examiner*, July 19, 2000, pp. B-5, B-8; Rebecca Buckman, "Microsoft Ready to Demonstrate a 'Me' Generation of Windows," *Wall Street Journal*, April 7, 2000, p. B3; and Edward C. Baig, "'Millennium': Cleaner Windows?" *USA Today*, January 10, 2000, p. 3D.

7. Scott Hillis, "Microsoft to Release Windows Test Version," *San Jose Mercury News*, July 28, 2000, p. 5C.

8. Steve Hamm, Peter Burrows, and Andy Reinhardt, "Is Windows Ready to Run E-Business?" *Business Week*, January 24, 2000, pp. 154–160.

9. Jim Kerstetter, "Novell Bets the Farm," *Business Week*, May 29, 2000, p. 96.

10. Stephen H. Wildstrom, "Windows 2000: Worth the Wait," *Business Week*, February 21, 2000, p. 22; Hamm, Burrows, and Reinhardt, 2000; John Markoff, "Microsoft Facing a Skeptical Market with Windows 2000," *New York Times*, February 14, 2000, pp. C1, C6; Ellis Booker, "Users Look Forward to Deploying Windows 2000," *InternetWeek*, February 18, 2000 (online); and David Bank, "Microsoft and Market Expect Slower Start for Windows 2000 Sales," *Wall Street Journal*, February 14, 2000, p. B8.

11. Hamm, Burrows, and Reinhardt, 2000; and Lawrence M. Fisher, "Sun Plans to Start Shipping Operating System Next Month," *New York Times*, January 27, 2000, p. C12.

12. Henry Norr, "Riding the Web Wave," *San Francisco Chronicle*, May 29, 2000, pp. B1, B2; and Henry Norr, "BSD Dates Back to Late '70s," *San Francisco Chronicle*, May 29, 2000, p. B2.

13. Irving Wladawsky-Berger, quoted in Deborah Solomon, "Could Linux Outdo Windows?" *USA Today*, March 9, 2000, pp. 1B, 2B.

14. Stanley Holmes, "Companies Form Group to Promote Linux," *Los Angeles Times*, March 10, 2000, pp. C2, C4; Sam Jaffe, "Will Linux Investors Be Left Out in the Cold?" *Business Week*, March 2000, pp. 178–180; David Kirkpatrick, "Dell to Wintel: Your Hegemony Is Over," *Fortune*, February 21, 2000, pp. 50, 52; Lawrence M. Fisher, "Looking for a New Life in Linux," *New York Times*, February 14, 2000, p. C4; Lawrence M. Fisher, "A View that Needs No Windows," *New York Times*, February 6, 2000, sec. 3, p. 2; and Bloomberg News, "Linux-Related Shares Soar on News of Revised Version," *Los Angeles Times*, February 3, 2000, p. C7.

15. David Rynecki, "Is Palm's IPO Really the One to Catch?" *Fortune*, February 7, 2000, pp. 213–214; Roy Furchgott,

"Using the Net to Soup Up Your Palm Top Computer," *Business Week*, November 8, 1999, pp. 165–166; Walter S. Mossberg, "A Pilot Rival Organizes Your Life, Then Morphs into Something Else," *Wall Street Journal*, September 16, 1999, p. B1; and Kevin Maney, "Palm Has the Whole World in Its Hand," *USA Today*, September 14, 1999, pp. 1B, 2B.

16. Edward C. Baig, "4 PDAs View to Be Your Assistant," *USA Today*, December 15, 1999, p. 3D.

17. Henry Norr, "Microsoft's Pocket PC Hits Store Shelves Next Month," *San Francisco Chronicle*, March 6, 2000, pp. B1, B3, B7.

18. Autumn de Leon, "Portable Computing at Hand," *Time*, November 29, 1999, p. 96.

19. Michel Marriott, "Newcomer in the OS Ranks," *New York Times*, February 10, 2000, p. D8.

20. Larry Blasko, "Utilities Do the Dirty—but Necessary—Work," *San Jose Mercury News*, January 31, 1999, p. 1F.

21. Jay Greene, " Windows of Opportunity," *Business Week*, July 3, 2000, pp. 140 – 144.

22. Greene, 2000; Jay Greene and Jim Kerstetter, "Bill's New Web Menu," *Business Week*, June 5, 2000, pp. 104–106; David Kirkpatrick, "The New Face of Microsoft," *Fortune*, February 7, 2000, pp. 87–96; Jon Swartz and Patrick McMahon, "Microsoft Unveils 'Dot-Net' Next-Generation Software Vision," *USA Today*, June 23, 2000, p. 6B; Jon Swartz, "Ballmer Bets All on Dot.Net," *USA Today*, June 26, 2000, p. 4B; and Dori Jones Yang, "Microsoft Breaks Up Windows All by Itself," *U.S. News & World Report*, July 3, 2000, pp. 33–34.

23. Greene and Kerstetter, 2000; and Jim Kerstetter and Peter Burrows, "H-P's E-Speak: Good Products, Botched Marketing," *Business Week*, July 3, 2000, p. 144;

24. David P. Hamilton, "Sun Microsystems to Formally Announce Jini Technology, Partners to Support It," *Wall Street Journal*, January 25, 1999, p. B6; David Einstein, "Sun Micro's Chief Scientist Prepares to Let the Jini Out of the Bottle," *San Francisco Chronicle*, January 25, 1999, pp. B1, B2; and Susan Gregory Thomas, "Home Network," *U.S. News & World Report*, August 10, 1998, pp. 57–59.

25. Henry Norr, "Turning Down Time into Cash," *San Francisco Chronicle*, June 19, 2000, pp. G1, G3.

26. *Microsoft Computer Dictionary*, 4th ed. (Redmond, WA: Microsoft Press, 1999), p. 147.

27. Dori Jones Yang, "Is the New Strategy a Big Mistake?" *U.S. News & World Report*, July 3, 2000, p. 34; John Markoff, "Critic Sees Flaws in Microsoft Strategy," *New York Times*, June 19, 2000, p. C2; and David Gelenter, "The Second Coming—A Manifesto," Edge, June 15–19, 2000, *www.edge.org*.

28. Dori Jones Yang, "Ganging Up on Wintel," *U.S. News & World Report*, June 12, 2000, p. 45; James Lardner, "Computers for Technophobes," *U.S. News & World Report*, May 22, 2000, pp. 58–60; Peter H. Lewis, "The Web without Microsoft," *New York Times*, November 11, 1999, pp. D1, D3; and Peter Burrows, "Bypassing Windows," *Business Week*, February 15, 1999, p. 35.

29. Randall E. Stross, "Will the Internet Break Windows?" *Wall Street Journal*, January 17, 2000, p. A18.

30. Anita Hamilton, "Scheduling Snafu," *Time*, May 10, 1999, p. 96.

31. Paul Davidson, "Online Software Catches On," *USA Today*, February 4, 2000, pp. 1B, 2B.

32. Davidson, 2000.

33. Luc Hatlestad, "Small Beginnings," *Red Herring*, March 2000, pp. 208–212; and Seth Schiesel, "AT&T to Invest $250 Million in Application Services Sector," *New York Times*, January 27, 2000, p. C11.

34. J. William Gurley, "The New Market for 'Rentalware,'" *Fortune*, May 10, 1999, p. 142.

35. Gary Bloom, quoted in Lawrence M. Fisher, "Software Evolving into a Service Rented Off the Net," *New York Times*, December 20, 1999, p. C37.

36. Stephen H. Wildstrom, "What Does a Freshman Need?" *Business Week*, August 7, 2000, p. 22.

37. Richard A. Seigel, "Palmtop Computers Prove Useful in Class" [letter], *Chronicle of Higher Education*, February 4, 2000, p. B10.

38. "Business Schools Struggle to Impose Laptop Etiquette," *San Francisco Chronicle*, April 20, 2000, p. B3; reprinted from *New York Times*.

39. Mark Babin(e)ck, "Lecture Notes Hit the Web," *San Francisco Chronicle*, February 1, 2000, p. C4; Goldie Blumenstyk, "Colleges Object as Companies Put Class Notes on Web Sites," *Chronicle of Higher Education*, September 17, 1999, p. A41; "Net Notes Trump Boring Lecture" [editorial] and Peter Wood, "Web Can't Supply Class Context" [opposing view], *USA Today*, September 15, 1999, p. 15A; and Jacques Steinberg, "Free College Notes on Web: Aid to Learning, or Laziness?" *New York Times*, September 9, 1999, pp. A1, A18.

40. Todd Gitlin, quoted in Dora Straus, "Lazy Teachers, Lazy Students" [letters], *New York Times*, September 12, 1999, sec. 4, p. 18.

41. Jean Richardson, quoted in Wendy R. Leibowitz, "At Yale's Demand, a Web Site Drops Lecture Notes from the University's Classes," *Chronicle of Higher Education*, March 17, 2000, pp. A49–A50.

42. Tanya Schevitz, "Web Sites Snag Students," *San Francisco Chronicle*, October 18, 1999, pp. A1, A13.

43. Jamie Horwitz, American Federation of Teachers, reported in Marco R. della Cava, "Blackboard Jungle Turns Ugly Online," *USA Today*, May 8, 2000, p. 3D.

44. Tanya Schevitz, "Prof Fights Web Trash Talk," *San Francisco Chronicle*, April 6, 2000, pp. A1, A14.

45. Lucia Perillo, "E-Mail and the Law of Unintended Consequences," *Chronicle of Higher Education*, March 3, 2000, p. A64.

Bookmark It! Box

a. Janet Rae-Dupree, "Help for Hard Drives," *San Jose Mercury News*, April 28, 1997, pp. 1E, 3E.

b. Harry Goldblatt, "Calling Doctor Hard Drive," *Fortune*, June 9, 1997, p. 140.

Chapter 5

1. Michael S. Malone, "The Tiniest Transformer," *San Jose Mercury News*, September 10, 1995, pp. 1D, 2D; excerpted from *The Microprocessor: A Biography* (New York: Telos/Springer Verlag, 1995).

2. Malone, 1995.

3. Hados Dembo, "The Way Things Were," *Wall Street Journal*, November 16, 1992, pp. R16–R17.

4. Laurence Hooper, "No Compromises," *Wall Street Journal*, November 16, 1992, p. R8.

5. "Computers, Then and Now," *Consumer Reports*, May 2000, p. 10.

6. Data from PC Data Inc., cited by Gary McWilliams, "Reversing Course, Home-PC Prices Head Higher," *Wall Street Journal*, January 13, 2000, pp. B1, B4; and John Simons, "Cheap Computers Bridge Digital Divide," *Wall Street Journal*, January 27, 2000, p. A22.

7. Stephen C. Miller, "Another Way to Get on Line without Buying a Computer," *New York Times*, October 14, 1999, p. D3; Ashley Dunn, "Pumped-Up Appliance or No-Frills Computer? i-Opener a Shortcut to Net," *Los Angeles Times*, March 9, 2000, pp. C1, C7; and Amy Harmon, "Courtesy of Amateurs, a $99 Personal Computer," *New York Times*, March 18, 2000, pp. B1, B2.

8. Associated Press, "Playing Games and Doing Banking on PlayStation Console," *San Francisco Chronicle*, March 31, 2000, p. C5.

9. Henry Norr, "The NC—with a Twist," *San Francisco Chronicle*, February 28, 2000, pp. B1, B2.

10. Henry Norr, "Why Thin Computing Is In," *San Francisco Chronicle*, February 28, 2000, p. B2.

11. Henry Norr, "Intel Introduces 2 Gigahertz Chip at Trade Show," *San Francisco Chronicle*, August 23, 2000, pp. D1, D7; Henry Norr, "Do You Need the Fastest Processor?" *San Francisco Chronicle*, April 17, 2000, pp. E1, E2, E3; Peter H. Lewis, "With 2 New Chips, the Gigahertz Decade Begins," *New York Times*, March 9, 2000, pp. D1, D3; Henry Norr,

"Intel Ships Fast Chip, Catches Up with AMD," *San Francisco Chronicle*, March 9, 2000, pp. B1, B4; and Edward C. Baig, "Gunning the Engine on 1-Gig PCs," *USA Today*, March 22, 2000, p. 3D.

12. John Markoff, "IBM to Show a Breakthrough in Chip Making," *New York Times*, April 3, 2000, pp. C1, C2; and Edward Iwata, "IBM Boasts of Better, Faster Chip," *USA Today*, April 3, 2000, p. 1B.

13. Henry Norr, "Pentium 4 Is the Chip Reinvented," *San Francisco Chronicle*, August 21, 2000, pp. B1, B4.

14. Baig, 2000; and Walter S. Mossberg, "Mossberg's Mailbox," *Wall Street Journal*, April 13, 2000, p. B9.

15. "In the Chips," *Reno Gazette-Journal*, July 17, 2000, pp. 1E, 4E.

16. "No Flash in the Pan," *Business Week*, June 12, 2000, p. 40D.

17. Ken Hawk, "Peripheral Issues: The USB," *Reno Gazette-Journal*, March 13, 2000, p. 3E.

18. Mossberg, 2000.

19. Becky Waring, "Taking the Desktop with You," *San Francisco Chronicle*, April 13, 2000, pp. D1, D3; Walter Mossberg, "Mossberg's Mailbox," *Wall Street Journal*, April 6, 2000, B10.

20. Stephen H. Wildstrom, "Kiss the Floppy Good-bye," *Business Week*, June 12, 2000, p. 34.

21. N'Gai Croal, "One Drive that Fits All," *Newsweek*, July 3, 2000, p. 38B.

22. Edward Baig, "Be Happy, Film Freaks," *Business Week*, May 26, 1997, pp. 172–173.

23. "2 New DVD Players Can Surf the Internet," *San Francisco Chronicle*, April 6, 2000, p. 3B.

24. Keith L. Alexander, "DVD Sales Energize Home Video Market," *USA Today*, February 28, 2000, pp. C1, C6.

25. Edmund Sanders, "Makers of Smart Cards Are Betting Big on U.S.," *Los Angeles Times*, February 28, 2000, pp. C1, C6.

26. Suzanne Kantra Kirschner, "Memories to Go," *Popular Science*, May 2000, pp. 88–90.

27. Paul M. Eng, Robert D. Hof, and Hiromi Uchida, "It's a Whole New Game: The Hards vs. the Cards," *Business Week*, June 8, 1992, pp. 101–103.

28. Henry Norr, "IBM to Unveil Breakthrough in Chip Speed," *San Francisco Chronicle*, May 22, 2000, p. E2; and John Markoff, "IBM to Show a Breakthrough in Chip Making," *New York Times*, April 3, 2000, pp. C1, C2.

29. Michio Kaku, "What Will Replace Silicon?" *Time*, June 19, 2000, pp. 98–99.

30. Gordon Moore, quoted in "Gordon Moore Q&A," *Time*, June 19, 2000, p. 99.

31. Otis Port, "Chips for the Post-PC Era," *Business Week*, March 27, 2000, pp. 96–104.

32. Philip J. Longman, "The Next Big Thing Is Small," *U.S. News & World Report*, July 3, 2000, pp. 30–33; Tim Appenzeller, "The Chemistry of Computing," *U.S. News & World*

Report, May 1, 2000, p. 56; and Ron Southwick, "Nanotechnology, the Study of Minute Matter, Becomes a Big Priority in the Budget," *Chronicle of Higher Education*, March 31, 2000, pp. A38–A39.

33. Alex Dominguez, "Speed of Light Pushed Faster," *Reno Gazette-Journal*, July 20, 2000, p. 3A.

34. John S. McNeil, "The Wet and Wild Future of Computers," *U.S. News & World Report*, February 14, 2000, p. 52; Tom Abate, "Scientists Develop Bionic Chip," *San Francisco*, February 26, 2000, pp. D1, D2; and Vincent Kiernan, "DNA-Based Computers Could Race Past Supercomputers, Researchers Predict," *Chronicle of Higher Education*, November 28, 1997, pp. A23–A24.

35. David P. Hamilton, "Quantum Computers Aren't Just Theoretical," *Wall Street Journal*, August 17, 2000, p. A1; Sara Robinson, "Gauging the Limits of Quantum Computing," *San Francisco Chronicle*, March 7, 2000, p. D3; "In the Chips," 2000; and Kau, 2000.

36. Kaku, 2000; Appenzeller, 2000; and John Markoff, "A New Era in Technology for Computers," *New York Times*, March 23, 2000, pp. C1, C24.

37. "Pack on Those Gigabytes," *Business Week*, May 15, 2000, p. 38B.

38. Henry Norr, "Tiny Drive Grows in Capacity," *San Francisco Chronicle*, June 20, 2000, p. C2; and Rachel K. Sobel, "Big Blue Thinks Small," *U.S. News & World Report*, March 27, 2000, p. 46.

39. Larry Armstrong, "Holographic Memories Get Easier to Write," *Business Week*, October 13, 1997, p. 126.

40. John Markoff, "Tiny Magnets May Form Basis for Computing Breakthrough," *New York Times*, January 27, 1997, p. C2.

41. Robert Birge, quoted in Amal Kumar Naj, "Researchers Isolate Bacteria Protein that Can Store Data in 3 Dimensions," *Wall Street Journal*, September 4, 1997, p. B4.

42. Lee Bruno, "The Age of Petabytes," *Red Herring*, March 2000, pp. 266–267.

43. George Gilder, "The End Is Drawing Nigh," *Forbes ASAP*, April 3, 2000, pp. 171–172.

44. Walter S. Mossberg, "How to Buy a Laptop: Some Basic Guidelines in a Dizzying Market," *Wall Street Journal*, October 21, 1999, p. B1.

45. Stephen H. Wildstrom, "How to Shop for a Laptop," *Business Week*, April 3, 2000, p. 25.

46. "The New Laptops," *Consumer Reports*, May 2000, pp. 12–16; Walter S. Mossberg, "Buying Your Next PC? Get the Most Memory, Not the Fastest Chip," *Wall Street Journal*, April 6, 2000, p. B1; and Bill Howard, "Notebook PCs," *PC Magazine*, August 1999, pp. 154–155.

Bookmark It! Box

a. Philip Robinson, "When the Power Fails," *San Jose Mercury News*, December 17, 1995, pp. 1F, 6F.

Chapter 6

1. Lorrie Grant, "Let Your Fingers Do Shopping . . . in Store," *USA Today*, July 28, 1999, p. 3B.
2. Salina Khan, "Kiosks Offer Maps, Hotel, Dining Info," *USA Today*, April 4, 2000, p. 5B.
3. "Coming to an ATM Near You: Movie Previews," *Reno Gazette-Journal*, May 2, 2000, pp. 1E, 6E; Marc Gunther, "Take Your $20, and a Coupon," *Fortune*, April 3, 2000, p. 48; and "Cash Crop," *New York Times Magazine*, August 15, 1999, p. 23.
4. Carol Jouzaitis, "Step Right Up, and Pay Your Taxes and Tickets," *USA Today*, October 2, 1997, p. 4A.
5. Edward C. Baig, "Tapping Out the Quick Keys to a Comfortable Keyboard," *USA Today*, March 29, 2000, p. 3D.
6. Bruce Headlam, "Spills Are Not a Problem for a Flexible Keyboard," *New York Times*, April 6, 2000, p. D3.
7. Becky Waring, "Microsoft Builds a Magical Mouse," *San Francisco Chronicle*, March 30, 2000, p. B3.
8. Claudia H. Deutsch, "There's Gold in Those Old Photos in the Attic," *New York Times*, June 30, 1997, p. C6.
9. Carrie Kirby, "Personal Web Cams for Class of 2000," *San Francisco Chronicle*, June 20, 2000, p. A18; and Joshua Quittner, "Joshing Online," *Time*, June 5, 2000, p. 98.
10. Becky Waring, "Computers Still Don't Know Their Master's Voice," *San Francisco Chronicle*, February 7, 2000, pp. C1, C3.
11. Susan M. Fulton, "Speak Softly, Carry a Big Chip," *New York Times*, March 30, 2000, pp. D1, D6.
12. Anita Hamilton, "Broadcasting from New Ork," *Time*, April 17, 2000, p. 80.
13. Philip Robinson, "Flat Panels: Are They Worth It?" *Reno Gazette-Journal*, March 13, 2000, p. 3E.
14. Verne Kopytoff, "Earful of Internet," *San Francisco Chronicle*, May 25, 2000, pp. B1, B4.
15. Stephen C. Miller, "Could That Be a Human Voice Coming Out of the Computer?" *New York Times*, November 25, 1999, p. D3.
16. Aimee Phan and Dan Vergano, "Wonder Is One of Many Looking into Vision Chips," *USA Today*, December 7, 1999, p. 7D.
17. Peter S. Sinton, "Building Independence," *San Francisco Chronicle*, July 26, 2000, pp. D1, D3; Otis Port, "Now, Electronic 'Eyes' for the Blind," *Business Week*, January 31, 2000, pp. 56–57; David Perlman, "Blind Man 'Sees' Shapes in Artificial Vision Test," *San Francisco Chronicle*, January 17, 2000, pp. A1, A8; and Anne Eisenberg, "Blind People with Eye Damage May Someday Use Chips to See," *New York Times*, June 24, 1999, p. D15.
18. Steve Kaye, quoted in Elisa Williams, "Making Your Home Work," *San Francisco Examiner*, October 8, 1995, pp. B-1, B8; reprinted from *Orange County Register*.
19. Kevin Maney, "Cellphones Could Be Secret Decoder Rings of Future," *USA Today*, June 21, 2000, p. 3B.
20. Dean Takahashi, "TV Interactive to Unveil 'Smart' Paper to Control Computers, Market Products," *Wall Street Journal*, September 9, 1996, p. B7D.
21. Daniel Sorid, "Giving Computers a Sense of Touch," *New York Times*, March 23, 2000, p. D11; and Marina Chicurel, "Once More with Feeling," *Stanford*, March/April 2000, pp. 70–73.
22. Jeff Nesmith, "Lots of Uses, Commercial and Otherwise, for Fast-Growing Electronic Nose," *San Francisco Chronicle*, April 28, 2000, p. A15; and Peter Wayner, "As Plaine as the 'Nose' on Your Chip," *New York Times*, July 8, 1999, p. D11.
23. John Carey, "Building a 'Canary on a Chip,'" *Business Week*, December 16, 1996, p. 130.
24. Edward Rothstein, "The Disklavier Is a Piano with Optical Sensors. What Would Chopin Have Said?" *New York Times*, September 29, 1997, p. C4.
25. Mike Allen, "Time Runs Out for Free Parking," *San Jose Mercury News*, April 25, 1999; reprinted from *New York Times*.
26. Constance L. Hays, "Variable-Price Coke Machine Being Tested," *New York Times*, October 28, 1999, pp. C1, C4.
27. Otis Port, "A Needle with Supersharp Sensors," *Business Week*, July 22, 1996, p. 59.
28. Steve Sternberg, "Device that Helps Quadriplegics Use Hands Wins Approval," *USA Today*, August 19, 1997, p. 1A.
29. Roger Matus, quoted in Kelly Zito, "Voices of the Future," *San Francisco Chronicle*, February 7, 2000, pp. C1, C3.
30. Tom Stein, "Computers May Ease Life's Vexations," *San Francisco Chronicle*, July 26, 1999, p. B3.
31. Daniel Goleman, "Laugh and Your Computer Will Laugh with You, Someday," *New York Times*, January 7, 1997, pp. B9, B14.
32. Tom Stein, "Computers May Ease Life's Vexations," *San Francisco Chronicle*, July 26, 1999, p. B3.
33. Malcom W. Browne, "How Brain Waves Can Fly a Plane," *New York Times*, March 7, 1995, pp. B1, B10; Don Clark "Mind Games: Soon You'll Be Zapping Bad Guys Without Lifting a Finger," *Wall Street Journal*, June 16, 1995, p. B12; Tom Abate, "Marin Investor Bets on Impulse," *San Francisco Chronicle*, July 2, 1995, pp. B-1, B-7; and Malcolm W. Browne, "Neuron Talks to Chip and Chip to Nerve Cell," *New York Times*, August 22, 1995, pp. B1, B9.
34. Mary Madison, "Mind Control for Computers," *San Francisco Chronicle*, December 2, 1995, pp. A1, A13.
35. "Fidler to Kent State," *Quill*, December 1996, p. 8; and Kevin Maney, "High-Tech Tablets: Next Step for Newspapers?" *USA Today*, June 5, 1997, p. 6B.
36. Paul M. Eng, "Web Surfing That'll Give You Whiplash," *Business Week*, April 7, 1997, p. 136C.
37. Phillip Robinson, "Home Theater No Longer Just for the Rich and Famous," *San Jose Mercury News*, November 30, 1997, p. 9S.
38. Kevin Maney, "High-Tech Tablets: Next Step for Newspapers?" *USA Today*, June 5, 1997, p. 6B.
39. Lawrence B. Johnson, "PC Makers Are Focusing on Fine-Tuning the Sound," *New York Times*, December 11, 1995, p. C3.
40. Dean Takahashi, "Blending Math and Music to Make PCs Sound Better," *Wall Street Journal*, March 4, 1997, pp. B1, B5.
41. Harry Somerfield, "3-D Is Headed for Home Television," *San Francisco Chronicle*, March 2, 1994, p. Z-5.
42. Charles Petit, "Read 3-D Debuts in Silicon Valley," *San Francisco Chronicle*, August 30, 1996, pp. A21, A26.
43. Diana Hembree and Ricardo Sandoval, "The Lady and the Dragon," *Columbia Journalism Review*, August 1991, pp. 44–45.
44. Bureau of Labor Statistics, cited in Ellen Neuborne, "Workers in Pain, Employers Up in Arms," *USA Today*, January 9, 1997, pp. 1B, 2B.
45. Edward Felsenthal, "An Epidemic or a Fad? The Debate Heats Up Over Repetitive Stress," *Wall Street Journal*, July 14, 1994, pp. A1, A4.
46. Bobby McGill, "Melding with Your Mouse," *San Francisco Examiner*, July 25, 1999, pp. B-5, B-6.
47. Felsenthal, 1994.
48. Ilana DeBare, "Eyestrain a Bulging Problem," *San Francisco Chronicle*, July 14, 1997, pp. B1, B2; and Jane E. Brody, "Reading a Computer Screen Is Different from Reading a Book, and Has Different Effects on Eyes," *New York Times*, August 7, 1997, p. B6.
49. Independent Expert Group on Mobile Phones, reported in Shawn Young, "Researchers Suggest Limiting Kids' Cellphone Use," *USA Today*, May 12, 2000, p. 1B.
50. George Carlo, in John Tuohy, "Cellphone Researcher's Call: Be Cautious," *USA Today*, August 1, 2000, p. 1D; and Claudia Kalb and Karen Springen, "Is Your Cell Really Safe?" *Newsweek*, August 7, 2000, p. 63.
51. Marty Jerome, "Boot Up or Die," *PC Computing*, April 1998, pp. 172–86.

Bookmark It! Box

a. Philip Robinson, "Organizing and Editing Digital Photos Can Be a Snap," *San Jose Mercury News*, November 23, 1997, pp. 1F, 5F; Richard Folkers, "Pixelated Photography," *U.S. News & World Report*, May 12, 1997, pp. 77–78; Mike Langberg, "HP a Digital Photography Hot-Shot," *San Jose Mercury News*, May 25, 1997, p. 4E; "Instant Images," *Fortune, Technology Buyer's Guide*, Winter 1997, pp. 185–87; and Philip Robinson, "What to Shoot for in a Digital Camera," *San Jose Mercury News*, November 16, 1997, pp. 1F. 4F.
b. Susan Gregory Thomas, "A Photo Lab on Your Desk," *U.S. News & World Report*, November 25, 1996, pp. 104–106.
c. Stewart Alsop, "Digital Photography Is the Next Big Thing," *Fortune*, August 4, 1997, pp. 220–21.
d. Mike Langberg, "Kodak's Online Photograph Venture Bold but Still Fuzzy," *San Jose Mercury News*, September 7, 1997, pp. 1E, 2E.

Chapter 7

1. Michel Marriott, "WebTV Puts It All Together," *New York Times*, December 30, 1999, pp. D2, D5.
2. Bruce Haring, "Step Right Up for the Next Push in Remote Control," *USA Today*, September 15, 1999, p. 7D.
3. Mike Snider, "Waves of Interference Slow Digital TV," *USA Today*, October 27, 1999, p. 7D; Neil Gross, Richard Siklos, and Heidi Dawley, "HDTV: You're Not Going to Like This Picture," *Business Week*, October 25, 1999, p. 50; and Andrew J. Glass, "HDTV Still Stationed at the Starting Gate," *San Francisco Chronicle*, September 27, 1999, p. E5.
4. "What Does 'Digital' Mean in Regard to Electronics?" *Popular Science*, August 1997, pp. 91–94.
5. Simon Romero, "Weavers Go Dot-Com, and Elders Move In," *New York Times*, March 28, 2000, pp. A1, A4.
6. Marcia Vickers, "Don't Touch that Dial: Why Should I Hire You?" *New York Times*, April 13, 1997, sec. 3, p. 11.
7. International Data Corp., cited in "Do Dirty Dishes, Dust Bunnies Violate OSHA Rules?" [editorial], *USA Today*, January 7, 2000, p. 14A.
8. Study by John J. Heldrich Center for Workforce Development, Rutgers University, cited in Kirstin Downey Grimsley, "How to Get Permission to Telecommute," *San Francisco Chronicle*, February 16, 2000, p. B3.
9. David Kline, quoted in W. James Au, "The Lonely Long-Distance Worker," *PC Computing*, February 2000, pp. 42–43.
10. David Leonhardt, "Telecommuting to Pick Up as Workers Iron Out Kinks," *New York Times*, December 20, 1999, p. C6.
11. Maryanne Murray Buechner, "Superconnected," *Time*, March 22, 1999.
12. Mike Romano, "Brave New Home," *U.S. News & World Report*, April 5, 1999, pp. 60–62.
13. Dori Jones Yang, "A Boob Tube with Brains," *U.S. News & World Report*, March 13, 2000, pp. 42–43.

14. Hank Hogan, "HDTV Is Here, but Can You Afford It?" *High-Tech Careers*, June/July 2000, pp. 17–20.

15. Neil Hickey, "The Digital Newsroom: Ready or Not," *Columbia Journalism Review*, February 2000, p. 56.

16. Peter Suciu, "HDTV: Getting the Big Picture," Fox News, May 11, 2000, www.foxnews.com.

17. Hank Kahrs, quoted in Evan Ramstad, "Works in Progress," *Wall Street Journal*, September 11, 1997, pp. 1E, 4E.

18. Carolyn Nielson, "GPS Ready to Take Off," *San Francisco Examiner*, June 4, 1995, pp. B-5, B-6.

19. David Perlman, "Satellite Network Captures Volcano Drama in Hawaii," *San Francisco Chronicle*, February 17, 1997, p. A4.

20. Anthony Ramirez, "Cheap Beeps: Across Nation, Electronic Pagers Proliferate," *New York Times*, July 19, 1993, pp. A1, C2.

21. "Who Needs a Cell Phone?" *Consumer Reports*, February 1997, pp. 10–15.

22. Ramstad, 1997.

23. James Kim, "Scrambling for the Sky," *USA Today*, June 18, 1997, pp. 1B, 2B.

24. Jonathan Marshall, "Speedy New Technology Begins Race to Market," *San Francisco Chronicle*, February 18, 1997, pp. B1, B2.

25. Karen Alexander, "Bluetooth: Next Big Wave in Electronics?" *Los Angeles Times*, August 14, 2000, pp. C1, C4.

26. Howard Banks, "The Law of the Photon," *Forbes*, October 6, 1997, pp. 66–73.

27. Jonathan Marshall, "Photonics Industry Is Changing Communications," *San Francisco Chronicle*, July 29, 1997, pp. C1, C7.

28. Otis Port, "A Lens that Tricks the Light Fantastic," *Business Week*, July 21, 1997, pp. 118–19.

29. Marshall, "Photonics Industry Is Changing Communications," 1997.

30. Yahoo!, cited in Del Jones, "Cyber-porn Poses Workplace Threat," *USA Today*, November 27, 1995, p. B1.

31. Lawrence J. Magid, "Be Wary, Stay Safe in the On-line World," *San Jose Mercury News*, May 15, 1994, p. 1F.

32. Peter H. Lewis, "Limiting a Medium without Boundaries," *New York Times*, January 15, 1996, pp. C1, C4.

33. John M. Broder, "Making America Safe for Electronic Commerce," *New York Times*, June 22, 1997, sec. 4, p. 4; Margaret Mannix and Susan Gregory Thomas, "Exposed Online," *U.S. News & World Report*, June 23, 1997, pp. 59–61; and Noah Matthews, "Shareware," *San Jose Mercury News*, October 12, 1997, p. 4F.

34. Survey by Equifax and Louis Harris & Associates, cited in Bruce Horovitz, "80% Fear Loss of Privacy to Computers," *USA Today*, October 31, 1995, p. 1A.

35. Fred Vogelstein, "Is It Sharing or Stealing?" *U.S. News & World Report*, June 12, 2000, pp. 38–40.

36. Lawrence Magid, "Kids Need to Respect Copyrights," *San Francisco Examiner*, August 20, 2000, pp. C-5, C-7.

37. Michel Marriott, "For Extra Cheese, Ctrl + Pizza," *New York Times*, February 10, 2000, pp. D1, D8; Walter S. Mossberg, "A Simple New Gadget Lets You Go Online without Using a PC," *Wall Street Journal*, January 27, 2000, p. B1.

38. Howie Hunger, quoted in John Yaukey, "Small, Simple PCs Take the Day," *Reno Gazette Journal*, May 8, 2000, p. 1E.

39. Roy Furchgott, "Web to Go—Sort of," *Business Week*, February 14, 2000, pp. 144–145.

40. Dori Jones Yang, "A Boob Tube with Brains," *U.S. News & World Report*, March 13, 2000, pp. 42–43; David Lieberman, "Malone: Cable Firms Must Tune in to Interactive TV," *USA Today*, April 17, 2000, p. 10B.

41. Kenneth Terrell, "PlayStation 2: Hold Your Fire—and Your Dollars," *U.S. News & World Report*, May 8, 2000, pp. 66–67; Kenneth Terrell, "Sega Takes Its Play Online," *U.S. News & World Report*, May 8, 2000, pp. 66–67; Jay Greene, Irene Kunii, and Janet Rae-Dupree, "Get Ready to Rumble," *Business Week*, March 20, 2000, p. 48; Dean Takahashi, "Sega Will Give Away Dreamcast Players to Lure Subscribers to the Web," *Wall Street Journal*, April 4, 2000, pp. B1, B4; Steven Levy, "Here Comes PlayStation 2," *Newsweek*, March 6, 2000, pp. 54–59; N'Gai Croal, "The Art of the Game," *Newsweek*, March 6, 2000, pp. 60–62; and Kelly Zito, "Sega's Got Game on the Web," *San Francisco Chronicle*, April 4, 2000, pp. C1, C5.

42. Kevin Maney, "Without Neat Plug-in Modules, Visor Suffers," *USA Today*, April 24, 2000, p. 7B; N'gai Croal, "A PC in Every Pocket," *Newsweek*, May 1, 2000, pp. 521–53; Henry Norr, "Technicolor Palm," *San Francisco Chronicle*, February 24, 2000, pp. C1, C4; Stephen H. Wildstrom, "Loosening Palm's Grip," *Business Week*, May 1, 2000, p. 28; and Joshua Quittner, "PCs? Forget 'Em!" *Time*, May 8, 2000, p. 105.

43. Chris O'Malley, "Two-Way Sharing," *Popular Science*, November 1999, p. 58.

44. "E-Mail When You're Far from Home," *Business Week Frontier*, March 27, 2000, p. F.6.

45. Guy Richardson, "Feeling Wired Because You're Not Wired," *Reno Gazette-Journal*, March 13, 2000, p. 3E.

46. Michel Marriott, "Out of the Mouths of Babes, Wirelessly," *New York Times*, March 23, 2000, pp. D1, D6.

Bookmark It! Box

a. Catherine Greenman, "You Want to Be in Pictures? Will You Settle for a Web Page?" *New York Times*, March 16, 2000, p. D9; Chris O'Malley, "Prefab Home Pages," *Popular Science*, September 1999, p. 56; and Deborah Kong, "Building a Web Site," *San Jose Mercury News*, April 1999, pp. 1F, 2F.

b. Thomas E. Weber, "There's No Place Like Home: A Reporter Builds a Web Page," *Wall Street Journal*, February 16, 1998, p. B22; Jakob Niesen, "Make Your Site User-Friendly," *FSB*, Spring 2000, p. e8; and Deborah Kong, "Building a Web Site," *San Jose Mercury News*, April 11, 1999, pp. 1F, 2F.

Chapter 8

1. Carole A. Lane, quoted in Leslie Miller, "You Are a Database and Access Abounds," *USA Today*, June 9, 1997, p. 6D.

2. Robert D. Atkinson and Randolph H. Court, *The New Economy Index: Understanding America's Economic Transformation* (Washington, D.C.: Progressive Policy Institute, 1999).

3. See J. Quittner, "Tim Berners-Lee," *Time*, March 29, 1999, pp. 193–194.

4. Forrester Research Inc., reported in "E-Commerce: It's Clicking" [editorial], *Business Week*, January 11, 1999, p. 154.

5. Odyssey, reported in S. Lohr, "Survey Suggests Consumers Are Taking to E-Commerce," *New York Times*, March 22, 1999, p. C4.

6. Sarah E. Hutchinson and Stacey C. Sawyer, *Computers, Communications, and Information: A User's Introduction*, rev. ed. (Burr Ridge, IL: Irwin/McGraw-Hill, 1998), pp. E1.1–E1.3.

7. Jeff Bezos, quoted in K. Southwick, interview, October 1996, www.upside.com.

8. D. Levy, "On-line Gamble Pays Off with Rocketing Success," *USA Today*, December 24, 1998, pp. 1B, 2B.

9. Kara Swisher, "Why Is Jeff Bezos Still Smiling?" *Wall Street Journal*, April 24, 2000, pp. B1, B10.

10. Jonathan Berry, John Verity, Kathleen Kerwin, and Gail DeGeorge, "Database Marketing," *Business Week*, September 5, 1994, pp. 56–62.

11. Cheryl D. Krivda, "Data-Mining Dynamite," *Byte*, October 1995, pp. 97–103.

12. Karen Watterson, "A Data Miner's Tools," *Byte*, October 1995, pp. 91–96.

13. Watterson, 1995.

14. Watterson, 1995.

15. Watterson, 1995.

16. Sarah Reese Hedberg, "The Data Gold Rush," *Byte*, October 1995, pp. 83–88.

17. Krivda, 1995.

18. Edmund X. DeJesus, "Data Mining," *Byte*, October 1995, p. 81.

19. Forester Research, cited in Denise Caruso, "Taking Stock of the Differences Between the Consumer Internet Market and Its Business-to-Business Cousin," *New York Times*, February 28, 2000, p. C5; and Gartner Group, cited in William J. Holstein, "Rewiring the 'Old Economy,'" *U.S. News & World Report*, April 10, 2000, pp. 38–40.

20. Don Tapscot, "Virtual Webs Will Revolutionize Business," *Wall Street Journal*, April 24, 2000, p. A38.

21. Carolyn Said, "Online Middlemen," *San Francisco Chronicle*, April 10, 2000, pp. C1, C3; Holstein, 2000; Claudia H. Deutsch, "Another Economy on the Supply Side," *New York Times*, April 8, 2000, pp. B1, B4; and Kelly Zito, "Online Exchange for Shops," *San Francisco Chronicle*, March 9, 2000, pp. B1, B4.

22. Tapscot, 2000.

23. John Chambers, quoted in Del Jones and Beth Belton, "Cisco Chief: Virtual Close to Hit Big," *USA Today*, October 12, 1999, p. 3B.

24. "Connecting the Dot-Coms," *Newsweek*, April 17, 2000, p. 68G.

25. James A. Larson, *Database Directories* (Upper Saddle River, NJ: Prentice Hall PTR, 1995).

26. David M. Kroenke, *Database Processing*, 5th ed. (Upper Saddle River, NJ: Prentice Hall, 1995), p. 271.

27. William Safire, "Art vs. Artifice," *New York Times*, January 3, 1994, p. A11.

28. Cover, *Newsweek*, June 27, 1994; and cover, *Time*, June 27, 1994.

29. Jonathan Alter, "When Photographs Lie," *Newsweek*, July 30, 1990, pp. 44–45.

30. Fred Ritchin, quoted in Alter, 1990.

31. Robert Zemeckis, cited in Laurence Hooper, "Digital Hollywood: How Computers Are Remaking Movie Making," *Rolling Stone*, August 11, 1994, pp. 55–58, 75.

32. Woody Hochswender, "When Seeing Cannot Be Believing," *New York Times*, June 23, 1992, pp. B1, B3.

33. Bruce Horowitz, "Believe Your Eyes? Ads Bend Reality," *USA Today*, April 24, 2000, pp. 1B, 2B.

34. Penny Williams, "Database Dangers," *Quill*, July/August 1994, pp. 37–38.

35. Lynn Davis, quoted in Williams, 1994.

36. Doug Rowan, quoted in Ronald B. Lieber, "Picture This: Bill Gates Dominating the Wide World of Digital Content," *Fortune*, December 11, 1995, p. 38.

37. Kathy Rebello, "The Ultimate Photo Op?" *Business Week*, October 23, 1995, p. 40.

38. Steve Lohr, "Huge Photo Archive Bought by Software Billionaire Gates," *New York Times*, October 11, 1995, pp. A1, C5.

39. Don Clark, "Bill Gates's Corbis Gains Sole Rights to Artist's Works," *Wall Street Journal*, April 2, 1996, p. B9.

40. Kathryn Rambo, quoted in Ramon G. McLeod, "New Thieves Prey on Your Very Name," *San Francisco Chronicle*, April 7, 1997, pp. A1, A6.

41. Rambo, quoted in T. Trent Gegax, "Stick 'Em Up? Not Anymore. Now It's Crime by Keyboard," *Newsweek*, July 21, 1997, p. 14.

42. McLeod, 1997.

Bookmark It! Box

a. Stephen Manes, "Time and Technology Threaten Digital Archives . . .," *New York Times*, April 7, 1998, p. B15.

b. Mike Snider, "Obsolescence: The No. 1 Built-In Feature," *USA Today*, October 29, 1997, p. 6D.

c. Laura Tangley, "Whoops, There Goes Another CD-ROM," *U.S. News & World Report*, February 16, 1999, pp. 67–68.

d. Marcia Stepanek, "From Digits to Dust," *Business Week*, April 20, 1998, pp. 128–130.

e. Manes, 1998.

Chapter 9

1. Katie Hafner, "We're Not All Connected, Yet," *New York Times*, January 27, 2000, pp. D1, D9.

2. Michael Dertouzos, "The Net Revolution Spawns a 'Fast Caste,'" *Los Angeles Times*, January 20, 2000, p. A15.

3. Survey by Inktomi Corp. and NEC Research Institute, reported in Ashley Dunn, "It's a Very Wide Web: 1 Billion Pages' Worth," *Los Angeles Times*, January 20, 2000, p. C7.

4. Alan Murray, "Trying to Make World Safe for E-Commerce," *Wall Street Journal*, November 29, 1999, p. A1.

5. Jube Shiver Jr., "Web Offers Few Riches for Poor," *Los Angeles Times*, March 16, 2000, pp. C1, C7; Katie Hafner, "A Credibility Gap in the Digital Divide," *New York Times*, March 5, 2000, sec. 4, p. 4; Janet Kornblum, "Poor Aren't Profiting, Non-profit Group Says," *USA Today*, March 14, 2000, p. 3D; and Jube Shiver Jr., "Big Bandage for a Narrowing Internet Gap," *Los Angeles Times*, January 29, 2000, pp. A1, A12.

6. We are grateful to Prof. John Durham for contributing these ideas.

7. John Allen Paulos, "Smart Machines, Foolish People," *Wall Street Journal*, October 5, 1999, p. A26.

8. Arthur M. Louis, "Nasdaq's Computer Crashes," *San Francisco Chronicle*, July 16, 1994, pp. D1, D3.

9. Stephen H. Wildstrom, "Oh, for a Bug-free Browser," *Business Week*, November 1, 1999, p. 24.

10. Henry K. Lee, "UC Student's Dissertation Stolen with Computer," *San Francisco Chronicle*, January 27, 1994, p. A15.

11. Thomas J. DeLoughry, "2 Students are Arrested for Software Piracy," *Chronicle of Higher Education*, April 20, 1994, p. A32.

12. John T. McQuiston, "4 College Students Charged with Theft Via Computer," *New York Times*, March 18, 1995, p. 38.

13. Janet Rae-Dupree and Richard J. Newman, "A Twisted Kind of Love," *U.S. News & World Report*, May 15, 2000, p. 24; Brad Stone, Mark Hosenball, and Stefan Theil, "Bitten by Love," *Newsweek*, May 15, 2000, pp. 42–43; and Lev Grossman et al., "Attack of the Love Bug," *Time*, May 15, 2000, pp. 49–56.

14. Edward Iwata, "Mutating Computer Virus Hits," *USA Today*, May 19, 2000, p. 1A.

15. David Carter, quoted in Associated Press, "Computer Crime Usually Inside Job," *USA Today*, October 25, 1995, p. 1B.

16. Bryan Burrough, "Invisible Enemies," *Vanity Fair*, June 2000, pp. 172 – 177, 208 – 214.

17. Steven Bellovin, cited in Jane Bird, "More Than a Nuisance," *The Times* (London), April 22, 1994, p. 31.

18. Eugene Carlson, "Some Forms of Identification Can't Be Handily Faked," *Wall Street Journal*, September 14, 1993, p. B2.

19. Justin Matlkick, "Security of Online Markets Could Well be at Stake," *San Francisco Chronicle*, September 16, 1997, p. A21.

20. John Holusha, "The Painful Lessons of Disruption," *New York Times*, March 17, 1993, pp. C1, C5.

21. The Enterprise Technology Center, cited in "Disaster Avoidance and Recovery Is Growing Business Priority," special advertising supplement in *LAN Magazine*, November 1992, p. SS3.

22. Justin Matlkick, "Security of Online Markets Could Well be at Stake," *San Francisco Chronicle*, September 16, 1997, p. A21.

23. John Schwartz, "Encryption Export Bars Relaxed," *San Francisco Chronicle*, July 18, 2000, p. A3; reprinted from the *Washington Post*.

24. John Holusha, "The Painful Lessons of Disruption," *New York Times*, March 17, 1993, pp. C1, C5.

25. The Enterprise Technology Center, cited in "Disaster Avoidance and Recovery Is Growing Business Priority," special advertising supplement in *LAN Magazine*, November 1992, p. SS3.

26. Micki Haverland, quoted in Fred R. Bleakley, "Rural County Balks at Joining Global Village," *Wall Street Journal*, January 4, 1996, pp. B1, B2.

27. David Ensunsa, "Proposed Cell-Phone Pole Faces Challenge," *The Idaho Statesman*, June 23, 1995, p. 4B.

28. Survey by Center for Energy and Climate Solutions, reported in Christine Y. Chen and Greg Lindsay, "Will Amazon(.com) Save the Amazon?" *Fortune*, March 20, 2000, p. 224.

29. Norm Alster and William Echikson, "Are Old PCs Poisoning Us?" *Business Week*, June 12, 2000, pp. 78–80; and William J. Holstein, "Take My Personal Computer—Please!" *U.S. News & World Report*, June 5, 2000, p. 51.

30. Andrew Kupfer, "Alone Together," *Fortune*, March 20, 1995, pp. 94–104.

31. Lutz Erbring, coauthor of Stanford University survey of 4113 people about Internet impact on daily activities, quoted in Joellen Perry, "Only the Cyberlonely," *U.S. News & World Report*, February 28, 2000, p. 62.

32. Dru Sefton, "Teen Girls Feel the Net Effect," *USA Today*, March 14, 2000, p. 3D; and Elizabeth Weise, "A Circle Unbroken by Surveys," *USA Today*, February 22, 2000, p. 3D.

33. Amitai Etzioni, quoted in Perry, 2000.

34. Tom Verdin, "Rapid Growth of Cyber-Gambling Prompts Calls for Regulation," *San Francisco Chronicle*, March 9, 2000, p. A9.

35. Survey by Microsoft Corp., reported in Don Clark and Kyle Pope, "Poll Finds Americans Like Using PCs, but May Find Them to Be Stressful," *Wall Street Journal*, April 10, 1995, p. B3.

36. Survey by Concord Communications, reported in Matt Richtel, "Rage Against the Machine: PCs Take Brunt of Office Anger," *New York Times*, March 11, 1999, p. D3.

37. Jonathan Marshall, "Some Say High-Tech Boom Is Actually a Bust," *San Francisco Chronicle*, July 10, 1995, pp. A1, A4.

38. Eleena de Lisser, "One-Click Commerce: What People Do Now to Goof Off at Work," *Wall Street Journal*, September 24, 1999, pp. A1, A8.

39. Surfwatch Checknet, cited in Keith Naughton, Joan Raymond, Ken Shulman, and Diane Struzzi, "CyberSlacking," *Newsweek*, November 29, 1999, pp. 62–65.

40. STB Accounting Systems 1992 survey, reported in Del Jones, "On-line Surfing Costs Firms Time and Money," *USA Today*, December 8, 1995, pp. 1A, 2A.

41. Steven Levy, "A Window on Their World," *Newsweek*, February 15, 1999, p. 61.

42. Dan Gillmor, "Online Reliability Will Carry a Price," *San Jose Mercury News*, July 18, 1999, pp. 1E, 7E.

43. Paul Saffo, quoted in Laura Evenson, "Pulling the Plug," *San Francisco Chronicle*, December 18, 1994, "Sunday" section, p. 53.

44. Daniel Yankelovich Group report, cited in Barbara Presley Noble, "Electronic Liberation or Entrapment," *New York Times*, June 15, 1994, p. C4.

45. Joseph Weber, Andy Reinhardt, and Rich Miller, "Tight Labor? Tech to the Rescue," *Business Week*, March 20, 2000, pp. 36–37; Louis Uchitelle, "Productivity Finally Shows the Impact of Computers," *New York Times*, March 12, 2000, sec. 3, p. 4; Jennifer Reingold, Marcia Stepanek, and Diane Brady, "Why the Productivity Revolution Will Spread," *Business Week*, February 14, 2000, pp. 112–118; Laura D'Andrea Tyson, "Though It's a New Economy, It's Got Some Old Flaws," *Business Week*, January 10, 2000, p. 32; and George Hager, "Worker Output Spurts, Helps Control Inflation," *USA Today*, December 8, 1999, p. 1B.

46. Stephen S. Roach, "Working Better or Just Harder?" *New York Times*, February 14, 2000, p. A27.

47. Jeremy Rifkin, "Technology's Curse: Fewer Jobs, Fewer Buyers," *San Francisco Examiner*, December 3, 1995, p. C-19.

48. Michael J. Mandel, "Economic Anxiety," *Business Week*, March 11, 1996, pp. 50–56; Bob Herbert, "A Job Myth Downsized," *New York Times*, March 8, 1996, p. A19; and Robert Kuttner, "The Myth of a Natural Jobless Rate," *Business Week*, October 20, 1997, p. 26.

49. Stewart Brand, in "Boon or Bane for Jobs?" *The Futurist*, January-February 1997, pp. 13–14.

50. Paul Krugman, "Long-Term Riches, Short-Term Pain," *New York Times*, September 25, 1994, sec. 3, p. 9.

51. Department of Commerce Survey, cited in "The Information 'Have Nots' [editorial]," *New York Times*, September 5, 1995, p. A12.

52. Beth Belton, "Degree-based Earnings Gap Grows Quickly," *USA Today*, February 16, 1996, p. 1B.

53. Alan Kruger, quoted in LynNell Hancock, Pat Wingert, Patricia King, Debra Rosenberg, and Alison Samuels, "The Haves and the Have-Nots," *Newsweek*, February 27, 1995, pp. 50–52.

54. Scott Carlson, "High-Speed Network Will Serve Universities for 3 More Years," *Chronicle of Higher Education*, April 21, 2000, p. A49.

55. Laurence H. Tribe, "The FCC vs. the Constitution," *Wall Street Journal*, September 5, 1997, p. A8.

56. John M. Broder, "Let It Be," *New York Times*, June 30, 1997, pp. C1, C9.

57. *A Framework for Global Electronic Commerce*, quoted in Steven Levy, "Bill and Al Get It Right," *Newsweek*, July 7, 1997, p. 80.

58. Jim Hornthal, quoted in Jon Schwartz, "Clinton Advocates Net Self-Rule," *San Francisco Chronicle*, July 2, 1997, pp. B1, B2.

59. Eli M. Noam, "An Unfettered Internet? Keep Dreaming," *New York Times*, July 11, 1997, p. A2.

60. Mike France, "What's in a Name.com? Plenty," *Business Week*, September 6, 1999, pp. 86 – 90.

61. Shelly Freierman, "For Life-long Learning, Click Here," *New York Times*, August 25, 1997, p. C6.

62. Dan Carnevale, "How to Procter from a Distance," *Chronicle of Higher Education*, December 12, 1999, pp. A47, A48.

63. Robert L. Johnson, "Extending the Reach of 'Virtual' Classrooms," *Chronicle of Higher Education*, July 6, 1994, pp. A19–A23.

64. Glenn R. Jones, *Cyberschools: An Education Renaissance* (Jones Digital Century, 1997).

65. Pam Dixon, *Virtual College* (Princeton, NJ: Peterson's, 1996).

66. Sarah Carr, "As Distance Education Comes of Age, the Challenge Is Keeping the

Students," *Chronicle of Higher Education*, February 11, 2000, pp. A39–A41.

67. Bruce Weber, "Notes from Cyberclass," *New York Times*, January 3, 1999, sec 4A, pp. 15, 46.

68. Jeff K. Powers, quoted in Carr, 2000.

69. Carr, 2000.

70. Brent Muirhead, "Looking at Net Colleges," *USA Today*, November 12, 1999, p. 14A.

71. University of Illinois, *The Online Pedagogy Report*, reported in P. J. Huffstutter and Robin Fields, "A Virtual Revolution in Teaching," *Los Angeles Times*, March 3, 2000, pp. A1, A7.

72. HuffStutter and Fields, 2000.

73. Student reported by Ann Fisher, "Readers Weigh In on Virtual Work and School," *Fortune*, June 21, 1999, p. 200.

74. Faith Bremmer, "On-line College Classes Get High Marks Among Students," *USA Today*, November 16, 1998, p. 16E.

75. Joseph B. Walter, quoted in William H. Honon, "Northwestern University Takes a Lead in Using the Internet to Add Sound and Sight to Courses," *New York Times*, May 28, 1997, p. A17.

76. Richard Clark, quoted in Bremmer, 1998.

Bookmark It! Box

a. Nate Stulman, "The Great Campus Goof-Off Machine," *New York Times*, March 15, 1999, p. A25.

b. Robert Kubey, "Internet Generation Isn't Just Wasting Time" (letters), *New York Times*, March 21, 1999, sec. 4, p. 14.

c. Marco R. della Cava, "Are Heavy Users Hooked or Just Online Fanatics?" *USA Today*, January 16, 1996, pp. 1A, 2A.

d. Kenneth Hamilton and Claudia Kalb, "They Log On, but They Can't Log Off," *Newsweek*, December 18, 1995, pp. 60 – 61; Kenneth Howe, "Diary of an AOL Addict," *San Francisco Chronicle*, April 5, 1995, pp. D1, D3.

e. Stella Yu, quoted in Hamilton and Kalb, 1995.

f. Jonathan Kandell, quoted in J. R. Young, "Students Are Unusually Vulnerable to Internet Addiction, Article Says," *Chronicle of Higher Education*, February 6, 1998, p. A25.

g. American Psychological Association, reported in R. Leibrock, "AOLaholic: Tales of an Online Addict," *Reno News & Review*, October 22, 1997, pp. 21, 24.

h. Hamilton and Kalb, 1995.

i. Keith J. Anderson, reported in Leo Reisberg, "10% of Students May Spend Too Much Time Online," *Chronicle of Higher Education*, June 16, 2000, p. A43.

j. R. Sanchez, "Colleges Seek Ways to Reach Internet-Addicted Students," *San Francisco Chronicle*, May 23, 1996, p. A16; reprinted from the *Washington Post*.

k. Sanchez, 1996.

l. P. Belluck, "The Symptoms of Internet Addiction," *New York Times*, December 1, 1996, sec. 4, p. 5.

m. Questionnaire adapted from chart, "Characteristics of 'Internet Dependent' Students," from Keith J. Anderson, Rensselaer Polytechnic Institute, in Reisberg, 2000.

n. Ben Gose, "A Dangerous Bet on Campus," *Chronicle of Higher Education*, April 7, 2000, pp. A49–A51.

o. Jeremy Siegel, Wharton School, University of Pennsylvania, quoted in David Segal, "The Minefield of Internet Trading," *San Jose Mercury News*, September 6, 1998, pp. 1F, 6F; reprinted from the *Washington Post*.

Chapter 10

1. Stephen Cohen, quoted in Carolyn Lochhead, "The Engines of a New Economy," *San Francisco Examiner & Chronicle*, "Sunday" section, February 27, 2000, p. 2.

2. Brad DeLong, quoted in Lochhead, 2000.

3. Guy Richardson, "Time to Curl Up and Take a Napster," *Reno Gazette-Journal*, May 29, 2000, p. 4E.

4. Survey by Webnoize Inc., Cambridge, Mass., reported in Martin Peers, "Survey Studies Napster's Spread on Campuses," *Wall Street Journal*, May 15, 2000, p. B8.

5. Randall E. Stross, "Napster Nonsense," *U.S. News & World Report*, May 29, 2000, p. 49.

6. Lee Gomes, "Think Music Moguls Don't Like Sharing? Try Copying Software," *Wall Street Journal*, August 14, 2000, p. B1.

7. Jefferson Graham, "Record Giants Join the Rust to Download," *USA Today*, June 13, 2000, p. 1D.

8. Jeanne B. Pinder, "Fuzzy Thinking Has Merits When It Comes to Elevators," *New York Times*, September 22, 1993, pp. C1, C7.

9. Robert Benfer Jr., Louanna Furbee, and Edward Brent Jr., quoted in Steve Weinberg, "Steve's Brain," *Columbia Journalism Review*, February 1991, pp. 50–52.

10. Robert McGrough, "Fidelity's Bradford Lewis Takes Aim at Indexes with His 'Neural Network' Computer Program," *Wall Street Journal*, October 27, 1992, pp. C1, C21.

11. Michael Waldholz, "Computer 'Brain' Outperforms Doctors Diagnosing Heart Attack Patients," *Wall Street Journal*, December 2, 1991, p. B78.

12. Otis Port, "Computers that Think Are Almost Here," *Business Week*, July 17, 1995, pp. 68–72.

13. Kenneth Chang, "Aping Biology, Computer Guides Automated Evolution of a Robot," *New York Times*, August 31, 2000, pp. A1, A18; and Curt Suplee, "A First: Robots Created by Robots," *San Francisco Chronicle*, August 31, 2000, pp. A1, A15; reprinted from *Washington Post*.

14. Judith Anne Gunther, "An Encounter with AI," *Popular Science*, June 1994, pp. 90–93.

15. William A. Wallace, *Ethics in Modeling* (New York: Elsevier Science, Inc., 1994).

16. Laura Johannes, "Meet the Doctor: A Computer that Knows a Few Things," *Wall Street Journal*, December 18, 1995, p. B1.

17. Christine Borgman, quoted in Gary Chapman, "What the Online World Really Needs Is an Old-Fashioned Librarian," *San Jose Mercury News*, August 21, 1995, p. 3D; reprinted from *Los Angeles Times*.

18. Campus Computing Project 1997 survey, reported in Lisa Guernsey, "E-Mail Is Now Used in a Third of College Courses, Survey Finds," *Chronicle of Higher Education*, October 17, 1997, p. A30; and Edward C. Baig, "A Little High Tech Goes a Long Way," *Business Week*, November 10, 1997, p. E10.

19. Student Monitor LLC, cited in Danielle Sessa, "For College Students, Web Offers a Lesson in Discounts," *Wall Street Journal*, January 21, 1999, p. B7.

20. U.S. Department of Education, cited in Steve Rhodes, "Classrooms with Class—and Possibly Espresso Machines," *Newsweek*, December 14, 1998, p. 20.

21. *Does It Compute?*, study by Harold Wenglinsky, reported in Ethan Bronner, "Computers Help Math Learning, Study Finds," *New York Times*, September 30, 1998, p. A16.

22. Dulcie Leimbach, "Encouraging Creativity, without the Mess," *New York Times*, November 12, 1998, p. D12.

23. Kelly McCollum, "Web Site Lets Thousands of Schoolchildren Follow Research on Albatrosses," *Chronicle of Higher Education*, February 13, 1998, p. A32.

24. Kelly McCollum, "High School Students Use Web Intelligently for Research, Study Finds," *The Chronicle of Higher Education*, December 4, 1998, p. A25.

25. Kelly McCollum, "How a Computer Program Learns to Grade Essays," *Chronicle of Higher Education*, September 4, 1998, pp. A37–A38; and William H. Honan, "High Tech Comes to the Classroom," *New York Times*, January 27, 1999, p. A22.

26. Nanette Asimov, "Home-Schoolers Plug into the Internet for Resources," *San Francisco Chronicle*, January 29, 1999, pp. A1, A15.

27. Mary Beth Marklein, "Distance Learning Takes a Gigantic Leap Forward," *USA Today*, June 4, 1998, pp. 1D, 2D; and Godie Blumenstyk, "Leading Community Colleges Go National with New Distance-Learning Network," *Chronicle of Higher Education*, July 10, 1998, pp. A16–A17.

28. Rebecca Quick, "Software Seeks to Breathe Life into Corporate Training Classes," *Wall Street Journal*, August 6, 1998, p. B8.

29. April Lynch, "Bleeding Sailor Performs Self-Surgery Via E-Mail," *San Francisco Chronicle*, November 19, 1998, pp. A1, A10.

30. Kate Murphy, "Telemedicine Getting a Test in Efforts to Cut Costs of Treating Prisoners," *New York Times*, June 8, 1998, p. C3.

31. Rick Satava, quoted in Gary Taubes, "Surgery in Cyberspace," *Discover*, December 1994, pp. 85–94.

32. Denise Grady, "Software to Compute Women's Cancer Risk," *New York Times*, January 26, 1999, p. D4.

33. Rita Beamish, "Computers Now Helping to Screen for Troubled Teen-Agers," *New York Times*, December 17, 1998, p. D9.

34. Jamie Beckett, "Sorting Out Cybershrinks," *San Francisco Chronicle*, August 11, 1998, p. C3.

35. Vincent Kiernan, "Using the Web, Epidemiologist Aims to Improve Public Health in Developing Nations," *Chronicle of Higher Education*, January 30, 1998, pp. A21–A22.

36. John O'Neil, "Implanted Chip Offers Hope of Simplifying Drug Regimens," *New York Times*, February 2, 1999, p. D6.

37. Nancy Ann Jeffrey, "Hydraulics and Computers Help Artificial Limbs Get 'Smarter,'" *Wall Street Journal*, August 14, 1998, p. B1.

38. Associated Press, "Implant Transmits Brain Signals Directly to Computer," *New York Times*, October 22, 1998, p. G9.

39. Associated Press, "Dad Uses Internet to Find Wonder Drug for His Son," *San Francisco Chronicle*, May 29, 1998, p. A22.

40. Jim Hudak, quoted in Heather Green and Linda Himelstein, "A Cyber Revolt in Health Care," *Business Week*, October 19, 1998, pp. 154–56.

41. Tanya Schevitz, "Pregnant in Cyberspace," *San Francisco Chronicle*, July 29, 1998, pp. A13, A17.

42. Vincent Kiernan, "Internet-Based 'Collaboratories' Help Scientists Work Together," *Chronicle of Higher Education*, March 12, 1999, pp. A22–A23.

43. Dan Werthimer, quoted in Henry Fountain, "Download Data, in Search of Intelligent Life," *New York Times*, December 8, 1998, p. D3; and Kelly McCollum, "Berkeley Astronomers Enlist Internet Users to Seek Alien Life," *Chronicle of Higher Education*, December 11, 1998, p. A40.

44. Frank Vizard, "In Search of Cleopatra's Palace," *Popular Science*, May 1999, pp. 78–81.

45. David L. Wheeloer, "Archeologists Use Technology to Avoid Invasive Excavations," *Chronicle of Higher Education*, November 30, 1998, pp. A13–A14.

46. Mindy Sink, "Fading Indian Rock Art Saved, at Least in Database," *New York Times*, October 1, 1998, p. D10; and Sharon Begley, "Secrets of the Cave's Art," *Newsweek*," May 24, 1999, pp. 65–66.

47. Sreenath Sreenivasan, "Environmental Issues Are Clarified," *New York Times*, October 27, 1997, p. C6.

48. Alex Barnum, "Finding Polluters Close to Home," *San Francisco Chronicle*, April 21, 1998, p. C3.

49. Myron Magnet, "Who's Winning the Information Revolution," *Fortune*, November 30, 1992, pp. 110–117.

50. Tony Rutkowski, quoted in Patricia Schnaidt, "The Electronic Superhighway," *LAN Magazine*, October 1993, pp. 6–8.

51. Hal Lancaster, "Technology Raises Bar for Sales Job; Know Your Dress Code," *Wall Street Journal*, January 21, 1997, p. B1.

52. Ingred Wickelgren, "Treasure Maps for the Masses," *Business Week/Enterprise*, 1996, pp. ENT22–ENT24.

53. Frank Barnako, "Online Sales Seen Worth $2 Trillion in 2003," CBS MarketWatch, May 1, 2000.

54. Tim McCollum, "New Horizons in Communications," *Nation's Business*, August 1996, pp. 38–39; and Laura Casteneda and Jon Swartz, "Kicking Virtual Tires," *San Francisco Chronicle*, October 15, 1996, p. C4.

55. Robert D. Hof, "Will Shoppers Take to Cyber Groceries?" *Business Week*, February 23, 1998, p. 110R.

56. 1998 Andersen Consulting study, cited in "Market for On-Line Supermarkets," *USA Today*, January 6, 1999, p. 1B.

57. David Leonhardt, "The Meat and Potatoes of Online Shopping?" *Business Week*, December 7, 1998, p. 46; George Anders, "Co-Founder of Borders to Launch Online Megagrocer," *Wall Street Journal*, Feburary 22, 1999, pp. B1, B4; Lorrie Grant, "Soon, On-Line Grocers Will Deliver the Goods to Your Door," *USA Today*, March 8, 1999, p. 1B; and .George Anders, "Amazon.com Buys 35% Stake of Seattle Online Grocery Firm," *Wall Street Journal*, May 18, 1999, p. B8.

58. Brian Bremner, Joan Warner, and Jonathan Ford, "Hold It Right There, Citibank," *Business Week*, March 25, 1996, p. 176; and Adam Zagorin, "Cashless, Not Bankless," *Time*, September 23, 1996, p. 52.

59. Paul Saffo, quoted in Jared Sandberg, "CyberCash Lowers Barriers to Small Transactions at Internet Storefronts," *Wall Street Journal*, September 30, 1996, p. B6.

60. Stewart Alsop, "The First Powerhouse Bank of the Virtual World," *Fortune*, September 7, 1998, pp. 159–160.

61. Christine Dugas, "Direct Deposit Wins for Safety, Speed," *USA Today*, January 25, 1999, p. 5B.

62. Arthur M. Louis, "The Check's on the Net," *San Francisco Chronicle*, June 22, 1999, pp. C1, C3; Christine Dugas, "Bank Deal Puts Stamp on Bill Delivery on Line," *USA Today*, June 24, 1999, p. 1B; and Martha Brannigan, "Bill Payments Via the Internet Get a Big Boost," *Wall Street Journal*, January 28, 1999, pp. B1, B14.

63. Denise Caruso, "On-Line Day Traders Are Starting to Have an Impact on a Few Big-Cap Internet Stocks in What Some Call a 'Feeding Frenzy,'" *New York Times*, December 14, 1998, p. C3.

64. Rebecca Buckman and Aaron Lucchetti, "Electronic Networks Threaten Trading Desks on Street," *Wall Street Journal*, December 23, 1998, pp. C1, C15.

65. Fred Vogelstein, "A Virtual Stock Market," *U.S. News & World Report*, April 26, 1999, pp. 47–48.

66. Eileen Glanton, "Will Extended Hours Benefit Small Investors?" *San Francisco Examiner*, June 6, 1999, p. B-3.

67. "Net Catches Merrill Lynch, but On-Line Waters Remain Rough" editorial, *USA Today*, June 2, 1999, p. 14A.

68. Allan Sloan, "Long Live the Middleman," *Newsweek*, June 14, 1999, p. 46.

69. Peter H. Lewis, "Play It Again, RAM," *New York Times*, November 12, 1998, pp. D1, D3.

70. Benny Evangelista, "AOL Trying to Catch Internet Music Wave," *San Francisco Chronicle*, June 2, 1999, pp. B1, B12.

71. Charles Bermant, "Musicians Tap Rich Lode of Sheet Music, Sold and Shared," *New York Times*, April 23, 1998, p. D5.

72. Lee Gomes, "Free Tunes for Everyone!" *Wall Street Journal*, June 15, 1999, pp. B1, B4.

73. Deidre Pike, "Reno Musicians: It's Business, Even Artists Need to Pay Bills," *Reno Gazette-Journal*, May 29, 2000, pp. 1E, 3E.

74. Jon Pareles, "Musicians Want a Revolution Waged on the Internet," *New York Times*, March 8, 1999, pp. B1, B8.

75. Stross, 2000.

76. Peter Stack, "An Animated Future," *San Francisco Chronicle*, December 17, 1996, pp. E1, E3.

77. Peter Stack, "The Digital Divide," *San Francisco Chronicle*, May 19, 1999, p. E1.

78. Bruce Haring, "Digitally Created Actors: Death Becomes Them," *USA Today*, June 24, 1998, p. 8D.

79. David Ansen and Ray Sawhill, "The New Jump Cut," *Newsweek*, September 2, 1996, pp. 64–66.

80. Dan Johnson, "Politics in Cyberspace," *The Futurist*, January 1999, p. 14.

81. *The State of "Electronically Enhanced Democracy": A Survey of the Internet* (New Brunswick, NJ: Rutgers University, Douglass Campus, Walt Whitman Center, Department of Political Science, 1998).

82. "Minus a Daily Newspaper, City Turns Online," *San Francisco Chronicle*, September 8, 1998, p. A22; reprinted from *New York Times*.

83. Frank Eltman, "Government of the People, Via the Net," *San Francisco Chronicle*, May 31, 2000, p. D3.

84. Michael Cornfield, quoted in Jon Swartz, "Electronic Engineering," *San Francisco Chronicle*, March 27, 1999, pp. D1, D3.

85. Ann Gearan, "Clinton Announces Program to Link Government Websites," *Reno Gazette-Journal*, June 25, 2000, p. 3A.

86. Forrester Research, reported in Glenn R. Simpson, "Putting Government on the Web," *Wall Street Journal*, May 17, 2000, pp. B1, B4.

87. Simpson, 2000.

88. Jonathan Marshall, "Surfing the Internet Can Land You a Job," *San Francisco Chronicle*, July 17, 1995, pp. D1, D3.

89. Marshall, 1995.

90. Belle Wise, "Magna Cum Market—Popular Jobs for College Grads," *USA Today*, April 12, 2000, p. 7A.

91. Lance Hale, "Demand Rising in IT Profession," *Reno Gazette-Journal*, May 14, 2000, p. 1F.

92. Deborah Mathis, "Learn or Be Left Behind," *Reno Gazette-Journal*, January 1, 2000, p. 41G.

93. Richard Bolles, quoted in T. Minton, "Job Hunting Requires Eyes and Ears of Friends," *San Francisco Chronicle*, January 25, 1991, p. D5.

94. Resumix, cited in Margaret Mannix, "Writing a Computer-Friendly Resume," *U.S. News & World Report*, October 26, 1992, pp. 90–93.

95. David Rampe, "Cyberspace Resumes Fit the Modern Job Hunt," *New York Times*, February 3, 1997, p. C6.

96. Kathleen Pender, "Jobseekers Urged to Pack Lots of 'Keywords' into Resumes," *San Francisco Chronicle*, May 16, 1994, pp. B1, B4.

97. Howard Bennet and Chuck McFadden, "How to Stand Out in a Crowd," *San Francisco Examiner & Chronicle*, October 17, 1993, help wanted section, p. 29.

98. Richard Bolles, quoted in Sylvia Rubin, "How to Open Your Job 'Parachute' After College," *San Francisco Chronicle*, February 24, 1994, p. E9.

Bookmark It! Box

a. Mike Tucker, "Going, Going Gone in Cyberspace," *USA Today*, June 16, 1998, p. 5E.

b. Steve Mollman, "Sold!," *PC Computing*, February 2000, pp. 131–51; Ken Bensinger, "The Perils of Online Auctions," *Wall Street Journal*, March 5, 1999, pp. W1, W12; and Pam Black, "All the World's an Auction," *Business Week*, February 8, 1999, pp. 120–21.

c. Michael Liedtke, "Call for More eBay Vigilance," *San Francisco Chronicle*, June 26, 2000, pp. B1, B7; Deborah Kong, "Internet Auction Fraud Increases," *USA Today*, June 23, 2000, p. 3B; and Jim Carlton and Ken Bensinger, "Phony Bids Put eBay on Defensive," *Wall Street Journal*, May 24, 2000, pp. B1, B4.

d. "Tips for Buying at Auction Web Sites," *USA Today*, June 23, 2000, p. 3B; Bob Tedeschi, "As On-Line Auctions Move into Pricier Merchandise, Escrow Services Offer Those About to Be Scammed a Little Safety," *New York Times*, April 19, 1999, p. C4; Bensinger, 1999; Jamie Beckett, "Rise of the Online Middlemen," *San Francisco Chronicle*, January 14, 1999, pp. B1, B4; Roberta Furgert, "Going Once, Going Twice . . .," *Parade Magazine*, November 29, 1998, p. 24; and "Tip: Protect Yourself Before Paying for Items," *USA Today*, March 10, 1998, p. 2A.

Chapter 11

1. Jessica Lipnack and Jeffrey Stamps, *Virtual Teams: Reaching Across Space, Time, and Organization with Technology* (New York: John Wiley & Sons, 1997).

2. "Virtual Teams Transcend Time and Space," *The Futurist*, September-October 1997, p. 59.

3. Sue Shellenbarger, "Overwork, Low Morale Vex the Mobile Office," *Wall Street Journal*, August 17, 1994, pp. B1, B4.

4. Gary Webb, "Potholes, Not 'Smooth Transition,' Mark Project," *San Jose Mercury News*, July 3, 1994, p. 18A.

Bookmark It! Box

a. John C. Dvorak, "Avoiding Information Overload," *PC Magazine*, December 17, 1996, p. 87.

b. "Web Crawlers Don't Surf Alone," *USA Today*, June 22, 2000, p. 1A.

c. Maria Puente, "Multi-tasking to the Max," *USA Today*, April 25, 2000, pp. 1D, 1E.

d. Patricia Wen, "Expert Advice for Those Who Try Two Things at Once—'Chill Out,'" *San Francisco Chronicle*, May 27, 2000, p. A10.

e. Odette Pollar, "How to Avoid the Overload of Information," *San Francisco Examiner*, August 23, 1998, p. J-3.

f. Odette Pollar, "Coping When Time Is Not on Your Side," *San Francisco Examiner*, June 8, 1997, p. J-3.

Appendix A

1. Alan Freedman, *The Computer Glossary*, 6th ed. (New York: AMACOM, 1993), p. 370.

2. Michael Krantz, "Keeping Tabs Online," *Time*, November 10, 1997, pp. 81–82.

Index

Photo Credits

Page 1 R.Ian Lloyd; **2** (top) courtesy of Motorola Corporation; **2** (middle) courtesy of Intel; **3** Fujifotos/The Image Works; **4** R.Ian Lloyd; **5** courtesy of Motorola Corporation; **6** Inge Yspeert/Corbis; **9** (top) Tony O'Brien; **9** (bottom left) Tom Tracy/Photophile; **9** (bottom right) courtesy of Sun Microsystems; **10** (top middle) courtesy of Sony Corporation; **10** (top right) courtesy of Toshiba; **10** (middle left) courtesy of Palm Inc.; **10** (middle) courtesy of Sharp Electronics; **10** (middle right) courtesy of IBM; **10** (bottom) courtesy of Intel; **14** courtesy of Intel; **19** courtesy of Microsoft; **20** (left) courtesy of Unisys Archives; **20** (right) Mark Richards/PhotoEdit; **21** (top) Don Mason/The Stock Market; **21** (bottom) Eric Draper/AP/Wide World; **22** Fujifotos/The Image Works; **24** R. Ian Lloyd; **27** R. Ian Lloyd; **29** (top) Tom Tracy/Photophile; **29** (middle) courtesy of Intel; **29** (bottom) courtesy of Toshiba; **30** (top) courtesy of IBM; **30** (bottom) Tony O'Brien; **31** courtesy of Sony Corporation; **32** R. Ian Lloyd; **34** courtesy of CIDCO; **35** Richard T. Nowitz/Corbis; **45** courtesy of CIDCO; **75** Richard T. Nowitz/Corbis; **79** Richard T. Nowitz/Corbis; **87** Gary Braasch; **88** Mark Richards/PhotoEdit; **90** Gary Braasch; **109** Mark Richards/PhotoEdit; **119** Gary Braasch; **123** Gary Braasch; **125** Gary Braasch; **131** Gary Braasch; **135** Peter Menzel; **139** Karen Kasmauski/Matrix; **142** courtesy of Apple Computer; **143** courtesy of Apple Computer; **146** Lori Adamski-Peck; **148** (bottom) courtesy of Handspring; **151** Charles O'Rear; **163** courtesy of Handspring; **167** J. Kyle Keener; **168** J. Kyle Keener; **170** John S. Reid; **171** courtesy of IBM; **172** J. Kyle Keener; **173** courtesy of IBM Archives; **174** (top) J. Kyle Keener; **174** (middle and bottom) courtesy of Intel; **182** courtesy of Intel; **183** (top) J. Kyle Keener; **183** (middle) Brian Williams; **183** (bottom) ;Brian Williams; **187** (top) Will & Deni McIntyre/Photo Researchers Inc.; **187** (bottom) Tom Pantages; **188** (right) Philip Quirk/Wildlight Photo Agency; **188** (left) Lori Adamson-Peck; **189** Brian Williams; **192** Brian Williams; **193** John S. Reid; **195** Brian Williams; **196** (top) John S. Reid; **197** courtesy of IBM Corp.; **198** Coco McCoy/Rainbow; **199** courtesy of IBM Corp.; **200** Brian Williams; **203** (top) Michael Schmelling/AP/Wide World Photos; **203** (bottom) courtesy of IBM Corp.; **204** Seiko Instruments Inc./AP/Wide World Photos; **206** courtesy of IBM; **208** (top) J. Kyle Keener; **208** (bottom) David Hanover/Stone; **209** (top) J. Kyle Keener; **209** (bottom) Brian Williams; **210** (top and middle) courtesy of Intel; **210** Brian Williams; **211** (top) John S. Reid; **211** (middle) Brian Williams; **211** (bottom) courtesy of IBM Corp.; **214** (top) courtesy of IBM Corp.; **214** (bottom) Coco McCoy/Rainbow; **216** (top) Brian Williams; **216** (bottom) John S. Reid; **220** (top) David Young-Wolff/PhotoEdit;; **220** (bottom) Brian Williams; **221** (left) courtesy of Sun Microsystems; **221** (right) Michael Newman/PhotoEdit; **222** Robert A. Flynn, Inc.; **223** (left) David Young-Wolff/PhotoEdit; **223** (right) Wernher Krutein/Photovault; **225** Bill Sikes/AP/Wide World Photos; **226** (top) Wernher Krutein/Photovault; **226** (bottom) courtesy of LandWare Inc.; **227** (top) Marble FX by Logitech; **227**(top middle) courtesy of IBM; **227** (top right) Brian Williams; **227** (bottom left) courtesy of AT & T Global Info/Solution; **227** (bottom middle) E. Young/John Rottet/AP/Wide World Photos; **227** (bottom right) Myrleen Cate/PhotoEdit; **228** (top left) courtesy of Compaq Computer; **228** (top right) courtesy of Aqcess Technologies, Inc.,Irvine,CA; **228** (bottom) courtesy of FTG Data Systems; **229** (top) courtesy of CalComp Ultraslate; **229** (bottom) Melissa Farlow; **230** (top) Gary Braasch; **230** (bottom) Charles Gupton/Stock Boston; **231** (top) PhotoDisc; **231** (bottom) StephenWeistead/LWA/Stock Market; **232** Larry Mulvehill/Rainbow; **233** courtesy of Intel and Konica; **234** courtesy of IBM Corporation; **236** (top) Jay Dickman; **236** (bottom) Tom Burdete/USGS; **237** J. Kyle Keener; **239** (left) courtesy of Sun Microsystems; **239** (right) courtesy of ViewSonic Corp.; **242** (top) Brian Williams; **242** (bottom) Michael Newman/PhotoEdit; **243** (top left) Dan McCoy/Rainbow; **243** (top right) Xerox ColorgrafX Systems; **243** (bottom) Don Mason/Stock Market; **246** Robert A. Flynn, Inc.; **247** Charles O'Rear; **248** Dr. Irfan Essa/Georgia Tech; **253** (top left) courtesy of Kinesis; **253** (top right) Coco McCoy/Rainbow; **253** (bottom) Owen Franken/Stock Boston; **256** (top) Charles Gupton/Stock Boston; **256** (bottom) courtesy of CalComp Ultraslate; **257** (top) courtesy of Intel and Konica; **257** (middle) courtesy of CalComp Ultraslate; **257** (bottom) Wernher Krutein/Photovault; **258** (top) courtesy of ViewSonic Corp.; **258** (bottom) Michael Newman/PhotoEdit; **259** courtesy of Kinesis; **260** (top) courtesy of Aqcess Technologies, Inc., Irvine,CA; **260** (middle) Marble FX by Logitech; **260** (bottom) courtesy of IBM; **261** (top) Charles O'Rear; **261** (bottom) Brian Williams; **262** (top) E. Young/John Rottet/AP/Wide World Photos; **262** (bottom) courtesy of IBM Corporation; **265** courtesy of AT&T; **266** Hulton Getty/Stone; **268** Brian Williams; **269** Ian Shaw/Stone; **270** (top) courtesy of AT&T; **276** Hulton Getty/Stone; **277** Bruce Ayres/Stone; **278** Dilip Mehta/Contact Press Images; **282** (top) courtesy of AT&T; **282** (middle) courtesy of AT&T; **282** (bottom) Brian Williams; **283** (top) Brian Williams; **283** (bottom) UPI/Bettmann/Corbis; **285** (left) AFP/Corbis; **285** (right) Mark Richards/PhotoEdit; **293** Rudi MeisalVisum; **299** Ian Shaw/Stone; **301** (top) courtesy of AT&T; **301** (bottom left) courtesy of NeoPoint; **301** (bottom right) Reuters NewMedia Inc./Corbis; **303** courtesy of AT&T; **304** (top) courtesy of AT&T; **304** (bottom) UPI/Bettmann/Corbis; **305** (top) courtesy of AT&T; **305** (middle) AFP/Corbis; **305** (bottom) Brian Williams; **308** Dilip Mehta/Contact Press Images; **309** courtesy of AT&T; **311** Jeffrey Aaronson/Network Aspen; **313** Nick Kelsh; **314** Paul Higdon/*New York Times*; **315** Jeffrey Aaronson/Network Aspen; **316** Michael St.Maur Sheil/Corbis; **318** Brian Williams; **325** Nick Kelsh; **327** Jeffrey Aaronson/Network Aspen; **334** Elastic Reality Inc.; **335** Paul Higdon/*New York Times*; **336** courtesy of Princeton Video Inc.; **338** Jeffrey Aaronson/Network Aspen; **339** Jeffrey Aaronson/Network Aspen; **341** Jeffrey Aaronson/Network Aspen; **343** Elastic Reality Inc.; **345** Jeffrey Aaronson/Network Aspen; **347** Yoshida-Fujifotos/The Image Works; **348** (top) AFP/Corbis; **348** (middle) Yoshida-Fujifotos/The Image Works; **348** (bottom) PhotoEdit; **349** (bottom) James Balog; **349** (top) Photex/Corbis; **350** Yoshida-Fujifotos/The Image Works; **352** (top) Bettmann Corbis; **352** (bottom) AFP/Corbis; **353** Brian Williams; **356** Robert Daemmrich/Stone; **357** Fujifotos/The Image Works; **358** Hironori Miyata/Fujifotos/The Image Works; **359** (top)Yoshida-Fujifotos/The Image Works; **359** (bottom) Theo Westenberger; **362** Ricki Rosen; **364** Yoshida-Fujifotos/The Image Works; **366** Joel Sartore; **367** Christof Stacke/AP/Wide World Photos; **368** (top) PhotoEdit; **368** (bottom) James Balog; **370** Yoshida-Fujifotos/The Image Works; **371** (top) Yoshida-Fujifotos/The Image Works; **371** (middle) Hironori Miyata/Fujifotos/The Image Works; **371** (bottom) Bettmann Corbis; **372** AFP/Corbis; **373** Yoshida-Fujifotos/The Image Works; **375** Jeffrey Morgan; **376** (top) Rick Allen & Cindy Burnham/Nautilus Productions; **376** (middle) Kaku Karita; **376** (bottom) Jeffrey Morgan; **377** (top) Denise Rocco/U.C. Berkeley; **377** (middle right) Fritz Hoffmann; **377** (middle left) Torin Boyd; **377** (bottom) Anna Clopet/Corbis; **378** Jeffrey Morgan; **379** (top left) Rick Allen & Cindy Burnham/Nautilus Productions; **379** (top right) Lori Grinker/Contact Press Images; **379** (bottom) Lori Grinker/Contact Press Images; **381** (top left) Rudi Meisel/Visum; **381** (top right) Charles Gupton; **381** (bottom left) Karen Kasmauski; **381** (bottom right) Peter Menzel; **382** (top) Kaku Karita; **382** (middle left) Denise Rocco/U. C. Berkeley; **382** (middle) Mark Richards/PhotoEdit; **382** (middle right) Misha Erwitt; **382** (bottom left) Hank Morgan/Rainbow; **382** (bottom right) Spencer Grant/Stock Boston; **387** (all) Gary Braasch; **388** (top) Fujifotos/The Image Works; **388** (bottom) Jeffrey Morgan; **389** (top) Shelly Katz; **389** (bottom left) Charles O'Rear; **389** (bottom right) Rob Crandall/Stock Boston; **390** (top) David Modell; **390** (bottom) John S. Reid; **391** (top) Denise Rocco/U. C.Berkeley; **391** (bottom) Fritz Hoffmann; **392** Jeffrey Morgan; **393** Greg Girgard/Contact Press Images; **394** Philip Quirk/Wildlight Photo Agency; **396** Torin Boyd; **401** Jeffrey Morgan; **403** Jeffrey Morgan; **404** (top) Misha Erwitt; **404** (bottom) Rudi Meisel/Visum; **405** Jeffrey Morgan; **409** PhotoDisc; **423** PhotoDisc; **424** PhotoDisc; **425** PhotoDisc; **427** PhotoDisc; **428** PhotoDisc; **433** PhotoDisc; **437** PhotoDisc; **438** PhotoDisc; **444** PhotoDisc; **446** PhotoDisc.